THE ESTIMATION OF ANIMAL ABUNDANCE AND RELATED PARAMETERS

Books on cognate subjects

*A volume in "Griffin's Statistical Monographs and Courses"

Descriptive catalogue available from the Publishers

THE ESTIMATION OF
ANIMAL ABUNDANCE
and related parameters

G. A. F. SEBER

M.Sc. (Auckland), Ph.D. (Manchester), F.S.S.

Professor of Statistics, Auckland University
New Zealand

GRIFFIN  LONDON

CHARLES GRIFFIN & COMPANY LIMITED
42 DRURY LANE, LONDON WC2B 5RX

First published   1973

Medium Octavo, xii + 506 pages
34 line illustrations, 106 tables
ISBN: 0 85264 207 5

Set by E. W. C. Wilkins Ltd London N12 0EH
Printed in Great Britain by J. W. Arrowsmith Ltd Bristol BS3 2NT

PREFACE

Today ecology is one of the "in" subjects. With the steady growth of technology and its resulting problems of pollution, more and more attention is being directed towards our environment and our natural resources. This upsurge of interest in ecological problems is being stimulated by the use of statistical methods in much the same way that the physical sciences have been stimulated by theoretical developments in mathematics. Although mathematical theory tends to leap way ahead of mathematical practice, it is now widely recognised that many branches of mathematics have a vital role to play in the study of biological systems, no matter how complex these systems may be. I believe that the interaction between mathematics and biology is of benefit to the mathematician as well as to the biologist.

In ecological studies the first problem that is usually encountered is a census one. We do not know the sizes of the various wildlife populations in our study area, nor do we know how these populations are changing with time. However, over the past forty years a large body of techniques have been developed for estimating animal population numbers and related parameters such as the mortality and birth rates, and this book is an attempt to bring this material together. Originally I had intended to write a short monograph, but as I searched deeper into the literature my working bibliography began to grow at an alarming rate; the monograph eventually became a book.

In organising this large body of research material my biggest problem has been to classify the various techniques in such a way that an appropriate technique could be readily located in the text. Eventually I decided to classify the techniques according to the type of population studied and the nature of the sampling information available from the population. Populations are divided into two categories called "closed" and "open", depending on whether the population remains unchanged during the period of investigation, or changes through such processes as mortality, migration, etc. For the two types of population, one can obtain sample information from plot studies, capture-tag-recapture data, catch-effort statistics, and demographic data such as the sex ratio or age structure.

Before using a statistical technique it is important that the user be fully aware of the assumptions that must hold for the technique to be valid. For this reason I have paid considerable attention to underlying assumptions and, where possible, have given methods for testing their validity. Regression techniques are widely used throughout as they are fairly robust with regard to departures from the underlying assumptions, and can be examined graphically. Sometimes a less efficient but more robust method is to

v

be preferred to a highly efficient method requiring strong assumptions.

In an endeavour to make this book more useful for the field worker I have included at least one numerical example demonstrating each technique, and, apart from one or two isolated cases, these examples are based on "real-life" data. To some extent the choice of examples has been rather arbitrary in that the one chosen for a particular technique has sometimes been the first one I encountered in my reading. However, in general, I have tried to use examples in which the authors have made some attempt to discuss the underlying assumptions from a field point of view. I hope that future users of these techniques will adopt the same critical attitude.

I am very grateful to a number of authors who kindly sent me pre-publication copies of their articles. This has, to some extent, compensated for various delays caused by typing difficulties, inaccessibility of certain journals, etc. I would also express my debt to Dr E.G. White of Lincoln College, Christchurch, N.Z., not only for reading the manuscript, but also for helping me to update some of the references prior to publication. Finally I would like to thank Mr John Whale of Auckland University for writing several computer programs and calculating Table 7.6 on page 321.

G.A.F. SEBER

ACKNOWLEDGMENTS

For permission to reproduce certain published tables, thanks are due to the Editors of the *Annals of the Institute of Statistical Mathematics* (**A4**), *Annals of Mathematical Statistics* (**A5**), *Biometrics* (Tables 7.5, 9.3, 9.4; Fig. 7.3, 7.4), *Biometrika* (**A3**; Tables 4.18, 9.1, 12.2), *Journal of the Fisheries Research Board of Canada* (**A2**; Fig. 6.2), *Journal of Wildlife Management* (Table 7.3; Fig. 9.1–9.3), and *Transactions of the American Fisheries Society* (**A6**; Tables 3.1, 3.2, 10.9; Fig. 3.1–3.6).

CONTENTS

CONTENTS

CONTENTS

CONTENTS

CONTENTS

CONTENTS

$$f(x) = \frac{1}{x} - \frac{1}{\exp(x) - 1}.$$

$$A_K(S) = \sum_{k=0}^{K} kS^k \sum_{k=0}^{K} S^k.$$

CHAPTER 1

PRELIMINARIES

1.1 INTRODUCTION

During the past thirty years there has been a growing realisation of the importance of sound statistical technique in the analysis of ecological data. This change of emphasis from qualitative to quantitative methods, as reflected for example in the bibliography of Schultz [1961] and in recent issues of the *Journal of Animal Ecology, Journal of Wildlife Management*, etc., has led to a greater use of statistical models and a greater appreciation of the need for careful experimental design.

The ecologist has also recognised the importance of obtaining data in the field from "natural" or free-ranging populations as opposed to data from "artificial" or laboratory populations. So often the population changes that occur in the laboratory give little indication as to what happens in the natural state, particularly when the animal under investigation is capable of exhibiting certain social tendencies in captivity. But the study of natural populations is not easy, as the experimenter is faced with the conflicting requirements of finding out as much about the population as possible and, at the same time, leaving the population undisturbed. Because of this lack of control over natural populations, statistical models should be used with caution, as they all depend on the validity of certain underlying assumptions which, in some cases, may be difficult to investigate. Some models are very sensitive to departures from their underlying assumptions, and it is therefore essential that with the development of a new model, there should be a corresponding development of procedures for testing the assumptions and for investigating the "robustness" of the model with regard to departures from the assumptions. Generally, the stronger the assumptions underlying a particular population model, the more "powerful" is the model *when valid* (e.g. estimates have smaller variances). However, a low-powered but robust model requiring few assumptions is often more useful than a complex high-powered model requiring many assumptions.

As a first step in understanding the structure and dynamics of a natural population it is essential to know something about the population size and related parameters such as the birth- and death-rates, etc. at given points in time. Since 1930, following Lincoln's [1930] work on banded waterfowl and the series of papers by Jackson [1933, 1937, 1939, 1940, 1944, 1948] on the tsetse fly, there has been a growing interest in methods of estimating population parameters, and today the literature is extensive. Much of the pioneering work has been done by fishery scientists, culminating in the monumental volumes of Beverton and Holt [1957] and Ricker [1958]. The importance of such methods, as for example in pest control and wildlife management, has led research workers of widely differing interests to enter the fray, and today the whole subject is expanding rapidly.

This book is an attempt to systematise the growing body of literature according to types of statistical models used and, where possible, to discuss in some detail the assumptions underlying the models. In checking my coverage of the literature I have been helped considerably by a number of reviews, especially De Lury [1954], Scattergood [1954], Ricker [1958: fish], Mosby* [1963: wildlife management], Southwood [1966: insects], and in particular Cormack [1968].

The remainder of this chapter is devoted to introducing the notation and outlining a number of statistical methods which are used throughout this book but are not usually dealt with in statistics textbooks.

1.2   NOTATION AND TERMINOLOGY

1.2.1   Notation

One of the difficulties in writing this book has been the choice of a suitable notation which could be maintained fairly consistently throughout. The international notation adopted by the F.A.O. for fishery research was found to be unsuitable and I finally opted for a "mnemonic" notation. For example, N and n denote the *number* of individuals in population and sample; M and m refer to the number of *marked* (or tagged members) of the population and sample; s represents the number of *samples,* etc. Where there has been a choice of symbols available I have endeavoured to choose those symbols which would help the reader in consulting the original articles. Occasionally, for the sake of clarity, I have found a change of notation necessary, and to avoid ambiguity I have made each chapter self-contained as far as the notation is concerned. In most cases the convention of distinguishing between a random variable and its observed value has not been followed because of notational difficulties.

Some statistical symbols are required: $E[y]$, $\sigma[y]$, $V[y]$ $(= \sigma^2[y])$, $C[y]$ $(= \sigma[y]/E[y])$ will represent the mean, standard deviation, variance, and coefficient of variation, respectively, of the random variable y; $\text{cov}[x, y]$ is the covariance of two random variables x and y, and $E[x|y]$, $V[x|y]$ are the mean and variance of x conditional on fixed y. The normal distribution with mean θ and variance σ^2 is represented by $\mathcal{N}[\theta, \sigma^2]$, and the random variable with a *unit normal* distribution $(\theta = 0, \sigma^2 = 1)$ will usually be denoted by z. The symbols z_α and $t_k[\alpha]$ will represent the 100α per cent upper tail values for z and the t-distribution with k degrees of freedom, respectively; thus

$$\Pr[z \geq z_\alpha] \;=\; \Pr[t_k \geq t_k[\alpha]] \;=\; \alpha,$$

---

*A revised edition of this book is now available (see Giles [1969].)

where Pr stands for Probability. The chi-squared distribution with k degrees of freedom is denoted by χ_k^2.

Occasionally the symbols $O[N]$ and $o[N]$ will be used: if g is a function of N, then $g[N] = O[N]$ if there exists an integer N_1 and a positive number A such that, for $N > N_1$, $|g(N)/N| < A$; $g(N) = o[N]$ if $\lim_{N \to \infty} \{g(N)/N\} = 0$. Roughly speaking, $O[N]$ means "of the same order of magnitude as N when N is large", while $o[N]$ means "of smaller order of magnitude than N when N is large".

In one or two chapters (especially Chapter 11) it is necessary to introduce vectors and matrices; these will be denoted by bold-face roman type. For example,

$$\mathbf{x} = [(x_i)] \quad \text{and} \quad \mathbf{\Sigma} = [(\sigma_{ij})]$$

represent respectively the column vector $\mathbf{x}$ with ith element x_i and the matrix $\mathbf{\Sigma}$ with i, jth element σ_{ij}.

If the vector $\mathbf{x}$ has a multivariate normal distribution with mean vector $\boldsymbol{\theta}$ and variance–covariance (dispersion) matrix $\mathbf{\Sigma}$, we shall write $\mathbf{x}$ is $\mathfrak{N}[\boldsymbol{\theta}, \mathbf{\Sigma}]$.

All logarithms, written log x, will be to the base e unless otherwise stated.

1.2.2 Terminology

The size of an animal population in a given area will be determined by the processes of *immigration* (or movement into the area), *emigration* (or movement out of the area), *total mortality*, and *recruitment*.

TOTAL MORTALITY. In dealing with exploited populations we shall usually subdivide total mortality into mortality due to exploitation and natural mortality, i.e. mortality due to natural processes such as predation, disease, climatic conditions: contrary to some authors, emigration is not included here under "mortality". We shall also distinguish between *mortality rate* and *instantaneous mortality rate* as follows:

Let ϕ_t be the probability that an animal survives for the period of time $[0, t]$, then if N_0 animals are alive at time zero we would expect $N_0\phi_t$ ($= N_t$ say) to be alive at time t. The proportion ϕ_t, sometimes expressed as a percentage, is called the *survival rate* over period t, and $1 - \phi_t$ is called the *mortality rate* over period t. If, however, the mortality may be regarded as a Poisson process with parameter μ — that is, the probability that an animal dies in the time-interval $(t, t+\delta t)$ is $\mu\delta t + o(\delta t)$ — then (e.g. Feller [1957]: Chapter 17),

$$\phi_t = e^{-\mu t}, \tag{1.1}$$

$$\frac{dN_t}{dt} = -\mu N_t, \tag{1.2}$$

and the parameter μ is called the *instantaneous mortality rate*.

MEAN LIFE EXPECTANCY. Let Y be the time at which a member of N_0 dies. Then

$$
\begin{aligned}
F[y] &= \Pr[Y \leqslant y] \\
&= 1 - \Pr[Y > y] \\
&= 1 - \Pr[\text{animal survives until time } y] \\
&= 1 - \exp(-\mu y),
\end{aligned}
$$

and Y has probability density function

$$f(y) = F'(y) = \mu e^{-\mu y}, \quad (y \geqslant 0).$$

Therefore the *mean life expectancy* is

$$
\begin{aligned}
E[Y] &= \int_0^\infty \mu y e^{-\mu y} \, dy \\
&= 1/\mu \\
&= -1/\log \phi_1.
\end{aligned}
\tag{1.3}
$$

RECRUITMENT. By recruitment we shall refer to those animals born into the population or, where applicable, those animals which grow into the catchable part of the population. In fishery research, recruitment sometimes denotes those fish which grow into the class of *legally* catchable fish. Thus we do not treat immigration as a component of recruitment.

OPEN AND CLOSED POPULATIONS. A population which remains unchanged during the period of investigation (i.e. the effects of migration, mortality and recruitment are negligible) is called a *closed* population. If the population is changing due to one or more of the above processes operating, then the population is said to be *open*.

1.3   SOME STATISTICAL METHODS

1.3.1   Maximum-likelihood estimation

Let $x_1, x_2, \ldots, x_n$ be a random sample from the probability (or probability density) function $f(x, \theta)$, and let

$$L(\theta) = \prod_{i=1}^{n} f(x_i, \theta)$$

be the likelihood function. Then, provided the function f has certain reasonable properties, $\hat{\theta}$, the maximum-likelihood estimate of θ, is a solution of

$$\frac{\partial \log L(\theta)}{\partial \theta} = 0,$$

and as $n \to \infty$, $\hat{\theta}$ is asymptotically $\mathcal{N}[\theta, \sigma_{\hat{\theta}}^2]$ where

$$\sigma_{\hat{\theta}}^2 = \left\{ -E\left[\frac{\partial^2 \log L(\theta)}{\partial \theta^2} \right] \right\}^{-1}. \tag{1.4}$$

Replacing θ by $\hat{\theta}$ leads to the estimate $\hat{\sigma}$, say, of σ_θ, and an approximate large-sample $100(1-\alpha)$ per cent confidence interval for θ is given by

$$\hat{\theta} \pm z_{\alpha/2}\hat{\sigma}. \tag{1.5}$$

In a number of standard situations (e.g. when x is binomial or Poisson) a more accurate confidence interval is (θ_1, θ_2), where θ_1 and θ_2 are the appropriate roots of the following equation in θ:

$$(\hat{\theta}-\theta)^2 = \sigma_\theta^2 z_{\alpha/2}^2. \tag{1.6}$$

For a 95 per cent confidence interval, $\alpha = 0\cdot05$ and $z_{0\cdot025} = 1\cdot96$.

COEFFICIENT OF VARIATION. The coefficient of variation of $\hat{\theta}$ is asymptotically

$$C[\hat{\theta}] = \sigma_\theta/\theta$$

which can be estimated by

$$\hat{C} = \hat{\sigma}/\hat{\theta}.$$

Here $\hat{C}$ is related to the proportional width of the interval (1.5) and is therefore a useful measure of the "accuracy" of $\hat{\theta}$.

BIAS. Occasionally an expression for the bias can be calculated, so that

$$E[\hat{\theta}] = \theta + b_\theta.$$

The estimate $\hat{b}$ of the bias term should be included in the interval $\hat{\theta} - \hat{b} \pm 1\cdot96\hat{\sigma}$ if it is more than 10 per cent of the magnitude of $\hat{\sigma}$ (Cochran [1963: 12–15]). We shall call b_θ/θ the *proportional bias* of $\hat{\theta}$.

SEVERAL PARAMETERS. Let $x_1, x_2, \ldots, x_n$ be a random sample from $f(x, \boldsymbol{\theta})$, where $\boldsymbol{\theta}$ is now a vector of parameters $\theta_1, \theta_2, \ldots, \theta_r$. Then, when f has certain reasonable properties, $\hat{\boldsymbol{\theta}}$, the vector of maximum-likelihood estimates, is a solution of the r equations

$$\frac{\partial \log L(\boldsymbol{\theta})}{\partial \theta_i} = 0, \quad (i = 1, 2, \ldots, r),$$

and is asymptotically distributed as the multivariate normal distribution $\mathcal{N}[\boldsymbol{\theta}, \mathbf{V}_\theta^{-1}]$, where $\mathbf{V}_\theta$ is the r-by-r matrix with i, jth element

$$-E\left[\frac{\partial^2 \log L(\boldsymbol{\theta})}{\partial \theta_i \partial \theta_j}\right]. \tag{1.7}$$

The matrix $\mathbf{V}_\theta$ is sometimes called the *information matrix*.

MOMENT ESTIMATES. The above maximum-likelihood theory is applicable to more general situations than those stated above. For example, the x_i may have different distributions or the x_i may not be independent but have a joint multinomial distribution. In these cases, if the number of random variables equals the number of unknown parameters, then the maximum-likelihood estimates can usually be obtained by equating each random variable to its expected value and solving the resulting equations for the unknown parameters. This method of estimation is called *moment estimation* and the estimates *moment estimates*.

1.3.2   Estimating a mean

WEIGHTED MEAN. Let x_i $(i = 1, 2, \ldots, n)$ be n independent random variables with known variances σ_i^2 and common mean θ. For the class of unbiased estimates of θ of the form

$$\bar{x}_w = (\sum_{i=1}^{n} w_i x_i)/(\sum_{i=1}^{n} w_i),$$

it is readily shown that $\bar{x}_w$ has minimum variance when w_i is proportional to $1/\sigma_i^2$. In particular, if $w_i \sigma_i^2 = a$, say, for $i = 1, 2, \ldots, n$, then

$$V[\bar{x}_w] = a/\sum w_i \qquad (1.8)$$

and, by setting $y_i = x_i - \theta$, it is readily shown that

$$v[\bar{x}_w] = \frac{\sum w_i (x_i - \bar{x}_w)^2}{(n-1) \sum w_i}$$

is an unbiased estimate of this minimum variance.

UNWEIGHTED MEAN. If the variances σ_i^2 are unknown then we can simply use the sample mean

$$\bar{x} = \sum x_i/n$$

as our estimate of θ. In this case it transpires that

$$v[\bar{x}] = \frac{\sum (x_i - \bar{x})^2}{n(n-1)} \qquad (1.9)$$

is an unbiased estimate of $V[\bar{x}]$.

A similar estimate of $V[\bar{x}]$ can also be obtained when the x_i's are not independent but correlated. Suppose that

$$\mathrm{cov}\,[x_i, x_j] = \begin{cases} \sigma_{ij} & j = i + 1 \\ 0 & j > i + 1 \end{cases}$$

so that the (unknown) non-zero covariances are $\sigma_{12}, \sigma_{23}, \ldots, \sigma_{n-1,n}$. Then

$$V[\bar{x}] = \frac{1}{n^2} \{ \sum_{i=1}^{n} \sigma_i^2 + 2 \sum_{i=1}^{n-1} \sigma_{i,i+1} \}$$

$$= \frac{1}{n^2} \{A + 2B\}, \text{ say,}$$

and the problem reduces to finding estimates of A and B. Let

$$S_1^2 = \sum_{i=1}^{n} (x_i - \bar{x})^2$$

and

$$S_2^2 = \sum_{i=1}^{n} (x_{i+1} - x_i)^2,$$

6

where $x_{n+1} = x_1$, then

$$E[S_1^2] = \frac{(n-1)}{n} A - \frac{2B}{n},$$

$$E[S_2^2] = 2A - 2B$$

and, using moment estimation, unbiased estimates of A and B are

$$\hat{A} = \frac{nS_1^2 - S_2^2}{n-3}$$

and

$$\hat{B} = \hat{A} - \tfrac{1}{2}S_2^2.$$

Therefore an unbiased estimate of $V[\bar{x}]$ is given by

$$v[\bar{x}] = \frac{\hat{A} + 2\hat{B}}{n^2} = \frac{3S_1^2 - S_2^2}{n(n-3)}. \tag{1.10}$$

We note in passing that $\hat{A} > 0$ since

$$nS_1^2 = \tfrac{1}{2} \sum_i \sum_j (x_i - x_j)^2 > S_2^2.$$

If the x_i's actually have different means θ_i then

$$E[v[\bar{x}]] = V[\bar{x}] + (3C_1 - C_2)/[n(n-3)],$$

where $C_1 = \sum_{i=1}^{n} (\theta_i - \bar{\theta})^2$ and $C_2 = \sum_{i=1}^{n} (\theta_{i+1} - \theta_i)^2.$

1.3.3   The delta method

A useful method used repeatedly in this book for finding approximate means, variances and covariances is demonstrated by the following examples.

MEAN. Let x_i be a random variable with mean θ_i ($i = 1, 2, \ldots, n$) and suppose we wish to find the mean of some function $g(x_1, x_2, \ldots, x_n)$ ($= g(\mathbf{x})$ say). Then using the first few terms of a Taylor expansion about $\boldsymbol{\theta}$, we have

$$g(\mathbf{x}) \approx g(\boldsymbol{\theta}) + \sum_{i=1}^{n} (x_i - \theta_i) \frac{\partial g}{\partial x_i} + \sum_{i=1}^{n} \sum_{j=1}^{n} \frac{(x_i - \theta_i)(x_j - \theta_j)}{2!} \frac{\partial^2 g}{\partial x_i \partial x_j},$$

where all partial derivatives are evaluated at $\mathbf{x} = \boldsymbol{\theta}$. Therefore, taking expected values,

$$E[g(\mathbf{x})] \approx g(\boldsymbol{\theta}) + b,$$

where $b = \sum_{i=1}^{n} \sum_{j=1}^{n} \frac{1}{2} \operatorname{cov}[x_i, x_j] \frac{\partial^2 g}{\partial x_i \partial x_j}$

$$= \sum_{i=1}^{n} \frac{1}{2} V[x_i] \frac{\partial^2 g}{\partial x_i^2} + \sum_{i<j} \sum \operatorname{cov}[x_i, x_j] \frac{\partial^2 g}{\partial x_i \partial x_j}.$$

VARIANCE. If we ignore the bias b and neglect quadratic terms in the above Taylor expansion, then

$$V[g(\mathbf{x})] \approx E[\{g(\mathbf{x}) - g(\mathbf{\theta})\}^2]$$

$$\approx \sum_{i=1}^{n} V[x_i] \left(\frac{\partial g}{\partial x_i}\right)^2 + 2 \sum\sum_{i<j} \text{cov}\,[x_i,\,x_j] \left(\frac{\partial g}{\partial x_i}\right)\left(\frac{\partial g}{\partial x_j}\right).$$

Example: Let $x_1, x_2, \ldots, x_n$ have a multinomial distribution

$$f(x_1, x_2, \ldots, x_n) = \frac{N!}{\prod_{i=1}^{n} x_i!} \prod_{i=1}^{n} p_i^{x_i}, \quad (\sum_{i=1}^{n} p_i = 1);$$

then, defining $q_i = 1 - p_i$,

$$E[x_i] = Np_i \quad (= \theta_i),$$
$$V[x_i] = Np_i q_i$$

and

$$\text{cov}\,[x_i, x_j] = -Np_i p_j \quad (i \neq j).$$

Suppose we wish to find the asymptotic variance of

$$g(\mathbf{x}) = \frac{x_1 x_2 \cdots x_r}{x_{r+1} x_{r+2} \cdots x_s} \quad (s \leqslant n).$$

Then, using the Taylor approximation,

$$g(\mathbf{x}) - g(\mathbf{\theta}) \approx \sum_i (x_i - \theta_i) \left[\frac{\partial g}{\partial x_i}\right]_{\mathbf{x}=\mathbf{\theta}},$$

and hence

$$\frac{g(\mathbf{x}) - g(\mathbf{\theta})}{g(\mathbf{\theta})} \approx \sum_{i=1}^{r} \left(\frac{x_i - \theta_i}{\theta_i}\right) - \sum_{i=r+1}^{s} \left(\frac{x_i - \theta_i}{\theta_i}\right).$$

Squaring both sides and taking expected values, we have

$$V[g(\mathbf{x})] \approx E[\{g(\mathbf{x}) - g(\mathbf{\theta})\}^2]$$

$$\approx [g(\mathbf{\theta})]^2 \left\{ \sum_{i=1}^{s} \left(\frac{V[x_i]}{\theta_i^2}\right) + 2\,\frac{\text{cov}\,[x_1, x_2]}{\theta_1 \theta_2} + \cdots + \right.$$

$$\left. + 2\,\frac{\text{cov}\,[x_{r+1}, x_{r+2}]}{\theta_{r+1}\theta_{r+2}} + \cdots - 2\,\frac{\text{cov}\,[x_1, x_{r+1}]}{\theta_1 \theta_{r+1}} - \cdots \right\}.$$

Now

$$\frac{V[x_i]}{\theta_i^2} = \frac{Np_i(1-p_i)}{N^2 p_i^2} = \frac{1}{\theta_i} - \frac{1}{N},$$

$$\frac{\text{cov}\,[x_i, x_j]}{\theta_i \theta_j} = -\frac{Np_i p_j}{N^2 p_i p_j} = -\frac{1}{N},$$

and therefore

$$V[g(\mathbf{x})] \approx [g(\boldsymbol{\theta})]^2 \left\{ \sum_{i=1}^{s} \theta_i^{-1} - \frac{s}{N} - \frac{2}{N} \left[\binom{r}{2} + \binom{s-r}{2} \right] - r(s-r) \right\}$$

$$= \frac{[g(\boldsymbol{\theta})]^2}{N} \{ \sum_{i=1}^{s} p_i^{-1} - (s-2r)^2 \},$$

a result given by Seber [1967]. We note that the second term in the above expression vanishes when $s = 2r$.

COVARIANCE. The delta method can also be used for finding an approximate formula for the covariance of two functions $g(\mathbf{x})$ and $h(\mathbf{x})$. Thus

$$\text{cov} [g(\mathbf{x}), h(\mathbf{x})] \approx E[\{g(\mathbf{x}) - g(\boldsymbol{\theta})\}\{h(\mathbf{x}) - h(\boldsymbol{\theta})\}]$$

$$\approx E\left[\left\{ \sum_i (x_i - \theta_i) \frac{\partial g}{\partial x_i} \right\} \left\{ \sum_j (x_j - \theta_j) \frac{\partial h}{\partial x_j} \right\} \right]$$

$$= \sum_i \sum_j \text{cov} [x_i, x_j] \frac{\partial g}{\partial x_i} \frac{\partial h}{\partial x_j} .$$

AN EXACT FORMULA. If x and y are independent random variables then we have the exact relation (Goodman [1960])

$$V[xy] = (E[x])^2 V[y] + (E[y])^2 V[x] + V[x]V[y]. \qquad (1.11)$$

Some generalisations of this result are given by Bohrnstedt and Goldberger [1969].

1.3.4 Conditional variances

Let x and y be a pair of random variables. Then from Kendall and Stuart [1968: 191] we have

$$E[x] = \underset{y}{E} \{E[x \mid y]\}$$

and

$$V[x] = \underset{y}{E} \{V[x \mid y]\} + \underset{y}{V} \{E[x \mid y]\}, \qquad (1.12)$$

where E, etc. denotes taking the expected value with respect to the distri-
$\quad y$
bution of y. We note that if $E[x \mid y]$ does not depend on y then the second term of (1.12) is zero and, by the delta method of **1.3.3**,

$$V[x] = \underset{y}{E} \{V[x \mid y]\}$$

$$= \underset{y}{E} \{g(y)\}, \text{ say}$$

$$\approx g(\theta)$$

$$= \{V[x \mid y]\}_{y=\theta} \qquad (1.13)$$

where $\theta = E[y]$.

1.3.5 Regression models

1 Weighted linear regression

Consider the regression line

$$Y_i = \beta_0 + \beta x_i + e_i \quad (i = 1, 2, \ldots, n), \tag{1.14}$$

where the x_i are constants, the e_i are random variables independently distributed as $\mathcal{N}[0, \sigma^2/w_i]$, the weights w_i are known, and β_0, β and σ^2 are unknown parameters. Then the weighted least-squares estimates $\hat{\beta}_0$ and $\hat{\beta}$ are found by minimising $\Sigma \, w_i e_i^2$ with respect to β_0 and β. Thus

$$\hat{\beta} = \Sigma \, w_i(Y_i - \bar{Y})(x_i - \bar{x})/\Sigma \, w_i(x_i - \bar{x})^2$$

and

$$\hat{\beta}_0 = \bar{Y} - \hat{\beta}\bar{x},$$

where $\bar{Y} = \Sigma \, w_i Y_i / \Sigma \, w_i$ and $\bar{x} = \Sigma \, w_i x_i / \Sigma \, w_i$.

Also,

$$V[\hat{\beta}] = \sigma^2/\Sigma \, w_i(x_i - \bar{x})^2$$

and an unbiased estimate of σ^2 is

$$\hat{\sigma}^2 = \Sigma \, w_i(Y_i - \bar{Y} - \hat{\beta}(x_i - \bar{x}))^2/(n-2).$$

A $100(1-\alpha)$ per cent confidence interval for β can be obtained in the usual manner from the t-distribution, namely

$$\hat{\beta} \pm t_{n-2}[\alpha/2](\hat{\sigma}^2/\Sigma \, w_i(x_i - \bar{x})^2)^{\frac{1}{2}}.$$

The assumptions underlying the above model can be investigated by examining the weighted residuals $\sqrt{w_i}(Y_i - \bar{Y} - \hat{\beta}(x_i - \bar{x}))$, using the methods of Draper and Smith [1966: Chapter 3].

CASE 1. When $\beta_0 = 0$ in (1.14) the least-squares estimate of β is now

$$\tilde{\beta} = \Sigma \, w_i Y_i x_i / \Sigma \, w_i x_i^2.$$

The corresponding confidence interval for β is

$$\tilde{\beta} \pm t_{n-1}[\alpha/2](\tilde{\sigma}^2/\Sigma \, w_i x_i^2)^{\frac{1}{2}}$$

where

$$(n-1)\tilde{\sigma}^2 = \Sigma \, w_i(Y_i - \tilde{\beta}x_i)^2$$

$$= \Sigma \, w_i Y_i^2 - (\Sigma \, w_i Y_i x_i)^2/(\Sigma \, w_i x_i^2)$$

CASE 2. When $\beta = 0$ in (1.14) the least-squares estimate of β_0 is

$$\tilde{\beta}_0 = \Sigma \, w_i Y_i / \Sigma \, w_i = \bar{Y}$$

with confidence interval

$$\tilde{\beta}_0 \pm t_{n-1}[\alpha/2](\tilde{\sigma}_0^2/\Sigma \, w_i)^{\frac{1}{2}}.$$

Here

$$(n-1)\tilde{\sigma}_0^2 = \Sigma \, w_i(Y_i - \tilde{\beta}_0)^2$$

$$= \Sigma \, w_i Y_i^2 - (\Sigma \, w_i Y_i)^2/(\Sigma \, w_i).$$

2   *Weighted multiple linear regression*

A generalisation of (1.14) is the multiple linear regression model

$$\mathbf{Y} = \mathbf{X\beta} + \mathbf{e},$$

where $\mathbf{e}$ has the multivariate normal distribution $\mathcal{N}[\mathbf{0}, \sigma^2\,\mathbf{B}]$, $\mathbf{X}$ is a known n-by-r matrix of rank r, $\mathbf{B}$ is a known n-by-n positive-definite matrix, and β and σ^2 are unknown parameters. From Draper and Smith [1966: 78] the weighted least-squares estimate of β, obtained by minimising

$$(\mathbf{Y} - \mathbf{X\beta})'\mathbf{B}^{-1}(\mathbf{Y} - \mathbf{X\beta})$$

with respect to β, is

$$\hat{\beta} = (\mathbf{X'B^{-1}X})^{-1}\mathbf{X'B^{-1}Y}.$$

The variance–covariance matrix of this estimate is

$$V[\hat{\beta}] = \sigma^2(\mathbf{X'B^{-1}X})^{-1}$$

and σ^2 is estimated by

$$\hat{\sigma}^2 = (\mathbf{Y'B^{-1}Y} - \mathbf{Y'B^{-1}X}\hat{\beta})/(n - r).$$

Multiple confidence intervals for linear combinations of the form $\mathbf{h'\beta}$ can be obtained in the usual manner (Miller [1966]).

3   *A special model*

Consider the linear regression model

$$Y_i = \gamma(\theta - x_i) + e_i, \quad (i = 1, 2, \dots, n), \tag{1.15}$$

where the e_i and x_i are defined as for (1.14) above (though for most applications in this book $w_i = 1$). Then, using the notation of section 1 above, the weighted least-squares estimates of γ and θ are

$$\hat{\gamma} = -\hat{\beta},$$
$$\hat{\theta} = \hat{\beta}_0/\hat{\gamma} = \bar{x} + (\bar{Y}/\hat{\gamma}),$$

and a $100(1-\alpha)$ per cent confidence interval for γ is

$$\hat{\gamma} \pm t_{n-2}[\alpha/2]\,(\hat{\sigma}^2/\Sigma\,w_i(x_i - \bar{x})^2)^{\frac{1}{2}}. \tag{1.16}$$

A confidence interval for θ can be obtained using a technique due to Fieller [1942] as follows. Now

$$\frac{E[\bar{Y}]}{E[\hat{\gamma}]} = \theta - \bar{x} \quad (= \delta \text{ say})$$

and, since $\operatorname{cov}[\bar{Y}, Y_i - \bar{Y}] = 0$,

$$\operatorname{cov}[\bar{Y}, \hat{\gamma}] = 0,$$
$$V[(\bar{Y} - \delta\hat{\gamma})] = V[\bar{Y}] + \delta^2 V[\hat{\gamma}]$$
$$= \sigma^2\left[\frac{1}{\Sigma\,w_i} + \frac{\delta^2}{\Sigma\,w_i(x - \bar{x})^2}\right]$$
$$= \sigma^2 v, \text{ say,}$$

and $\bar{Y} - \delta\hat{\gamma}$ is $\mathcal{N}[0, \sigma^2 v]$. Also $(n-2)\hat{\sigma}^2/\sigma^2$ is distributed as χ^2_{n-2} and is independent of both $\bar{Y}$ and $\hat{\gamma}$. Therefore

$$t = \frac{\bar{Y} - \delta\hat{\gamma}}{\hat{\sigma}\sqrt{v}}$$

has the t-distribution with $n-2$ degrees of freedom, and a $100(1-\alpha)$ per cent confidence interval for δ is given by

$$t^2 \leqslant t^2_{n-2}[\alpha/2], \quad \text{or } d_1 \leqslant \delta \leqslant d_2,$$

where d_1 and d_2 are the roots of the quadratic

$$d^2 \left(\hat{\gamma}^2 - \frac{\hat{\sigma}^2 t^2_{n-2}[\alpha/2]}{\Sigma\, w_i(x_i - \bar{x})^2}\right) - 2d\bar{Y}\hat{\gamma} + \left(\bar{Y}^2 - \frac{\hat{\sigma}^2 t^2_{n-2}[\alpha/2]}{\Sigma\, w_i}\right) = 0. \quad (1.17)$$

The corresponding confidence interval for θ is $(d_1 + \bar{x}, d_2 + \bar{x})$.

Writing (1.17) in the form

$$d^2 (\hat{\gamma}^2 - A) - 2d\bar{Y}\hat{\gamma} + \bar{Y}^2 - B = 0,$$

we see that it has real unequal roots when

$$A\bar{Y}^2 + B\hat{\gamma}^2 - AB > 0.$$

Sufficient conditions for this are $\hat{\gamma}^2 \geqslant A$ or $\bar{Y}^2 \geqslant B$.

When n is large, an approximate expression for the variance of $\hat{\theta}$ can be derived using the delta method, thus

$$\begin{aligned}
V[\hat{\theta}] &= V[\hat{\theta} - \bar{x}] \\
&= V[\bar{Y}/\hat{\gamma}] \\
&\approx \frac{\sigma^2}{\gamma^2}\left[\frac{1}{\Sigma\, w_i} + \frac{(\theta - \bar{x})^2}{\Sigma\, w_i(x_i - \bar{x})^2}\right].
\end{aligned} \quad (1.18)$$

Assuming $\hat{\theta}$ to be approximately normally distributed, an approximate 95 per cent confidence interval for θ is

$$\hat{\theta} \pm 1\cdot96\, \hat{V}^{\frac{1}{2}}, \quad (1.19)$$

where $\hat{V}$ is $V[\hat{\theta}]$ with θ, γ and σ^2 replaced by their estimates. However, for the applications considered in this book n will usually be small (< 10), so the assumption that $\hat{\theta}$ is approximately normal is open to question. Therefore (1.17) will be used in all applications.

1.3.6   Goodness-of-fit tests

1   Binomial distribution

Let $x_1, x_2, \dots, x_n$ be a random sample from the binomial distribution

$$f(x) = \binom{N}{x} p^x q^{N-x}, \quad (q = 1-p, \quad x = 0, 1, \dots, N).$$

Suppose that x_i takes a value x with frequency f_x ($\sum_x f_x = n$), then it is readily shown that the maximum-likelihood estimate of p is

$$\hat{p} = \bar{x}/N = \sum_{x=0}^{n} x f_x/(nN)$$

and the expected frequencies, E_x, are given by

$$E_x = n \binom{N}{x} \hat{p}^x \hat{q}^{N-x}, \quad x = 0, 1, \ldots, N.$$

Since the joint distribution of the random variables f_x is multinomial with $N + 1$ categories, the goodness-of-fit statistic for testing the appropriateness of the binomial model is

$$T_1 = \sum_{x=0}^{N} (f_x - E_x)^2/E_x,$$

which is approximately distributed as χ^2_{N-1} when n is large.

An alternative test statistic can be obtained by putting the data in the form of a contingency table, namely

x_1	x_2	$\ldots$	x_n	$\sum x_i$
$N - x_1$	$N - x_2$	$\ldots$	$N - x_n$	$nN - \sum x_i$
N	N	$\ldots$	N	nN

and carrying out a test for homogeneity. The test statistic is then

$$T_2 = \sum_{i=1}^{n} \frac{(x_i - N\hat{p})^2}{N\hat{p}\hat{q}}$$

$$= \frac{\sum_{i=1}^{n} (x_i - \bar{x})^2}{\bar{x}[1 - \bar{x}/N]}$$

$$= \frac{\sum_{x=0}^{N} f_x(x - \bar{x})^2}{\bar{x}[1 - \bar{x}/N]}$$

the so-called *Binomial Index of Dispersion*, and it is asymptotically distributed as χ^2_{n-1}. We note that $T_2/(n-1)$ is effectively based on comparing the observed variance estimate $\sum (x_i - \bar{x})^2/(n-1)$ with $N\hat{p}\hat{q}$, an estimate of the expected variance under a binomial model.

2   Poisson distribution

Let $x_1, x_2, \ldots, x_n$ be a random sample from a Poisson distribution

$$f(x) = e^{-\lambda} \frac{\lambda^x}{x!} \quad x = 0, 1, 2, \ldots .$$

Then the maximum-likelihood estimate of λ is $\hat{\lambda} = \bar{x}$ and the expected frequencies are

$$E_x = n e^{-\hat{\lambda}} \cdot \frac{\lambda^x}{x!}.$$

Usually the expected frequencies are pooled for $x \geq X$ so as to ensure that $n - \sum_{x=0}^{X-1} E_x \ (= E_{X+}$ say) is greater than about 5 (though a value as small as 1 can usually be tolerated if $X \geq 4$), and the goodness-of-fit statistic

$$T_1 = \sum_{x=0}^{X-1} \left\{ \frac{(f_x - E_x)^2}{E_x} \right\} + \frac{(f_{X+} - E_{X+})^2}{E_{X+}}$$

is then approximately distributed as χ^2_{X-1}.

Alternatively, we can use the *Poisson Index of Dispersion*

$$T_2 = \sum_{i=0}^{n} (x_i - \bar{x})^2 / \bar{x} = \sum_x f_x (x - \bar{x})^2 / \bar{x},$$

which is asymptotically distributed as χ^2_{n-1}. Since the mean of a Poisson variable equals its variance, $T_2 / (n-1)$ can be regarded as a statistic for comparing the observed variance estimate with the estimate, $\bar{x}$, of the expected variance under a Poisson model.

In general, T_2 will provide a more sensitive test than T_1 (Kendall and Stuart [1973: 599]), though when the underlying distribution is not Poisson a comparison of the f_x and E_x may give some idea as to the form of departure from Poisson. Also T_2 can be used for quite small values of n ($n > 20$; or if $\bar{x} > 1$, $n > 6$: Kathirgamatamby [1953]), while T_1 requires a much larger sample size n in order to ensure that $E_x \geq 5$ for several values of x. The statistic T_2 is further discussed in Rao and Chakravarti [1956] who suggest that the chi-squared approximation to the distribution of T_2 is a good one when $\bar{x} > 3$; they also give a useful table for carrying out an exact small-sample test based on the statistic Σx_i^2.

3   *Multinomial distribution with N unknown*

Let $y_1, y_2, \ldots, y_k$ have a multinomial distribution

$$f_1(y_1, y_2, \ldots, y_k) = \frac{N!}{\left(\prod_{i=1}^{k} y_i! \right)(N-r)!} \left(\prod_{i=1}^{k} p_i^{y_i} \right) Q^{N-r}, \qquad (1.20)$$

where $r = \sum_{i=1}^{k} y_i$, $Q = 1 - \sum_{i=1}^{k} p_i$.

We wish to test the hypothesis H that $p_i = p_i(\boldsymbol{\theta})$ ($i = 1, 2, \ldots, k$), where $p_i(\boldsymbol{\theta})$ is a function of t unknown parameters $\theta_1, \theta_2, \ldots, \theta_t$. When N is known, we can test H using the standard multinomial goodness-of-fit statistic

$$T_1 = \sum_{i=1}^{k} \left\{ \frac{(y_i - N\tilde{p}_i)^2}{N\tilde{p}_i} \right\} + \frac{(N - r - N\tilde{Q})^2}{N\tilde{Q}},$$

where $\widetilde{p}_i = p_i(\widetilde{\theta})$, $\widetilde{Q} = 1 - \sum_{i=1}^{k} \widetilde{p}_i$, and $\widetilde{\theta}$ is the maximum-likelihood estimate of θ for (1.20).

When N is unknown, it transpires that we can work with the conditional multinomial distribution (cf. 1.3.7(2) for the method of derivation)

$$f_2(y_1, y_2, \ldots, y_k \mid r) = \frac{r!}{\prod_{i=1}^{k} y_i!} \prod_{i=1}^{k} \left(\frac{p_i}{1-Q}\right)^{y_i} \qquad (1.21)$$

and use

$$T_2 = \sum_{i=1}^{k} \frac{[y_i - r\hat{p}_i/(1-\hat{Q})]^2}{r\hat{p}_i/(1-\hat{Q})}$$

$$= \sum_{i=1}^{k} \frac{(y_i - \hat{N}\hat{p}_i)^2}{\hat{N}\hat{p}_i},$$

where $\hat{N} = r/(1-\hat{Q})$, $\hat{p}_i = p_i(\hat{\theta})$, and $\hat{\theta}$ is the maximum-likelihood estimate of θ for (1.21). It can be shown, using, for example, the methods of Birch [1964] or Darroch [1959: 347], that when H is true, T_2 is asymptotically distributed as χ^2_{k-t-1} as $N \to \infty$.

By solving the equations $\partial \log f_1/\partial \theta_j = 0$ $(j = 1, 2, \ldots, t)$ and $\Delta \log f_1 = 0$ (where Δ denotes the first difference with respect to N), we find that when N is unknown, $\hat{\theta}$ and $\hat{N}$ are the maximum-likelihood estimates of θ and N for the model (1.20).

1.3.7 Some conditional distributions

1 Poisson variables

If x_1 and x_2 are independent Poisson random variables with means θ_1 and θ_2 respectively, then it is readily shown that the distribution of x_1 conditional on $y = x_1 + x_2$ is binomial, namely

$$f(x_1 \mid y) = \binom{y}{x_1} p^{x_1} q^{x_2}$$

where $p = \theta_1/(\theta_1 + \theta_2)$. Conversely, if x_1 and y are a pair of random variables such that the conditional distribution of x_1 given y is binomial with parameters y and P, and y is Poisson with mean λ, then the unconditional distribution of x_1 is Poisson with mean λP (cf. Feller [1957: 160, Example 27]).

2 Multinomial variables

Let $x_1, x_2, \ldots, x_k$ have a multinomial distribution

$$f(x_1, x_2, \ldots, x_k) = \frac{n!}{\prod_{i=1}^{k} x_i!} \prod_{i=1}^{k} p_i^{x_i}, \quad (\textstyle\sum_i p_i = 1);$$

then the joint marginal distribution of x_1 and x_2 is

$$f_1(x_1, x_2) = \frac{n!}{x_1! \, x_2! \, (n - x_1 - x_2)!} \, p_1^{x_1} p_2^{x_2} (1 - p_1 - p_2)^{n - x_1 - x_2}.$$

If $y = x_1 + x_2$, then y has probability function

$$f_2(y) = \binom{n}{y} (p_1 + p_2)^y (1 - p_1 - p_2)^{n - y}$$

and

$$
\begin{aligned}
f(x_1 \mid y = x_1 + x_2) &= \Pr[X_1 = x_1 \mid Y = y] \\
&= \frac{\Pr[X_1 = x_1 \text{ and } Y = y]}{\Pr[Y = y]} \\
&= \frac{\Pr[X_1 = x_1 \text{ and } X_2 = x_2]}{\Pr[Y = y]} \\
&= \frac{f_1(x_1, x_2)}{f_2(y)} \\
&= \binom{y}{x_1} \left(\frac{p_1}{p_1 + p_2}\right)^{x_1} \left(\frac{p_2}{p_1 + p_2}\right)^{x_2}.
\end{aligned}
\tag{1.22}
$$

1.3.8   Iterative solution of equations

1   One unknown

Suppose we wish to find a root θ_0 of the equation $h(\theta) = 0$. Then, starting with a trial solution $\theta_{(1)}$, a number of iterative procedures are available in which the ith step of the iteration takes the form

$$\theta_{(i+1)} = \theta_{(i)} - h(\theta_{(i)})/k_i. \tag{1.23}$$

For example, if $\theta_{(i)} - \theta_0$ is small, we have the Taylor approximation

$$
\begin{aligned}
0 &= h(\theta_0) \\
&\approx h(\theta_{(i)}) + (\theta_0 - \theta_{(i)}) h'(\theta_{(i)}),
\end{aligned}
$$

where $h'(\theta_{(i)})$ is the first derivative of $h(\theta)$ evaluated at $\theta = \theta_{(i)}$. Rearranging this last equation, we have

$$\theta_0 \approx \theta_{(i)} - h(\theta_{(i)})/h'(\theta_{(i)}), \tag{1.24}$$

so that one method of iteration is to set $k_i = h'(\theta_{(i)})$ in (1.23), the so-called Newton–Raphson method. Another possibility is to choose

$$k_i = \frac{h(\theta_{(i)}) - h(\theta_{(i-1)})}{\theta_{(i)} - \theta_{(i-1)}};$$

this is known as the method of "false positions". When θ_0 is only required to the nearest integer, a useful discrete analogue of the Newton–Raphson method, used for example by Robson and Regier [1968], is to set

$$k_i = \nabla h(\theta_{(i)}) = h(\theta_{(i)}) - h(\theta_{(i)} - 1).$$

Conditions for the convergence of such procedures are discussed in standard textbooks on numerical analysis (see also Kale [1961, 1962]).

2   Several unknowns

Suppose that we have r equations

$$h_i(\boldsymbol{\theta}) \;=\; 0 \quad (i = 1, 2, \dots , r)$$

in r unknowns $\theta_1, \theta_2, \dots , \theta_r$. Then the multivariate analogue of the Newton–Raphson method (1.24) is

$$\boldsymbol{\theta}_{(i+1)} \;=\; \boldsymbol{\theta}_{(i)} - \mathbf{H}_{(i)}^{-1}\mathbf{h}(\boldsymbol{\theta}_{(i)}), \qquad (1.25)$$

where

$$\mathbf{H}_{(i)} \;=\; \left[\left(\frac{\partial h_i(\boldsymbol{\theta})}{\partial \theta_j}\right)\right]_{\boldsymbol{\theta}=\boldsymbol{\theta}_{(i)}}$$

and

$$\mathbf{h}(\boldsymbol{\theta}_{(i)}) \;=\; \begin{bmatrix} h_1(\boldsymbol{\theta}_{(i)}) \\ h_2(\boldsymbol{\theta}_{(i)}) \\ \cdots \\ h_r(\boldsymbol{\theta}_{(i)}) \end{bmatrix}.$$

3   Maximum-likelihood equations

ONE PARAMETER. Let $x_1, x_2, \dots , x_n$ be a random sample from a probability (or probability density) function $f(x, \theta)$ and let

$$L(\theta) \;=\; \prod_{i=1}^{n} f(x_i, \theta)$$

be the likelihood function. Then $\hat{\theta}$, the maximum-likelihood estimate of θ, is in general a solution of

$$h(\theta) \;=\; \partial \log L(\theta)/\partial \theta \;=\; 0, \qquad (1.26)$$

which can be solved by the methods of section 1 above.

If f has certain reasonable properties, then $\hat{\theta}$ is asymptotically $\mathfrak{N}[\theta, \sigma_{\hat{\theta}}^2]$, where

$$\sigma_{\hat{\theta}}^2 \;=\; \left\{ - E\left[\frac{\partial^2 \log L(\theta)}{\partial \theta^2}\right]\right\}^{-1}$$

can be estimated by

$$v \;=\; \left\{ -\left[\frac{\partial^2 \log L(\theta)}{\partial \theta^2}\right]\right\}^{-1}_{\theta=\hat{\theta}}. \qquad (1.27)$$

Since $-1/h'(\theta_{(i)})$ tends to v as $\theta_{(i)}$ approaches $\hat{\theta}$, we find that v can be calculated from the last iteration of the Newton–Raphson method.

Some of the problems involved in solving (1.26) are discussed by Barnett [1966]. He shows that the method of false positions is preferred when (1.26) has multiple roots.

SEVERAL PARAMETERS. If θ is now replaced by a vector $\boldsymbol{\theta} = [(\theta_i)]$ of r parameters, then $\hat{\boldsymbol{\theta}}$, the maximum-likelihood estimate of $\boldsymbol{\theta}$, is a solution of the r equations

$$h_i(\boldsymbol{\theta}) \ = \ \partial \log L(\boldsymbol{\theta})/\partial\theta_i \ = \ 0 \quad (i = 1, 2, \ldots, r).$$

Using the Newton–Raphson method (equation (1.25)) we have that

$$\mathbf{H}_{(i)} \ = \ [(\partial^2 \log L(\boldsymbol{\theta})/\partial\theta_i \partial\theta_j)]_{\boldsymbol{\theta}=\boldsymbol{\theta}_{(i)}}$$

and, as in the case of one parameter, the term $-\mathbf{H}^{-1}$ in the last step of the iteration provides an estimate of

$$- [(E[\partial^2 \log L(\boldsymbol{\theta})/\partial\theta_i \partial\theta_j])]^{-1} , \qquad\qquad (1.28)$$

the asymptotic variance–covariance matrix of $\hat{\boldsymbol{\theta}}$.

CHAPTER 2

DENSITY ESTIMATES FOR CLOSED POPULATIONS

2.1 ABSOLUTE DENSITY

2.1.1 Total count over whole area

Occasionally it is possible to count all the animals of a particular species in a given area by systematically traversing the whole area on foot or, in the case of large animals, by air. For example, territorial species such as birds can be counted in this manner as the animals tend to stay in their territories when observed, thus reducing the possibility of duplication in counting through movement in and out of adjacent areas. Animals which congregate in groups can often be photographed and counted later (Davis [1963]), while fish which must migrate through rivers during part of their life cycles can be counted individually using traps or weirs (Scattergood [1954]). Recently radar has been used for estimating bird densities (Dyer [1967], Myres [1969]) and echo-sounding for counting fish (Cushing [1964, 1968], Shiraishi and Furuta [1963]).

The accuracy of the total count will depend on the visibility of the animals; this in turn depends on a number of factors such as animal size, vegetation cover available, and activity of the animal at the particular time of the year. During the censusing the population should be disturbed as little as possible, to avoid both the duplication in counting mentioned above and the possibility of frightened animals moving out of the area altogether.

In an effort to carry out a total count some attempts at killing or live-trapping all the members of a given population have been recorded in the literature. For example, fish populations have been counted by draining ponds in order to check on the accuracy of other methods of estimation (Buck and Thoits [1965]). Also with some populations of small mammals it is possible to tag and release animals until only tagged animals are being caught. In this way virtually the whole population can be counted, provided all the animals are of trappable age and there is no evidence of trap avoidance. When such counts are to be converted to density estimates of so many per unit area, the problem arises as to just what area has been trapped. Usually one adds to the area enclosed by the traps a strip of width one-half the home range of the animal (Dice [1938]). Some authors subtract from the total population area all regions which are definitely uninhabitable, as far as the given species is concerned, to obtain a "maximal" density estimate. A method resembling direct enumeration, called the "calendar of captures" method, has been widely used in Poland (cf. Petrusewicz and Andrzejewski [1962], Andrzejewski [1963], Ryszkowski et al. [1966], Andrzejewski [1967]).

The home range, if it exists, can be estimated from the trap records of individuals (Southwood [1966: 261], Mohr and Stumpf [1966], Sanderson [1966], Kikkawa [1964: 276], Jennrich and Turner [1969]) or by using various tracking devices such as radio transmitters (Giles [1963], Adams and Davis [1967]), radioactive tracers (Miller [1957], Kaye [1961], Kikkawa [1964], Harvey and Barbour [1965]), smoked kymograph paper (Justice [1961], Sheppe [1965], Bailey [1968]), dyes (New [1958]), and even spools of thread (Dole [1965]); for general reviews see Spitz [1963] and Sanderson [1966]. A technique for the simultaneous estimation of home range and population density, due to MacLulich [1951], is described in **2.1.5.**

2.1.2   Total counts on sample plots

With some populations it is impracticable or impossible to count the animals over the whole area because of the disturbance caused or the number of personnel required. In this case a sampling scheme is required whereby total counts are made on randomly chosen sample areas. An estimate of total population size is then obtained by multiplying the average density per unit area, estimated from the sample areas, by the total area of the population. The main steps in devising such a sampling scheme are as follows:

(1) The size and shape of the sample area or "plot" should be determined and one's choice will be guided by such factors as the species under consideration, the terrain, the distribution of vegetation cover and the distribution of food supply. One can use either quadrats, a term often loosely applied to plots such as squares, rectangles, circles etc. (though strictly it applies to square plots), or transects, the term for strips running the length of the population area. Line transects, which are discussed in **2.1.3.**, are also being widely used in ecological research.

(2) The number of sample plots should be determined in advance, unless of course sequential methods are used (e.g. Southwood [1966: 43]). This number will depend on the precision required for the estimate of population density, and such factors as staff and time available for the sampling.

(3) The sample plots should be located by some random procedure so that valid sampling errors can be calculated. If there are marked differences in, for example, vegetation cover, one can divide up the total area into domains or "strata" of uniform habitat and then use stratified random sampling (Cochran [1963]). This stratification can be achieved using any factor related to population density: for example, in the study of ruffed grouse populations Ammann and Ryel [1963] used a stratification based on deer density. Another useful application of stratification is in sampling for soil animals. Here, very small plots are used, so that choosing the plots at random over the whole area may lead to an undesirable bunching of the plots (Healy [1962]). However, by using stratification and randomisation within each stratum a certain degree of coverage is achieved.

We note that for mobile species the number counted on a sample plot may not be the actual number present before the arrival of the investigator. Some

of the animals will hide while others may move away, so that the counts are
more of an index of population density than a measure of absolute density.
In big game studies aerial sampling can be used (Hunter and Yeager [1949],
Siniff and Skoog [1964]), but once again the observer may not be able to see
all the animals present. For example, the animals may tend to stay under
cover in certain weather conditions or at certain times of the day. Sometimes
the terrain is so rugged that planes cannot fly safely at the low levels
necessary for adequate observation. Examples of comparing aerial and
ground counts are given in Riordan [1948] and Gilbert and Grieb [1957].

We shall now discuss the above three steps in some detail with regard
to quadrats and strip transects: line transects will be dealt with in a
separate section.

1   Quadrats: simple random sampling

Quadrat sampling is particularly useful for dealing with ant and termite
mounds (Waloff and Blackith [1962], Sands [1965]) and animals which are
fairly stationary, e.g. soil animals (Murphy [1962]), some species of insects
(Southwood [1966: 108]), and birds which are readily detected (Kendeigh
[1944], Einarsen [1945: pheasants], Ammann and Baldwin [1960: wood-
peckers]). It has also been used for mobile animals such as small mammals
(Bole [1939], Turček [1958], Pelikan *et al.* [1964]) and deer (Trippensee
[1948]); in this case the size of the quadrat appears to be crucial. Siniff and
Skoog [1964] preferred quadrats to strips in aerial censusing.

SHAPE. Although circular and even hexagonal plots (e.g. Lloyd [1967])
have been used, squares and rectangles are generally the easiest shapes to
mark out or identify, particularly if large plots are required. Also the
sampling procedures for selecting such plots are usually simpler than those
for plots with curved sides. In comparing squares and rectangles, sometimes long narrow rectangles are more efficient than squares of the same
area (e.g. Clapham [1932]). However, the longer the rectangle, the greater
the length of boundary per unit area, so that there is a greater chance of
error per unit area in determining whether individuals on the edges of the
plot are inside or outside the boundary, and whether or not the plot population is closed during the counting. Although this error per unit area decreases as the shape of the plot approaches a circle, circular plots do not
fit together without leaving spaces. This complicates the sampling procedure unless the plot areas are very small compared to the total area, as
in soil sampling. Another possibility is the hexagonal, a compromise between a square and a circle as far as edge effect is concerned. However,
although hexagonals can be fitted together, the time taken to mark them out
precludes their use in most situations. Therefore, as a general recommen-
dation, square sample plots should be used for most populations.

SIZE. In determining the size of a quadrat we note that a density estimate from a large number of small quadrats will usually have a smaller

variance than that from a few large quadrats of the same total area (though the reverse can occur when the population is not randomly distributed). It is therefore recommended that the quadrat be as small as possible, bearing in mind that the edge effect mentioned above increases as the plot size decreases. For example, in populations of small mammals where a quadrat count is obtained by trapping out the quadrat area over several trap-nights, the smaller quadrats are more affected by animals drifting into them from adjacent areas (Bole [1939]). Also the quadrat should not be so small that the majority of quadrats are found to be empty. One working rule suggested by Greig-Smith [1964] is that there should not be more empty quadrats than quadrats with just one individual.

NUMBER OF PLOTS. Let

N = total population size,
A = area occupied by the population,
D = N/A (the average population density),
pA = area sampled,

and n = number of animals found in the sampling area.

We shall assume that the population is randomly distributed, i.e. the probability of finding an individual at any particular point in the population area is the same for all points, and the presence of one individual does not influence the position occupied by another. Then n is binomially distributed with probability function

$$f(n) = \binom{N}{n} p^n q^{N-n} \quad (q = 1-p)$$

and a natural estimate of N is

$$\hat{N} = n/p. \tag{2.1}$$

This is unbiased with variance

$$V[\hat{N}] = Nq/p$$

and coefficient of variation

$$C = \sqrt{\left(\frac{q}{Np}\right)}.$$

Rearranging this last equation, we have

$$p = 1/(1 + NC^2), \tag{2.2}$$

so that given a rough lower bound for N and a prescribed value of C, p can be determined. For example, one general rule often stated (e.g. Dice [1952: 33]) is that about 5–10 per cent of the population area should be sampled. This would mean that for $p = 0.1$, say, we could only obtain a precision of $C \leqslant 0.2$ if $N \geqslant 225$. Therefore it is not possible to obtain reasonably precise estimates for small populations without sampling a very high proportion of the population area. In fact, for $p < 0 \cdot 2$ we have $\sqrt{q} \approx 1$, and since n is an estimate of Np, C is estimated by $1/\sqrt{n}$. This means that the precision depends only on the total number of animals seen when less than 20 per cent of the population area is sampled.

When the population is not randomly distributed it is difficult to determine the precise effect of non-randomness on (2.2), though one or two comments may be made. If the animals tend to cluster (i.e. the population is "contagious"), as is often the case, $V[\hat{N}]$ will generally be greater than Nq/p and (2.2) will underestimate p. However, this effect will be small if the size of the quadrat is very much smaller than the average cluster size, or if the quadrat is large enough so that it generally includes one or more clusters (Greig-Smith [1964: 56–57]). In general, (2.2) will be satisfactory if the total area sampled is sufficiently large, so that the clustering tends to affect only the distribution and not the number of animals in the sampled area.

When p has been determined, the next step is to choose the quadrat area a and the number of sample plots s, subject to $sa = pA$. Normally one would endeavour to make s large as as possible, bearing in mind the time taken to map out a quadrat in the field and the possible edge effects of small quadrats.

SELECTING THE QUADRATS. Once s and a are determined, the quadrats must be chosen by some random procedure if the statistical error in our estimate of N is to be calculated. If a is very small compared with A, as in soil sampling where circular plots of less than 10 in. (260 cm) diameter are used, the simplest method for locating the centre of the plot (or one corner if square plots are used) is to read off a pair of numbers from a table of random numbers and use them as coordinates on a grid system (cf. Sampford [1962: Chapter 4]). However, if a is an appreciable fraction of A then the above method of random selection will lead to repeated overlapping of the sample quadrats. In this case the simplest method is to divide up the whole area into numbered quadrats of area a (ignoring incomplete quadrats on the boundary), and then use a table of random numbers to obtain a selection of s of these plots.

ESTIMATION. Let x_i be the number of animals seen in the ith quadrat $(i = 1, 2, \ldots, s)$; then $n = \Sigma \; x_i$ and, since $sa = pA$,

$$\hat{N} \; = \; n/p \; = \; \bar{x}A/a \; = \; \bar{x}S, \quad \text{say.} \tag{2.3}$$

This is the usual estimate of a total based on a simple random sample without replacement of s quadrats from a population of $S \; (= A/a)$ quadrats (Cochran [1963: Chapter 2]). Therefore $\hat{N}$ will still be an unbiased estimate of N, even when the population is not randomly distributed, though the formula for $V[\hat{N}]$ above will no longer be valid. In this case an unbiased estimate of the true variance is given by (Cochran [1963: 25])

$$\hat{\sigma}_N^2 \; = \; S^2 \, \frac{v}{s} \left(1 - \frac{s}{S} \right),$$

where

$$v = \sum_{i=1}^{s} (x_i - \bar{x})^2/(s-1).$$

Also an approximate $100(1-\alpha)$ per cent confidence interval for N is given by $\hat{N} \pm t_{s-1}[\alpha/2]\hat{\sigma}_N$: dividing this interval by A gives a confidence interval for the population density D.

PILOT SURVEY. In designing an experiment, if a rough estimate of N is not available, an alternative approach would be to choose a so that S is large and then carry out a pilot sample using s_0 quadrats, say. Calculating $\bar{x}$ and v from this sample, it is readily shown that $s - s_0$ further quadrats are required, where s is the smallest integer satisfying

$$\frac{1}{s} \leqslant \frac{1}{S} + \frac{C^2\bar{x}^2}{v}, \tag{2.4}$$

to give a prescribed coefficient of variation C.

VARIABLE PLOT SIZE. In some populations the total area may already be divided up into T plots of varying shapes and sizes by natural or man-made barriers such as watercourses, fences, etc. We then find that, provided the s plots are chosen at random, the above theory still holds but with S replaced by T. Obviously, when the population density per unit area is constant, so that the number of animals in a plot is proportional to plot size, the greater the variation in plot size the larger is v. This means that for a constant density and a given value of T, the most efficient method of sampling is to use plots of equal area.

TESTS OF RANDOMNESS. When the population is randomly distributed then x, the number of individuals in a sample plot of area δA, has a binomial distribution with parameters M and θ, where M is the maximum number of individuals the sample area can contain and θ is the probability that any one of the M possible "places" in δA is occupied. When M is large, θ small and $M\theta$ moderate, the distribution of x is best approximated by the Poisson distribution with mean $M\theta$, where $M\theta$, the expected number found in δA, is $D\delta A$ (cf. Greig-Smith [1964: 11–14] or, for a rigorous derivation, Dacey [1964a]). For animal populations which are randomly distributed, this Poisson model would normally be more appropriate than the binomial, as δA is usually very large compared with the area taken up by an individual, and D is usually small, so that the actual number of places occupied by individuals is only a small fraction of the M possible places. Therefore, if $x_1, x_2, \ldots, x_s$ are the numbers of animals found in s sample plots of equal area, a test of randomness amounts to testing the hypothesis that these observations come from a Poisson distribution. One can use either the standard chi-squared goodness-of-fit test or the Poisson Index of Dispersion (cf. **1.3.6** (2)). However, it is advisable to use both tests as the rare situation can arise in which one of the tests may detect evident non-randomness when the other fails to do so (Greig-Smith [1964: 68–70]). Also it should be stressed that the degree of non-randomness detected will depend on the size of the sample plot used (cf. Greig-Smith [1964: Chapter 3] and Kershaw [1964: Chapter 6]).

It seems that, in general, natural plant or animal populations are rarely distributed at random and that organisms are usually found to be clustered together more than a Poisson distribution would predict. For example, small mammals will tend to congregate in regions where there is shelter, thus avoiding open areas. This means that there will be more empty plots and more plots with many individuals than expected, so that the variance of x will be greater than its mean. When the Poisson distribution is not applicable a number of alternative distributions have been put forward to account for the spatial patterns observed (cf. Greig-Smith [1964], Southwood [1966], Pielou [1969], and King [1969] for references). In particular the negative binomial distribution

$$f(x) = \frac{k(k+1)\ldots(k+x-1)}{x!} P^x Q^{-x-k} \quad x = 0, 1, 2, \ldots$$

obtained by expanding $(Q-P)^{-k}$, where $Q = 1 + P$ and $k > 0$, has been found to be the most useful because of its flexibility. Since

$$E[x] = kP = \mu, \quad \text{say,}$$

and

$$V[x] = kPQ = \mu + \mu^2/k,$$

we see that this distribution can be fitted to a wide range of distributions, ranging from the Poisson with variance equal to the mean ($k = \infty$) to those in which the variance is much greater than the mean. For example, in insect populations k is often found to be about 2 (Southwood [1966: 25]). A general discussion on the negative binomial is given in Southwood, and some interesting examples of fitting data to Poisson and negative binomial distributions are given in Andrewartha [1961], Debauche [1962] and South [1965]; further applications to fish populations are given by Moyle and Lound [1960], Roessler [1965], and Houser and Dunn [1967]. A recent study on goodness-of-fit tests for the negative binomial is given in Pahl [1969].

If it is known from past experience that a population is scattered according to a negative binomial then from (2.3) we have

$$N = E[\hat{N}] = SE[\bar{x}] = SkP,$$

$$V[\hat{N}] = \frac{S^2}{s} V[x] = \frac{S^2}{s} kP(1+P) = \frac{N}{s}\left(S+\frac{N}{k}\right)$$

and

$$(C[\hat{N}])^2 = V[\hat{N}]/N^2 = \frac{1}{s}\left(\frac{S}{N}+\frac{1}{k}\right).$$

Rearranging this last equation yields

$$s = \frac{1}{C^2}\left(\frac{S}{N}+\frac{1}{k}\right),$$

so that if approximate estimates of N and k are available, s, the number of sample plots of area a, can be calculated to give a prescribed C. If a pilot sample is used then N/S is estimated by $\bar{x}$, and k can be estimated using one of the methods described in Southwood [1966: 26].

2   Quadrats: *stratified random sampling*

NUMBER OF PLOTS. Suppose now that the total population area is divided into J domains (or strata), each of area A_j ($j = 1, 2, \ldots, J$; $\Sigma A_j = A$), and let N_j be the number of animals in the jth domain ($\Sigma N_j = N$). If an area $p_j A_j$ is sampled in the jth domain and found to contain n_j animals, then if the animals are randomly distributed within each domain, an unbiased estimate of N is given by (cf. (2.1))

$$\widetilde{N} = \sum_j \hat{N}_j = \sum_j (n_j/p_j)$$

with variance

$$V[\widetilde{N}] = \sum_j N_j q_j/p_j .$$

To determine the optimum allocation of sampling effort we minimise $V[\widetilde{N}]$, subject to

$$\Sigma p_j A_j = \text{constant} \ (= pA, \text{ say})$$

and obtain

$$p_j = K\sqrt{(N_j/A_j)} = K\sqrt{D_j}, \quad \text{say},$$

where $K = pA/\sum_j \sqrt{(N_j A_j)}$.

If these values of p_j are used, we find that

$$V[\widetilde{N}] = \frac{\{\Sigma \sqrt{(N_j A_j)}\}^2}{pA} - N$$

and for a given coefficient of variation C we must have

$$p = \frac{\{\Sigma \sqrt{(N_j A_j)}\}^2}{AN(1+NC^2)} .$$

When rough estimates of the N_j are available, then for a given C, p — and hence the p_j — can be determined by the above equations.

We note that for optimum allocation we must choose p_j proportional to the square root of the population density D_j in the jth domain. In fact, even just relative density estimates (cf. **2.2**) can be used for determining the p_j. For example, if $p_j = K_1 \sqrt{R_j}$, where R_j is an estimate of relative density (the same units being used for each domain), and p is determined in advance (by the resources available for sampling), then

$$K_1 = pA/\Sigma A_j \sqrt{R_j}$$

and the p_j can be calculated.

If a proportional allocation of area is used, so that $p_j = p$, then

$$\widetilde{N} = \sum_j n_j/p$$

and p can be determined from (cf. (2.2))

$$p = 1/(1+NC^2).$$

ESTIMATION. Suppose that s_j plots are chosen at random from the T_j plots available in the jth domain and let x_{ij} be the count on the ith sample plot $(i = 1, 2, \ldots, s_j;\ \Sigma\ s_j = s,\ \Sigma\ T_j = T)$. We shall assume, for completeness, the general case where the plots may have different sizes. Then if

$$\overline{x}_j = \sum_{i=1}^{s_j} x_{ij}/s_j$$

is the average sample count for the jth domain, N is estimated by (cf. (2.3))

$$\widetilde{N} = \sum_j \hat{N}_j = \sum_j \overline{x}_j T_j$$

and an unbiased estimate of $V[\widetilde{N}]$ is given by

$$\widetilde{\sigma}_N^2 = \sum_j \hat{\sigma}_{N_j}^2 = \sum_j T_j^2 \frac{v_j}{s_j}\left(1 - \frac{s_j}{T_j}\right),$$

where $\quad v_j = \sum_i (x_{ij} - \overline{x}_j)^2/(s_j - 1).$

A $100(1 - \alpha)$ per cent confidence interval for N is given by $\widetilde{N} \pm t_k[\alpha/2]\widetilde{\sigma}_N$, where k, the number of degrees of freedom, is given by (Cochran [1963: 95])

$$k = \frac{\widetilde{\sigma}_N^4}{\displaystyle\sum_j \left(\frac{\hat{\sigma}_{N_j}^4}{s_j - 1}\right)}.$$

We note that the above theory does not depend on the assumption that the population is randomly distributed within each domain.

PILOT SURVEY. When estimates $\widetilde{N}$ and v_j are available from a pilot sample then, for a given s, the optimum value of s_j which minimises $V[\widetilde{N}]$ is estimated by (Cochran [1963: 97])

$$s_j = \frac{s T_j \sqrt{v_j}}{\Sigma\ T_j \sqrt{v_j}} = s r_j, \quad \text{say.}$$

With this choice of s_j it is readily shown that, for a given coefficient of variation C, we must have

$$s = \frac{\Sigma\ T_j^2 v_j r_j^{-1}}{\widetilde{N}^2 C^2 + \Sigma\ T_j v_j}. \tag{2.5}$$

If, however, proportional sampling is used, so that $s_j/T_j = s/T$, then we simply set $r_j = T_j/T$ in (2.5).

In some situations it may not be possible or practicable to use either optimum or proportional sampling. However, data from a pilot sample can still be used to design an experiment in which $\widetilde{N}$ has a coefficient of variation no greater than C. Calculating $\overline{x}_j$, v_j and $\hat{N}_j$ from the pilot sample, we choose s_j so that the coefficient of variation of the estimate of N_j will be no greater than C, i.e. from (2.4) we have

$$\frac{1}{s_j} \leq \frac{1}{T_j} + \frac{C^2\,\bar{x}_j^2}{v_j}.$$

Then, defining

$$b_j = \left[T_j^2\, \frac{v_j}{s_j} \left(1 - \frac{s_j}{T_j} \right) \right]^{\frac{1}{2}}$$

we have $b_j/\hat{N}_j \leq C$, and

$$\hat{C}[\tilde{N}] = \frac{(\Sigma\, b_j^2)^{\frac{1}{2}}}{\Sigma\, \hat{N}_j} < \frac{\Sigma\, b_j}{\Sigma\, \hat{N}_j} \leq \max_j \left\{ \frac{b_j}{\hat{N}_j} \right\} \leq C.$$

Some recent examples of stratified sampling applied to animal populations are given in Foote *et al.* [1958], Wiegert [1964], Siniff and Skoog [1964], Evans *et al.* [1966], and Sen [1970].

3   *Strip transects*

For a given proportion p of area sampled, strips will generally give more variable estimates of population size than quadrats because of the edge effects mentioned above in section *1* and the fact that strips running the full length of the area cover a greater variation in type of habitat. However, strip sampling is usually easier to carry out, so that the time saved can be devoted to covering a greater area. In particular, by simply working from a base-line running across the population area, a strip is more readily located than a quadrat. This is a considerable advantage when large quadrats are needed, as in the case of sparsely scattered populations. Also, quadrat sampling may be impracticable where the terrain is difficult or where the animals are very mobile (e.g. Brock [1954: reef fish populations], Pielowski [1969: European hare]).

Usually population areas are irregular in shape, so that strips running the full length of the area will vary in size. This means that as far as the estimation of N is concerned, the general theory of the previous section as applied to sampling with a variable plot size can be used here. A correction for the lack of visibility in immobile populations is given by Anderson and Pospahala [1970].

2.1.3   Line transect methods

1   *General theory*

In the line transect method, an observer walks a distance L across the population in non-intersecting and non-overlapping lines, counting n, the number of animals sighted, and recording one or more of the following statistics at the time of first sighting:

(a) radial distance r_i $(i = 1, 2, \ldots, n)$ from observer to animal,
(b) right-angle distance y_i from the animal sighted to the path of the observer, and
(c) angle of sighting θ_i from the observer's path to the point at which the animal was first sighted.

This technique seems to be most useful for populations in which the animals can only be seen when they are disturbed and flushed into the open; in this case r_i is called the "flushing" distance.

A number of different methods have been proposed for estimating the population density $D (= N/A)$ from the above data, and for a summary of the methods to date the reader is referred to Eberhardt [1968] and Gates [1969]. All the formulae are of the form $\hat{D} = n/(2Lw)$, where w is some measure of one-half the "effective width" of the strip covered by the observer as he moves down the line transect. The basic assumptions underlying the various models are:

(1) The animals are randomly and independently distributed over the population area, i.e. the probability of a given animal being in a particular region of area δA is $\delta A/A$.
(2) The sighting of one animal is independent of the sighting of another.
(3) No animal is counted more than once.
(4) When animals are seen through being flushed into the open, each animal is seen at the exact position it occupied when startled.
(5) The response behaviour of the population as a whole does not substantially change in the course of running a transect.
(6) The individuals are homogeneous with regard to their response behaviour, regardless of sex, age, etc.
(7) The probability of an animal being seen, given that it is a right-angle distance y from the line transect path (irrespective of which side of the path it is on), is a simple function $g(y)$, say, of y, such that $g(0) = 1$ (i.e. probability 1 of seeing an animal on the path).

Given the above assumptions, and using a slight generalisation of a method given by Gates *et al.* [1968], we shall now derive the joint distribution of the observed distances $y_1, y_2, \ldots, y_n$ and n, the number of sightings.

If an animal is on a particular line at right angles to the transect, then by assumptions (1) and (7),

$$\Pr[\text{animal in } (y, y+dy)] = 2L\,dy/A,$$
$$\Pr[\text{animal seen} \mid \text{animal in } (y, y+dy)] = g(y),$$

so that

$$\Pr[\text{animal seen in } (y, y+dy)] = 2Lg(y)dy/A$$

and the probability of sighting an animal from the transect is

$$P = \frac{2L}{A} \int_0^\infty g(y)dy = \frac{2Lc}{A}, \quad \text{say.} \tag{2.6}$$

Since

$$\Pr[\text{animal in } (y, y+dy) \mid \text{animal is seen}]$$

$$= \Pr[\text{animal seen in } (y, y+dy)]/\Pr[\text{animal is seen}]$$
$$= 2L g(y)dy/(AP)$$
$$= c^{-1}g(y)dy,$$

the probability density function (p.d.f.) for y is

$$f(y) = c^{-1}g(y), \tag{2.7}$$

and the joint p.d.f. of $y_1, y_2, \dots, y_n$ conditional on n is

$$f(y_1, y_2, \dots, y_n \mid n) = c^{-n} \prod_{i=1}^{n} g(y_i). \tag{2.8}$$

From assumption (2), the probability function of n is binomial, namely

$$f(n) = \binom{N}{n} P^n Q^{N-n}, \tag{2.9}$$

where $Q = 1 - P$, so that the joint p.d.f. of the y_i's and n is

$$f(y_1, y_2, \dots, y_n, n) = \binom{N}{n} P^n Q^{N-n} c^{-n} \prod_{i=1}^{n} g(y_i). \tag{2.10}$$

Once $g(y)$ is specified as a function of y and of certain unknown parameters, c and P can be derived in terms of these parameters and then (2.10) can be maximised to obtain maximum-likelihood estimates of N and the unknown parameters. It transpires that maximising (2.10) is equivalent to maximising (2.8) with respect to the unknown parameters and then, using these estimates in P, maximising (2.9) with respect to N, i.e. $\hat{N} = n/\hat{P}$. We shall now consider three different models for the function g.

2   *Exponential law for probability of detection*

Gates *et al.* [1968] choose

$$g(y) = \exp(-\lambda_1 y)$$

so that $c = 1/\lambda_1$, $P = 2L/(A\lambda_1)$, and show that the maximum-likelihood estimate of λ_1 (corrected for bias) is

$$\hat{\lambda}_1 = (n-1)/\Sigma y_i.$$

Hence $\hat{N}$, the maximum-likelihood estimate of N, is given by

$$\hat{N}_5 = n/\hat{P} = nA\hat{\lambda}_1/(2L)$$

and the population density is estimated by

$$\hat{D}_5 = (n-1)/(2\bar{y}L).$$

(The suffix 5 is introduced in preparation for a later comparison of density estimates.) Setting $\hat{c} = 1/\hat{\lambda}_1$, we also have

$$\hat{D}_5 = n/(2L\hat{c}). \tag{2.11}$$

We note in passing that the error in using infinity as the upper limit in (2.6) when the population area is strictly finite will in most cases be negligible. This is because A is usually large, so that the line transect is well away from the boundaries, i.e. $g(y)$ is approximately zero for y, the distance to the nearest boundary.

Gates et $al.$ [1968] prove that

$$E[\hat{N}_5] = N(1 - Q^{N-1})$$

and give an approximate expression for the variance, namely

$$V[\hat{N}_5] \approx \frac{N}{P} \left\{ Q + NPE\left[\frac{1}{n-2}\right]\right\}.$$ (2.12)

This means that for moderate N, $\hat{N}_5$ is almost unbiased, and a consistent estimate of the above variance approximation is

$$v_5 = \frac{n}{\hat{P}^2}\left[1 - \hat{P} + \frac{n}{n-2}\right].$$ (2.13)

We note that in applying the above theory y_i, L and A should be measured in the same units (e.g. miles (km) and square miles (sq km) respectively).

EXPERIMENTAL DESIGN. As in quadrat sampling, the first problem in designing a line transect survey is to determine in advance what L or n should be in order to obtain an estimate of N with a prescribed coefficient of variation C. From (2.13) C^2 is estimated by

$$\hat{C}^2 = \frac{1 - \hat{P}}{n} + \frac{1}{n-2},$$

and neglecting $\hat{P}$ (which is usually small), this quadratic in n can be solved for the larger root. For example, if $C = 0{\cdot}25$ then $n = 33$. Therefore, choosing $C < 0{\cdot}25$, $C^2 \approx 2/n$ which leads to a value of n approximately twice that required for quadrat sampling (p. 23: see also 12.1.2 (1) for a cost comparison). This means that line transect estimates are not very precise unless n is large. We note that if $\hat{P}$ is not negligible then ignoring it will lead to a conservative value of n, i.e. n will be larger than necessary.

If a preliminary pilot survey is carried out with length L, then Gates et $al.$ show that the total length, L_T say, required to give a prescribed C is given by

$$L_T = \frac{\hat{\lambda}_1 A(n-1)}{(C^2\hat{N}_5 + 1)(n-2)},$$

where $\hat{\lambda}_1$, $\hat{N}_5$ and n are obtained from the pilot survey. This expression can derived by noting that if n_T is the number of animals that would be observed for length L_T then, since $n_T > n$,

$$n/(n-2) > n_T/(n_T - 2)$$

and $n/(n-2)$ can be used as a conservative estimate of $NPE[1/(n_T - 2)]$ in (2.12).

UNDERLYING ASSUMPTIONS. Assumptions (1), (2) and (3) (p. 29) can be tested by dividing L into a large number of smaller segments of length l and noting n_l, the number seen in each segment. If P_l is the probability of sighting an animal from such a segment, then P_l will be very small, so that when assumptions (1)–(3) hold, n_l will have a Poisson rather than a binomial distribution with mean NP_l. Therefore one can test the goodness of fit of the values n_l to the Poisson distribution by using a standard chi-squared test or the Poisson Index of Dispersion (**1.3.6** (2)).

Assumption (2) is violated if the animals tend to flush in pairs, for example. However, according to Gates [1969], the estimate $\hat{N}_5$ will not be biased by this tendency, though the variance of $\hat{N}_5$ will be increased.

Assumption (4) can be relaxed where it is possible to assume that a certain constant proportion of the animals flee and are not seen on the approach of the observer. For example, some game birds run along the ground instead of flushing and may not be seen in heavy cover. Where such a constant proportion is known to exist and can be determined independently, one may substitute $P' = kP$ for P in the general theory and still obtain an approximately unbiased estimate of N.

Assumption (6) can be investigated by analysing the data from the different subgroups in the population separately. However, when the other assumptions hold, we find that even in the extreme case where every individual is from a different subgroup with regard to behaviour pattern, the true variance of $\hat{N}_5$ will be *smaller* than that given by (2.11) (Kendall and Stuart [1969: 127]).

To test whether $g(y)$ is exponential, we note from (2.7) that y is distributed as $\lambda_1 \exp(-\lambda_1 y)$. If n is large enough, we can therefore carry out a standard chi-squared goodness-of-fit test to see whether the observations y_i come from this negative exponential distribution (cf. p. 46).

Any inaccuracies in the estimation of the right-angle distances y_i will not bias $\hat{N}_5$, provided the errors are not biased; though the variance of $\hat{N}_5$ will be increased. Therefore, in view of the fact that some of the departures mentioned above do not bias $\hat{N}_5$ but tend to increase the variance, it is recommended that one should also calculate the less efficient estimate $\bar{N}_5$, the average of the estimates $\hat{N}_{5j}$, say $(j = 1, 2, \ldots, s)$, for s transects of length L/s. Then, from (1.9), a robust estimate of $V[\bar{N}_5]$ is given by

$$\frac{1}{s(s-1)} \sum_{j=1}^{s} (\hat{N}_{5j} - \bar{N}_5)^2.$$

Example 2.1   Ruffed grouse (*Bonasa umbellus*): Gates *et al.* [1968].

Estimated densities for ruffed grouse at the Cloquet Forest Research Centre, Minnesota, in the line transect surveys conducted there during the years 1950–58 are presented in Table 2.1. The King, Hayne and Webb estimates of density are also included for comparison (see section 5 for definitions).

TABLE 2.1
Estimated densities for a ruffed grouse population at Cloquet Forest Research Centre, Minnesota, 1950–58: from Gates *et al.* [1968: Table 1].

Date of survey	L (miles)	n (number seen)	$\hat{D}_1$ (King)	$\hat{D}_2$ (Hayne)	$\hat{D}_3$ (Webb)	$\hat{D}_4$ ± standard deviation (Gates *et al.*)
			Density estimate (birds per square mile)			
1950A	28·00	45	98·5	188·7	155·7	148·8 ± 30·6
1950B	30·75	35	79·1	167·8	84·0	121·6 ± 29·1
1951A	26·50	41	85·3	132·2	183·3	195·2 ± 43·2
1951B	31·40	27	59·8	97·2	114·4	122·4 ± 33·0
1954	34·25	8	10·9	13·3	21·3	17·1 ± 9·1
1955	34·25	8	17·2	50·5	30·3	28·8 ± 15·4
1956A	31·50	12	20·9	42·5	29·0	36·2 ± 15·2
1956B	31·50	8	12·6	19·4	36·9	41·5 ± 22·2
1957A	36·50	4	6·4	7·2	18·3	14·5 ± 12·4
1957B	34·30	9	15·9	25·1	23·7	30·6 ± 15·2
1958A	32·00	12	17·0	23·2	26·2	27·3 ± 11·4
1958B	31·65	17	21·4	37·3	36·0	48·5 ± 16·9

All surveys: 226

A goodness-of-fit test of the y_i observations to a negative exponential distribution was carried out using the data of all 7 years, and the observed and expected frequencies are given in Table 2.2 (y_i was not recorded for 2 birds, giving a total of 224 observations). Class intervals of 5 yards were used and, because of bias in reporting, those observations recording at exactly 5 yards were divided equally between the class above and below. A value $\hat{\lambda}_1 = 0 \cdot 126$ was calculated from the ungrouped data and used to derive the expected frequencies. The chi-squared goodness-of-fit test yielded a value of 2·295 which, with 6 degrees of freedom, indicates an excellent fit.

TABLE 2.2
Frequency distribution of 224 right-angle flushing distances (y) fitted to a negative exponential distribution: from Gates *et al.* [1968: Table 2].

y (yards)	Frequencies	
	Observed	Expected
0–5	103·5	104·6
5–10	61·5	55·8
10–15	28·5	29·8
15–20	11·5	15·9
20–25	9·0	8·3
25–30	3·5	4·5
> 30	6·5	5·2
Total	224·0	224·1

To test assumptions (1)–(3), the numbers n_l were recorded for each l = quarter-mile and a test of fit for the Poisson distribution was carried out. The observed and expected frequencies are given in Table 2.3, and pooling

TABLE 2.3

Distribution of ruffed grouse flushings per quarter-mile segment of line transect fitted to a Poisson distribution: from Gates et al. [1968: Table 3].

n_l	Frequencies	
	Observed	Expected
0	2450	2410·8
1	292	361·7
2	52	27·1
3	6	1·4
4	0	0·0
5	0	0·0
6	0	0·0
7	1	0·0
Total	2801	2801·0

the last 5 classes into one class gives a chi-squared value of 59·4 with 2 degrees of freedom. This is highly significant, thus suggesting a breakdown in the underlying assumptions. An examination of Table 2.3 indicates that there are either too many one-quarter mile segments with two or more birds flushing at the expense of single-bird flushes, or alternatively an excess of zeros. The authors suggest two possible explanations, not mutually exclusive, for the former and one possible explanation for the latter failure to follow a Poisson distribution, namely (i) the grouse have a tendency to flush in pairs (assumption (2) false), (ii) some grouse may fly ahead of the observer and be flushed twice in the same one-quarter mile segment (assumption (3) false), and (iii) some parts of the area may not be suitable for occupation (assumption (1) false).

3   *Power law for probability of detection*

Eberhardt [1968] suggests using the function

$$g(y) = \begin{cases} 1 - \left(\dfrac{y}{W}\right)^a & 0 \leqslant y \leqslant W \\ 0 & y > W \end{cases}$$

which leads to $c = Wa/(a+1)$ and the density function (cf. (2.7))

$$f(y) = \frac{(a+1)}{aW}\left(1 - \left(\frac{y}{W}\right)^a\right). \tag{2.14}$$

If a is known, then from

$$E[\bar{y}] \, (= E[y]) \; = \; \frac{W(a+1)}{2(a+2)} \tag{2.15}$$

we have the moment estimate $\hat{W} = 2\bar{y}(a+2)/(a+1)$, which leads to the estimate (cf. (2.11))

$$\hat{D}_8 \; = \; \frac{n(a+1)^2}{4L\bar{y}a(a+2)}.$$

Both $\hat{D}_5$ (the biased version $n/(2\bar{y}L)$) and $\hat{D}_8$ are derived as moment estimates by Eberhardt [1968], using a different method. However, his estimates differ from ours by a factor of two as he deals with only one side of the transect line (setting $\nu = 2DL$ in his paper leads to $\hat{D}_5$, $\hat{D}_8$ and $\hat{D}_9$ given below).

When a is unknown the maximum-likelihood estimates of W and a are y_M, the maximum observed value of y, and the solution of

$$\frac{n}{a(a+1)} + \sum_i \left\{ \left(\frac{y_i}{y_M}\right)^a \cdot \log\left(\frac{y_i}{y_M}\right) \cdot \frac{1}{1 - (y_i/y_M)^a} \right\} \; = \; 0.$$

As this equation is rather intractable it would seem preferable to use a simpler, though less efficient, estimate of a. For example, setting $W = y_M$ one can use (2.15) to obtain such an estimate.

4 All animals detected over a constant unknown distance

The simplest, and probably the least realistic, assumption one can make about $g(y)$ is that *all* animals are sighted for some unknown distance W out from the transect and none are sighted beyond this distance, i.e.

$$g(y) \; = \; 1, \quad 0 \leqslant y \leqslant W,$$

$c = W$, and y has the uniform distribution (cf. (2.7))

$$f(y) \; = \; 1/W, \quad 0 \leqslant y \leqslant W.$$

Since $E[y] = W/2$, c can be estimated by $2\bar{y}$, which leads to the density estimate (cf. (2.11))

$$\hat{D}_9 \; = \; n/(4L\bar{y}).$$

Alternatively c can be estimated by y_M, leading to

$$\hat{D}_{10} \; = \; n/(2Ly_M).$$

This estimate, suggested by Amman and Baldwin [1960], was found to be useful in the study of woodpecker populations, as virtually all the birds could be detected from feeding sounds.

5 Density estimates based on flushing distances

All the previous models can be used either when the animals are stationary and visibility is limited (e.g. mortality surveys of dead deer: Eberhardt [1968: 86–7]), or when the animals are mobile and have to be flushed into the open. In the latter situation one can also make use of the flushing distances r_i which are generally easier to measure than right-angle distances. For example, setting $\hat{D}_i = n/(2Lw_i)$ we have, as possibilities: $w_1 = \bar{r}$, the so-called "King method" (Leopold [1933]); $w_2 = n/\Sigma\ (1/r_i)$ (Hayne [1949a]); $w_3 = \bar{r}\sin\bar{\theta}$, where $\bar{\theta}$ is the average angle of sighting (Webb [1942]); and $w_4 = \bar{G}$, the geometric mean of the r_i (Gates [1969]).

Hayne examined the King method $\hat{D}_1$ and pointed out that it is based on three main assumptions: (i) in a population animals vary with regard to the distance at which they will flush upon the approach of an observer; (ii) the various classifications of animals (with regard to flushing distance) are scattered about over the study area in a random fashion, or at least random with respect to the path of the observer; and (iii) the average flushing distance $\bar{r}$ observed by the investigator is a good estimate of the average flushing distance for the whole population. There is also a fourth assumption implicit in his discussion, namely (iv) all animals flushed are actually seen by the observer. The first two assumptions are generally satisfied if there is adequate randomisation in the choice of line transect, though (ii) will not hold if there is any directional movement of animals in relation to the observer's path. For example, certain animals appear to avoid an observer so successfully that the flushing distance in this case must be considered greater than the range of visibility of the observer. Also any tendency for the animals to "drive", that is move down the line of travel ahead of the observer after being flushed, will increase the apparent population density.

The third assumption appears to be false, and Hayne demonstrated mathematically that w_1 will generally lead to an overestimate of effective strip width with a consequent underestimate of population density. This is readily seen by comparing the observed average flushing distance with the true average flushing distance for a strip centred on the line transect. Obviously animals in the strip at a perpendicular distance from the line greater than their flushing distance will not be flushed, so that the observed average will be based on fewer short flushing distances than the true average. To allow for this bias, Hayne suggested the modification w_2 which will generally give unbiased results, provided assumptions (i), (ii) and (iv) are satisfied (see also Overton and Davis [1969: 420–23]).

The King method was initially devised for censusing ruffed grouse populations. Webb [1942], however, felt that the method could not be applied to snowshoe hares because of the difference in flushing behaviour. When the grouse are flushed they make a considerable noise with their wings, so that the birds flushed are less likely to be overlooked. In flushing hares, however, Webb argued that the observer may see the hares that move ahead of him and those which remain motionless until he is very close, but will tend to miss those that move directly to the right or left. This means that the effective strip width w should be reduced and on intuitive grounds Webb proposed the modification w_3.

We note that w_1 and w_2 are the arithmetic mean and the harmonic mean respectively of the flushing distances, so that from Kendall and Stuart

[1969: 36] $w_2 \leqslant w_4 \leqslant w_1$, and $\hat{D}_1 \leqslant \hat{D}_4 \leqslant \hat{D}_2$.

Although Webb's estimate allows for the possibility of flushed animals not being seen, a general method for dealing with this problem is given by Gates [1969] as follows. We shall make assumptions (1)–(7) of p. 29 above, together with the following:

(8) $g(y) = \exp(-\lambda_1 y)$, i.e. the p.d.f. of y is

$$f(y) = \lambda_1 \exp(-\lambda_1 y). \tag{2.16}$$

(9) The conditional probability density function of the radial distance r, given the right-angle distance y, is

$$f(r \mid y) = \lambda_2 \exp[-\lambda_2(r-y)], \quad y \leqslant r < \infty, \ \lambda_2 > 0. \tag{2.17}$$

Then, using the transformation $y = r \sin \theta$, Gates shows that the joint distribution of r and θ (the angle of flushing), given that the animal is seen, is

$$f(r, \theta) = \lambda_1 \lambda_2 (\cos \theta) r \exp(-\alpha r), \quad 0 \leqslant \theta \leqslant \pi/2, \tag{2.18}$$

where $\alpha = (\lambda_1 - \lambda_2) \sin \theta + \lambda_2$.

CASE 1: $\lambda_1 = \lambda_2 = \lambda$. It follows from (2.18) that

$$f(r, \theta) = \lambda^2 r (\cos \theta) \exp(-\lambda r)$$

with marginal distributions

$$f_1(r) = \lambda^2 r \exp(-\lambda r) \tag{2.19}$$

and

$$f_2(\theta) = \cos \theta, \quad 0 \leqslant \theta \leqslant \pi/2. \tag{2.20}$$

By the same argument which led to equation (2.10) we find that the joint distribution of $r_1, r_2, \ldots, r_n$ and n is

$$f(r_1, r_2, \ldots, r_n, n) = \binom{N}{n} P^n Q^{N-n} \prod_{i=1}^{n} \{\lambda^2 r_i \exp(-\lambda r_i)\}. \tag{2.21}$$

Gates shows that the maximum-likelihood estimates of λ and N, after correction for bias, are

$$\hat{\lambda} = (2n-1)/\Sigma \, r_i$$

and

$$\hat{N}_7 = n/\hat{P} = nA\hat{\lambda}/(2L),$$

so that the estimate of population density is given by

$$\hat{D}_7 = (2n-1)/(2L\bar{r}).$$

Also, using the delta method, Gates proves that

$$V[\hat{N}_7] \approx N^2 E \left[\frac{1}{2n-2} \right] + \frac{N(1-P)}{P},$$

and a consistent estimate of this expression is

$$v_7 = \frac{n}{\hat{P}^2} \left[\frac{3n-2}{2n-2} - \hat{P} \right].$$

It can be shown that $v_7 < v_5$ (cf. equation (2.13)), so that radial distances are not only easier to measure, but when the above model is applicable, they also give a more efficient estimate of population size.

Using (2.19), it is readily shown that

$$\hat{\lambda}' = \frac{1}{n} \Sigma \frac{1}{r_i}$$

is also unbiased for λ, and the corresponding unbiased estimate of N is

$$\hat{N}_2 = \frac{An\hat{\lambda}'}{2L} = \frac{A}{2L} \Sigma \frac{1}{r_i},$$

which, expressed in terms of density, is Hayne's [1949a] estimate $\hat{D}_2$. However, $V[\hat{D}_2]$ does not exist for the present model, as $E[1/r^2]$ does not exist.

From (2.16), (2.17) and (2.19) we find that

$$f(y \mid r) = 1/r, \quad 0 \leqslant y \leqslant r$$

so that for given r, y has a uniform distribution. Katz, in an appendix to Hayne's 1949a paper, assumed this distribution to start with, and by writing $y = r \sin\theta$, proved that $\sin\theta$ is uniformly distributed on $[0, 1]$ and that θ has the probability density function (2.20). These two distributional properties of θ can be used as a check on the validity of the above model. For example, from (2.20) we have

$$E[\theta] = \frac{\pi}{2} - 1 \quad \text{radians}$$

$$= 32 \cdot 704°,$$

and we can test whether the sample mean $\bar{\theta}$ differs significantly from $32 \cdot 704°$. One can also carry out a goodness-of-fit test to see whether the sample values $\sin\theta_i$ ($i = 1, 2, \ldots, n$) come from a uniform distribution. A variety of tests of fit are available, and these are listed in many textbooks (e.g. Kendall and Stuart [1973], Keeping [1962]). When n is sufficiently large, the usual Pearson chi-squared goodness-of-fit statistic can be used, though the Kolmogorov statistic gives a more sensitive test and its exact distribution has been tabled. The validity of (2.19) can also be tested using a chi-squared goodness-of-fit test, and an example of this is given in Gates [1969].

CASE 2: $\lambda_1 \neq \lambda_2$. From (2.18) the marginal distributions of r and θ are

$$f_1(r) = \frac{\lambda_1\lambda_2}{\lambda_2 - \lambda_1} \{\exp(-r\lambda_1) - \exp(-r\lambda_2)\}$$

and

$$f_2(\theta) = \lambda_1\lambda_2 (\cos\theta)/\alpha^2. \tag{2.22}$$

Using the same argument that led to (2.10) and (2.21), the maximum-likelihood estimates $\hat{N}$, $\hat{\lambda}_1$ and $\hat{\lambda}_2$ are given by

$$\hat{N} = n A \hat{\lambda}_1 / (2L),$$

where $\hat{\lambda}_1$, $\hat{\lambda}_2$ are the solutions of

$$\frac{n\lambda_1}{\lambda_2(\lambda_2-\lambda_1)} = \sum_i \left\{ \frac{r_i \exp(-r_i\lambda_2)}{\exp(-r_i\lambda_1) - \exp(-r_i\lambda_2)} \right\}$$

$$\frac{n\lambda_2}{\lambda_1(\lambda_2-\lambda_1)} = \sum_i \left\{ \frac{r_i \exp(-r_i\lambda_1)}{\exp(-r_i\lambda_1) - \exp(-r_i\lambda_2)} \right\}$$

Unfortunately these equations, obtained from $\prod_i f_1(r_i)$, do not have simple solutions and must be solved iteratively (cf. **1.3.8**). Gates suggests obtaining initial values for the iterative process by using the moment estimates

$$\bar{r} = E[r] = (\lambda_1+\lambda_2)/(\lambda_1\lambda_2), \tag{2.23}$$

$$s^2 = V[r] = (\lambda_1^2+\lambda_2^2)/(\lambda_1^2\lambda_2^2), \tag{2.24}$$

where $s^2 = \Sigma(r_i - \bar{r})^2/(n-1)$.

Solving these two equations we find the estimates, if they exist, to be

$$\frac{\bar{r} \pm \sqrt{(2s^2 - \bar{r}^2)}}{\bar{r}^2 - s^2}.$$

Because of the symmetry of λ_1 and λ_2 in (2.23) and (2.24) this result has the unfortunate feature that the two estimates of λ are not distinguishable. However, one possible method for identifying these estimates is described below.

From (2.22) Gates shows that the average angle of flushing depends on the relative magnitudes of λ_1 and λ_2; thus

$$E[\theta] = \begin{cases} k\left[\dfrac{-\pi}{2\lambda_1} + \dfrac{2}{\sqrt{c}} \tan^{-1}\left(\dfrac{2\lambda_2-\lambda_1}{\sqrt{c}}\right) \right], & \text{for } \lambda_2 > \tfrac{1}{2}\lambda_1 \\[3mm] k\left[\dfrac{-\pi}{2\lambda_1} + \dfrac{1}{d} \log\left\{\dfrac{(\lambda_1-d)(\lambda_1-\lambda_2+d)}{(\lambda_1+d)(\lambda_1-\lambda_2-d)}\right\} \right], & \text{for } \lambda_2 < \tfrac{1}{2}\lambda_1 \end{cases}$$

where $c = 2\lambda_1\lambda_2 - \lambda_1^2$, $d = \sqrt{(-c)}$, $k = \lambda_1\lambda_2/(\lambda_1-\lambda_2)$ and $\lambda_1 \neq \lambda_2$. When λ_2 is greater or less than λ_1, it can be shown that $E[\theta]$ is respectively greater or less than $32\cdot704°$. Therefore any significant departure of $\bar{\theta}$ from $32\cdot704°$ would suggest that $\lambda_1 \neq \lambda_2$, and the direction of the departure would indicate which λ is the greater.

6   Comparison of estimators

Using computer simulation, Gates [1969] compared the estimates $\hat{D}_i$ ($i = 1, 2, \ldots, 5$) (cf. pp. 36 and 30 for definitions) with respect to bias, variance, and robustness towards certain departures from the underlying assumptions. The estimate $\hat{D}_7$ was derived after most of the computer calculations had been carried out, so that it was compared with the other estimators in a few cases only. His conclusions are summarised below.

BIAS. In this section we shall say that an estimator is "biased" or "unbiased" according to whether or not it is significantly different from the true population density. Gates found that when assumptions (1)–(9) were satisfied and $\lambda_1 = \lambda_2 = \lambda$, $\hat{D}_2$, $\hat{D}_5$ and $\hat{D}_7$ were generally unbiased while $\hat{D}_1$ and $\hat{D}_4$ were negatively biased (the differences usually being significant at the 1 per cent level of significance). The Webb estimator $\hat{D}_3$, although frequently biased, appeared to be less biased than $\hat{D}_1$ or $\hat{D}_4$. It was found that the relative bias of the various estimators did not depend on λ, though the the variances were directly proportional to λ.

VARIANCE. The Hayne estimator $\hat{D}_2$ consistently had a larger variance than the other estimators. When the estimators are ranked according to increasing size of variance we have the order $\hat{D}_1$, $\hat{D}_4$, $\hat{D}_5$, $\hat{D}_3$ and $\hat{D}_2$.

ROBUSTNESS. When there are unbiased errors in measuring distances it was found that the relative bias in all the estimators appeared to be unaffected, though the variances were increased as expected. $\hat{D}_2$ appeared to be particularly susceptible to measurement error, while the other estimators appeared to be less so.

Sensitivity to departure from the negative exponential distribution of assumption (8) was investigated by considering two alternative distributions: (i) a right-triangle type of distribution (cf. (2.14) with $a = 1$), and (ii) a half-normal distribution. It was found that all the estimators $\hat{D}_1$ to $\hat{D}_5$ appeared sensitive to such departures, and $\hat{D}_5$ consistently overestimated the true population density in both cases.

In some populations, as for example grouse, there is often the tendency for animals to flush in pairs, thus violating assumption (2) that animals act independently. Using simulated pairing, Gates found that such pairing did not change the relative bias of the estimators, and in particular $\hat{D}_2$, $\hat{D}_5$ and $\hat{D}_7$ were unbiased for the four computer runs based on 10, 20 and 30 per cent pairing. It was also found that the variances of the estimators increased as the percentage pairing increased.

As pointed out above, $E[\theta]$ is greater or less than $32 \cdot 704°$ when λ_2/λ_1 is greater or less than unity, respectively. Gates found that the estimators $\hat{D}_i$ ($i = 1, 2, 3, 4$), and in particular D_2, were sensitive to departures from the ratio $\lambda_2/\lambda_1 = 1$, while $\hat{D}_5$ was apparently insensitive. His results also verified the findings of Robinette *et al.* [1956] who concluded that when $E[\theta] > 40°$, the Hayne method is positively biased.

OTHER ESTIMATORS. Several density estimates have been derived by Yapp [1956] and Skellam [1958] for the more complex situation where the population is on the move. Unfortunately the requirement of knowing the velocities of the animals encountered rather precludes the use of these methods on practical grounds. However Gates *et al.* [1968] show that under certain reasonable conditions the two formulae proposed by Yapp reduce to the King estimator $\hat{D}_1$.

In conclusion it should be stressed that no matter which estimator is used, the data should be recorded separately for each segment of length l (= L/s) as described on p. 32. In this way the estimate and its theoretical variance estimate can be compared with the sample mean and variance estimate based on s repeated samples.

2.1.4   Density estimates based on the spatial distribution

1   Introduction

When a population is randomly distributed over the population area, one can use the distances of individuals from randomly chosen points (closest-individual techniques) or the distances between neighbours (nearest neighbour techniques) to calculate estimates of population density. Such techniques, although most suitable for stationary populations such as plants or trees (Greig-Smith [1964]), can be applied to conspicuous and relatively stationary animal populations (e.g. snails: Keuls *et al.* [1963]) or to well-marked colonies such as ant or termite mounds (Waloff and Blackith [1962], Blackith *et al.* [1963], Sands [1965]). They have also been applied to more mobile populations such as grasshoppers (Blackith [1958]), cricket frogs (Turner [1960a]), and quail coveys (Ellis *et al.* [1969]).

Before discussing these distance estimates of density in detail it should be emphasised that most of the estimates depend on the assumption of random distribution, so that tests of randomness must be carried out. However, it is hoped that further research will eventually show that some of the methods are robust with regard to certain departures from randomness: some useful methods for handling non-randomly distributed populations have been given by Dacey [1964b, 1965, 1966]. If the density is known (or estimated by other techniques), these distance methods can be used for detecting non-randomness (e.g. Holgate [1965], McLaren [1967], Pielou [1969: 115]).

Finally it should be pointed out that some of the published work using the above distance methods reveals a lack of understanding with regard to the background theory (Kendall and Moran [1963: 38]): this is particularly the case in the sampling methods used for choosing animals at random (see section 4 below).

2   *Closest-individual techniques: Poisson model*

UNBIASED ESTIMATION. We shall assume that:

(i) The population area is infinite with a constant population density D (restriction to a finite area A is considered later).

(ii) The number of individuals found in an area δA, chosen at random, has a Poisson distribution with parameter $D\delta A$ (cf. p. 24).

Suppose that a sample point is chosen at random and let X_1 be the distance to the nearest animal. Then from assumption (ii) we have (Eberhardt [1967])

$$F_1(x) = \Pr[X_1 \leqslant x]$$

$$= \Pr[\text{finding at least one individual in circular plot of radius } x]$$

$$= 1 - \Pr[\text{no individuals in plot of area } \pi x^2]$$

$$= 1 - \exp(-D\pi x^2), \qquad\qquad (2.25)$$

and the probability density function (p.d.f.) for X_1 is

$$f_1(x) = F_1'(x) = 2D\pi x \exp(-D\pi x^2). \tag{2.26}$$

If X_r is the distance to the rth nearest animal, then, from Eberhardt[1967],

$$F_r(x) = \Pr[X_r \leqslant x]$$

$$= \Pr[r \text{ or more animals in circular plot of area } \pi x^2]$$

$$= \sum_{i=r}^{\infty} \exp(-D\pi x^2) \frac{(D\pi x^2)^i}{i!}$$

and

$$f_r(x) = F_r'(x) = \frac{2(D\pi)^r}{(r-1)!} x^{2r-1} \exp(-D\pi x^2). \tag{2.27}$$

This probability density was derived independently by Morisita [1954], Moore [1954], and Thompson [1956] (see Dacey [1964a] for a review) and generalised to the case of n-dimensional space by Dacey [1963].

Setting $Y = \pi X_r^2$, then the probability density of Y has the simple form

$$g_r(y) = D^r y^{r-1} \exp(-Dy)/(r-1)!. \tag{2.28}$$

To estimate D, suppose that s such sample points are chosen at random and let Y_i ($i = 1, 2, \dots, s$) be the corresponding values of Y. If $Y_. = \Sigma Y_i$, then from Moore [1954] and Holgate [1964a] an unbiased estimate of D is the modified maximum-likelihood estimate

$$\hat{D} = (sr - 1)/Y_. \tag{2.29}$$

with variance

$$V[\hat{D}] = D^2/(sr - 2). \tag{2.30}$$

Since $2DY$ is distributed as χ^2_{2r}, $2DY_.$ is χ^2_{2rs}, and this statistic can be used for constructing confidence intervals for D. For example a 95 per cent confidence interval for D is given by $(c_1/(2Y_.), c_2/(2Y_.))$ where c_1 and c_2 are the lower and upper 2·5 per cent points of χ^2_{2rs} respectively. Generally $2rs$ will be greater than the degrees of freedom tabulated, and in this case a normal approximation is used: thus

$$z = \sqrt{(4DY_.)} - \sqrt{(4rs - 1)}$$

is approximately unit normal and

$$0.95 \approx \Pr\left[\frac{-1.96 + \sqrt{(4rs - 1)}}{\sqrt{(4Y_.)}} < \sqrt{D} < \frac{1.96 + \sqrt{(4rs - 1)}}{\sqrt{(4Y_.)}}\right].$$

Since, from (2.28),

$$E[Y^{-1}] = D/(r - 1),$$

Eberhardt [1967] suggests the unbiased estimate

$$\tilde{D} = (r - 1) \sum_{i=1}^{s} (Y_i^{-1}/s) = \sum_{i=1}^{s} \tilde{D}_i/s, \text{ say} \tag{2.31}$$

which is evidently equivalent to one proposed by Morisita [1957]. It can be shown that, for $r \geqslant 3$,

$$V[\tilde{D}] = V[\tilde{D}_i]/s = D^2/[s(r-2)],\qquad(2.32)$$

which, although greater than $V[\hat{D}]$ of (2.30), has an unbiased replicated-sample estimate (cf. (1.9))

$$v[\hat{D}] = \frac{\sum\limits_{i=1}^{s}(\tilde{D}_i - \tilde{D})^2}{s(s-1)}.\qquad(2.33)$$

FINITE POPULATION AREA. In practice the population area is finite, and to avoid the possibility of "edge effect" one should choose a subarea in the population away from the boundary and restrict sample points to this subarea only. If a sample point then falls on the edge of the subarea, one would expect the distribution of X_r to be the same as that for points within the subarea, provided that the edge of the subarea was at a sufficient distance from the population boundary. As a rough guide, this distance, x say, should be such that for $y = \pi x^2$, $g_r(y)$ is sufficiently small for the appropriate truncation of (2.28) to have a negligible effect. We note in passing that a common method of choosing a sample point is to divide up the total area into quadrats of equal area (allowing a "buffer" strip around the boundary), choose a quadrat (with replacement) using random numbers, and then select a point in the quadrat by reading off a pair of coordinates from the random numbers (cf. Sampford [1962: Chapter 4]).

COMPARING DENSITIES. A simple test for comparing the densities D_1 and D_2 of two areas is available from Moore [1954]. If s_j sample points are located in the jth area ($j = 1, 2$) and $Y_{.j}$ is the corresponding value of $Y_.$, then

$$F = (r_2 s_2 D_1 Y_{.1})/(r_1 s_1 D_2 Y_{.2})$$

has an F-distribution with $2r_1 s_1$ and $2r_2 s_2$ degrees of freedom respectively. When $D_1 = D_2$, F reduces to a simple statistic which can be entered in the the tables of the F-distribution. As in the F-test for equal variances, one can observe the convention of putting the larger value of $Y_{.j}/(r_i s_i)$ in the numerator.

EXPERIMENTAL DESIGN. We note that $V[\hat{D}]$ of (2.30) and the coefficient of variation of $\hat{D}$, $1/\sqrt{(sr-2)}$, depend on the product sr, so that we can, for example, use either sr samples of nearest individuals ($r = 1$) or s samples of rth nearest individuals. The choice of procedure would depend on a comparison of the time taken in the field to locate a random point and the time to search for the rth nearest animal. In mobile populations, however, it may only be possible to measure the distance to the nearest animal: this special case has been considered by Skellam [1952: 349], Clark and Evans [1954, 1955], Hopkins [1954], and Moore [1954].

When the main effort involved is in searching for the animals rather than the location of sample points or quadrat boundaries, Holgate [1964a] shows that the above method has approximately the same efficiency as quadrat sampling. For example, if quadrats of total area A_0 yield a count of n individuals, then, assuming n to have a Poisson distribution with mean DA_0 (cf. p. 24), D is estimated by $D' = n/A_0$ with variance D/A_0. To compare this with the closest-individual technique based on X_r, it is assumed that the experimenter searches in increasing circles until exactly r animals have been found. Then the expected total area covered for s sample points will be $E[Y_.]$ $(= sr/D)$. Therefore, setting $A_0 = E[Y_.]$, we find that

$$V[D'] = D/A_0 = D^2/(sr),$$

which is slightly less than $V[\hat{D}]$ of (2.30). However, this analysis does not apply to the common situation where animals are readily seen, and the main effort is involved in laying out the random quadrats or sample points, and making the measurements (for such a comparison see **12.1.2** (*1*)).

QUARTER METHOD. A modification of the above technique, called the "point-centred quarter method", has been discussed by Cottam *et al.* [1953], Cottam and Curtis [1956] and Morisita [1954]. The sample point is chosen at random, together with two perpendicular directions fixed in advance, and the distances Z_1, Z_2, Z_3, and Z_4 say, from the sample point to the nearest individual in each of the four quadrants are measured. Let

$$Y = \tfrac{1}{4} \pi (Z_1^2 + Z_2^2 + Z_3^2 + Z_4^2),$$

then, from Kendall and Moran [1963: 40], the $\tfrac{1}{2}\pi D Z_i^2$ are independently distributed as χ_2^2, so that $2DY$ is χ_8^2. If s sample points are used and $Y_. = \Sigma Y$, then $2DY_.$ is χ_{8s}^2. Therefore, setting $r = 4$ in the above general theory, we see that this quarter method is equivalent to looking for the fourth nearest individual. This accounts for the fact discovered empirically by Cottam and Curtis [1956: 457] that $4s$ sample points using closest-individual measurements ($r = 1$) are needed to give an estimate of D with approximately the same accuracy as s points using the quarter method (Cottam and Curtis actually use a less efficient method of estimation based on the mean distance $\bar{Z}$: it is readily shown that $E[\bar{Z}] = 1/\sqrt{D}$). We note that in the field it will usually be easier to locate the nearest individual in each of the four quadrants than to look for the fourth nearest individual, though either method could only be applied to relatively immobile animals.

APPROXIMATE METHODS. Approximate methods for obtaining point and interval estimates of D based on the untransformed distances X_r have also been given. From Thompson [1956] we have

$$E[X_r] = a_r/\sqrt{D},$$
$$V[X_r] = \sigma_r^2 = b_r/D,$$

where

$$a_r = \frac{(2r)!r}{(2^r r!)^2} \quad \text{and} \quad b_r = \frac{r}{\pi} - a_r^2.$$

If $\bar{X}_r$ is the mean of a sample of s values of X_r, then a moment estimate of D is

$$D^* = a_r^2 / \bar{X}_r^2.$$

Although the distribution of X_r is fairly skew for small values of r (cf. Thompson [1956]), $\bar{X}_r$ will be approximately normally distributed for large s by the Central Limit theorem. We can then obtain an approximate 95 per cent confidence interval for $\sqrt{D}$ as follows:

$$0\cdot95 \approx \Pr\left[-1\cdot96 < \frac{\bar{X}_r - E[\bar{X}_r]}{\sigma_r/\sqrt{s}} < 1\cdot96\right]$$

$$= \Pr\left[\frac{-1\cdot96\sqrt{(b_r/s)} + a_r}{\bar{X}_r} < \sqrt{D} < \frac{1\cdot96\sqrt{(b_r/s)} + a_r}{\bar{X}_r}\right].$$

When $r = 1$ we have $a_1 = \frac{1}{2}$ and

$$D^* = 1/(4\bar{X}_1^2),$$

an estimate first given by Clark and Evans [1954].

INDEX OF NON-RANDOMNESS. If $C[X_r]$ is the coefficient of variation of X_r, then from the previous paragraph we can define

$$
\begin{aligned}
I_r &= (C[X_r])^2 + 1 &\quad (2.34)\\
&= E[X_r^2]/(E[X_r])^2\\
&= \frac{r/(\pi D)}{a_r^2/D}\\
&= \frac{(2^r r!)^4}{(2r!)^2 \pi r},
\end{aligned}
$$

which is independent of D: for example, when $r = 1$, $I_1 = 4/\pi = 1\cdot27$. Since I_r is readily estimated from sample data, and increases with increasing tendency for aggregation, Eberhardt [1967] suggests that it may be a useful measure or index of "non-randomness".

TESTING UNDERLYING ASSUMPTIONS. Several methods are available for testing the validity of the Poisson assumptions. For example, if one of the above methods of estimation is used along with quadrat sampling, then quadrat counts will provide a test of fit to the Poisson distribution. It was noted on p. 25 that the two standard tests of fit, and in fact all tests of randomness based on quadrat counts, are dependent on the size of the quadrat used. Quadrats which are too small or too large with respect to the average size of the "patches" of individuals may fail to detect any non-randomness (Greig-Smith [1964: 28, 56–7]). As a rough guide it is suggested that the quadrat size should be such that the average number of animals per quadrat is not less than unity. In this case it appears that the Poisson Index of Dispersion (cf. 1.3.6 (2)) is satisfactory for as few as 6 sample quadrats (Kathirgamatamby [1953]).

When quadrat sampling is not used it is simpler to work with the transformed distances Y_i and calculate a standard chi-squared goodness of fit to
the probability density (2.28). In particular, when $r = 1$, (2.28) reduces to the
negative exponential distribution.

$$g_1(y) = D \exp(-Dy),$$

and the expected frequencies E_j, say, are readily calculated. For example,
with the class-interval $[a_{j-1}, a_j)$, $(j = 1, 2, \ldots, k; a_0 = 0, a_k = \infty)$,

$$E_j = s \int_{a_{j-1}}^{a_j} \hat{D} \exp(-\hat{D}y)dy$$

$$= s\{\exp(-\hat{D}a_{j-1}) - \exp(-\hat{D}a_j)\}.$$

If f_j is the observed frequency of the y values in the jth class interval, then
the goodness-of-fit statistic

$$\sum_{j=1}^{k} (f_j - E_j)^2/E_j$$

is approximately distributed as χ^2_{k-1} when s is large. We note that one
degree of freedom is not subtracted for the estimation of D as $\hat{D}$ is calculated
from the ungrouped data (Kendall and Stuart [1973: 447]).

Since the times between events for a Poisson time-process follow the
negative exponential distribution (cf. p. 4), a number of other test procedures,
summarised in Cox and Lewis [1966: 153 ff.], can be used here. For example,
if we transform from the Y_i to

$$Z_i = \sum_{r=1}^{i} Y_r / \sum_{r=1}^{s} Y_r,$$

then $Z_1, Z_2, \ldots, Z_{s-1}$ represent an ordered sample from the uniform distribution on $[0, 1]$, and the usual tests of the goodness of fit of the Z_i to this
distribution can be carried out (Seshadri et al. [1969]).

A good description of the various test methods is also given in Epstein
[1960a, b].

3   Closest-individual techniques: alternative models

BINOMIAL MODEL. Consider a population of N animals randomly distributed in a region of area A. Then if a sampling point is chosen at random in
the area, there is the possibility that the point may be closer to the boundary
than to the nearest individual. However, if sampling points are restricted to
an interior area sufficiently far from the boundaries to avoid the above
difficulty, then, using the same argument that led to (2.26), we have
(Eberhardt [1967])

$$F_1(x) = 1 - \left(1 - \frac{\pi x^2}{A}\right)^N$$

and

$$f_1(x) = \frac{2\pi N x}{A}\left(1 - \frac{\pi x^2}{A}\right)^{N-1}. \tag{2.35}$$

When $N \to \infty$, $A \to \infty$ in such a way that $D = N/A$ is constant, we find that (2.35) tends to (2.26) as expected.

Eberhardt shows that (2.27) now becomes

$$f_r(x) = \frac{2\pi N x \, (N-1)\,!}{A(N-r)\,!\,(r-1)\,!} \cdot \left(\frac{\pi x^2}{A}\right)^{r-1} \left(1 - \frac{\pi x^2}{A}\right)^{N-r} \qquad (2.36)$$

or, setting $Y = \pi X_r^2$,

$$g_r(y) = \frac{N\,!}{A(N-r)\,!\,(r-1)\,!} \cdot \left(\frac{y}{A}\right)^{r-1} \left(1 - \frac{y}{A}\right)^{N-r}, \qquad 0 \leqslant y \leqslant A,$$

which is beta-distribution. Once again $\tilde{D}$ of (2.31) is an unbiased estimate of D and, for $r \geqslant 3$,

$$V[\tilde{D}] = \frac{D(D - (r-1)/A)}{s(r-2)}.$$

For the case when a sampling point may be closer to the boundary than to the nearest animal we can use the following method, due to Craig [1953a]. Suppose that sampling is carried out in a quadrat of side a as shown in Fig. 2.1. Let Q be the point in the square with coordinates (u, v) with respect to the origin O and let $w = a - u$, $z = a - v$. Then the basis of

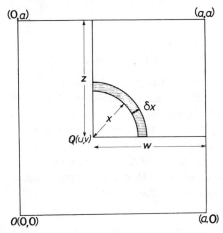

Fig. 2.1 Q is a point chosen at random in a quadrat of side a. The shaded circular strip of width δx contains the individual in the first quadrant with Q as centre which is nearest to Q.

Craig's method is to choose Q at random and, provided $X_1 \leqslant d_0 = $ minimum (w, z), to measure X_1, the distance to the nearest animal in the first quadrant with Q as centre. Now if there are n animals in the square, X_1 is the distance to the nearest animal if one animal lies in the curved shaded area of Fig. 2.1 and $n - 1$ animals lie outside the quarter circle with centre Q and radius X_1. Hence

$$\Pr\left[x < X_1 < x + \delta x \mid w, z\right] = \binom{n}{1} \frac{\pi x \delta x}{2a^2} \left(1 - \frac{\pi x^2}{4a^2}\right)^{n-1} \quad (x \leqslant d_0)$$

and

$$\Pr\left[x < X_1 < x + \delta x\right] = \int_x^a \int_x^a \frac{1}{a^2} \Pr\left[x < X_1 < x + \delta x \mid w, z\right] dw \, dz$$

$$= \Pr\left[x < X_1 < x + \delta x \mid w, z\right] \left(1 - \frac{x}{a}\right)^2 \quad (x \leqslant d_0),$$

since the above conditional probability does not depend functionally on w and z when $x \leqslant d_0$. Therefore the probability density function for X_1, given that Q is chosen at random and $x \leqslant d_0$, is given by

$$f(x) = \frac{n \pi x}{2a^2} \left(1 - \frac{\pi x^2}{4a^2}\right)^{n-1} \left(1 - \frac{x}{a}\right)^2.$$

If s points Q are chosen, k say of these points will have x's less than or equal to the distance from Q to the nearest of the right and upper boundaries of the square. For these k values of X_1 the likelihood function $\prod_i f(x_i)$ can be maximised to give a maximum-likelihood estimate of n, namely

$$\hat{n} = -k / \sum_{i=1}^k \log \left(1 - \frac{\pi x_i^2}{4a^2}\right) \tag{2.37}$$

with asymptotic variance (cf. (1.4))

$$V[\hat{n}] = n^2 / k.$$

The corresponding estimate of population density is $\hat{n}/a^2$ and, from a random sample of quadrats of side a, an average density estimate can be calculated.

One advantage of Craig's method is that it takes into account the problem of boundary points, so that the square could have a side along or near the boundary of the population area under consideration. This convenience, however, is obtained at the expense of $s - k$ sampling points. Defining $P[n]$ as the probability that a sampling point has an x value satisfying $x \leqslant d_0$, Craig gives the following table of $P[n]$ for different values of n:

n:	10	20	30	50	100	200
$P[n]$:	0·5058	0·6216	0·6804	0·7442	0·8134	0·8652

If a is sufficiently large so that $n > 50$, we see from this table that less than about one quarter of the sampling points will be wasted.

Blackith [1958] discusses Craig's model and points out that if $x_i/a < 0·1$ (which will often be the case for animal populations) we can approximate $-\log(1 - \epsilon)$ by ϵ in (2.37) and obtain the simpler expression

$$\hat{n} = 4ka^2 / (\pi \sum_i x_i^2).$$

However, he does not interpret the x_i correctly (Eberhardt [1967: 213]) which to some extent accounts for his inability to reconcile the models of Craig [1953a] and Clark and Evans [1954].

NEGATIVE BINOMIAL MODEL. Suppose that animals are not randomly distributed but are distributed according to a negative binomial law (cf. p. 25)

$$f(u) = \frac{k(k+1)\dots(k+u-1)}{u\,!}\, P^u (1+P)^{-u-k}, \qquad (2.38)$$

where $f(u)$ is the probability of finding $U = u$ animals in a randomly chosen plot of area πx^2. Then, setting

$$D\pi x^2 = E[U]\,(= kP)$$

we have

$$P = D\pi x^2/k.$$

If we assume that k, the "heterogeneity parameter," is independent of x, then X_1, the distance from a randomly chosen point to the nearest animal, has distribution function:

$$\begin{aligned} F_1(x) &= \Pr[X_1 \leqslant x] \\ &= 1 - \Pr[\text{no animals in } \pi x^2] \\ &= 1 - f(0) \\ &= 1 - (1+P)^{-k} \\ &= 1 - \left(1 + \frac{D\pi x^2}{k}\right)^{-k} \end{aligned}$$

and

$$f_1(x) = F_1'(x) = 2D\pi x \left(1 + \frac{D\pi x^2}{k}\right)^{-k-1}. \qquad (2.39)$$

In deriving (2.39) Eberhardt [1967: 209] points out that k will tend to vary with x, so that the above derivation may not be realistic. However, (2.39) may serve as a more useful distribution than either (2.35) or (2.26) for data displaying heterogeneity.

Using the same argument which led to (2.27), Eberhardt shows that X_r has probability density

$$f_r(x) = \frac{2D\pi x\,\Gamma(r+k)(D\pi x^2)^{r-1}k^k}{\Gamma(k)(r-1)!\,(k+D\pi x^2)^{r+k}}$$

which tends to (2.27) when $k \to \infty$. Setting $Y = \pi X_r^2$, we have

$$g_r(y) = \frac{\Gamma(r+k)(Dy)^{r-1}k^k D}{\Gamma(k)(r-1)!\,(k+Dy)^{r+k}}$$

which is related to the beta distribution. Using $g_r(y)$, it is readily shown that $\tilde{D}$ of (2.31) is still an unbiased estimate of D and, for $r \geqslant 3$,

$$V[\widetilde{D}] = \frac{D^2(1 + (r-1)/k)}{s(r-2)},$$

which tends to (2.32) as $k \to \infty$.

COMPARISON OF MODELS. When $r > 1$, $\widetilde{D}$ of (2.31) is an unbiased estimate of D for the Poisson, binomial and negative binomial models, so that it can be used for most animal populations along with the robust estimate of variance $v[\widetilde{D}]$ (cf. (2.33)). When $r = 1$, some indications as to which model is appropriate may be given by Eberhardt's index I_1 of (2.34). For example, Eberhardt gives the following values:

Poisson: $I_1 = 1{\cdot}27$

Negative binomial: $k = 10 \quad 2 \quad 1{\cdot}50 \quad 1{\cdot}10 \quad 1{\cdot}01$

$I_1 = 1{\cdot}31 \quad 1{\cdot}62 \quad 2{\cdot}00 \quad 5{\cdot}19 \quad 42$

and he evaluates the sample value of I_1 for published data from Clark and Evans [1954], Cottam and Curtis [1956], and Blackith [1958].

4   Nearest-neighbour techniques

If instead of choosing a sampling point at random, an animal is chosen at random and the distance X_r' to its nearest neighbour is measured, then provided the animals are randomly distributed, X_r' will have the same distribution as X_r, the distance from a sample point to the nearest animal. A number of methods for selecting an animal have been suggested, such as (a) numbering all the animals in the population and using a table of random numbers, (b) choosing a quadrat at random and then applying method (a) to the quadrat population, and (c) choosing a point at random and using the nearest animal. Method (a) is of course out of the question except in the situation when the population size is easily determined and one is primarily interested in spatial distribution (e.g. termite mounds). Even then, however, if the population is large, considerable labour is involved in identifying all the members of the population. Method (b), although not giving a proper random sample of nearest-neighbour distances, could perhaps be used along with quadrat sampling methods, provided the locating of all the animals in the quadrat did not disturb their pattern of distribution. Method (c) seems to be the most popular, no doubt because of its simplicity, but it also does not give a random sample (Cottam and Curtis [1956]). This method has been considered by Kendall and Moran [1963: 39] who show that if $\bar{X}_r'$ is the mean of a sample of rth nearest-neighbour measurements chosen by method (c) then $0{\cdot}8396\ \bar{X}_r'$ is an unbiased estimate of $1/\sqrt{D}$. A variation of the nearest-neighbour method, called the "random pairs" method, has been proposed (Cottam and Curtis [1949, 1955, 1956]), but this also uses method (c) for sampling. A helpful discussion on this sampling problem is given in Pielou [1969: 117].

In conclusion we see that on account of sampling difficulties and the present lack of theory for methods (b) and (c), the closest-individual techniques of the previous section are preferred to the above nearest-neighbour techniques for estimating population density. The same comments apply to a number of tests of randomness based on comparing closest-individual and nearest-neighbour measurements (Greig-Smith [1964: Chapter 3]). Batchelor and Bell [1970], however, give a method which utilises both types of measurement for estimating the density of a random or of a non-random population.

2.1.5 Simultaneous estimation of density and home range

MacLulich [1951] has given a trapping method for the simultaneous estimation of population density and home range. It is based on the observation that, given a sufficient number of traps and a reasonable trapping time (e.g. four trap-nights, though a longer period may be necessary: Sanderson [1950]), a line or quadrat of traps will catch most of the animals whose ranges of movement overlap the line or area enclosed by the traps.

Suppose that R is the mean diameter of the home range or range of movement of the animals during the trapping period. Then the effective trapping area of a rectangular quadrat of length a and width b, obtained by adding a strip of width $\frac{1}{2}R$ to the quadrat boundary (Dice [1938]) as in Fig. 2.2, is given by

$$(a+R)(b+R) - (1-\tfrac{1}{4}\pi)R^2.$$

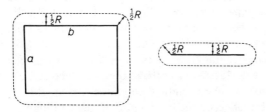

Fig 2.2   Effective trapping areas for a quadrat and a line of traps.

Assuming that the population density D is constant throughout the population area, and defining n_1, n_2 to be the number of animals resident in the effective trapping areas of two such quadrats of different areas, we have for $i = 1, 2$ the deterministic equation

$$n_i = D[(a_i + R)(b_i + R) - (1-\tfrac{1}{4}\pi)R^2] = DA_i, \quad \text{say.}$$

Therefore, setting $n_1/n_2 = A_1/A_2$ leads to the quadratic in R

$$\frac{\pi}{4}\left(1-\frac{n_1}{n_2}\right)R^2 + \left[(a_1+b_1) - \frac{n_1}{n_2}(a_2+b_2)\right]R + \left(a_1 b_1 - \frac{n_1}{n_2} a_2 b_2\right) = 0,$$

which can be solved for the positive root. Substituting this value of R in A_i, the population density is then given by

$$D = n_1/A_1 = n_2/A_2.$$

If live-traps and tagging are used, then the n_i can be estimated by the methods of Chapters 3 and 4 or by trapping until no new animals are caught. Alternatively, if breakback traps are used, n_i can be estimated by the removal methods of Chapter 7 or by trapping out the quadrat areas completely. Obviously the two quadrats should be of reasonably different areas so as to reduce the probability of the larger quadrat giving the smaller value of n_i through variations in population density.

Instead of using two quadrats the experimenter can use one quadrat and one straight line of traps of length L, say, for which the effective trapping area is (Fig. 2.2) $RL + \frac{1}{4}\pi R^2$. If the number of individuals in this area is n_1, then using such a line instead of the first quadrat above leads to the following quadratic in R:

$$\frac{\pi}{4}\left(1-\frac{n_1}{n_2}\right) R^2 + \left(L - \frac{n_1}{n_2}\,(a_2 + b_2)\right) R - \frac{n_1}{n_2}\, a_2 b_2 \;=\; 0.$$

Again, R is given by the positive root, and $D = n_2/A_2$. For three specified line–quadrat combinations MacLulich [1951] gives a graph for reading off the solution of the above quadratic for a given value of n_1/n_2. He also gives a number of practical recommendations for the spacing of the traps, and the reader is referred to his article for further details.

In conclusion we note that the two main assumptions underlying the above mathematical models are (i) the population has constant density, and (ii) the range of movement is not too variable, so that R is approximately constant over the population area. Obviously the best check on these assumptions is by replication, as any marked departures from (i) and (ii) will lead to wide fluctuations in the estimates of D and R. From replicate quadrat–quadrat or quadrat–line pairs, average estimates of D and R can be calculated along with the usual estimates of variance (cf. (1.9)). An interesting modification of the above method which uses a single quadrat and assumes a variable edge effect is given by Hansson [1969].

2.2 RELATIVE DENSITY

2.2.1 Direct methods

Often it is easier, and sometimes more appropriate, to measure population density in units other than area. For example, in the study of insect populations a useful concept is that of population intensity (Southwood [1966]) which is the number of animals per unit of habitat, e.g. per leaf, per tree, per host. Other units which are commonly employed are distance (e.g. animals seen per mile while driving or walking through the area on a prescribed route), time (e.g. animals per hour crossing a given path), and trapping effort (e.g. mice per trap-night, insects per sweep, pheasants killed per hunter-hour); some examples are given in Dice [1952: 39–42]. The roadside census method, in which counting is done from a car usually driven at constant speed, has the advantage that large areas can be quickly covered, and in the U.S.A. it is one of the few methods applicable on a state-wide basis. It has been applied for example to rabbits (Newman [1959], Wight [1959], Kline [1965]), small birds (Howell [1951]), woodcocks (Kozicky *et al.* [1954]), mourning doves (Foote *et al.* [1958]), pheasants (Fisher *et al.* [1947], Kozicky *et al.* [1952], Hartley *et al.* [1955]), blackbirds (Hewitt [1967]), and rhesus monkeys in India (Southwick and Siddiqi [1968]). Also some states of U.S.A. secure the cooperation of rural mail carriers to obtain relative estimates of their principal game species.

These relative estimates of population density are particularly useful in detecting changes in population density with time or in comparing populations in different areas. It should, however, be stressed that if any comparisons are to be made then the censusing should be carried out under as nearly identical conditions as possible. For example, an increase in roadside counts may be due to a genuine increase in population size or to an increase in activity of the animals. As so many factors such as time of day, weather, food supply, and vegetation cover can affect activity (Davis [1963]), a blind application of one of the above methods without a careful study of daily and seasonal activity of the species concerned could give a completely false picture. Newman [1959], for example, using regression analysis, examined the factors which could possibly influence a winter roadside count of cottontails. In comparing counts for different times of the year and different localities, regression and analysis-of-variance methods have proved useful (e.g. Hartley *et al.* [1955], Schultz and Byrd [1957], Schultz and Muncy [1957], Schultz and Brooks [1958], Sen [1970]): see Overton and Davis [1969] for a helpful discussion.

If absolute estimates of population density are also available from time to time from other methods, then the conversion factors thus provided can be used to convert the relative estimates to absolute estimates. Howell [1951], for example, calculated a measure of conspicuousness for various bird species by determining the ratio of those birds seen on the roadside count to the absolute number known to be in the area. These ratios could then be used to convert roadside counts to absolute densities. However, such conversion factors should be used with caution (Dice [1952: 42]) as they will vary with time and place, and even if applicable will give a density estimate with a large coefficient of variation. (See also Gates and Smith [1972].).

An ingenious method for converting roadside counts of territorial birds into absolute density estimates has been suggested by Hewitt [1967]. A census route is driven at approximately 15 miles per hour, and on the first trip the investigator tallies the birds seen and describes their locations in relation to roadside landmarks on a continuously operating tape recorder. The route is then rerun immediately, at similar speed and with a continuous playback of the recording, so the investigator can tally and distinguish "marked" birds seen on the first trip and new "unmarked" birds. The Petersen or Schnabel methods of Chapters 3 and 4 can then be used to estimate the population size. When suitable landmarks for locating the birds are absent, a sensitive resettable odometer can be used for reading off bird locations to the nearest 0·01 mile, say (Harke and Stickley [1968]).

2.2.2 Indirect methods*

Sometimes it is not easy to observe an animal, and its presence has to be inferred by some sign such as dens, burrows, nests, houses, tracks, faeces, dead individuals, songs, calls, shed antlers, etc. (Scattergood [1954: 279], Davis [1963]). Such measures of animal abundance, usually called population indices, are generally more unreliable than the direct methods described above. However, in studying population trends the more methods that can be used and compared the better. For secretive animals an index is sometimes all that is available.

Before a particular index is used it should be studied carefully to see how it is affected by changing conditions: here regression and analysis-of-variance techniques can often be used (e.g. Kozicky [1952], Carney and Petrides [1957], Cohen *et al.* [1960], Smith [1964], Progulske and Duerre [1964], Gates [1966]). Generally it is not possible to examine the whole population area for animal signs, so that sampling methods such as those described in **2.1.2** must be used. For example, Foote *et al.* [1958] used stratified random sampling in choosing routes for collecting data on the call counts of mourning doves.

A useful index for some populations is the number of dwellings per unit area (e.g. Leedy [1949: pheasants], Reid *et al.* [1966: mountain pocket gophers]). However, unless it is possible to distinguish between used and unused dwellings, the number of dwellings will not necessarily reflect the population size.

Large animals can often be detected by their tracks, and in snow these tracks may be distinguishable from the air. Small mammals can be detected by their tracks on smoked kymograph paper (Justice [1961], Sheppe [1965], Marten [1971]), or by runways that they build up.

Faecal pellets have been used as a basis for a population index in big game (Neff [1968]), pheasants (McClure [1945]), mice (Emlen *et al.* [1957]), ruffed grouse (Dorney [1958]), rabbits (Taylor and Williams [1956]), and snowshoe hares (Adams [1959]). Indices based on the call counts of birds are widely used, and the following references cover some of the problems involved in using such indices: woodcocks (Kozicky *et al.* [1954]); mourning doves (Lowe [1956], Foote *et al.* [1958], Cohen *et al.* [1960]); pheasants (Davis [1963: 101], Nelson *et al.* [1962], Gates [1966], Martinson and Grondahl [1966]); and ruffed grouse (Petraborg *et al.* [1953], Dorney *et al.* [1958], Gullion [1966]). In particular, Kozicky *et al.* [1954] use an analysis-of-variance technique for analysing the data with respect to years and routes, and they give a method for estimating the number of routes required to detect a given percentage change in the count index.

Sometimes an index can be converted to a measure of absolute density

---

*For a useful review see Overton and Davis [1969: 427–32].

if a reliable conversion factor of signs per animal is available, e.g. calls per bird (Gates and Smith [1972]), pellet groups per deer (Rogers *et al.* [1958]). A useful technique given by Davis [1963] for estimating absolute density is to calculate the index before and after a known number of animals are removed. For example, suppose that a population of size N occupying a given area yields an index I_1 for the area (e.g. I_1 is the total number of calls or pellet groups). If the index is I_2 after the removal of n animals from the area, and the number of signs per animal remains constant, we have the relation

$$\frac{I_1}{N} = \frac{I_2}{N-n} \left(= \frac{I_1 - I_2}{n} \right)$$

or
$$N = nI_1/(I_1 - I_2).$$

This method is further exploited in Chapter 9 (cf. (9.22)).

FREQUENCY INDEX. One other index worthy of consideration is the so-called "frequency index" (Dice [1952], Scattergood [1954: 279], Davis [1963]), which is the proportion, $\hat{p}$ say, of the sample units which contain at least one or more animals (or signs). This index is particularly useful when it is difficult to count the number of individuals in a unit (e.g. fleas per rat) but easy to determine the presence or absence of individuals. We note that $\hat{p}$ will depend on the size of the sampling unit, and the same size unit must be used if comparisons are to be made. If quadrat sampling is used and the quadrat is too big, $\hat{p}$ will be unity, and obviously no comparison can be made between two populations that are present in 100 per cent of the quadrats, even if the densities are widely different. To overcome this problem, quadrats of several sizes should be used and a practical way of doing this (Davis [1963]) is to divide a large quadrat into smaller-size quarters for recording data. The data can then be analysed at any time in the future by quarters, halves or wholes. If several species are to be compared it is recommended that the quadrat size should be such that $\hat{p}$ is about 0·8 for the more important species (Fisher [1954: 62]).

Suppose that the population area consists of S quadrats with a proportion $p\ (= 1-q)$ occupied by animals. If n quadrats from a sample of s quadrats are found to contain animals, then from Cochran [1963], n has a hypergeometric distribution, $\hat{p} = n/s$ is an unbiased estimate of p, and

$$V[\hat{p}] = \frac{pq\ (S-s)}{s\ (S-1)}$$

with unbiased estimate $v[\hat{p}] = \frac{\hat{p}\hat{q}}{s-1} \left(1 - \frac{s}{S} \right).$

If the sampling fraction s/S is less than 0·1 it can be ignored and a knowledge of S is then not required for $v[\hat{p}]$. The question of constructing a confidence interval for p is discussed in Cochran [1963: Chapter 3]. By redefining p, the above theory can be extended to the case of counting the number of plots with t or less animals (cf. Gerrard and Chiang [1970]).

When the animals are randomly distributed, the frequency index can be converted to an estimate of the absolute population density D. In this case the number of animals in a quadrat of area a follows a Poisson distribution with parameter Da (p. 24), and D can be estimated from

$$\hat{q} = \exp(-\hat{D}a). \tag{2.40}$$

From the delta method (**1.3.3**) we have

$$V[\hat{D}] \approx V[\hat{p}]/(qa)^2.$$

We note that if the population is not randomly distributed and the animals tend to group together more than as predicted by the Poisson distribution (which will usually be the case), then the probability of finding a given quadrat empty will be greater than $\exp(-Da)$ and $\hat{D}$ will underestimate the true density. When a non-random distribution is known or suspected, Dice [1952] suggests that the best procedure is to reduce the quadrat size or sampling unit until only rarely will more than one individual be found in any quadrat. In this case $\hat{D}a$ will be small, so that, expanding the exponential in (2.40),

$$\hat{p} = 1 - \hat{q} \approx \hat{D}a, \quad \text{or} \quad \hat{D} = n/(sa).$$

However, this is simply the same as saying that n is a good approximation to the total count on a region of area sa when the probability that a quadrat contains more than one individual is small. For a further discussion on the question of non-random distribution see Gerrard and Chiang [1970].

The frequency indexes of two or more populations can be compared by means of a contingency table, and details are given, for example, in Greig-Smith [1964: 37–41]. It should, however, be stressed that a difference in frequency indexes does not necessarily imply a difference in population densities. For example, two populations with the same average density will give different p values if the animals in one population tend to be clustered together more than in the other. Obviously the greater the clustering the greater the number of empty quadrats and the smaller the value of p.

Finally it is noted that if the presence or absence of several species is recorded for each quadrat, one can test for association between species (Greig-Smith [1964: Chapter 4], Pielou [1969: Chapter 13]).

Example 2.2   Grey fox (*Urocyon cinereoargenteus*): Wood [1959]

Two basic methods for obtaining information on the relative density of grey fox populations were considered, namely trapping with standardised trap-lines and track-counts. The trap-lines consisted of pairs of traps or "stations" spaced at equal intervals along the line. In looking for a suitable method of standardising the trap-line sampling, such questions as how to trap, where to trap, how long to trap, and how to space the stations were considered. For example, it was found that the highest catches occurred when the trap-lines were set along primitive roads, and a spacing of 0·2 mile between stations yielded the highest proportion of "positive" stations, i.e. stations which recorded at least one capture. A trapping period of 7 days was adopted as it was found that lines run for 8, 9 or 10 days caught over 90 per cent of their total catch by the 7th day, while lines run for 11–15 days caught over 80 per cent by the 7th day.

Two variations of the track-count method were used. The first procedure, called the scent-post method, was conducted in the same manner as the above trapping method except that no traps were set. At each trap site there was a raked area of about 3 feet in diameter with a clump of grass containing a scent lure in the centre or near one edge. Animals visiting the lure had to cross part of the raked area, and their visit was recorded by the track left in the soft soil. The second procedure, called the random track-count method, required no field preparation. Undisturbed sandy, dusty or muddy areas that would record the track of a passing animal were randomly selected and checked for the presence or absence of fox signs.

Several indices of relative density were considered for the trap-line data. However, it was found that the frequency index $\hat{p}$, the observed proportion of positive stations, gave the simplest means for comparing data from different areas. This index could also be applied to data from either of the track-count methods.

In Table 2.4 the observed numbers of stations catching 0, 1, 2, 3 or 4 foxes in a period of 7 days are compared with the expected number of stations

TABLE 2.4
Comparison of the number of observed and of expected stations that captured 0, 1, 2, 3 or 4 grey foxes: from Wood [1959: Table 5].

No. of foxes	Observed stations (O)	Expected stations (E)	$\dfrac{(O-E)^2}{E}$
0	3771	3702·6	1·264
1	524	647·0	23·383
2	97	55·2	31·652
3	15	3·2	
	17	3·4	54·400
4	2	0·2	
Total	4409	4408·2	110·699

calculated on the assumption that catches follow a Poisson distribution. The chi-squared goodness-of-fit statistic (cf. **1.3.6**(2)) yields the high value of 110·699 (2 degrees of freedom), thus indicating that the distribution of foxes was not random and that some form of social behaviour was affecting the numbers caught (Davis [1963: 98]). However, the observed number of stations catching zero foxes does not differ significantly from the expected number. We also find that the average catch of

$$[1(524) + 2(97) + 3(15) + 4(2)]/4409 = 0·175$$

foxes per station is close to

$$\log (1/\hat{q}) = \log (4409/3771) = 0 \cdot 16$$

(cf. equation (2.40) with a equivalent to one station). This suggests that $\hat{q}$ (and hence $\hat{p}$) is a suitable index for comparing population densities. For example, in Table 2.5 we have a 2-by-2 contingency table for comparing the

TABLE 2.5

Comparison of the number of positive stations (i.e. stations catching at least one fox) for summer and winter seasons.

	Summer	Winter	Total
Positive stations	16	14	30
Zero-catch stations	60	27	87
Total	76	41	117
$\hat{p}$	0·210	0·341	$\chi^2_1 = 1·76$

frequency indexes for two seasons. The value of chi-squared is not significant, so that there is no evidence of a seasonal difference.

In conclusion it is noted that the two track-count methods were not developed early enough for extensive use in the above population study. However, they showed promise as a means of obtaining density indexes, and the random track-count could obviously be used in areas lacking a sufficient number of primitive roads.

BOUNDED COUNTS METHOD. Population units (whether animals or signs) are sometimes not easy to count because of mobility or lack of distinguishability, and a person counting units on an area or sample plot may not be sure that he has counted all the units. If no units are counted twice and repeated counts are possible, Regier and Robson [1967] suggest the following "bounded counts" method based on the theory of Robson and Whitlock [1964].

Let N be the true number of units and let N_m, N_{m-1} be the largest and second largest counts obtained, respectively. Then N can be estimated by

$$\hat{N} = N_m + (N_m - N_{m-1}) = 2N_m - N_{m-1}$$

and an approximate $100(1-\alpha)$ per cent confidence interval for N is $N_m < N < [N_m - (1-\alpha)N_{m-1}]/\alpha$. Regier and Robson give several possible applications of this method to fresh-water fish populations. For example, it can be used in stream censuses by a number of divers obtaining independent counts under conditions similar to those described by Northcote and Wilke [1963]. It could also be applied in counting migrating fish-runs from a number of vantage points by equally perceptive enumerators or mechanical devices, and in small ponds through which large seines may be drawn at least twice during an interval when the population is closed.*

If s independent counts are made, then the bias of $\hat{N}$ is of order $1/s^2$. For cases when more than two counts are made, Robson and Whitlock [1964] derive further corrections to reduce the bias.

---

*A further example is given by Overton and Davis [1969: 426].

CHAPTER 3

CLOSED POPULATION: SINGLE MARK RELEASE

3.1 ESTIMATION

3.1.1 Hypergeometric model

A simple method, which we shall call the "Petersen method", for estimating N, the number of animals in a closed population, is the following. A sample of n_1 animals is taken from the population, the animals are marked or tagged for future identification and then returned to the population. After allowing time for marked and unmarked to mix, a second sample of n_2 animals is then taken and it is found that m_2 are marked. Assuming that the proportion of marked in the second sample is a reasonable estimate of the unknown population proportion, we can equate the two and obtain an estimate of N. Thus:

$$\frac{m_2}{n_2} = \frac{n_1}{\hat{N}}, \quad \text{or} \quad \hat{N} = \frac{n_1 n_2}{m_2},$$

the so-called "Petersen estimate" (or "Lincoln Index": cf. Le Cren [1965]). As this estimate is widely used in ecological investigations we shall now discuss the above method in some detail.

The first step is to decide what assumptions must hold if $\hat{N}$ is to be a suitable estimate of N. These are usually listed as follows:

(a) The population is closed (cf. **1.2.2**), so that N is constant.
(b) All animals have the same probability of being caught in the first sample.
(c) Marking does not affect the catchability of an animal.
(d) The second sample is a simple random sample, i.e. each of the $\binom{N}{n_2}$ possible samples has an equal chance of being chosen.
(e) Animals do not lose their marks in the time between the two samples.
(f) All marks are reported on recovery in the second sample.

We note that these assumptions are not mutually exclusive. For example, (d) will depend on the validity of (b) and (c), as any variation in the catchability of the animals, whether natural or induced by the handling and marking, will lead to a non-random second sample. This point is discussed further in **3.2.2**.

When assumptions (a), (d), (e) and (f) are satisfied, the conditional distribution of m_2, given n_1 and n_2, is the hypergeometric distribution

$$f(m_2 \mid n_1, n_2) = \binom{n_1}{m_2}\binom{N-n_1}{n_2 - m_2} \Big/ \binom{N}{n_2}, \tag{3.1}$$

where $m_2 = 0, 1, 2, \ldots$, minimum (n_1, n_2). The properties of $\hat{N}$ with respect to this distribution have been discussed fully by Chapman [1951]. He shows that although $\hat{N}$ is a best asymptotically normal estimate of N as $N \to \infty$, it is biased, and the bias can be large for small samples. However, when $n_1 + n_2 \geqslant N$, his modified estimate

$$N^* = \frac{(n_1 + 1)(n_2 + 1)}{(m_2 + 1)} - 1$$

is exactly unbiased, while if $n_1 + n_2 < N$ we have, to a reasonable degree of approximation (Robson and Regier [1964]),

$$E[N^* \mid n_1, n_2] = N - Nb,$$

where $b = \exp\{-(n_1 + 1)(n_2 + 1)/N\}$. Defining

$$\mu = n_1 n_2 / N = E[m_2 \mid n_1, n_2],$$

Robson and Regier recommend that in designing a Petersen type experiment it is essential that $\mu > 4$, so that b is small (in this case less than $0 \cdot 02$). They also state that if $m_2 \geqslant 7$ in a given experiment, then we are 95 per cent confident that $\mu > 4$. This means that for 7 or more recaptures we can be 95 per cent confident that the bias of N^* is negligible.

Chapman [1951] shows that N^* not only has a smaller expected mean square error than $\hat{N}$ for values encountered in practice, but it also appears close to being a minimum variance unbiased estimate over the range of parameter values for which it is almost unbiased. Using what is essentially a Poisson approximation to (3.1), he shows that the variance of N^* is approximately given by

$$V[N^* \mid n_1, n_2] = N^2(\mu^{-1} + 2\mu^{-2} + 6\mu^{-3}). \tag{3.2}$$

If the expected number of recaptures μ is small, this variance is large, and Chapman [1951: 148] concludes that "sample census programmes in which the expected number of tagged members in the sample is much smaller than 10 may fail to give even the order of magnitude of the population correctly." For example, when $\mu = 10$, (3.2) yields a standard deviation of $0 \cdot 36N$.

An estimate, V^* say, of the variance of N^* is obtained by simply replacing N by N^* in (3.2). However, an approximately unbiased estimate has been given by Seber [1970a], namely

$$v^* = \frac{(n_1 + 1)(n_2 + 1)(n_1 - m_2)(n_2 - m_2)}{(m_2 + 1)^2(m_2 + 2)},$$

which has a positive proportional bias of order $\mu^2 \exp(-\mu)$. It can be shown that v^* is exactly unbiased when $n_1 + n_2 \geqslant N$. (In fact, when $n_1 + n_2 \geqslant N$, N^* and v^* are both *unique* unbiased estimates because of the completeness of m_2.)

The coefficient of variation of N^* is approximately given by

$$C[N^*] = 1/\sqrt{\mu},$$

and if a rough estimate of N is available before the experiment, n_1 and n_2 can be chosen beforehand to give a desired value of C; this question of experimental design is discussed more fully in **3.1.5**. We note that an estimate of C is obtained by replacing μ by m_2, giving $C = 1/\sqrt{m_2}$. This means that the "accuracy" of N^* is almost solely dependent on the number of recaptures m_2.

3.1.2 Bailey's binomial model

Using a binomial approximation to the hypergeometric distribution (3.1), we have (Bailey [1951, 1952]):

$$f(m_2 \mid n_1, n_2) \approx \binom{n_2}{m_2} \left(\frac{n_1}{N}\right)^{m_2} \left(1 - \frac{n_1}{N}\right)^{n_2 - m_2}, \qquad (3.3)$$

and the maximum-likelihood estimate of N is the Petersen estimate $\hat{N}$. Bailey shows that $\hat{N}$ is biased with respect to this binomial distribution and suggests the modification

$$\hat{N}_1 = n_1 (n_2 + 1)/(m_2 + 1)$$

which has a proportional bias of order $\exp(-\mu)$. The variance of this estimate may be estimated, with a positive proportional bias of order $\mu^2 \exp(-\mu)$, by

$$v_1 = \frac{n_1^2(n_2 + 1)(n_2 - m_2)}{(m_2 + 1)^2(m_2 + 2)}.$$

If the sampling fraction n_2/N is sufficiently small for one to be able to ignore the complications of sampling without replacement then $\hat{N}_1$ can be used instead of N^*, though in practice there will often be little difference in the two estimates. Obviously (3.3) will hold exactly for the less common situation when random sampling *with* replacement is used (cf. Example 3.9, p. 110, where the animals are merely observed and not actually captured: this point is discussed further in **12.1.2** (3)). However, there is one other practical situation where (3.3) may be more appropriate than (3.1). We saw above that $\hat{N}$ is an intuitively reasonable estimate when the sample proportion of marked in the second sample faithfully reflects the population proportion of marked. This means that $\hat{N}$ can still be used even when assumption (d) is false and the second sample is a systematic rather than a random sample, provided that (i) there is uniform mixing of marked and unmarked so that the proportion (n_1/N) of marked throughout the population is constant, and (ii) given that a certain location in the population area is sampled, all animals at that location, whether marked or unmarked, have the same probability of being caught. When (i) and (ii) are satisfied, the probability that an animal is found to be marked, given that it is caught in the second sample, is n_1/N, and the binomial model (3.3) applies. The question of systematic sampling is discussed further on p. 82.

An alternative model to the above hypergeometric and binomial models, using a distribution derived by Skellam [1948] on the assumption that the probability of capture is not constant but follows a beta distribution, has been given by Eberhardt [1969a] (cf. (4.34) in **4.1.6** (4)).

3.1.3   Random sample size

In practice it is not always possible to fix n_2 in advance as the sample size may depend on the effort or time available for sampling. However, when n_2 is regarded as a random variable rather than a fixed parameter, N^* is still approximately unbiased, since

$$E[N^* \mid n_1] = \underset{n_2}{E}\, E[N^* \mid n_1, n_2]$$

$$\approx \underset{n_2}{E}\, [N]$$

$$= N,$$

and using a similar argument it is readily shown that v^* is an approximately unbiased estimate of $V[N^* \mid n_1]$. Also, from **1.3.4** we have

$$V[N^* \mid n_1] = \underset{n_2}{E}\, \{V[N^* \mid n_1, n_2]\} + \underset{n_2}{V}\{E[N^* \mid n_1, n_2]\}$$

$$\approx \underset{n_2}{E}\, \{V[N^* \mid n_1, n_2]\}$$

$$\approx \{V[N^* \mid n_1, n_2]\}_{n_2 = E[n_2 \mid n_1]}.$$

This means that for large N, the only difference between $V[N^* \mid n_1]$ and $V[N^* \mid n_1, n_2]$ is that in the former, n_2 is replaced by $E[n_2 \mid n_1]$. Since in practice one would estimate $E[n_2 \mid n_1]$ by n_2 in a variance formula, there is therefore little difference between treating n_2 as a fixed parameter or as a random variable as far as estimation is concerned. But it can be argued that once n_2 is known, we are only interested in the distribution of m_2 given n_1 and n_2, and that $f(m_2 \mid n_1, n_2)$ is then the appropriate distribution, irrespective of whether n_2 is fixed or random.

3.1.4   Confidence intervals

We now turn our attention to the problem of finding confidence intervals for N. As $N \to \infty$, N^* is asymptotically normally distributed, so that an approximate 95 per cent confidence interval for N is given by

$$N^* \pm 1 \cdot 96 \sqrt{v^*}.$$

However, according to Ricker [1958], $1/N^*$ is more symmetrically distributed and more nearly normal than N^*, so that in general it is better to base confidence intervals on the probability distribution of m_2. This hypergeometric distribution (3.1) has been tabled, and exact confidence limits for $p = n_1/N$ when N is known and n_1 unknown are available (e.g. Chung and De Lury [1950]). Unfortunately no such tables are available for the case when N is unknown and n_1 known, so that approximate methods have to be used. For various values of n_1, n_2 and N the hypergeometric distribution can be satisfactorily approximated by the Poisson, binomial and normal distributions (Chapman [1948], Lieberman and Owen [1961]). But the choice of which approximation to use when N is unknown still needs further investigation, so that the following recommendations should be regarded as a general guide only.

Let $\hat{p} = m_2/n_2$; then when $\hat{p} < 0.1$ and $m_2/n_1 < 0.1$, the Poisson approximation is recommended using m_2 as the "entering" variable in appropriate tables. For example, a confidence interval for μ ($= n_1 n_2/N$) can be read off from tables such as Pearson and Hartley [1966: 227, $m_2 \leqslant 50$], Crow and Gardner [1959: $m_2 \leqslant 300$] or from a graph (Adams [1951: $m_2 \leqslant 50$]). For $m_2 \leqslant 50$, however, it is simpler to use a table specially prepared by Chapman [1948], giving the shortest 95 per cent confidence intervals for N/λ where $\lambda = n_1 n_2$ (for example, when $m_2 = 10$, Chapman's interval is 5 per cent shorter than the "equi-tail" interval of Pearson and Hartley). This table is reproduced in **A1**, and we demonstrate its use with the following example. Suppose $n_1 = 1000$, $n_2 = 500$, $m_2 = 20$, then $\hat{p} = 0.04$, $m_2/n_1 = 0.02$ and the Poisson approximation is appropriate. Using m_2 as the entering variable, a 95 per cent confidence interval for N/λ is $(0.030\ 04, 0.0773)$, and multiplying these limits by $n_1 n_2$ gives the corresponding interval for N, namely $(15\ 020, 38\ 650)$.

When $\hat{p} < 0.1$ and $m_2 > 50$ we can use a normal approximation given by Cochran [1963: 87] to obtain a 95 per cent confidence interval for p, namely

$$\hat{p} \pm \{1.96[(1-f)\hat{p}(1-\hat{p})/(n_2-1)]^{\frac{1}{2}} + 1/(2n_2)\} \qquad (3.4)$$

which can be inverted to give a confidence interval for N. Here $f(= n_2/N)$, the unknown "sampling fraction", can be neglected if its estimate $f = m_2/n_1$ is less than 0.1; also $1/(2n_2)$, the correction for continuity, will often be negligible. For example, if $n_1 = 2000$, $n_2 = 1000$, $m_2 = 80$, then $\hat{p} = 0.08$, $\hat{f} = 0.04$ and the interval for p is

$$0.08 \pm 1.96[(0.08)(0.92)/1000]^{\frac{1}{2}},$$

or $(0.0632, 0.0968)$. The corresponding interval for N is $(20\ 700, 31\ 600)$. Neglecting f when $f > 0.1$ will lead to conservative confidence intervals, i.e. intervals which are overwide. This is not serious, however, as the assumptions given in **3.1.1** are never exactly true and variances have a habit of being larger than predicted by theory!

When $N > 150$, $n_1 > 50$ and $n_2 > 50$, m_2 is approximately normal (Robson and Regier [1964]), and a more accurate method than the one above is to solve the following cubic equation in N:

$$\frac{\left(m_2 - \dfrac{n_1 n_2}{N}\right)^2}{n_2 \cdot \dfrac{n_1}{N}\left(1 - \dfrac{n_1}{N}\right)\left(\dfrac{N - n_2}{N - 1}\right)} = 1.96^2.$$

The two largest roots then give an approximate 95 per cent confidence interval for N (Chapman [1948]); a graphical method of solution is discussed in Schaefer [1951].

If $\hat{p} > 0 \cdot 1$ we can use either the binomial approximation (3.3) or the normal approximations mentioned above. A rough guide as to the smallest values of n_2 for which the normal approximation (3.4) is applicable is given by the following table reproduced from Cochran [1963: 57]:

$\hat{p}$ (or $1 - \hat{p}$)	0·5	0·4	0·3	0·2	0·1
n_2	30	50	80	200	600

For example, if $\hat{p} = 0 \cdot 3$ (or $0 \cdot 7$), the normal approximation can be used if $n_2 > 80$. When the normal approximation is not applicable, a binomial confidence interval for p can be obtained from the Clopper–Pearson charts in Pearson and Hartley [1966: 228–229] and Adams [1951], or from extensive binomial tables such as those of the Harvard Computation Laboratory [1955]. If, for example, $m_2 = 18$, $n_2 = 60$ then $\hat{p} = 0 \cdot 3$, and the first Clopper–Pearson chart (with $n = 60$) gives $(0 \cdot 190, 0 \cdot 433)$ as the 95 per cent confidence interval for p.

3.1.5   Choice of sample sizes

1   Prescribed accuracy

It was mentioned in **3.1.1** that given a rough estimate of N, the product $n_1 n_2$ can be determined to give a prescribed coefficient of variation $C[N^*]$; useful tables for doing this are given in Davis [1964]. However, although accuracy is often measured in terms of coefficient of variation, a more appropriate definition of the accuracy A of an estimate is given by Robson and Regier [1964] as follows.

Let $(1 - \alpha)$ be the probability that the Petersen estimate $\hat{N}$ will not differ from the true population size by more than $100A$ per cent. Then

$$1 - \alpha \leqslant \Pr\left[- A < \frac{\hat{N} - N}{N} < A\right],$$

where α and A are to be chosen by the experimenter. Three standard levels for α and A are suggested:

(i) $1 - \alpha = 0 \cdot 95$, $A = 0 \cdot 50$; recommended for preliminary studies or management surveys where only a rough idea of population size is needed.

(ii) $1 - \alpha = 0 \cdot 95$, $A = 0 \cdot 25$; recommended for more accurate management work.

(iii) $1 - \alpha = 0 \cdot 95$, $A = 0 \cdot 10$; recommended for careful research into population dynamics.

When $\hat{N}$ is multiplied or divided by some other estimate (e.g. multiplied by weight to obtain total biomass or divided by another population estimate to obtain probability of survival) a relatively high accuracy is required of each estimate if the variance of the product or ratio is to be reasonably small. In this situation, level (iii) would then be most appropriate.

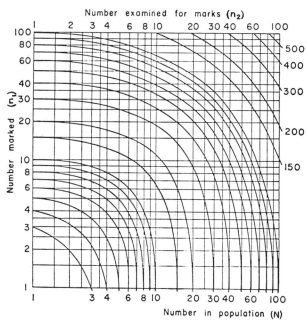

Fig. 3.1   Sample sizes when $1 - \alpha = 0.95$, $A = 0.5$ and $N \leqslant 500$; recommended for preliminary studies or management surveys. (From Robson and Regier [1964].)

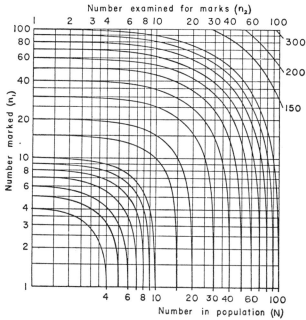

Fig. 3.2   Sample sizes when $1 - \alpha = 0.95$, $A = 0.25$ and $N \leqslant 300$; recommended for management studies. (From Robson and Regier [1964].)

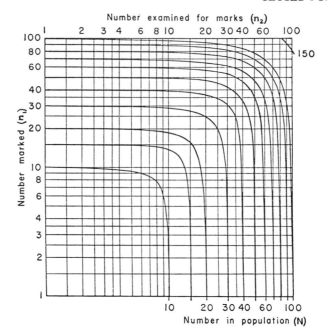

Fig. 3.3   Sample sizes when $1 - \alpha = 0 \cdot 95$, $A = 0 \cdot 10$ and $N \leqslant 150$; recommended for research. (From Robson and Regier [1964].)

Robson and Regier give charts (Fig. 3.1–3.6) for determining n_1 and n_2 from a rough estimate of N for the above three levels of α and A. They point out that for the values of n_1 and n_2 given by these charts the bias mentioned in **3.1.1** is negligible (less than about 1 per cent). The first three charts ($N \leqslant 100$) were calculated using tables of the hypergeometric distribution given by Lieberman and Owen [1961], while the second three charts ($N > 100$) were obtained using a normal approximation to the hypergeometric. In this latter case only part of the range of the combinations of n_1 and n_2 for each N are drawn, as the relationships determining the charts (equations (3.5) and (3.6)) are symmetric for n_1 and n_2. Thus, for a given α, sample sizes $n_1 = 300$, $n_2 = 600$ will yield an estimate $\hat{N}$ of the same accuracy A as sample sizes $n_1 = 600$, $n_2 = 300$. To illustrate the use of the tables suppose that N is estimated to be 10 000 and we require $1 - \alpha = 0 \cdot 95$, $A = 0 \cdot 25$. Then from Fig. 3.5 we obtain possible pairs such as (6000, 50), (4000, 100) and (1000, 600),

If, for $N > 100$, we wish to use values of α and A other than those mentioned above, charts for n_1 and n_2 can be drawn using the normal approximation as follows. Given α and A, we solve

$$1 - \alpha = \phi\left(\frac{A\sqrt{D}}{1 - A}\right) - \phi\left(\frac{-A\sqrt{D}}{1 + A}\right) \tag{3.5}$$

for D, where $\phi(z)$ is the cumulative unit normal distribution, and then, using an estimate of N, plot n_1 and n_2 subject to the constraint

TABLE 3.1

Values of D satisfying equation (3.5) for selected α and A: from Robson and Regier [1964: Table 2].

$1 - \alpha$	A	D
0.75	0.50	4.75
0.90	0.50	14.8
0.90	0.25	45.5
0.95	0.50	24.4
0.95	0.25	69.9
0.95	0.10	392
0.99	0.10	695
0.99	0.01	66 300

TABLE 3.2

Combinations of A and $1 - \alpha$ satisfying (3.5) for selected D: from Robson and Regier [1964: Table 3].

D	A	$1 - \alpha$
24.4	0.88	0.99
24.4	0.50	0.95
24.4	0.25	0.79
24.4	0.10	0.38
24.4	0	0
69.9	0.39	0.99
69.9	0.25	0.95
69.9	0.10	0.60
69.9	0	0
392	0.14	0.99
392	0.10	0.95
392	0.05	0.75
392	0	0

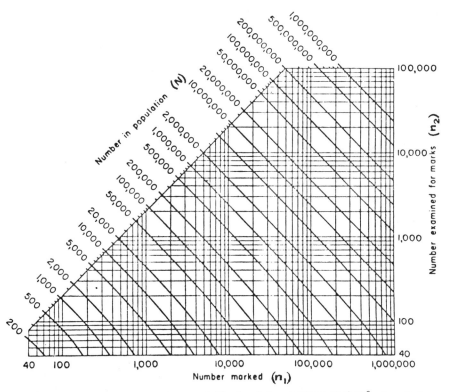

Fig. 3.4 Sample sizes when $1 - \alpha = 0.95$, $A = 0.5$ and $200 \leqslant N \leqslant 10^9$: from Robson and Regier [1964].

67

$$\frac{n_1 n_2 (N-1)}{(N-n_1)(N-n_2)} = D. \qquad (3.6)$$

Robson and Regier give two useful tables (Tables 3.1 and 3.2) for handling (3.5); Table 3.1 gives the solution D for various combinations of $1 - \alpha$ and A, and Table 3.2 gives various combinations of $1 - \alpha$ and A satisfying (3.5)

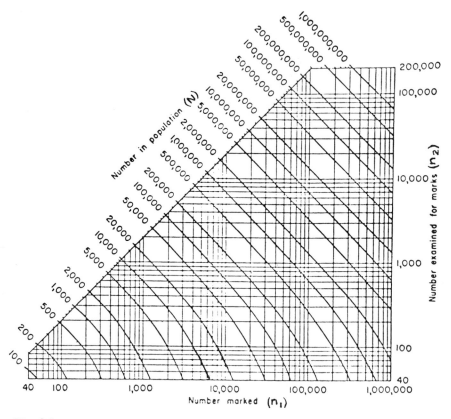

Fig. **3.5**   Sample sizes when $1 - \alpha = 0.95$, $A = 0.25$ and $100 \leqslant N \leqslant 10^9$: from Robson and Regier [1964].

for the values of D used in the construction of Fig. 3.4–3.6, namely $D = 24\cdot4$, $69\cdot9$, 392. For example, given $1 - \alpha = 0\cdot99$, $A = 0\cdot10$, then from Table 3.1, $D = 695$. If N is estimated to be 1000, n_1 and n_2 are the solutions of

$$\frac{n_1 n_2 (1000-1)}{(1000-n_1)(1000-n_2)} = 695;$$

a particular solution is $n_1 = n_2 = 455$.

   In conclusion it is noted that sample sizes less than 40 are not graphed in Fig. 3.4–3.6 as the normal approximation is not sufficiently accurate in this case. The lines for smaller sample sizes could have been derived using the Poisson or binomial approximations to the hypergeometric, but the authors did not consider it worth the effort. As demonstrated below, experiments utilising markedly unequal sample sizes (either large n_1 and very small n_2 or vice versa) would almost certainly be much more costly than if more nearly equal-sized samples were used. For a further discussion on the accuracy of the Petersen estimate see **12.1.2**(*3*).

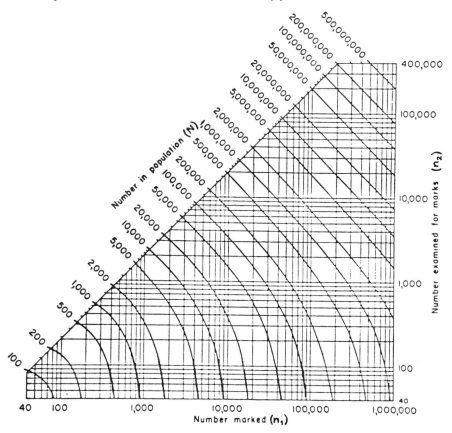

Fig. 3.6 Sample sizes when $1 - \alpha = 0.95$, $A = 0.10$ and $100 \leqslant N \leqslant 5 \times 10^8$: from Robson and Regier [1964].

2   Optimum allocation of resources

MINIMUM α WITH LIMITED FUNDS. Suppose c_1 and c_2 are the costs per animal of catching—marking and catching—examining and let f represent the known fixed or overhead costs, so that the total cost of the experiment is

$$c_T = c_1 n_1 + c_2 n_2 + f. \qquad (3.7)$$

If funds are limited and the accuracy A is predetermined, then an optimum allocation would be to choose n_1 and n_2 which maximise $1 - \alpha$, subject to c_T being fixed. Robson and Regier [1964] show that the solution of this is

$$\frac{c_1 n_1}{c_2 n_2} = \frac{N - n_2}{N - n_1}, \qquad (3.8)$$

69

and given estimates of c_1, c_2 and N, equations (3.7) and (3.8) can be solved for n_1 and n_2. When n_1 and n_2 are small relative to N, then (3.8) becomes $c_1 n_1 \approx c_2 n_2$, so that an equal division of resources between catching–marking and catching–examining is optimal.

MINIMUM COSTS FOR FIXED α AND A. If funds are not restricted, so that both α and A can be fixed in advance, then one would wish to choose n_1 and n_2 so that c_T is minimised. Given rough estimates of N and c_1/c_2, Robson and Regier show that the solution is obtained by consulting the appropriate curve in Fig. 3.1–3.6 and choosing the pair (n_1, n_2) satisfying (3.8). If a combination of α and A other than one of the three given above is required, then one must solve equations (3.6) and (3.8), where D is calculated from (3.5).

ALTERNATIVE METHODS OF CAPTURE. If two or more alternative methods of capture are available and α and A are predetermined, then the problem of choosing the least costly method arises. Suppose we compare two methods with sample sizes n_{i1}, n_{i2} and costs c_{i1}, c_{i2} ($i = 1, 2$), where, for the ith method, n_{i1} and n_{i2} are optimal in the sense of the previous paragraph. Then, assuming the same fixed overhead cost for each method, Robson and Regier show that

$$\frac{\text{cost for 2nd method}}{\text{cost for 1st method}} \approx \sqrt{\left(\frac{c_{21} c_{22}}{c_{11} c_{12}}\right)} = r, \quad \text{say.}$$

For example, if the second method of capture costs twice as much per captured animal as the first method, and both methods utilise the same marking technique, then we can write

$$c_{11} = a + b, \; c_{12} = a; \quad c_{21} = 2a + b, \; c_{22} = 2a;$$

and

$$r = \sqrt{\left(\frac{4a + 2b}{a + b}\right)}.$$

3.2 VALIDITY OF ASSUMPTIONS

3.2.1 Assumption (a): closed population

If the assumption of constant population size is to hold, the experiment should be carried out over a short period of time, in fact, ideally at a single point in time. For this reason the Petersen method is often called a "point census".

Departures from assumption (a) can occur in a number of ways, and we shall now discuss these in detail.

1 Accidental deaths

If there are d accidental deaths through the process of catching and marking the first sample, the general theory of **3.1** remains unchanged, provided that n_1 now refers to the number returned *alive* to the population. $\hat{N}$ and N^* are estimates of $N - d$.

2   Natural mortality

Suppose that mortality is taking place in the time between the two samples, and let N be the size of the population when the first sample is released. Robson [1969] points out that when assumptions (d), (e) and (f) of **3.1.1** are true and the only departure from (a) is due to mortality, the hypergeometric model (3.1) still holds, provided the mortality process is such that the deaths constitute a *simple random sample* of unknown size. This follows from the fact that the survivors will also constitute a simple random sample of which the second sample is a random subsample (assumption (d)). Since a random subsample of a random sample is itself a random sample, the second sample will still represent a simple random sample from the original population. Chapman [1952: 300, 1954: 5] also demonstrates this feature of the Petersen method by using a binomial model for the mortality process.

When the deaths do not constitute a simple random sample, the Petersen estimate can still be used, provided the marked and unmarked have the same average probability ϕ of surviving up till the time of the second sample. This can be seen intuitively from the equation

$$E\left[\frac{m_2}{n_2}\middle| n_1\right] \approx \frac{\phi n_1}{\phi N} = \frac{n_1}{N}.$$

We note that mortality is often selective with regard to size or age of the animal. However, if the first sample is a simple random sample, the more "vulnerable" individuals will be proportionately represented in both marked and unmarked populations, thus ensuring that marked and unmarked have the same average survival probabilities. This point is discussed further on p. 87

To examine the effect of variable mortality in the marked portion, suppose that there are various subcategories in the population with numbers $X, Y, \ldots, W$, where $N = X + Y + \ldots + W$. For the *marked* members in these categories let $\phi_x, \phi_y, \ldots, \phi_w$ be the respective survival probabilities and let $p_{2x}, p_{2y}, \ldots,$ be the probabilities of recapture in the second sample. Then, using the suffix x to denote membership of category X, we have

$$E[m_{2x}/n_{1x}] = \phi_x p_{2x}$$

and we can test the hypothesis $H_0 : \phi_x p_{2x} = \ldots = \phi_w p_{2w}$ using a standard chi-squared test based on the contingency Table 3.3 (Dunnet [1963: 92],

TABLE 3.3

| | Subcategory | | | | |
	X	Y	...	W	Total
Recaptured	m_{2x}	m_{2y}	...	m_{2w}	m_2
Not recaptured	$n_{1x} - m_{2x}$	$n_{1y} - m_{2y}$	...	$n_{1w} - m_{2w}$	$n_1 - m_2$
Number released	n_{1x}	n_{1y}	...	n_{1w}	n_1

Robson and Regier [1968]). When $\phi_x = \phi_y = \ldots = \phi_w$, a test of H_0 is a test that the second sample is random with respect to the marked individuals in the various categories. Conversely, when the second sample is random so that $p_{2x} = p_{2y} = \ldots = p_{2w}$, this test is a test of constant survival probability for marked members.

3   Catchable population

It should be noted that N may sometimes refer to the catchable portion of the population only, and not to the whole population. For example, in sampling foraging ants, Ayre [1962] obtained a Petersen estimate of 109 for the population of an anthill known to be in the region of 3000. This enormous bias was explained by the fact that only a small proportion of the ants ever go foraging, so that the majority remain uncatchable.

If an approximately unbiased estimate $(\hat{p}_c)$ of the catchable proportion (p_c) of the population is available then we can estimate the total population N_T by

$$\hat{N}_T = N^*/\hat{p}_c.$$

If N^* and $\hat{p}_c$ are based on separate sampling experiments, as will usually be the case, they are statistically independent. Therefore, using the delta method (**1.3.3**), we have

$$E[\hat{N}_T] = E[N^*]E[1/\hat{p}_c]$$

$$\approx N\left(\frac{1}{p_c} + \frac{V[\hat{p}_c]}{p_c^3}\right)$$

$$= N_T\left(1 + \frac{V[\hat{p}_c]}{p_c^2}\right)$$

and

$$V[\hat{N}_T] \approx \frac{V[N^*]}{p_c^2} + \frac{N_T^2}{p_c^2}V[\hat{p}_c].$$

If binomial sampling is used to obtain $\hat{p}_c$ then $V[\hat{p}_c] = p_c q_c/n$, where n is the number of animals investigated for catchability.

Assuming $\hat{N}_T$ to be approximately normally distributed, an approximate confidence interval for N_T can be calculated in the usual manner. However, unless $\hat{p}_c$ and N^* are accurate estimates, this interval may be too wide to be of much practical use.

4   Recruitment

Sometimes the time-lapse between the two samples is sufficient to allow the recruitment of younger animals into the catchable part of the population. These new recruits would tend to decrease the proportion of marked in the second sample, and the Petersen estimate $\hat{N}$ would over-estimate the initial population size. In the situation where there is recruitment but no mortality, $\hat{N}$ will be a valid estimate of the population number at the time when the second sample is taken. However, as pointed out by Robson and Regier [1968], when both recruitment and mortality occur, $\hat{N}$ will overestimate both the initial and final population size. This is seen mathematically by noting that if there are r recruits in the population at the time of the second sample, we have the approximate relation

$$E \left[\frac{m_2}{n_2} \mid n_1 \right] \approx \frac{\phi n_1}{\phi N + r},$$

or

$$E[\hat{N} \mid n_1] \approx N + r\phi^{-1},$$

where $N + r\phi^{-1}$ is greater than N and $\phi N + r$ ($0 < \phi < 1$). By enlarging the definition of r to include *permanent* immigrants, and redefining ϕ as the average probability that an animal in the population at the release of the first sample is alive and still in the population at the time of the second sample, then, provided ϕ is the same for marked and unmarked, the above comments apply to a population in which there is immigration and emigration also.

If an animal becomes immediately catchable as soon as it reaches a certain age, then an age analysis of the second sample would provide an estimate of the ratio of recruits to non-recruits. Using this ratio and a Petersen estimate of non-recruits (from the proportion of marked in the non-recruited members of the second sample) we could then obtain an estimate of total recruits. But the process of age determination is usually time-consuming and the "threshold" age for recruitment is not usually well defined, so that partial recruitment may occur over a range of younger ages. Usually the more readily available information such as length or weight is used to classify the individual, and such data can be used for carrying out the following two tests of recruitment.

CHI-SQUARED TEST. If the individual measurements are not actually recorded and the animals are simply allotted to particular size-classes, we can test for recruitment as follows. Let $X, Y, \ldots, W$ denote both the classes and the numbers in the classes at the beginning of the experiment, and suppose that X increases to $X + r_x$, etc. through recruitment. Then if the second sample is random with respect to within (but not necessarily between) classes, we have

$$E \left[\frac{m_{2x}}{n_{2x}} \right] = \underset{n_{1x}}{E} \; E \left[\frac{m_{2x}}{n_{2x}} \mid n_{1x} \right]$$

$$= \underset{n_{1x}}{E} \left[\frac{n_{1x}}{X + r_x} \right] = p_{1x}, \quad \text{say},$$

and the hypothesis $H_0 : p_{1x} = p_{1y} = \ldots = p_{1w}$ can be tested using a standard chi-squared statistic based on the contingency Table 3.4 (Robson and Regier [1968]). When there is no recruitment, so that $r_x = 0$ for each class, then this test will be a test that the first sample is random with respect to size-class. Conversely, if the first sample is a simple random sample then $E[n_{1x}/X]$ will

TABLE 3.4

	Size-class X	Size-class Y	...	W	Total
Marked	m_{2x}	m_{2y}	...	m_{2w}	m_2
Unmarked	$n_{2x} - m_{2x}$	$n_{2y} - m_{2y}$	...	$n_{2w} - m_{2w}$	$n_2 - m_2$
Total	n_{2x}	n_{2y}	...	n_{2w}	n_2

be the same for each class and, since there is no recruitment in the classes with larger animals, a test of H_0 will then amount to a test of $r_x = 0$ for *all* the classes. In this latter situation the test will be unaffected by mortality, provided that the survival probabilities are the same for marked and un-marked. This follows from the simple relationship

$$E\left[\frac{m_{2x}}{n_{2x}}\right] = E\left[\frac{\phi_x n_{1x}}{\phi_x X + r_x}\right]$$

$$= E\left[\frac{n_{1x}}{X + r_x \phi_x^{-1}}\right],$$

where ϕ_x is the average survival probability for class X. Finally it is noted that the above test can still be used, even when animals grow from one class into another with unknown overlap during the course of the experiment. The reason for this is that although size-classes are usually determined from the second sample, X may be regarded as the "conceptual" population, existing at the time of the first sample, which grows into the required class: n_{1x} will then be unknown.

NON-PARAMETRIC TEST. When individual size measurements are recorded, Robson and Flick [1965] suggest the following non-parametric method for detecting and eliminating recruits. To simplify the notation we shall drop the suffix 2 from m_2 and n_2 and define $u = n_2 - m_2$. In the follow-ing discussion the word "length" will denote some readily available measure-ment of size.

We shall assume that n_1 is sufficiently large and the first sample sufficiently random for the length distribution of animals to be the same for both marked and unmarked. Suppose that the lengths of the m recaptures are $L_1 < L_2 < ... < L_m$ and let u_i ($i = 1, 2, ... , m + 1$) be the number of unmarked animals caught in the second sample with lengths L in the interval $L_{i-1} \leqslant L < L_i$ ($L_0 = 0$, $L_{m+1} = \infty$). If the second sample is random with respect to mark status and length, the probability that the length of an unmarked animal falls into any one of the above $m + 1$ length-classes is $1/(m + 1)$, and the expected value of each u_i will be $u/(m + 1)$. However, if recruitment has occurred in the shorter size range, the observed u_i for the intervals in this range will be greater than expected. Thus, if recruitment in the length-class $[0, L_1)$ has occurred, u_1 will be significantly larger than $u/(m+1)$. To determine the significance of u_1 we calculate the tail probability

$$\Pr[U_1 \geqslant u_1] = \binom{u+m-u_1}{m} \bigg/ \binom{u+m}{m}, \tag{3.9}$$

where U_1 is the random variable taking the values u_1, and compare this with the significance level α (usually $\alpha = 0\cdot05$). For example, if $u = 110$, $m = 50$ and $u_1 = 8$, then

$$u_1 > u/(m+1) = 2\cdot2$$

and

$$\Pr[U_1 \geqslant 8] = \binom{160-8}{50} \bigg/ \binom{160}{50}$$

$$= \frac{152!\ 110!}{102!\ 160!}.$$

Using Stirling's approximation for large factorials, namely

$$\log K! \approx \tfrac{1}{2} \log (2\pi) + (K + \tfrac{1}{2}) \log K - K,$$

we have, after some simplification,

$$\log \Pr[U_1 \geqslant 8] \approx (152\cdot5) \log 1\cdot52 + (110\cdot5) \log 1\cdot 10$$
$$- (102\cdot5) \log 1\cdot02 - (160\cdot5) \log 1\cdot60$$
$$= -3\cdot0821,$$

or

$$\Pr[U_1 \geqslant 8] \approx 0\cdot046.$$

Since this probability is just less than $0\cdot05$, we reject the hypothesis of no recruitment in the length-class $[0, L_1)$ at the 5 per cent level of significance.

The same procedure can now be applied to the second length-class $[L_1, L_2)$ by eliminating L_1 and the class of u_1 animals from the data. Thus we compare u_2 with its expected value $(u-u_1)/m$ and if this difference is significant we compare u_3 with $(u-u_1-u_2)/(m-1)$, and so on. Proceeding in this stepwise fashion through the larger classes, the recruits, if any, will dwindle in number until the rth step, say, is reached when the recruits no longer make a significant contribution. Thus u_r will not be significantly greater than $(u-a_{r-1})/(m-r+2)$, where $a_r = u_1 + u_2 + \ldots + u_r$; for this step the tail probability is

$$\Pr[U_r \geqslant u_r \mid u_1, u_2, \ldots, u_{r-1}] = \frac{\binom{n-a_r-r+1}{m-r+1}}{\binom{n-a_{r-1}-r+1}{m-r+1}}.$$

This would suggest that the remaining sample of $u_{r+1} + u_{r+2} + \ldots + u_{m+1}$ unmarked animals is free of recruits, and that the average

$$\bar{u}_{r+1} = (u_{r+1} + \ldots + u_{m+1})/(m+1-r)$$

is therefore an estimate of the number of unmarked non-recruits that should occur between every adjacent pair of marked recaptures. Hence the estimated number of unmarked non-recruits in the second sample is $(m+1)\bar{u}_{r+1}$ and the modified Petersen estimate of N becomes

$$N^* = \frac{(n_1+1)[\bar{u}_{r+1}(m+1)+m+1]}{m+1} - 1$$

$$= (n_1+1)(\bar{u}_{r+1}+1) - 1.$$

In evaluating the mean and variance of N^* we run into the difficulty of r being a random variable. This same problem arises, for example, in fitting a polynomial regression where the degree of the final polynomial obtained is strictly a random variable. However, as with the regression problem, treating r as though it were a constant would not seem unreasonable and would perhaps lead to a slight underestimate of $V[N^*]$. Now under the assumption of non-recruitment after the rth class, we have

$$E[\bar{u}_{r+1}\,|\,m,\,r,\,a_r] = \frac{u-a_r}{m+1-r}$$

leading to

$$E[N^*\,|\,m,\,r,\,a_r,\,n_1] = \frac{(n_1+1)(n-r-a_r+1)}{(m-r+1)} - 1$$

$$= \frac{(n_1+1)(n'+1)}{(m'+1)} - 1, \text{ say,}$$

where m' and n' are simply the values of m and n obtained by truncating the second sample at length L_r. If this truncation successfully eliminates recruits we would then expect

$$E[N^*\,|\,n_1] \approx N.$$

Also

$$V[N^*\,|\,n_1] = (n_1+1)^2\,V[\bar{u}_{r+1}],$$

where the variance of $\bar{u}_{r+1}$ can be estimated robustly from the replicated u's (cf. (1.9)), namely from

$$v[u_{r+1}] = \frac{1}{(m+1-r)(m-r)}\sum_{i=1}^{m+1-r}(u_{r+i}-\bar{u}_{r+1})^2.$$

Robson and Flick [1965] point out that as recruitment will generally tend to decrease with increasing body-length, there will be a decrease in the probability of detecting these recruits. Also this decrease in detectability is further accentuated by a decrease in the length interval between marked animals as the test progresses from the lower tail toward the centre of the length distribution, and also by the decrease in sample size resulting from the successive removal of the intervals tested. To overcome this difficulty we require some method of pooling intervals as the number of recruits falls off.

The authors mention that a further need for combining intervals arises when length measurements are sufficiently crude to permit ties to occur. In particular, if several marked animals are recorded as having the same body-length, then the resulting degenerate intervals must be combined to include all unmarked animals having that same recorded length.

It transpires that the optimal pooling procedure is simply to combine adjoining intervals giving a new total interval and a new total number of unmarked. If the first k intervals are combined, then a test for recruitment in this total interval has a tail probability of

$$T[s_k] = \Pr[S_k \geqslant s_k]$$

$$= \sum_{r=0}^{k-1} \binom{s_k+r-1}{r}\binom{u+m-s_k-r}{m-r}\bigg/\binom{u+m}{m} \qquad (3.10)$$

where $s_k(=u_1+u_2+\ldots+u_k)$ is the number of unmarked in the total interval. As k gets large, (3.10) becomes computationally awkward, and for small values of s_k the recursive relation

$$\Pr[S_k = s_k] = \frac{(s_k+k-1)(u-s_k+1)}{s_k(u+m-s_k-k+1)}\Pr[S_k = s_k-1], \qquad (3.11)$$

where $\Pr[S_k = 0] = \binom{u+m-k}{u}\bigg/\binom{u+m}{u}$,

is useful ($u-s_k-1$ should be replaced by $u-s_k+1$ in equation (3) of Robson's and Flick [1965]; the correct equation has been used in their calculations). When u is much greater than m, so that

$$\frac{m+1-k}{kn} \approx 0 \quad \text{and} \quad \frac{u(n+1)}{n^2} \approx 1,$$

Robson and Flick suggest the incomplete beta approximation

$$\Pr[S_k \geqslant s_k] = \frac{\Gamma(m+1)}{\Gamma(k)\,\Gamma(m+1-k)}\int_{p}^{1} t^{k-1}(1-t)^{m-k}\,dt$$

$$= \sum_{i=0}^{k-1}\binom{m}{i}p^i(1-p)^{m-i}, \qquad (3.12)$$

where $p = (s_k+k-1)/n$, and they give a useful table indicating the accuracy of the approximation. We note that (3.12) can be evaluated from standard binomial tables.

Suppose now that the first k_1 intervals, the second k_2 intervals ... are combined to give new length-classes $[0, L_{k_1})$, $[L_{k_1}, L_{k_2})$, etc. with unmarked numbers s_{k_1}, s_{k_2}, etc. Then if

$$\Pr[S_{k_1} \geqslant s_{k_1}] < \alpha,$$

we reject the hypothesis of no recruitment in the length-class $[0, L_{k_1})$ at the α level of significance and proceed to consider s_{k_2}. Dropping the data in the first interval from the sample, we now evaluate the tail probability (cf. (3.10))

$$T[s_{k_2}; s_{k_1}] = \Pr[S_{k_2} \geqslant s_{k_2} \mid S_{k_1} = s_{k_1}]$$

$$= \sum_{r=0}^{k_2-1} \binom{s_{k_2} + r - 1}{r} \binom{u' + m' - s_{k_2} - r}{m' - r} \bigg/ \binom{u' + m'}{m'}, \quad (3.13)$$

where $m' = m - k_1$ and $u' = u - s_{k_1}$ are the "new" values of m and u. We again reject the hypothesis of no recruitment in $[L_{k_1}, L_{k_2})$ if the above probability is less than α. This process can then be repeated for s_{k_3}, s_{k_4}, etc. until non-significance is achieved; the number of non-recruits in the second sample can then be estimated as before.

One of the problems in combining adjacent intervals is to determine the best sequence $k_1, k_2, \dots$. Since there is a practical possibility that all the recruits are shorter than the shortest non-recruit, it would seem reasonable to use $k_1 = 1$. Also, because of the steady reduction in numbers of recruits between successively larger marked animals, the k-sequence should be increasing, so that $1 = k_1 \leqslant k_2 \leqslant \dots$. Unfortunately the optimum sequence can only be determined if the length–frequency distributions are known for both recruits and non-recruits, although as the statistics S_{k_1}, S_{k_2}, etc. are virtually independent, the k sequence could perhaps be determined sequentially by regression methods. For example, Robson and Flick suggest extrapolating the regression of u_i on i for $i = 2, 3, \dots, k_2$ to indicate the number k_3 of intervals which must be next combined in order to achieve the relation

$$u_{k_1 + k_2 + k_3} \approx \frac{u - s_{k_1} - s_{k_2} - s_{k_3}}{m - k_1 - k_2 - k_3 + 1}.$$

Further research needs to be done on such methods of finding a suitable k-sequence.

If one wishes to combine intervals still further (e.g. the first $k_1 + k_2$ intervals), then, as mentioned above, the optimal procedure is simply to use the sum $S_{k_1} + S_{k_2}$ ($= S_{k_1 + k_2}$) and evaluate the tail probability

$$T[s_{k_1 + k_2}] = \Pr[S_{k_1 + k_2} \geqslant s_{k_1 + k_2}].$$

However, to avoid this additional computation Robson and Flick suggest a number of approximate procedures such as using $T[s_{k_1}] + T[s_{k_2}; s_{k_1}]$ with a significance level of $(2\alpha)^{1/2}$, or

$$T[s_{k_1}] + T[s_{k_2}; s_{k_1}] + T[s_{k_3}; s_{k_2}, s_{k_1}]$$

with a significance level of $(6\alpha)^{1/3}$ if three groups are pooled.

Example 3.1 Brook trout (*Salvelinus fontinalis*): Robson and Flick [1965]

In 1962 a number of brook trout were captured with trap nets from an experimental pond and were returned with jaw tags. The following year 70 marked and 165 unmarked trout were caught from the same pond, again using trap nets, and Table 3.5 gives the length-frequency distribution of these fish. The catchable trout in 1962 were presumed to be of age I+ and

TABLE 3.5

Length–frequency distributions of marked and unmarked brook trout in a trap-netted sample: from Robson and Flick [1965: Table 2].

Length-class (in.)	Marked	Unmarked	Length-class (in.)	Marked	Unmarked
< 9·7	0	54	11·4	2	6
		*	11·5	5	8
9·7 (L_1)	1	1			*
9·8	0	4	11·6	3	2
9·9	0	7	11·7	3	3
10·0	0	8	11·8	4	2
10·1 (L_2)	1	1			*
10·2	0	2	11·9	7	1
		*	12·0	6	2
10·3 (L_3)	1	4			*
10·4 (L_4, L_5)	2	7	12·1	4	3
10·5 (L_6, L_7)	2	4	12·2	4	2
10·6 (L_8)	1	2	12·3	2	1
10·7 (L_9, L_{10})	2	4			*
		*	12·4	1	0
10·8 (L_{11})	1	5	12·5	4	1
10·9 (L_{12})	1	5	12·6	0	2
11·0 ($L_{13}, ..., L_{15}$)	3	1	12·7	2	0
		*	> 12·7	0	12
11·1 ($L_{16}, ..., L_{19}$)	4	3			*
11·2 (L_{20})	1	4	Total	70	165
11·3 ($L_{21}, ..., L_{23}$)	3	4			
		*			

* Intervals used in testing for recruits.

the 1963 sample was then expected to include these same fish now at age II+ plus the new recruits of age I+.

The first few rows of Table 3.5 indicate that recruitment clearly occurred; also the number with length greater than 12·7 in. seems significantly high, but, for simplicity of exposition, we shall neglect this fact for the moment.

The k-sequence was arbitrarily chosen as $k_i = i$, subject to any necessary modifications due to ties. Thus the length-intervals are $[0, L_1), [L_1, L_3),$ $[L_3, L_6), [L_6, L_{11}),$ etc.; the interval $[L_6, L_{11})$ includes 5 rather than 4 lengths because 2 marked fish have the same length of 10·7 in. The k-sequence is now 1, 2, 3, 5, 5, 8, 7, 10, 13 and 10. We recall that u_i is the number of unmarked animals with lengths in $[L_{i-1}, L_i)$ and the various tests are as follows.

(i) $u = 165$, $m = 70$, $u_1 = 54$. From equation (3.9)

$$\Pr[U_1 \geqslant 54] = \binom{165 + 70 - 54}{70} \Bigg/ \binom{165 + 70}{70}$$

$$= 0 \cdot 2098 \times 10^{-9}.$$

(ii) $s_2 = u_2 + u_3 = 20 + 3 = 23$.
$u' = 165 - 54 = 111$, $m' = 70 - 1 = 69$.
From equation (3.13) we have

$$\Pr[S_2 \geqslant 23] = \left[\binom{22}{0} \binom{69 + 111 - 23}{69} + \binom{23}{1} \binom{69 + 111 - 24}{68} \right] \Bigg/ \binom{69 + 111}{69}$$

$$= \frac{156 \, ! \, 111 \, !}{88 \, ! \, 180 \, !} [157 + 23(69)]$$

$$= 0 \cdot 6160 \times 10^{-4}.$$

(iii) $s_3 = u_4 + (u_5 + u_6) = 4 + 7 = 11$.
$u' = 111 - 23 = 88$, $m' = 69 - 2 = 67$.

$$\Pr[S_3 \geqslant 11] = \left[\binom{10}{0} \binom{144}{67} + \binom{11}{1} \binom{143}{66} + \binom{12}{2} \binom{142}{65} \right] \Bigg/ \binom{155}{67}$$

$$= 0 \cdot 030 \, 04.$$

(iv) $s_5 = (u_7 + u_8) + u_9 + (u_{10} + u_{11}) = 4 + 2 + 4 = 10$, $u' = 77$, $m' = 64$.
$\Pr[S_5 \geqslant 10] = 0 \cdot 147 \, 72$.

(v) $s_5 = u_{12} + u_{13} + (u_{14} + u_{15} + u_{16}) = 5 + 5 + 1 = 11$, $u' = 67$, $m' = 59$.
$\Pr[S_5 \geqslant 11] = 0 \cdot 086$.

(vi) $s_8 = u_{17} + \ldots + u_{24} = 11$, $u' = 56$, $m' = 54$.
$\Pr[S_8 \geqslant 11] = 0 \cdot 246 \, 05$.

(vii) $s_7 = u_{25} + \ldots + u_{31} = 14$, $u' = 45$, $m' = 46$.
$\Pr[S_7 \geqslant 14] = 0 \cdot 033 \, 06$.

(viii) $S_{10} = u_{32} + \ldots + u_{41} = 7$, $u' = 31$, $m' = 39$.
To calculate $\Pr[S_{10} \geqslant 7]$ it is simpler to use the recursive relation (3.11) rather than (3.13). Thus

$$\Pr[S_{10} = 0] = \binom{31 + 39 - 10}{31} \Bigg/ \binom{31 + 39}{31} = \binom{60}{31} \Bigg/ \binom{70}{31},$$

$$\Pr[S_{10} = 1] = \frac{(1 + 10 - 1)(31 - 1 + 1)}{1(31 + 39 - 1 - 10 + 1)} \Pr[S_{10} = 0]$$

$$= \frac{10(31)}{60} \Pr[S_{10} = 0], \text{ etc.}$$

Hence

$$\Pr[S_{10} \geqslant 7] = 1 - \sum_{i=0}^{6} \Pr[S_{10} = i]$$

$$= 0{\cdot}628\ 96.$$

All the higher length-intervals fail to show a significant recruitment until the last interval is reached, for which the present procedure provides no test. This last interval has been ignored for purposes of exposition, but, as the authors point out, there is obviously an excessive number of unmarked trout longer than 12·7 in. (322 mm). If it had been suspected before the experiment that recruitment was possible at both ends of the length distribution then the appropriate test procedure would have been to alternate from one tail to the other, progressing toward the centre as far as possible from each end.

It is noted that the probabilities of the above tests fluctuate considerably, with (i), (ii), (iii) and (vii) being significant at the 5 per cent level of significance. These erratic results could be due to the arbitrary choice of $k_i = i$, and further pooling of the intervals is suggested. Thus, starting with the interval [9·7, 10·2), we again follow the rule of combining one, then two, then three intervals, etc., and using the approximate method of adding tail probabilities described above we obtain

$$0{\cdot}6160 \times 10^{-4} < 0{\cdot}05$$
$$0{\cdot}030\ 04 + 0{\cdot}147\ 72 < 0{\cdot}3162\ (= [2(0{\cdot}05)]^{1/2})$$
$$0{\cdot}0806 + 0{\cdot}2461 + 0{\cdot}0331 < 0{\cdot}6694\ (= [6(0{\cdot}05)]^{1/3}),$$

thus obtaining significant recruitment up to 11·5 inches (292 mm).

3.2.2 Assumptions (b), (c) and (d)

1 Practical considerations

VARIABLE CATCHABILITY. One of the crucial assumptions underlying the theory of 3.1.1 is assumption (d) that the second sample is a simple random sample (though we shall see below in section 2 that the theory still holds under alternative assumptions). Strictly speaking, such a random sample can only be obtained by numbering the animals $1, 2, \ldots, N$ and using a table of random numbers to select n_2 animals. However, in practice, if all the animals have the same catchability, we can approximate to a random sample by arranging that every point of the population area has the same probability·of being sampled (e.g. using random number pairs as coordinates) and that all points selected are sampled with the same effort. If a more even coverage is required one can use stratified random sampling, whereby the population area is divided into equal subareas and one or more points are allotted at random within each subarea (e.g. Leslie et al. [1953: 139]). Unfortunately the requirement of constant probability of capture may not hold, either because of an inherent variation in catchability or because catching and handling in the first sample affect future catchability. Very often the probability of capture will vary between various subgroups defined by age, sex, species, etc. (Kikkawa [1964: 260, 290], Pucek [1969]). For example,

certain subgroups may be more mobile and have a different habitat preference
(e.g. Corbet [1952: male and female dragonflies]) while others may have
certain bait and trap preferences (e.g. Chitty and Shorten [1946: bait indiffer-
ence shown by some rats], Dobzhansky *et al.* [1956: differential attraction
of species of Drosophila to different kinds of yeast]). Strandgaard [1967]
found age and sex variations in the sighting and trapping of roe deer.

In fisheries, catchability usually varies with the size of the fish, and
considerable research has been carried out on such problems as "gear
selectivity" and "length-selection" curves (cf. International Commission for
Northwest Atlantic Fisheries [1963], Lagler [1968]). When recruitment and
mortality are negligible (or any mortality is random with respect to size, cf.
p. 71) and the marked members are individually identifiable, Robson [1969]
suggests testing the randomness of the second sample with respect to size
by partitioning the first sample into that portion which is ultimately recaptured and the portion which is not. A Mann–Whitney (Wilcoxon) rank–sum
test (e.g. Keeping [1962]) comparing these two subsamples with respect to
body size will then be a test for a monotonic relation between body size and
probability of capture in the second sample. Alternatively, if the marked
members are allotted to size-classes one can carry out the goodness-of-fit
test described on p. 71 (Table 3.3).

When there is variation in the inherent catchabilities of the individuals
and the first sample is not random, assumption (b) will be false, and the
more catchable individuals will be caught in the first sample. This means
that for the second sample the marked will in general be more catchable than
the unmarked, and assumption (d) will be false. Unfortunately, apart from a
careful choice of catching method and preliminary studies – for example, on
activity, feeding habits, "length-selection" curves, etc. – little can be done
to overcome this problem of variable catchability in the first sample. However,
we shall see in section 2 below that the bias in the Petersen estimate $\hat{N}$ due
to variation in catchability can be reduced by using different trapping methods
for the two samples.

If the catchability is constant within certain well-defined subgroups and
there are sufficient recaptures from each subgroup, the numbers in each subgroup should be estimated separately (e.g. Gunderson [1968]). The question
of pooling data from different subgroups is discussed in **3.2.5** and **11.1.4**.

SYSTEMATIC SAMPLING. It was noted in **3.1.2** that the Petersen estimate
can still be used, even when assumption (d) is false and a systematic rather
than a random sample is taken, provided there is uniform mixing of marked
and unmarked and all animals are equicatchable in the second sample. But
in many populations uniform mixing is unlikely because of territorial behaviour and the presence of well-defined home ranges. Another situation
where it is difficult to obtain a uniform mixing is when the population is not
randomly distributed throughout the population area and the animals are
relatively immobile. For example, Hancock [1963] suggested that the

excessive variability in the monthly returns of marked whelks may have been due to the random distribution of marked individuals among essentially non-randomly distributed unmarked individuals. Banks and Brown [1962] found the same lack of dispersal of the marked individuals in the study of Sunn Pest populations (cf. Example 3.10, p. 115).

It would seem that where possible the experimenter should aim for a random sample rather than rely on the assumption of uniform mixing. However, in many population studies it is helpful to arrange the release of the first sample so that mixing can take place as much as possible. For example, one can divide up the total area into subareas, sample each subarea with the same effort, and then release the marked animals back into the same area from which they were taken. If the catchability is- independent of subarea one would hope that this method produces roughly the same proportion of marked in each subarea. To check this, one can carry out a standard goodness-of-fit test to see whether the proportions of marks recorded in the second sample from the various subareas are significantly different (cf. **3.4.1** (*1*)).

CATCHING AND HANDLING. Departures from assumption (c) that trapping and marking do not affect catchability can be minimised if the following points are observed by the experimenter:

(i) *Type of trap.* It is essential that a trapping method be used which will not harm the animal in any way. For example, in small mammal populations such expedients as placing the traps under cover of vegetation, drugging the bait, and visiting traps frequently can reduce trap mortality (Buckner [1957]). But in spite of careful trapping technique the shock of actually being trapped may have considerable effect on the animal (e.g. Guthrie *et al.* [1967: squirrels], Keith *et al.* [1968: snowshoe hares], Bouck and Ball [1966: rainbow trout]).

If several types of trap are available one would endeavour to choose the type which is most efficient, as the accuracy of the Petersen estimate increases with n_1 and n_2 (cf. Pucek [1969]). Also to increase trap efficiency, the bait or lure should be selective for the species under investigation and some consideration should be given to the spacing and distribution of the traps (cf. Andrzejewski *et al.* [1966]).

(ii) *Method of handling.* Care is needed in handling the captured animals so that they quickly recover on their return to the population. For example, a major problem in much fishery work is the high mortality of tagged fish immediately after release, and recent evidence indicates that tagging and marking can have a greater effect on fish than is apparent (Ricker [1958: Chapter 3], North Atlantic Fish Marking Symposium [1963: 7–13], Paulik [1963a: 41–2], Clancy [1963], Shetter [1967]).

Another problem arises with small mammals and birds where "trap addiction" or "trap shyness" can alter an animal's pattern of behaviour after it has been caught for the first time. Numerous examples of this are given in the literature, and the following are a selection: rats and voles (Chitty and

Shorten [1946], Tanaka [1951], Bailey [1968]), house mice (Young *et al.*
[1952]), squirrels (Evans [1951], Flyger [1959], Nixon *et al.* [1967]), rabbits
(Geis [1955], Huber [1962], Eberhardt *et al.* [1963], Edwards and Eberhardt
[1967]), small mammals (Morris [1955], Tanaka [1956, 1963a], Getz [1961],
Pucek [1969: 421]) and bluetits (Taylor [1966]). Obviously there is a need
for more research into methods of measuring the response of individuals to
traps (e.g. Sealander *et al.* [1958], Balph [1968], Bailey [1968], Bailey
[1969]). One way of detecting trap response, which is discussed later in
Chapter 4, is by analysing the frequencies of recaptures of individuals from
multiple-recapture experiments (cf. **4.1.5**(2) and **4.1.6**); another method, which
can be used when the second sample is taken in stages, is discussed at the
end of this chapter on p. 129.

The effect of trap shyness can be minimised by prebaiting the traps for
a suitable period of time before the census, thus allowing the animals to get
used to the presence of the traps (Chitty and Kempson [1949], Tanaka [1970]).
This, however, does not always work (e.g. Young *et al.* [1952], Crowcroft and
Jeffers [1961], Balph [1968]), and of course trap addiction is not necessarily
helped by prebaiting. Sometimes trap addiction can be reduced by altering
the trap positions throughout the trapping period, thus, for example, prevent-
ing individuals from building their runaways up to the mouth of the trap
(Brown [1954]). Another way of minimising the effect of trap response is to
use a different trapping method for taking the second sample. For instance,
animals can be live-trapped for marking and then shot for recapture or, if the
mark is conspicuous, sight of the mark itself could be the means of "recapture" (e.g. Dunnet [1957: rabbits], Flyger [1959: squirrels], Strandgaard [1967:
deer]). In the latter case, observing the animals and noting the proportion of
marked amounts to sampling with replacement, so that Bailey's binomial
model (**3.1.2**) is appropriate.

In some circumstances the tag itself may affect the longevity and behaviour of the animals. For example, jaw tags on fish can interfere with
feeding and thus affect growth rate, while Petersen disc tags can make the
fish more vulnerable to gill nets through the net catching under the disc
(Ricker [1958]). Newly emerged insects may be more sensitive to the toxic
substances used in paint markers than older insects, and labels attached to
their wings may interfere with blood circulation (Southwood [1966: 58]).

Another aspect of marking particularly relevant to insects is that the
presence of conspicuous marks may well destroy an animal's natural camou-
flage and make it more or less liable to predation. Also when animals are
sampled by a method which relies on the sight of the collector then the
marked may tend to be collected more than the unmarked (Southwood [1966:
58]). On the other hand, if the tags are not conspicuous enough they may be
overlooked, particularly if one relies on huntsmen, fishermen, farmers, etc.,
to return the tags.

In general the effect of a particular marking method should be checked
by both laboratory and field experiments where possible. A comprehensive

review of methods of catching and handling wild animals is given in Taber and Cowan [1969].

(iii) *Method of release.* Animals often show a high level of activity immediately on release, and efforts should be made to minimise this. For example, birds and insects could perhaps be restrained from flying immediately by covering them with small cages until the effects of handling wear off, while tagged fish could be held in tanks so as to reduce and measure initial tagging mortality (e.g. North Atlantic Fish Marking Symposium [1963: 8]). If the animals have a rhythm of activity during the day they could perhaps be released during an inactive period (Southwood [1966: 74]). Where an animal is released may have some effect on its future behaviour, particularly if released in strange surroundings (Flyger [1960: 369]).

2   Theoretical analysis of catchability

For the jth member of the population ($j = 1, 2, \ldots, N$), let x_j be the probability that it is caught in the first sample, let y_j be the conditional probability that it is caught in the second sample given that it is caught in the first, and let z_j be the conditional probability that it is caught in the second, given that it is not caught in the first. Assuming that the population represents a random sample of N triples (x_j, y_j, z_j) with regard to the species as a whole and the particular trapping methods used, then (x_j, y_j, z_j) may be regarded as a random observation from a trivariate probability density function, $f(x, y, z)$ say. Then from Seber [1970a] we have that the conditional distribution of m_2 given n_1 and n_2 is approximately binomial, i.e.

$$f(m_2 \,|\, n_1, n_2) \approx \binom{n_2}{m_2} P^{m_2} (1-P)^{n_2 - m_2}, \qquad (3.14)$$

where $P = \dfrac{n_1}{(N - n_1)k + n_1}$,

$k = \dfrac{(p_3 - p_{13})p_1}{p_{12}(1 - p_1)}$,

and $p_1 = E[x]$, $p_3 = E[z]$, $p_{13} = E[xz]$ and $p_{12} = E[xy]$; all expectations being with respect to $f(x, y, z)$. From (3.14) it can be shown that

$$E[N^* \,|\, n_1, n_2] = \frac{n_1 + 1}{P} \{1 - (1 - P)^{n_2 + 1}\} - 1$$

$$\approx n_1 / P$$

$$= (N - n_1)k + n_1,$$

or

$$E[N^* - n_1 \,|\, n_1, n_2] \approx (N - n_1)k, \qquad (3.15)$$

so that N^* is an approximately unbiased estimate of N if and only if $k = 1$. Seber [1970a] also shows that V^* and v^* of **3.1.1** are still satisfactory estimates of the "true" variance (i.e. the variance of N^* with respect to (3.14)), even when $k \neq 1$.

ASSUMPTION (c) TRUE. In studying this special case it is helpful to make the transformation

$$k = (B-p_1)/(1-p_1),\tag{3.16}$$

where $B = p_1(p_3-p_{13}+p_{12})/p_{12}$.

Obviously $k = 1$ if and only if $B = 1$ and $k < B$ when $B < 1$. Now when assumption (c) is true, i.e. marking does not affect catchability, we have $y_j = z_j$ ($j = 1, 2, \ldots, N$), so that $p_{13} = p_{12}$, $p_3 = p_2$ and

$$1 - B = 1 - (p_1p_2/p_{12})\tag{3.17}$$

$$= \text{cov}\,[x, y]/E[xy].$$

This equation was first considered by Junge [1963] and its implications discussed in some detail by him. He pointed out that $B = 1$ if and only if x and y are uncorrelated; a positive correlation will lead to an underestimate of N (as $k < B < 1$) and a negative correlation to an overestimate. We note that if a correlation exists we would generally expect it to be positive. This may account for the persistent underestimation observed by Buck and Thoits [1965] who checked on several estimates of fish population numbers by draining the ponds containing the populations.

We conclude from the above that variation in catchability due to, say, trapping selectivity could exist for both samples without introducing bias if the sources of selectivity in the two samples were independent. This leads support to the statement by a number of authors that bias due to difference in catchability can be reduced by using a different sampling method for each sample (Buck and Thoits [1965], Waters [1960]). For example, Ricker [1958: 96] argues that if fish for marking are trapped while the second sample is obtained by angling, it would seem unlikely for a similar sampling bias to be present in both gears. Junge [1963] gives an interesting example on river tagging of migrating salmon. In this situation it may be possible to tag non-selectively with respect to size but not with respect to time, e.g. sampling effort may vary with time. On the other hand, the recovery process on the spawning grounds may be selective with respect to fish size and spawning area, so that if the time of passage past the tagging site is uncorrelated with fish size and area of spawning, no bias is introduced.

Junge also discusses the sensitivity of B to variation in catchability by studying the special case of $y_j = bx_j$ ($b > 0$) when the correlation is unity. We find that

$$B = 1 - V[x]/(E[x])^2 \leqslant 1$$

with equality if and only if $V[x] = 0$ or x is constant (i.e. assumption (b) is true). He shows that if the range of x is $[c, d]$ ($0 \leqslant c < d \leqslant 1$) then, provided $w = c/d$ is not too small, B will be near unity and insensitive to the shape of $f(x)$, the probability density function of x. For example, if $f(x)$ is the uniform distribution, then

$$B = 3(1+w)^2/4(1+w+w^2)$$

and $B = 1$, $\frac{27}{28}$, $\frac{27}{31}$, $\frac{3}{4}$ when $w = 1$, $\frac{1}{2}$, $\frac{1}{5}$ and 0. On the other hand, if $c = 0$, B may be significantly less than unity; if in addition $f(x)$ is concentrated near $x = 0$ (i.e. probabilities of capture are near zero for a substantial proportion of the population) Junge shows that B could be much smaller still (see also **4.1.6**(4) and Table 4.20). Since $k < B$ when $B < 1$ these effects will be accentuated in the value of k, e.g. for the extreme case when $f(x)$ is the uniform distribution on $[0, 1]$ we have $p_1 = \frac{1}{2}$, $B = \frac{3}{4}$ and $k = \frac{1}{2}$.

It was noted by Junge that x and y are uncorrelated if either x or y is constant, i.e. $B = 1$ if at least one of the two samples is a random sample. In particular, if the first sample is random, the second sample need not be random and in fact could be highly selective, provided the selectivity was independent of mark status (assumption (c)). This fact is also noted by Robson [1969] who points out that when assumption (c) is true, the hypergeometric model given by (3.1) still holds if either sample is random because of the symmetry of n_1 and n_2 in the formula.

When mortality is taking place and just the second sample is random we saw in **3.2.1**(2) that (3.1) is valid only when the mortality process is random. However, if just the first sample is random, (3.1) will still hold even if the mortality is non-random, provided that it is independent of mark status. This follows from the fact that when the first sample is random, the tags are distributed randomly through every possible subgroup or category existing in the population and therefore throughout any portion of the population subsequently removed for investigation. Since most mortality tends to be selective, Robson therefore suggests that "the most effective plan for the two-sample experiment consists of a determined effort to obtain a random sample for marking and then exploiting the habits of the creature to obtain a large, if selective, sample in the recapture stage."

If the first sample is random with respect to body size and there is no recruitment, we would expect the marked and unmarked portions of the second sample to represent random subsamples from the same size-distribution. Therefore, as suggested by Robson [1969], a Mann–Whitney (Wilcoxon) rank sum test (e.g. Keeping [1962]) would provide a test of randomness with respect to body size. Alternatively the goodness-of-fit test given on p. 74 (Table 3.4) could be used to test the randomness of the first sample with respect to size or any other category.

ASSUMPTION (b) TRUE. When x is constant we find that $p_{13} = p_1 p_3$, $p_{12} = p_1 p_2$ and $k = p_3/p_2$. This means that $k = 1$ if and only if the average probability of capture of the marked in the second sample is the same as the average for the unmarked.

CONCLUSION. We see from the above discussion that if assumption (c) is true and $y_j = b x_j$, then $k = 1$ if and only if assumption (b) is true; a test for this case based on taking two samples from a known (e.g. marked) population is given below in section 3. Conversely, if assumption (b) is true then $k = 1$ if and only if assumption (c) is true. In practice there may be departures from both assumptions, and a general method for testing $k = 1$ based on multiple recaptures is given in **4.1.3**(*3*). A model due to Skellam [1948], based on a beta distribution for the probability of capture, is used by Eberhardt [1969] to estimate N: this is discussed in **4.1.6**(*4*).

3   Test for constant catchability

We shall now consider the problem of testing assumption (b), given that assumptions (a), (c), (e) and (f) are true, by taking three samples and using the first sample as an identifiable population (or else by taking two samples from a population of known size).

Suppose that the m_2 tagged animals in the second sample are given another tag (if individually numbered tags are not used for the first sample) and the second sample then returned to the population. If a third sample of size n_3 is now taken, then on the basis of the tagging information obtained and assuming that catching and tagging do not affect future catchability, Cormack [1966] gives two procedures, which we discuss below, for testing the hypothesis of constant catchability. The choice of procedure depends on which of two methods is used for catching n_1. In the first method, sample one is captured by a different technique from that used for samples two and three. We then require an additional assumption that the first sample is a random sample of probabilities of capture with respect to the *latter* catching technique. This will imply that the distribution of probability of capture over the first sample with regard to the latter catching technique is the same as over the entire population. For the second method the same catching technique is used for all three samples and no further assumptions are necessary.

In the following discussion we note that n_1, n_2 and n_3 play the same role as N, n_1, n_2 in the theoretical analysis of the previous section above.

METHOD 1. The probability p_j $(j = 1, 2, \ldots, n_1)$ that the jth member of the first sample is captured in a later sample will be proportional to its inherent catchability and to the intensity of sampling or sampling effort. Assuming the sampling effort f to be the same for each individual, we therefore define $c_j = p_j/f$ as the catchability of the jth individual for the particular catching method used. We shall assume the c_j's to be a random sample of size n_1 from a probability density function $g(c)$ with moments μ_r' about the origin and moments μ_r about the mean. If we standardise c_j so that the domain of g is $[0, 1]$ then f will be uniquely determined and $0 \leqslant f \leqslant 1$. Cormack points out that the catchability of an animal may be regarded as the probability with which it places itself in a position where the experimenter is able to catch it, and the sampling intensity is then the probability that an animal in this position will be caught. Alternatively, if c_j is not standardised we can regard it as the probability that one unit of sampling effort catches the ith individual. Then, considering f as the number of units of effort expended on the sample $(f > 1)$, and assuming units of effort to be additive, $p_j = fc_j$ as before.

Let f_2 and f_3 be the sampling efforts for samples two and three respectively. Then if x_j and y_j are the probabilities that the jth member of sample one is caught in samples two and three respectively, and assumption (c) is true, we have $x_j = f_2 c_j$, $y_j = f_3 c_j$, and hence $y_j = b x_j$. Let m_{10} be those individuals caught in the first sample only, m_{12} those caught in both samples one and two only, m_{13} those caught in both samples one and three only, and m_{123} those caught in all three samples. Then Cormack shows that the joint probability function of m_{12}, m_{13} and m_{123} is given by

$$\frac{n_1!}{m_{12}!\, m_{13}!\, m_{123}!\, m_{10}!} \, [\alpha_2(1-\alpha_3 \lambda)]^{m_{12}} [\alpha_3(1-\alpha_2 \lambda)]^{m_{13}} [\alpha_2 \alpha_3 \lambda]^{m_{123}}$$

$$\times \, [1 - \alpha_2 - \alpha_3 + \alpha_2 \alpha_3 \lambda]^{m_{10}}, \tag{3.18}$$

where $\alpha_2 = f_2 \mu_1'$, $\alpha_3 = f_3 \mu_1'$ and $\lambda = \mu_2'/(\mu_1')^2$. To test the hypothesis of constant catchability it is sufficient to test whether the variance of c is zero. This is equivalent to testing $H_0 : d = 0$, where d, the square of the coefficient of variation, is given by

$$d = \mu_2/(\mu_1')^2$$
$$= \lambda - 1. \tag{3.19}$$

Now, for the multinomial distribution (3.18) the maximum-likelihood estimates of α_2, α_3 and λ are simply the moment estimates

$$\hat{\alpha}_1 = (m_{12} + m_{123})/n_1 ;$$
$$\hat{\alpha}_2 = (m_{13} + m_{123})/n_1 ,$$

and

$$\hat{\lambda} = n_1 m_{123}/[(m_{12} + m_{123})(m_{13} + m_{123})].$$

Therefore, writing $\hat{d} = \hat{\lambda} - 1$, we have, from the delta method, the asymptotic expressions

$$E[\hat{d}] = d + d(1+d)/n_1 \tag{3.20}$$

and

$$V[\hat{d}] = \frac{d+1}{n_1 \alpha_2 \alpha_3} [1 - (\alpha_2 + \alpha_3)(d+1) + \alpha_2 \alpha_3 (d+1)(2d+1)]. \tag{3.21}$$

The asymptotic bias and variance of $\hat{d}$ can be estimated as usual by replacing each unknown parameter by its estimate; thus

$$\hat{V}[\hat{d}] = \frac{m_{123}^2 \, n_1^2}{(m_{12}+m_{123})^2 \, (m_{13}+m_{123})^2} \left(\frac{1}{m_{123}} - \frac{1}{n_1} - \frac{m_{12}+m_{13}}{(m_{12}+m_{123})(m_{13}+m_{123})} \right) \tag{3.22}$$

Under the null hypothesis H_0 we have, assuming approximate normality, that $\hat{d}$ is $\mathcal{N}(0, (1-\alpha_2)(1-\alpha_3)/(n_1 \alpha_2 \alpha_3))$. Therefore a one-sided test of H_0 (one-sided since $d \geqslant 0$) is given by the statistic

$$z = \hat{d}\sqrt{\{n_1 \hat{\alpha}_2 \hat{\alpha}_3/[(1-\hat{\alpha}_2)(1-\hat{\alpha}_3)]\}}, \tag{3.23}$$

which is approximately distributed as the unit normal when H_0 is true: if z is negative we accept $d = 0$ as the most reasonable hypothesis. It is readily seen that the power of this test will be maximised when f_2, f_3 and n_1 are as large as possible.

In conclusion we note from Cormack that the above experiment does not have to be carried out on a free-ranging population. As the interest is in the physiological or psychological behaviour of the animal with regard to one particular sampling technique, it is possible that the assumptions necessary for the method could still be satisfied by a controlled population of size n_1 living within fixed boundaries.

METHOD 2. Suppose now that the sampling technique is the same for all three samples and let c_j be the catchability of the jth individual in the *total* population ($j = 1, 2, \ldots, N$). If the sampling intensity for the first sample is f_1 then the probability that the jth individual is caught in the first sample is $f_1 c_j$. Hence

$$\Pr[\text{in } n_1 \text{ and } c_j < c < c_j + \delta c_j] = \Pr[\text{in } n_1 \mid c_j < c < c_j + \delta c_j]$$
$$\times \Pr[c_j < c < c_j + \delta c_j]$$
$$\approx f_1 c_j \, g(c_j) \, \delta c_j \, ,$$

$$\Pr[\text{in } n_1] = \int_0^1 f_1 c \, g(c) \, dc = f_1 \mu_1'$$

and

$$\Pr[c_j < c < c_j + \delta c_j \mid \text{in } n_1] \approx (1/\mu_1') c_j g(c_j) \delta c_j .$$

Therefore, in effect, subsequent sampling for n_2 and n_3 is from a marked population of size n_1 with catchability density function

$$h(c) = (1/\mu_1') c \, g(c) .$$

Hence $\hat{d}$ defined in Method 1 is now an estimate of D, the square of the co-efficient of variation for the density $h(c)$. This means that d must be replaced by D in (3.20) and (3.21); the statistic z (3.23) now provides a test for $D = 0$ rather than for $d = 0$.

From Cormack [1966] it can be shown that

$$D = [\mu_3' \mu_1' / (\mu_2')^2] - 1$$
$$= \frac{1}{(1+d)^2} \left[\frac{\mu_3}{(\mu_1')^3} + d(1-d) \right] , \tag{3.24}$$

so that $\hat{d}$ cannot be used for estimating d without some knowledge of μ_3 and μ_1'. However, if we can make the additional assumption that the original density $g(c)$ is symmetrical, then $\mu_3 = 0$, $\mu_1' = \frac{1}{2}$, and solving (3.24), d can be estimated by

$$\hat{d}_1 = \frac{1 - 2\hat{d} - (1 - 8\hat{d})^{\frac{1}{2}}}{2(1+\hat{d})} ,$$

where $\hat{d}$ is the estimate of D. Using the delta method, the asymptotic variance of $\hat{d}_1$ is given by

$$V[\hat{d}_1] = \frac{(1+d)^6}{(1-3d)^2} \, V[\hat{d}] ,$$

where d is estimated by $\hat{d}_1$ and $V[\hat{d}]$ by (3.22). Now $d = 0$ implies both $D = 0$ and $V[\hat{d}_1] = V[\hat{d}]$, so that the statistic for testing $d = 0$ is the same as before, namely (3.23) but with $\hat{d}$ replaced by $\hat{d}_1$. Since from (3.24) $D < d$ when $g(c)$ is symmetrical, we have $\hat{d} < \hat{d}_1$ and a situation could arise in which $\hat{d}_1$ is significantly different from zero but $\hat{d}$ is not. This means that we must be clear which of the two methods is appropriate to a given experimental situation and choose the correct statistic, $\hat{d}$ or $\hat{d}_1$, for testing $d = 0$.

COMPARING THE METHODS. For a certain class of alternatives (namely $g(c)$ is a beta distribution) Cormack compares the relative powers of the two methods given above and notes that for moderate deviations from the null hypothesis of constant catchability the first method is the more powerful. However, as the alternative hypothesis comes closer to the null hypothesis, the second method eventually becomes the more powerful. Unfortunately, as with the Petersen method, both tests will be rather insensitive unless n_1 is large and a significant proportion of the first sample is captured in the second and third samples. For example, Table 3.6 gives the size n_1 required for different intensities $f_1 \, (= f_2 = f_3)$, to obtain a 90 per cent probability of disproving the null hypothesis at the 5 per cent level of significance when $g(c)$ is actually the uniform distribution on $[0, 1]$. Thus, for a recapture rate of $12\frac{1}{2}$ per cent we require $n_1 = 4133$ for the first method.

TABLE 3.6
The size (n_1) of the first sample required to discover a uniform distribution for different sampling intensities (f_1): from Cormack [1966: Table 1].

f_1	0·25	0·5	1
% recaptured $(100\mu_1' f_1)$	$12\frac{1}{2}\%$	25%	50%
n_1 (method 1)	4133	730	84
n_1 (method 2)	8485	1559	174

MORTALITY PRESENT. Suppose we relax the assumption of a closed population to the extent of allowing mortality. Let ϕ_1, ϕ_2 be the probabilities of survival for a tagged animal between the first two and the second two samples respectively. To allow for the estimation of the ϕ_i, we require additional information provided by releasing a further r_2 tagged animals (in addition to the m_2) into the population after the second sample and noting the recaptures, m_{23} say, from this group in the third sample.

Under the assumptions of method 1, and assuming the members of r_2 to represent a random sample of catchabilities from $g(c)$, the joint probability of m_{12}, m_{13}, m_{23}, m_{123} is given by

$$\frac{n_1!}{m_{12}! \, m_{13}! \, m_{123}! \, m_{10}!} \, [\phi_1 \alpha_2 (1 - \phi_2 \alpha_3 \lambda)]^{m_{12}} [\phi_1 (1 - \alpha_2 \lambda) \phi_2 \alpha_3]^{m_{13}}$$

$$\times \, (\phi_1 \alpha_2 \phi_2 \alpha_3 \lambda)^{m_{123}} \, \theta^{m_{10}} \binom{r_2}{m_{23}} (\phi_2 \alpha_3)^{m_{23}} (1 - \phi_2 \alpha_3)^{r_2 - m_{23}},$$

where $\theta = 1 - \phi_1 \alpha_2 - \phi_1 \phi_2 \alpha_3 (1 - \alpha_2 \lambda)$.

In the above probability function we have four independent observations but five parameters, and we find that ϕ_2 and α_3 cannot be estimated separately; only the product $\phi_2 \alpha_3$ is estimable.

The maximum-likelihood estimates (which are also the moment estimates in this case) are

$$\hat{\phi}_1 = (m_{13} + m_{123}) r_2 / (m_{23} n_1),$$

$$\hat{\alpha}_2 = m_{23}(m_{12} + m_{123}) / [r_2 (m_{13} + m_{123})],$$

$$\hat{\phi}_2 \hat{\alpha}_3 = m_{23} / r_2,$$

and

$$\hat{\lambda} = m_{123} r_2 / [m_{23}(m_{12} + m_{123})].$$

Setting $\hat{d} = \hat{\lambda} - 1$ as before, we have approximately

$$E[\hat{d}] = d + \frac{\lambda}{r_2}\left(\frac{1}{\phi_2 \alpha_3} - 1\right)$$

and

$$V[\hat{d}] = \frac{\lambda}{n_1 \phi_1 \alpha_2 \phi_2 \alpha_3}\,[1 + \lambda(\phi_1 \alpha_2 n_1 r_2^{-1} - \phi_2 \alpha_3 - \phi_1 \alpha_2 \phi_2 \alpha_3 n_1 r_2^{-1})].$$

To test the hypothesis $d = 0$ we again use a one-tailed test based on the statistic $z = (\hat{d} - \hat{b})/\hat{\sigma}$, where

$$\hat{b} = \hat{\lambda}\left(\frac{1}{m_{23}} - \frac{1}{r_2}\right)$$

and

$$\hat{\sigma}^2 = \frac{r_2^2 m_{123}^2}{m_{23}^2 (m_{12} + m_{123})^2}\left\{\frac{1}{m_{123}} + \frac{1}{m_{23}} - \frac{1}{(m_{12} + m_{123})} - \frac{1}{r_2}\right\}.$$

In testing $d = 0$ there is unfortunately a very considerable loss in power through having to estimate ϕ_1. Cormack [1966] states that even when the death-rate is actually zero ($\phi_1 = \phi_2 = 1$), five to ten times (depending on the sampling intensity) the number of tagged animals $n_1 + r_2$ are required to give the same discriminatory power as the test for a closed population. Because of this lack of sensitivity and the need for such large numbers of tagged animals Cormack suggests that the experiment should be arranged so that the possibility of death can be neglected.

If information on ϕ_1 is required we can use $\hat{\phi}_1$ and the approximate variance formula

$$V[\hat{\phi}_1] \approx \frac{\phi_1}{n_1 r_2 \phi_2 \alpha_3}\,[r_2(1 - \phi_1 \phi_2 \alpha_3) + n_1 \phi_1 (1 - \phi_2 \alpha_3)]$$

to obtain an approximate confidence interval for ϕ_1. We note that if f_3/f_2 ($= \alpha_3/\alpha_2$) is known, we can obtain the estimates

$$\hat{\alpha}_3 = \hat{\alpha}_2 f_3/f_2$$

and

$$\hat{\phi}_2 = m_{23}/(r_2 \hat{\alpha}_3).$$

When $\phi_1 = 1$ and f_3/f_2 is known, the second release of r_2 animals is unnecessary as the parameters α_2, ϕ_2 and λ can be estimated from the joint multinomial distribution of m_{12}, m_{13}, m_{123} and m_{10}. In this case the maximum-likelihood estimates are again the moment-estimates, and their asymptotic means and variances can be derived using the delta method. Also the hypothesis $\phi_2 = 1$ can be tested using a standard goodness-of-fit statistic.

3.2.3   Assumption (e): no loss of tags

If animals lose their marks or tags, the observed recaptures will be smaller than expected and N^* will overestimate N. Therefore considerable thought should be given to the choice of tag, and some experiments should be carried out either before or during the census period to check for tag losses or tag deterioration. The type of tag chosen will depend on such factors as the species studied, the information required by the tag, time and personnel available for tagging, and the method of tag return — whether by hunter, fisherman or research worker. Obviously tags should be durable so that they are not lost through the effects of weather or physical changes in the animal, such as moulting.

1   Types of marks and tags

There are a number of useful review articles on marking methods, e.g. mammals (Taber [1956]), birds (Cottam [1956]), amphibians and reptiles (Woodbury [1956]), fish (Ricker [1956], North Atlantic Fish Marking Symposium [1963], Stott [1968]), insects (Dobson [1962], Gangwere et al. [1964], Southwood [1966] and White [1970]) and for general references Taber and Cowan [1969] and Southwood [1966]. The following is a brief list of the main methods of marking and tagging.

PAINTS. Slow-drying oil paints, quick-drying cellulose paints and lacquers, and reflecting paints for night detection, are particularly useful for the study of insect populations.

DYES. Dyes in solution or powder form have been used for a wide variety of populations. They can be applied externally, and in this respect automatic marking without capturing is sometimes possible (Taber and Cowan [1963]), or they can be applied internally by injecting into the tissues. For relative estimates of population density, dyes can also be introduced into food, giving marked faeces (New [1958, 1959]).

TAGS. Various tags, bands and rings can be attached externally, and small internal tags can sometimes be used, particularly when the recapturing is done by trained personnel.

MUTILATION. Mutilation methods have been widely used for vertebrates, e.g. fin clipping or punching for fish, toe clipping and fur clipping for small mammals.

MUTATIONS. Occasionally mutants which are readily distinguished from the normal species can be introduced into the population to act as the marked sample n_1. This method has been used for studying insect populations (Dobson [1962]).

FLUORESCENT SOLUTIONS. These can for example be sprayed onto insects and later detected with an ultraviolet lamp.

RADIOACTIVE ISOTOPES. Despite the obvious problems of cost and danger to personnel, radioactive isotopes are being widely used for population studies; a full discussion is given in Giles [1963], Southwood [1966] and Peterle [1969].

RADIO TRACKING. With the development of miniature transistorised radio transmitters, radio tracking using transmitter "tags" is now being used for many species (Giles [1963])*. Transmitters can be used not only for locating animals irrespective of cover, weather or time of day, but they can also be used for relaying physiological data such as body temperature, pulse-rate etc. (e.g. Slater [1963], Stoddart [1970]). Recently transmitters have been used for studying prey—predator movements (Mech [1967]). This approach is a step towards solving the problem of obtaining biological data from free-ranging animals rather than from animals in captivity where behaviour patterns may be very different. Some recent examples of radio tracking, together with further references, are given in Sanderson [1966], *Journal of Wildlife Management* [1967], Adams and Davis [1967], Heezen and Tester [1967], Knowlton *et al.* [1968], and Hessler *et al.* [1970]; ultrasonic transmitters have also been useful for studying fish movements (e.g. Henderson *et al.* [1966]).

PARASITES. These can sometimes be used as "natural" tags (e.g. Sindermann [1961]).

2   *Estimating tag loss*

One simple method of detecting tag loss is to give all the n_1 animals in the first sample two types of tags and then to note those recaptures with just one tag and those with both tags intact (Beverton and Holt [1957], Gulland [1963]). Denoting the two types of tag by A and B, we define:

$$\pi_i = \text{probability that a tag of type } i \text{ is lost by the time of the second sample } (i = A, B),$$

$$\pi_{AB} = \text{probability that both tags are lost,}$$

$$m_i = \text{number of tagged animals in the second sample, with tag } i \text{ only } (i = A, B),$$

$$m_{AB} = \text{number of tagged animals in the second sample with both tags,}$$

and m_2 = members of n_1 caught in n_2.

Assuming that the tags are independent of each other ($\pi_{AB} = \pi_A \pi_B$), the

---

* See also Brander and Cochran [1969].

joint probability function of m_A, m_B, m_{AB} and m_2 is given by

$$f(m_A, m_B, m_{AB}, m_2 \mid n_1, n_2) = f(m_A, m_B, m_{AB} \mid m_2)f(m_2 \mid n_1, n_2)$$

where

$$f(m_A, m_B, m_{AB} \mid m_2) = \frac{m_2!}{m_A! \, m_B! \, m_{AB}! \, m_0!} [(1-\pi_A)\pi_B]^{m_A}$$

$$\times [\pi_A(1-\pi_B)]^{m_B} [(1-\pi_A)(1-\pi_B)]^{m_{AB}} [\pi_A\pi_B]^{m_0},$$

$$m_0 = m_2 - m_A - m_B - m_{AB},$$

and $f(m_2 \mid n_1, n_2)$ is given by equation (3.1). The maximum-likelihood estimates of N, m_2, π_A and π_B (which are also the moment estimates) are given by

$$\hat{N}_{AB} = n_1 n_2 / \hat{m}_2,$$

$$m_A = \hat{m}_2(1-\hat{\pi}_A)\hat{\pi}_B,$$

$$m_B = \hat{m}_2\hat{\pi}_A(1-\hat{\pi}_B),$$

$$m_{AB} = \hat{m}_2(1-\hat{\pi}_A)(1-\hat{\pi}_B),$$

which have solutions

$$\hat{\pi}_A = m_B/(m_B + m_{AB}),$$

$$\hat{\pi}_B = m_A/(m_A + m_{AB}),$$

and

$$\hat{m}_2 = (m_A + m_{AB})(m_B + m_{AB})/m_{AB}$$

$$= c(m_A + m_B + m_{AB}), \quad \text{say}.$$

This means that the observed recaptures $m_A + m_B + m_{AB}$ must be corrected by a factor

$$c = \left[1 - \frac{m_A m_B}{(m_A + m_{AB})(m_B + m_{AB})}\right]^{-1}$$

to give an estimate of the actual number of recaptures m_2. For large samples, $\hat{N}_{AB}$ is an approximately unbiased estimate of N, and defining $\hat{N} = n_1 n_2 / m_2$, we have from **1.3.4**

$$V[\hat{N}_{AB} \mid n_1, n_2] = \underset{m_2}{E} \{V[\hat{N}_{AB} \mid n_1, n_2, m_2]\} + \underset{m_2}{V} \{E[\hat{N}_{AB} \mid n_1, n_2, m_2]\}$$

$$\approx \underset{m_2}{E} \{V[\hat{N}_{AB} \mid n_1, n_2, m_2]\} + V\{\hat{N} \mid n_1, n_2\},$$

which, by the delta method (and equation 3.2), is approximately equal to

$$\frac{N^3}{n_1 n_2} \pi_A\pi_B\left[2 + \frac{1}{(1-\pi_A)(1-\pi_B)}\right] + \frac{N^3}{n_1 n_2}\left[1 + \frac{2N}{n_1 n_2} + 6\left(\frac{N}{n_1 n_2}\right)^2\right]. \quad (3.25)$$

In some situations the only information recorded is the number of tags for each tagged individual, so that just the numbers m_{AB} and m_C ($= m_A + m_B$) are available. For this case we can still estimate m_2 if we can assume that $\pi_A = \pi_B$ ($= \pi$ say). We then have

$$f(m_C, m_{AB} \mid m_2) = \frac{m_2!}{m_C! \, m_{AB}! \, m_0!} \, [2\pi(1-\pi)]^{m_C} [(1-\pi)^2]^{m_{AB}} [\pi^2]^{m_0},$$

and the maximum-likelihood estimates (and moment estimates) of m_2 and π are

$$\tilde{m}_2 = (m_C + 2m_{AB})^2 / 4m_{AB}$$

and

$$\hat{\pi} = m_C / (m_C + 2m_{AB}).$$

Setting

$$\tilde{N}_{AB} = n_1 n_2 / \tilde{m}_2,$$

this estimate is asymptotically unbiased, and its asymptotic variance is the same as (3.25) but for the first expression which is replaced by

$$\frac{\cdot N^3 \pi^2}{n_1 n_2 (1-\pi)^2}.$$

When information on m_A and m_B is available for the whole or perhaps a part of the second sample, we can test the hypothesis $\pi_A = \pi_B$ as follows. Let $\pi_B = k\pi_A$; then, from (**1.3.7**(2)), the conditional probability function of m_A given $m_A + m_B$ is given by

$$f(m_A \mid m_A + m_B) = \binom{m_A + m_B}{m_A} p^{m_A} q^{m_B}, \tag{3.26}$$

where $p = \dfrac{k(1 - \pi_A)\pi_A}{k(1 - \pi_A)\pi_A + \pi_A(1 - k\pi_A)}$ and $q = 1 - p.$

Testing $k = 1$ is therefore equivalent to testing $p = \frac{1}{2}$ for the above binomial distribution (3.26).

We see from **3.2.1**(2) that the above theory will still hold when there is mortality, provided that the deaths constitute a random sample and the second sample is random. It is also noted that when tag losses are small, which will be the case for many populations, the above variance terms involving π_A, π_B and π will be negligible, so that the effect of tag loss on the variance can be neglected. For example, when π_A and π_B are both less than 0·1 the contribution of the first expression to (3.25) is less than about $3\frac{1}{2}$ per cent. In such cases, therefore, the general theory of **3.1** can still be used but with m_2 replaced by $\hat{m}_2$ or $\tilde{m}_2$.

In conclusion we note that the above theory can be extended to the case of more than two marks by defining π_A as the probability that a particular mark is lost and π_B as the probability of losing at least one of the other marks. Whatever the marking method used, it is recommended (Southwood [1966: 72]) that one should follow the policy of Michener *et al.* [1955], and use a system in which all individuals bear the same number of marks (e.g. if colour marking is used one can allocate a colour to zero in the units, tens, hundreds position). An interesting coding technique for insects based on this system is given by White [1970].

3.2.4   Assumption (f): all tags reported

When there are incomplete tag returns the observed value of m_2 will be too small and N^* will overestimate N. This problem arises when tags are returned by hunters, commercial fishermen, local inhabitants, etc. who may or may not be interested in the experiment. It is found that the percentage return is usually related to such factors as the training of observers, size of reward (Bellrose [1955], Atwood and Geis [1960], Paulik [1961]), publicity given to the experiment, and ease of visibility of the tag. However, if the second sample can be classified into two categories, one which has a known reported ratio of unity or nearly so (e.g. through inspection by special observers), and the other with an unknown reported ratio, then Paulik's [1961] method described below can be used to test whether the unknown ratio is significantly less than unity.

Let $n_2 = n_{2a} + n_{2b}$, where the suffixes a and b denote the two categories respectively. Let m_{2a}, m_{2b} be the number of recaptures in the two groups and let r_{2a}, r_{2b} be the number of recaptures actually reported (i.e. $r_{2a} = m_{2a}$). Then if n_2 is large, the tag ratio n_1/N small, and the tag ratio the same for both groups, we can use the Poisson approximation to the hypergeometric distribution (cf. **3.1.4**) and assume that the recaptures m_{2i} in each group have independent Poisson distributions with parameters $n_1 n_{2i}/N$ ($i = a, b$). If ρ is the (constant) probability that a member of m_{2b} is reported, then the conditional probability function of r_{2b} given m_{2b} is

$$f(r_{2b} \mid m_{2b}) = \binom{m_{2b}}{r_{2b}} \rho^{r_{2b}} (1-\rho)^{m_{2b}-r_{2b}} \tag{3.27}$$

and, from **1.3.7**(1), the unconditional probability function of r_{2b} is Poisson with parameter $n_1 n_{2b} \rho/N$. As r_{2a} and r_{2b} are independent Poisson variables it then follows from **1.3.7**(1) that

$$f(r_{2a} \mid r) = \binom{r}{r_{2a}} p^{r_{2a}} q^{r_{2b}}, \tag{3.28}$$

where $r = r_{2a} + r_{2b}$ and $p = n_{2a}/(n_{2a} + \rho n_{2b})$. Therefore an estimate of ρ is given by

$$r_{2a}/r = n_{2a}/(n_{2a} + \hat{\rho} n_{2b}),$$

or

$$\hat{\rho} = \left(\frac{r_{2b}}{n_{2b}}\right) \bigg/ \left(\frac{r_{2a}}{n_{2a}}\right),$$

and a test of $H_0 : \rho = 1$ against the one-sided alternative $\rho < 1$ is equivalent to testing $p = n_{2a}/n_2$ ($= p_0$ say) against $p > p_0$ for the binomial model (3.28). We note that an estimate of m_2, the actual number of tagged individuals recaptured, is given by

$$\hat{m}_2 = m_{2a} + (r_{2b}/\hat{\rho}) = m_{2a} n_2/n_{2a}$$

and

$$\hat{N} = n_1 n_2 / \hat{m}_2 = n_1 n_{2a} / m_{2a}.$$

This means that we base the Petersen estimate on the recapture data for which we have a 100 per cent reporting rate.

In fishery research, if the tag ratio remains constant during the entire season, the above scheme is very flexible. For example, all the fish landed during the first part of the season may be inspected and none of those during the latter part of the season, or vice versa. However, if the tag ratio varies, suppose that the season can be divided into intervals for which the tag ratio is constant in each interval. Then, provided a constant fraction of the catch in each interval is inspected, i.e. $n_{2b}/n_{2a} (= \gamma)$ is constant, Paulik shows that (3.28) still holds but with $p = (1+\rho\gamma)^{-1}$; r and r_{2a} now refer to the season's total.

If the tag ratio n_1/N is not small, the m_{2i} may be more appropriately represented by a binomial law (cf. **3.1.4**), namely

$$f_1(m_{2i}) = \binom{n_{2i}}{m_{2i}} \left(\frac{n_1}{N}\right)^{m_{2i}} \left(1-\frac{n_1}{N}\right)^{n_{2i}-m_{2i}}.$$

Assuming (3.27), the above equation then leads to

$$f_2(r_{2i}) = \binom{n_{2i}}{r_{2i}} \left(\frac{\rho_i n_1}{N}\right)^{r_{2i}} \left(1-\frac{\rho_i n_1}{N}\right)^{n_{2i}-r_{2i}}, \quad (i = a, b), \quad (3.29)$$

where $\rho_a = 1$ and $\rho_b = \rho$. Therefore testing H_0 is now equivalent to a test for homogeneity in the 2-by-2 table:

r_{2a}	r_{2b}	r
$n_{2a} - r_{2a}$	$n_{2b} - r_{2b}$	$n_2 - r$
n_{2a}	n_{2b}	n_2

As the alternative hypothesis is one-sided, the usual test is modified slightly by only rejecting H_0 when $r_{2a}/n_{2a} > r_{2b}/n_{2b}$, and the chi-squared statistic is significant at the 2α level of significance, where α is the size of the test. When the r_{2i} and n_{2i} are small, Fisher's exact test (Keeping [1962]) should be used.

It is important not only to detect incomplete reporting after the experiment, but also to decide before the experiment how many tags should be released and how much of the sample should be inspected to be reasonably sure of detecting non-reporting of a given magnitude. Paulik shows that such information can be obtained by examining the power of the test of H_0 for particular alternatives $\rho < 1$. When the smaller of rp and rq is greater than 5 (preferably greater than 10: Raff [1956: 296]), and the correction for continuity is used, the normal approximation to the binomial distribution (3.28) can be used. In this case the most powerful one-sided test of H_0 is to reject H_0 when $r_{2a} \geq d$, where

$$d = z_\alpha \sqrt{(rp_0 q_0)} + \tfrac{1}{2} + rp_0$$

(or the nearest integer greater than the right-hand side). Here z_α satisfies $\Pr[z \geqslant z_\alpha] = \alpha$, where z has a unit normal distribution. If β is the Type II error for a given alternative ρ, i.e. $1 - \beta = \Pr[r_{2a} \geqslant d \mid \rho]$, then the number of recoveries needed to test H_0 at pre-set values of α and β for different values of ρ can be expressed as

$$r = (z_\beta\sqrt{(pq)} + z_\alpha\sqrt{(p_0 q_0)})^2/(p-p_0)^2 ,$$

where $\quad p = \dfrac{n_{2a}}{n_{2a} + \rho n_{2b}} = \dfrac{p_0}{p_0 + \rho(1-p_0)}.$

Paulik has tabulated r as a function of β, ρ and p_0 for $\alpha = 0\cdot 10,\ 0\cdot 05,\ 0\cdot 01$, and these tables are reproduced in **A2**. He points out that if the binomial model (3.28) is more appropriate than the Poisson model which led to (3.29), then the value of r obtained from the tables is conservative in that the true power of the test exceeds $1 - \beta$.

To demonstrate the use of Paulik's table, suppose that a biologist wants to be 99 per cent sure of detecting a non-reporting of 20 per cent or more of the tags recovered by a commercial fishery. From past experience he knows that at least 30 per cent of the tags released will be recovered. If $\alpha = 0\cdot 10$ and 25 per cent of the total catch is to be inspected, how many tags should be released? Entering Table **A2** for $\alpha = 0\cdot 10$, $\rho = 0\cdot 80$, $p_0 = 0\cdot 25$ and $1 - \beta = 0\cdot 99$, we find that $r = 1340$. Now

$$E[r] = E[m_{2a}] + E[r_{2b}]$$

$$= \frac{n_1 n_{2a}}{N} + \frac{n_1 n_{2b}\rho}{N}$$

$$= \frac{n_2}{N}[p_0 + \rho q_0]n_1$$

and substituting the above values (with $n_2/N \approx m_2/n_1$),

$$1340 = 0\cdot 30[0\cdot 25 + (0\cdot 80)(0\cdot 75)]n_1$$

or $n_1 = 5255$ tags liberated. If, however, the 30 per cent recovery estimate had been based on the percentage of tags actually reported then $n_2/N \approx 0\cdot 30/\rho$ and $n_1 = 4204$.

The above method of Paulik's is a very flexible one and can be applied to a number of different experimental situations. In particular it can be used for the case when p_0 refers to the proportion of the total *effort* inspected rather than the proportion of the total catch. For example, in the above numerical application p_0 could refer to the percentage of traps to be inspected, or even to the percentage of fishermen who agree to attend a special course of instruction and cooperate fully in the experiment. A theoretical justification for this extension of the above theory, along with several examples, is given in **8.3.1**.

We note that incomplete reporting can arise either through (i) tags being accidentally overlooked or (ii) tags being deliberately withheld. For larger animals a tag can usually be designed so that the chance of it being overlooked is negligible. In fisheries, however, where large numbers are rapidly handled and tags are small, this source of error can be a real problem. The second source of error can usually be minimised by extensive advertising and the offering of an "adequate" reward for a returned tag. One obvious method of separating the two types of incomplete reporting is by planting tagged fish in the catches (Margetts [1963]).

In conclusion we see that the above methods of Paulik will still apply even if natural mortality is operating, provided that tagged and untagged have the same mortality rate, so that the tag ratio remains constant throughout.

3.2.5 Populations with sub-categories

1 Pooling

It will frequently happen that the population being investigated can be divided into several sub-categories according to sex, age, size, species, etc. If the Petersen method is used and the appropriate assumptions are satisfied for the population as a whole, the question arises as to whether it is better to estimate N by pooling the data or by estimating the size of each subgroup separately and then summing the estimates. This question is particularly crucial when the number of recaptures in each subgroup is small. By considering just two subgroups, Chapman [1951: 151] shows that the former method, namely pooling, is to be preferred. That is, if $N = X + Y$ (where X and Y are the subgroup sizes), $n_1 = n_{1x} + n_{1y}$, etc., and the assumptions of **3.1.1** are satisfied for the *whole* population, then

$$N^* = \frac{(n_1 + 1)(n_2 + 1)}{(m_2 + 1)} - 1$$

has smaller variance than

$$X^* + Y^* = \frac{(n_{1x} + 1)(n_{2x} + 1)}{(m_{2x} + 1)} + \frac{(n_{1y} + 1)(n_{2y} + 1)}{(m_{2y} + 1)} - 2.$$

However, if we wish to estimate X, say, Chapman suggests the estimate

$$\tilde{X} = \frac{n_{1x} + n_{2x} - m_{2x}}{n_1 + n_2 - m_2} \cdot N^*$$

$$= \tilde{P}_x N^*, \quad \text{say,}$$

where $P_x = X/N$. When both samples are random, so that $E[n_{1x} \mid n_1] = n_1 P_x$, $E[n_{2x} \mid n_2] = n_2 P_x$, etc. then

$$E[\tilde{X} \mid n_1, n_2] = \underset{m_2}{E} \; E[\tilde{X} \mid n_1, n_2, m_2]$$

$$= P_x \underset{m_2}{E} \; [N^* \mid n_1, n_2]$$

$$\approx P_x N$$

$$= X$$

and $\widetilde{X}$ is almost unbiased (exactly unbiased when $n_1 + n_2 \geq N$). The asymptotic variance of $\widetilde{X}$ is given by

$$V[\widetilde{X} \mid n_1, n_2] = P_x^2 N^2 \left[\left(\frac{N}{n_1 n_2} \right) + 2 \left(\frac{N}{n_1 n_2} \right)^2 + 6 \left(\frac{N}{n_1 n_2} \right)^3 \right] + \frac{P_x(1-P_x)N^2}{n_1 + n_2 + 1}$$

and, estimating N by the Petersen estimate $\hat{N}$ in the above square bracket, we have the variance estimate

$$v[\widetilde{X}] = \widetilde{X}^2 \left[\frac{1}{m_2} + \frac{2}{m_2^2} + \frac{6}{m_2^3} \right] + \frac{\widetilde{X}(N^* - \widetilde{X})}{n_1 + n_2 + 1} .$$

To compare the efficiencies of $\widetilde{X}$ and X^* we note that the variance of X^* can be estimated by (cf. (3.2))

$$v[X^*] = X^{*2} \left[\frac{1}{m_{2x}} + \frac{2}{m_{2x}^2} + \frac{6}{m_{2x}^3} \right].$$

Since $m_2 > m_{2x}$ and the second term of $v[\widetilde{X}]$ is relatively small for large n_1 and n_2, a comparison of the above two estimates would indicate that $\widetilde{X}$ will generally have a smaller variance than X^*. In fact, m_{2x} may be so small (e.g. less than 7) that X^* not only has a comparatively large variance but also has considerable bias.

We note that all the above formulae still apply when there are more than two subgroups. The more subgroups there are, the smaller the ratios m_{2x}/m_2 and $v[\widetilde{X}]/v[X^*]$ will be.

2   *Variable catchability*

As pointed out in **3.2.2**, the catchability of individuals may vary between subgroups because of such factors as mobility, bait or trap preference, etc. It is therefore of interest to investigate the effect of this variation on our estimates for the special case when the probability of capture is constant within each subgroup for both samples. Let p_{ix} be the probability that a member of X is caught in the ith sample ($i = 1$, 2) and let X, Y, $Z, \ldots$ denote both the respective subgroups and their sizes. Then since, for a given subgroup, the probability of capture in the second sample is constant irrespective of mark status, assumption (c) that marking does not affect catchability is satisfied. Therefore, in the notation of **3.2.2**(2), the pair of random variables (x_j, y_j) takes values (p_{1x}, p_{2x}), (p_{1y}, p_{2y}), etc. with probabilities X/N, Y/N, etc., so that·

$$p_1 = E[x_j] = (p_{1x} X + p_{1y} Y + \ldots)/N$$

and

$$p_{12} = E[x_j y_j] = (p_{1x} p_{2x} X + p_{1y} p_{2y} Y + \ldots)/N.$$

Hence from (3.14), (3.15) and (3.16):

$$E[(N^* - n_1) \mid n_1, n_2] \approx \frac{(B - p_1)(N - n_1)}{(1 - p_1)} ,$$

where $1 - B = 1 - (p_1 p_2 / p_{12})$

$$= 1 - \frac{(p_{1x} X + p_{1y} Y + \ldots)(p_{2x} X + p_{2y} Y + \ldots)}{N(p_{1x} p_{2x} X + p_{1y} p_{2y} Y + \ldots)}$$

$$= \frac{\Sigma(p_{1x} - p_{1y})(p_{2x} - p_{2y}) XY}{N(p_{1x} p_{2x} X + p_{1y} p_{2y} Y + \ldots)}$$

and Σ denotes summation over all pairs of subgroups (X, Y), (X, Z), (Y, Z), etc. Now for any pair (X, Y) we would expect $(p_{1x} - p_{1y})$ and $(p_{2x} - p_{2y})$ to have the same sign. Therefore N^* will generally underestimate N, though the bias will be small if the p_{1x}, p_{1y}, etc. are not too different. This of course follows directly from the general theory of p. 86, where it was noted that a positive correlation between the probabilities of capture in the two samples leads to underestimation.

When $p_{ix} \neq p_{iy}$ $(i = 1, 2)$, N^* is biased and $\tilde{P}_x$ is no longer a satisfactory estimate of P_x; in this case X^* should be used instead of $\tilde{X}$. However, when the probabilities of capture are the same for the first sample (assumption (b) true) and different for the second, we have $p_{1x} = p_{1y} = \ldots$, $B = 1$, and N^* is approximately unbiased. This means that $\tilde{X}$ can now be used but with P_x estimated by n_{1x}/n_1. The only change in $V[\tilde{X} \mid n_1, n_2]$ above is to replace the second expression by $P_x (1 - P_x) N^2 / n_1$. For the opposite situation, when the probabilities of capture are the same for the second sample only, we simply estimate P_x by n_{2x}/n_2, and the corresponding variance term is $P_x(1 - P_x) N^2 / n_2$. We recall that when mortality and recruitment are negligible, and tagging does not affect catchability, the goodness-of-fit tests mentioned on pp. 71 and 73 are tests of the hypotheses $p_{2x} = p_{2y} = \ldots = p_{2w}$ and $p_{1x} = p_{1y} = \ldots = p_{1w}$ respectively. In conclusion we note that populations can also be stratified geographically and temporally; a detailed discussion of this problem is given in Chapter 11.

3.2.6 Independent estimates of migration, mortality and recruitment

Suppose that a_m and a_u are the net additions to the numbers of marked and unmarked between the two samples and let $a = a_m + a_u$ ($a_m \leqslant 0$, though a_u may be positive or negative). Then, if $\hat{a}_m$ and $\hat{a}_u$ are unbiased estimates of a_m and a_u respectively, the Petersen estimate of N, the initial population size, is

$$\hat{N} = \frac{n_1 + \hat{a}_m}{\hat{p}} - \hat{a}, \tag{3.30}$$

where $\hat{a} = \hat{a}_m + \hat{a}_u$ and $\hat{p} = m_2 / n_2$. This estimate is asymptotically unbiased, and when $\hat{a}_m$, $\hat{a}_u$ and $\hat{p}$ are independent estimates (as will usually be the case) we have, using the delta method,

$$V[\hat{N} \mid n_1, n_2] \approx p^{-2}\{(N+a)^2 V[\hat{p} \mid n_1, n_2] + (1-p)^2 V[\hat{a}_m] + p^2 V[\hat{a}_u]\}, \tag{3.31}$$

where $p = E[\hat{p}]$.

If the second sample is drawn randomly without replacement so that the hypergeometric model (3.1) is applicable, then

$$v[\hat{p}] = \frac{\hat{p}\hat{q}}{n_2} \left(1 - \frac{n_2}{\hat{N} + \hat{a}}\right)$$

is an approximately unbiased estimate of $V[\hat{p}]$. In this situation we also find that the modified estimate

$$N^* = \frac{(n_1 + \hat{a}_m + 1)(n_2 + 1)}{m_2 + 1} - \hat{a} - 1$$

is still approximately unbiased (exactly unbiased when $n_1 + \hat{a}_m + n_2 \geqslant N + a$).

When subsampling is used, p may be estimated by the methods of **3.4**.

Example 3.2   Large-mouth bass (*Micropterus salmoides*): Paulik and Robson [1969]

In a mark-recapture experiment to estimate the number of large-mouth bass in a reservoir, 3000 bass were seined from the population for marking a month before the fishing season. A subsample of 500 marked bass were retained in a live-pen for observation and the remaining 2500 bass were returned to the population. It was found that 100 of the marked bass in the pen died in the first three weeks and none died in the fourth week, while in a control group of unmarked bass held in the same live-pen there were no mortalities. This suggests that the actual marking process was responsible for the mortality among the marked, and that marking mortalities were complete before the season started: we shall assume this to be true for the whole population. During the following fishing season 2300 bass were inspected and 400 were found to be marked.

Let θ be the proportion of marked bass surviving until the fishing season; then θ is estimated by $\hat{\theta} = 400/500 = 0\cdot80$. Now $n_1 = 3000 - 500 = 2500$, $\hat{a}_m = -n_1(1-\hat{\theta}) = -500$, $a_u = -500$ (those marked fish kept in the live-pen), $\hat{p} = 400/2300 = 0\cdot1739$, $n_1 + \hat{a}_m = n_1\hat{\theta} = 2000$, and from (3.30)

$$\hat{N} = \frac{2000(2300)}{400} + (500 + 500) = 12\ 500.$$

As a first step in estimating the variance of $\hat{N}$ we note that

$$\frac{n_2}{\hat{N} + \hat{a}} = \frac{m_2}{n_1 + \hat{a}_m} = \frac{400}{2000} = \frac{1}{5}$$

and

$$v[\hat{p}] = \frac{0\cdot1739(0\cdot8261)}{2300} \cdot \frac{4}{5} = 0\cdot000\ 500.$$

To derive $V[\hat{a}_m]$ we assume that $\hat{\theta}$ is a binomial random variable with parameters 500 and θ, i.e.

$$\begin{aligned}
V[\hat{a}_m] &= n_1^2 V[1-\hat{\theta}] \\
&= n_1^2 V[\hat{\theta}] \\
&= n_1^2 \theta(1-\theta)/500,
\end{aligned}$$

103

which is estimated by

$$v[\hat{a}_m] = 2500^2 \, \hat{\theta}(1-\hat{\theta})/500 = 2000.$$

Since a_u is known, $V[a_u] = 0$, so that from (3.31) $V[\hat{N} \mid n_1, n_2]$ is finally estimated by

$$v[\hat{N}] = \left(\frac{2300}{400}\right)^2 \{(11\ 500)^2(0 \cdot 000\ 050\ 0) + (0 \cdot 8261)^2(2000)\}$$

$$= 263\ 830.$$

Hence

$$\sqrt{v\ [\hat{N}]} = 514$$

and the approximate 95 per cent confidence interval $\hat{N} \pm 1.96\sqrt{v[\hat{N}]}$ is (11 500, 13 500).

3.3   EXAMPLES OF THE PETERSEN METHOD

In this section we shall consider seven applications of the Petersen method to biological populations; the first two examples are included for their historical interest.

Example 3.3   Population of France: Laplace [1783].

Laplace used the Petersen method for estimating the total population size N from a register of births for the whole country (the "marked" individuals n_1). His second "sample" consisted of a number of parishes of known total population size n_2, and m_2 was the number of births recorded for these parishes.

Example 3.4   North American Ducks: Lincoln [1930].

Lincoln trapped and banded large numbers of ducks before the annual dispersal from the breeding grounds. At every shooting season after a release he consistently received from the shooters about 12 per cent of the bands. From this he inferred that if 12 per cent of the banded had been shot, 12 per cent of the unbanded ducks also met the same fate. Estimating the total kill for a particular year at 5 million ducks, he used the Petersen method as follows to obtain an estimate of the total duck population N for the North American continent.

Let n_1 = number banded, n_2 = total kill, and m_2 = number of returned bands. Then $m_2/n_1 = 0 \cdot 12$ and

$$\hat{N} = n_2 \Big/ \left(\frac{m_2}{n_1}\right)$$

$$= 5 \times 10^6 \times 100/12$$

$$\approx 42 \text{ million birds.}$$

We note that this estimate depends on the assumption of a 100 per cent band return from the shooters.

Example 3.5   Underground ant (*Lasius flavus*): Odum and Pontin [1961].

It was found that radioactive tagging with P^{32} is a satisfactory method of tagging large numbers of ants in a colony of *Lasius flavus*. The population was effectively constant in size throughout the experiment as (i) the workers were relatively long-lived, (ii) the individuals newly emerged from the pupae could be recognised by their lighter colour and therefore eliminated from the counts, and (iii) there was little or no movement of individuals from one colony to another. The method of sampling on both occasions was to place a flat stone on top of the mound and then carefully lift the stone after a suitable period of time. It was found that the ants construct extensive galleries directly under the stone where warmth from the sun is favourable for the development of the pupae. During the summer the workers move through these chambers in a continuous stream to and from underground tunnels leading to the aphids clustered on roots of nearby plants. By gently lifting the stone, large numbers of individuals could therefore be removed for tagging and then conveniently returned to the mound with a minimum disturbance of the colony. It was stressed that a sufficient period of time should be allowed (5–10 days in this case) for the proper mixing of tagged and untagged before the stones are again lifted for the second sample. But this period should not be too long (no greater than 10 days) as the short half-life of P^{32} and the rapid biological turnover leads to the eventual "loss" of the tag.

For mass tagging, a dipping device was used to soak the ants thoroughly in a P^{32} solution. As soon as they had dried out on filter-paper and were beginning to move about, the paper was placed next to the hill and the ants allowed to crawl back under the stone on their own accord. Any individuals which appeared to be abnormal or injured were removed, so that only active, uninjured ants were returned. It was found that mortality during the tagging process was less than 1 per cent. A number of preliminary laboratory experiments carried out over 6 days seemed to indicate that tagging did not affect subsequent behaviour of the ants. Although in some cases untagged ants picked up a small amount of radioactivity, this "secondary" tagging was so much less than the "primary" tagging that there was never any doubt as to which individuals had received the primary tag.

For illustrative purposes only, I have arbitrarily selected the results of just six experiments from a table given by the authors and renumbered the colonies studied in order of magnitude of the number of recaptures, m_2. The data are set out in Table 3.7. We recall from **3.1.1** that

$$N^* = \frac{(n_1 + 1)(n_2 + 1)}{(m_2 + 1)} - 1,$$

$$v^* = \frac{(n_1 + 1)(n_2 + 1)(n_1 - m_2)(n_2 - m_2)}{(m_2 + 1)^2 (m_2 + 2)}$$

is an almost unbiased estimate of $V[N^*]$, and that

TABLE 3.7

Recapture data for six ant colonies: modified from Odum and Pontin [1961: Table 1].

Colony	n_1	n_2	m_2	N^*	$\sqrt{v^*}$	$100\,\hat{C}$
1	500	149	7	9 393	3 022	38
2	500	159	11	6 679	1 761	30
3	500	189	17	5 287	1 133	24
4	1 000	243	21	11 101	2 184	22
5	500	437	68	3 179	324	12
6	600	321	89	2 149	176	11

$$\hat{C} = 1/\sqrt{m_2}$$

is an estimate of the coefficient of variation of N^*. To obtain confidence intervals for N we calculate $\hat{p} = m_2/n_2$ and $\hat{f} = m_2/n_1$, as a guide to the choice of approximation mentioned in **3.1.4**. Using the rules outlined there, we find that the Poisson approximation is appropriate to the first four colonies ($\hat{p}$ and $\hat{f}$ both less than 0·1) and the normal approximation (3.4) for the last two colonies. The 95 per cent confidence limits thus calculated are set out in Table 3.8, and comparing these intervals with those based on $N^* \pm 1\cdot96\sqrt{v^*}$ we notice that the latter are too narrow for small $\hat{p}$. In particular, the upper limits are too low, thus reflecting the skewness of the distribution of N^* when p is small.

TABLE 3.8

95 per cent confidence intervals for the sizes of six ant colonies (workers only): data from Table 3.7.

$\hat{p}$	$\hat{f}$	Confidence interval	$N^* \pm 1\cdot96\sqrt{v^*}$
0·047	0·014	(4 179, 23 021)	(3 470, 15 316)
0·069	0·022	(3 522, 13 118)	(3 228, 10 130)
0·089	0·034	(3 177, 8 930)	(3 066, 7 508)
0·086	0·021	(7 049, 17 715)	(6 820, 15 382)
0·156	0·136	(2 671, 4 032)	(2 545, 3 815)
0·277	0·148	(1 860, 2 591)	(1 803, 2 495)

It should be emphasised that the point and interval estimates of N are for the number of workers only (the "catchable" population) and not for the whole population in the colony. For a further study of the above marking technique cf. Stradling [1970].

Example 3.6   Snowshoe hares (*Lepus americanus*): Green and Evans [1940]

The aim of this investigation was to study the fluctuations in abundance

of snowshoe hares on the Lake Alexander area, Minnesota, over a period of
years from 1932 to 1939, using a series of Petersen estimates. The trapping
season extended from October or November to April or May of the following
spring, and trapping activities were discontinued during the warmer months
of the year as hares would not then enter traps in appreciable numbers,
probably because desirable food was abundant. In each year the major part
of the trapping season was utilised to band as many new hares as possible,
and the final two and a half weeks of the season were used to retrap the
entire area.

The animals were tagged with numbered metal ear-bands which were
clipped on with special pliers. Although the bands made only a minute per-
foration they seemed to stay on throughout the animal's entire life and
caused no irritation. The sampling method consisted of dividing up the area
into "stations" and then operating each station for six trap-days, either by
using six traps for one day or more commonly three traps for two days. Some
of the experimental results for the whole Lake Alexander Area are given in
Table 3.9: the normal approximation (3.4) in **3.1.4** was used to calculate the
confidence intervals for N.

TABLE 3.9

**95 per cent confidence intervals for the size of a snowshoe hare
population, Lake Alexander, Minnesota: data from Green and Evans [1940].**

Year	n_1	n_2	m_2	N^*	$\hat{p}$	$\hat{f}$	Confidence interval
1932–3	948	421	167	2383	0·3976	0·176	(2153, 2685)
1933–4	876	329	154	1866	0·4681	0·176	(1689, 2098)
1934–5	659	272	105	1699	0·3860	0·159	(1494, 1991)

The population of hares was not strictly closed as there was a natural
mortality, particularly over the winter months. However, as mentioned on
page 71 this would not affect the Petersen estimate if the mortality rate was
the same for tagged and untagged. As the hares lived in a markedly restricted
range, seldom travelling more than $\frac{1}{8}$ mile from the point where they were
first trapped, migration to and from the area was considered to be negligible.
Usually a hare trapped at one station was recaptured at the same station or
an adjoining one, even when a year or two had elapsed between the two
trappings. As already mentioned, tag losses were considered negligible, and
trapping did not seem to affect an animal's behaviour with regard to future
trapping.

Systematic rather than random sampling was used, so that for the
Petersen estimate to apply, the tagged proportion must be constant over the
population area before the second sample is taken (cf. **3.1.2**). Since the
hares had restricted home ranges, then, provided that all the animals were

equicatchable, this uniformity would have been achieved by systematically trapping the whole area with a uniform trapping effort. The uniform effort would ensure that a fixed proportion of animals in any subarea would be caught for tagging. But in this particular investigation this was not done as there were more traps and more trapping effort used on a particular subarea which the authors called the Mile Area. The authors felt that this departure from the assumptions of the Petersen method led to an underestimate of the total population by about 5 to 10 per cent.

Example 3.7   Climbing cutworms: Wood [1963]

Population estimates of climbing cutworms are difficult to obtain as the larvae feed only at night and then conceal themselves in the litter layer during the day. It was found that the sweep net method (cf. Southwood [1966]) provided a satisfactory measure of the relative abundance of cutworms between fields or from year to year. The Petersen method was also used to estimate the number of insects per unit area in a small field ($\frac{1}{4} - \frac{1}{2}$ acre) of blueberry plants. The field was swept systematically at night and all larvae captured were taken to the laboratory for examination, where they were painted or sprayed with a solution of a fluorescent compound and then redistributed systematically over the field on the same night; any specimen injured in collecting or handling was discarded. On the following two or three evenings the field was again swept and marked specimens were detected by passing an ultraviolet lamp over the collected larvae. The results of the experiment are given in Table 3.10, and the confidence limits were obtained using the Poisson approximation (cf. **3.1.4**). We note that for $m_2 < 10$ the intervals are very wide.

TABLE 3.10

95 per cent confidence intervals for the density of a climbing cutworm population: data from Wood [1963: Table 1].

Year	n_1	n_2	m_2	N^*	$\hat{p}$	Confidence interval	Density per square foot
1956	125	107	8	15 119	0·074	(7 035, 34 240)	(0·06, 0·29)
1958	75	143	6	15 633	0·042	(6 435, 41 613)	(0·08, 0·52)
1959	93	114	3	27 024	0·026	(7 803, 130 405)	(0·07, 1·30)
1961	200	122	14	16 481	0·115	(9 321, 29 524)	(0·09, 0·30)
1962	1 000	1 755	41	41 850	0·023	(30 414, 57 915)	(1·73, 3·29)

Wood felt that by and large the assumptions underlying the Petersen model were satisfied and in particular the method of marking had no adverse effect on the insects. Although the fluorescent marker did not wash off, it was found that the proportion of marked larvae decreased with each successive night of collecting. This apparent loss of marked insects was attributed to loss of markings through moulting. Since the longer the interval the greater the possibility of moulting, it was decided to limit the counts to those taken on the first recapture date (24 hours after marking).

Example 3.8   Redpolls (*Acanthis linaria*): Nunneley [1964]

In 1962 a flock of redpolls frequented a banding station in Massachusetts, U.S.A., from March 3 to April 5, and 122 of them were banded. Three categories of information about the flock were sought: (i) the lengths of stay for individuals and the flock, (ii) the changing population size produced by the invasion movement from day to day, and (iii) information on whether the number of birds passing through the station was much larger than the number entering traps. I shall concern myself with just (ii) as a means of illustrating the Petersen method; the recapture data in Table 3.11 are taken from Table 4 of Nunneley's article.

TABLE 3.11

95 per cent confidence intervals for the size of a redpoll population: data from Nunneley [1964: Table 4].

Date (March)	n_1	n_2	m_2	N^*	$\hat{p}$	Confidence interval
3	—	7	0	—	—	—
6	6	12	5	14	0·417	(8, 40)
8	11	16	9	19	0·563	(14, 37)
13	13	45	9	63	0·200	(37, 133)
19	33	39	23	56	0·590	(45, 78)
22	24	43	18	57	0·419	(42, 88)
25	25	28	17	41	0·607	(32, 62)
27	13	6	3	24	0·500	(15, 108)
28	12	17	11	19	0·647	(14, 31)

Of the 122 banded there were 37 males, 74 females and 11 of unknown sex. As there appeared to be negligible differences in the arrival, average length of stay and recapture behaviour of the two sexes, the data for the two sexes were pooled and the redpolls treated as a single population. The assumption of a closed population (assumption (a)) seemed reasonable for two reasons. First, by confining tagging and recapturing to a single day, natural mortality and new additions were considered negligible over this short period. Secondly, banded birds appeared to be confined to the vicinity of the banding station. This meant that they had a very narrow feeding range, so that over a day, losses through emigration would be small. Tag losses would be negligible, and comparing captures per day with both length of stay and number of repeats suggested that trap shyness or trap addiction was negligible. The 95 per cent confidence intervals in Table 3.11 were calculated using the binomial approximation (**3.1.4**).

Example 3.9   Roe-deer (*Capreolus capreolus* L.): Andersen [1962]

In most European countries, the roe-deer is a widely distributed and popular animal, even though it causes considerable damage to forests. For example, in Denmark about 25 000 roe-deer are killed each year by more than 100 000 licensed sportsmen. With such a high annual kill it is imperative that deer management and shooting policy should be based on a sound knowledge of roe-deer biology. In particular, Andersen mentions the danger of using purely subjective methods in estimating roe-deer density as roe-deer are extremely good at escaping detection.

The aim of Andersen's work was to analyse both the size and composition of a herd which filled to carrying capacity the two woods of an experimental game farm (Kalø estate), using the Petersen method. The tag used was a leather collar studded with different coloured plastic buttons and bearing a copper plate giving a serial number and address. As a special precaution, a metal ear-tag supplying the relevant information was attached at the same time so as to take care of the rare cases when the collar was lost. From experience it was felt that the collar did not hamper the deer or expose them to danger, and there was no evidence to indicate that tagged animals behaved differently from untagged.

The traps were fenced enclosures and there was a period of prebaiting to allow the deer time to become used to the traps. Trapping took place in January and February, and the trapped animals were released immediately after tagging. Many entered the traps again, sometimes on the next day, in which case the collars were read and the animals released. The total number n_1 of animals tagged formed the first sample. During the following months, March, April and May, one man had the job of combing the woods and their immediate surroundings with binoculars, and noting all the individuals observed; this total n_2 formed the "recaptured" sample. Recording was not equally easy at all times as light-coloured buttons showed up better during dusk hours or in dense woodland, and it is here that observer bias could possibly have affected the data as doubtful records were excluded. For example, if the observer sees a group of, say, five deer together he must be able to "read" all five; if that is not possible Andersen recommends that they should all be disregarded. To achieve adequate sampling of the whole area, a number of observation posts were established at strategic points. In addition the observer stalked the deer, creating the least possible disturbance, and took special precautions to cover the entire woodland and to vary the time of his visit to different parts of the wood.

Concerning the assumption of constant N which underlies the Petersen method the following observations were made. Although the estate was fenced, public roads crossed the perimeter and the animals used them as routes to pass in and out of the area. However, it was felt that during the trapping and observation periods (autumn and winter respectively) migration took place to a very limited extent. As far as emigration out of the area was concerned, the marked deer were easy to spot, and all forest owners and sportsmen in the neighbourhood knew of the experiments. In particular a reward of approximately \$1·50 was paid to sportsmen who handed in a collar. As far as mortality is concerned, a few died through trapping; although the natural mortality during the experimental period was not estimated, it would not seriously affect the Petersen estimate if it was small and proportionately the same for tagged and untagged (cf. p. 71).

During the trapping two age-classes were distinguished, i.e. fawns born during the preceding summer and adults. It appeared that the fawns were more easily trapped than the adults and that about 75 per cent of the population had to be trapped before there was stability in the estimate of the fawn—adult ratio. The data for fawns and adults were then pooled as the percentage of adults in both n_1 and m_2 was approximately the same (57 per cent and 55 per cent respectively).

It is noted that members of the population could be "recaptured" (observed) several times, so that n_2 was much larger than N. Here the sampling was effectively with replacement, so that Bailey's binomial model (3.1.2) could be used, provided that the method of observing was random, or at least representative, at all times. For example, if a particular subarea contained a higher proportion than average of marked deer and that subarea was observed more often than other areas, then n_2 would contain too many tagged animals and N would consequently be underestimated.

The following numbers were obtained: $n_1 = 74$, $n_2 = 462$ and $m_2 = 340$, so that from 3.1.2

$$\hat{N}_1 = \frac{74(463)}{341} = 100 \quad \text{and} \quad \hat{p} = \frac{340}{462} = 0·736.$$

Using the normal approximation to the binomial distribution, an approximate 95 per cent confidence interval for $p = n_1/N$ is

$$\hat{p} \pm 1·96\,[\hat{p}(1-\hat{p})/n_2]^{\frac{1}{2}}.$$

Inverting this gives the corresponding confidence interval for N, namely (95, 106).

For a further study of this deer population the reader is referred to Strandgaard [1967].

3.4   ESTIMATION FROM SEVERAL SAMPLES

Let n_1 animals be captured, marked, and released throughout the population and let $p = n_1/N$, where N is the total population size. Suppose that the total population area can be divided up into K subareas of which k are selected at random for further sampling. Let

N_i = number of animals in ith subarea ($i = 1, 2, \ldots, K$),

$\bar{N}$ = N/K,

n_{1i} = number of marked animals in the ith subarea,

$\{n_{1i}\}$ = set of n_{1i} $(i = 1, 2, \ldots, K)$,

$\{n_{1i}\}_k$ = sample set of n_{1i} $(i = 1, 2, \ldots, k)$,

n_{2i} = number of animals caught in the ith subarea $(i = 1, 2, \ldots, k)$,

m_{2i} = number of marked in n_{2i},

p_i = n_{1i}/N_i,

$\hat{p}_i$ = m_{2i}/n_{2i},

f_1 = k/K,

and f_{2i} = n_{2i}/N_i.

We note that

$$p = \left(\sum_{i=1}^{K} N_i\, p_i \right) \Big/ \sum_{i=1}^{K} N_i.$$

3.4.1 Ratio estimate

With the above experimental set-up we effectively have two-stage sampling, in which we choose k unequal size-units or clusters at the first stage and then subsample from each unit, noting the proportion $\hat{p}_i$ of marked in each subsample. Therefore, if the cluster sizes N_i were known, p could be estimated by the ratio-type estimator (cf. Cochran [1963: 300])

$$\left(\sum_{i=1}^{k} N_i\, \hat{p}_i \right) \Big/ \left(\sum_{i=1}^{k} N_i \right).$$

However, as the N_i are unknown, one possibility is to use weights proportional to the sample sizes n_{2i}, so that p is estimated by

$$\hat{p} = \left(\sum_{i=1}^{k} n_{2i}\, \hat{p}_i \right) \Big/ \left(\sum_{i=1}^{k} n_{2i} \right)$$

$$= \left(\sum_{i=1}^{k} m_{2i} \right) \Big/ \left(\sum_{i=1}^{k} n_{2i} \right)$$

$$= m_2/n_2 \quad \text{say.}$$

It is noted that

$$E[\hat{p} \mid \{n_{2i}\}, \{n_{1i}\}_k] = \left(\sum_{i=1}^{k} n_{2i}\, p_i \right) \Big/ n_2 \qquad (3.32)$$

and we now consider two cases where $\hat{p}$ is unbiased and approximately unbiased, respectively.

1   Constant mark ratio

When $p_i \rightarrow p$ $(i = 1, 2, \ldots, k)$ the right-hand side of (3.32) reduces to p and $\hat{p}$ is unbiased. In this case $V[\hat{p} \mid \{n_{1i}\}]$, the variance of $\hat{p}$, can be estimated by (ignoring finite population corrections due to sampling without replacement) —

$$v_1[\hat{p}] = \hat{p}(1-\hat{p})/(n_2 - 1).$$

This estimate was used, for example, by Welch [1960]. The hypothesis H that $p_i = p$ can be tested using the standard goodness-of-fit statistic

$$\sum_{i=1}^{k} \frac{(m_{2i} - n_{2i}\hat{p})^2}{n_{2i}\,\hat{p}(1-\hat{p})},$$

which, when H is true, is approximately distributed as chi-squared with $k - 1$ degrees of freedom. When H is true and n_2 is large (say greater than 30), $\hat{p}$ is asymptotically normal with mean p and variance estimate $v_1[\hat{p}]$. An approximate confidence interval for p (and hence for N) can then be calculated in the usual manner.

2   Constant sampling effort

In general, if H is not true $\hat{p}$ will be biased. If, however, the same sampling effort is used within each subarea and the expected fraction caught (θ say) is proportional to the sampling effort, then $E[n_{2i}\,|\,n_{1i}] = E[n_{2i}] = N_i\theta$ and $E[m_{2i}\,|\,n_{1i}] = n_{1i}\,\theta$. Hence, for large n_{2i},

$$E[\hat{p}\,|\,\{n_{1i}\}_k] \approx \frac{\sum\limits_{i=1}^{k} E[m_{2i}\,|\,n_{1i}]}{\sum\limits_{i=1}^{k} E[n_{2i}\,|\,n_{1i}]}$$

$$= \left(\sum_{i=1}^{k} n_{1i}\right)\Big/\left(\sum_{i=1}^{k} N_i\right). \qquad (3.33)$$

Since the k subareas are chosen at random, (3.33) represents a ratio estimate of p with respect to the first stage in the sampling (cf. Cochran [1963: 29]). Therefore, taking expectations with respect to this first stage, we have, for large k,

$$E[\hat{p}\,|\,\{n_{1i}\}] = \underset{k}{E}\, E[\hat{p}\,|\,\{n_{1i}\}_k]$$

$$\approx \left(\sum_{i=1}^{K} n_{1i}\right)\Big/\left(\sum_{i=1}^{K} N_i\right)$$

$$= n_1/N = p.$$

Thus $\hat{p}$ is approximately unbiased for large n_{2i} and large k.

Noting that

$$\theta = \frac{E[n_{2i}]}{N_i} = \frac{E[n_2]}{\sum\limits_{i=1}^{k} N_i} \approx \frac{E[n_2]}{k\bar{N}} = \frac{E[\bar{n}_2]}{\bar{N}},$$

where $\bar{n}_2 = n_2/k$, $N_i/\bar{N}$ can be estimated by $n_{2i}/\bar{n}_2$. Therefore, from Cochran [1963: 300–2 and particularly 324, Ex. 11.5, with $M_i = N_i$] an approximate estimate of $V[\hat{p}\,|\,\{n_{1i}\}]$ is given by

$$v_2[\hat{p}] = \frac{1 - f_1}{k(k-1)} \sum_{i=1}^{k} \left(\frac{n_{2i}}{\bar{n}_2}\right)^2 (\hat{p}_i - \hat{p})^2 +$$

$$+ \frac{f_1(1-f_2)}{kn_2} \sum_{i=1}^{k} \left(\frac{n_{2i}}{\bar{n}_2}\right) \cdot \frac{n_{2i}}{n_{2i}-1} \cdot \hat{p}_i(1-\hat{p}_i), \qquad (3.34)$$

where f_2 ($\approx \bar{n}_2/N$) can either be ignored, or estimated using $n_1/\hat{p}$ as an estimate of N.

When $k = K$, i.e. $f_1 = 1$, we have the special case of stratified random sampling. The expression (3.33) then reduces to p and the first term in (3.34) is zero.

When f_1 is small the second term in (3.34) can be neglected, and $V[\hat{p} \mid \{n_{1i}\}]$ is now estimated by

$$v_3[\hat{p}] = \frac{k}{(k-1)n_2^2} \sum_{i=1}^{k} (m_{2i} - n_{2i}\hat{p})^2.$$

Using the delta method, the variance of the Petersen estimate $\hat{N} = n_1/\hat{p}$ is approximately given by

$$n_1^2 p^{-4} V[\hat{p} \mid \{n_{1i}\}],$$

which can then be estimated by

$$
\begin{aligned}
v[\hat{N}] &= n_1^2 \hat{p}^{-4} v_3[\hat{p}] \\
&= \frac{\hat{N}^2 k}{m_2^2(k-1)} \sum_{i=1}^{k} (m_{2i} - n_{2i}\hat{p})^2.
\end{aligned}
\tag{3.35}
$$

From the relationship

$$\Sigma(m_{2i} - n_{2i}\hat{p})^2 = m_2^2 \left[\frac{\Sigma(m_{2i} - \bar{m}_2)^2}{m_2^2} - \frac{2\Sigma(m_{2i} - \bar{m}_2)(n_{2i} - \bar{n}_2)}{m_2 n_2} + \frac{\Sigma(n_{2i} - \bar{n}_2)^2}{n_2^2} \right],$$

it can be shown that (3.35) is identical with a formula used by Banks and Brown [1962] in Example 3.11 below.

3.4.2 Mean estimates

An alternative estimate of p is the average

$$\hat{\bar{p}} = \sum_{i=1}^{k} \hat{p}_i/k$$

and from the general theory of Cochran [1963: 272–8, setting $y_i' = \hat{p}_i/k$, $Y_i' = p_i/k$, $\pi_i = k/K$ etc.] an unbiased estimate of the variance is given by

$$v[\hat{\bar{p}}] = \frac{(1-f_1)}{k(k-1)} \sum_{i=1}^{k} (\hat{p}_i - \hat{\bar{p}})^2 + \frac{f_1}{k^2} \sum_{i=1}^{k} (1 - f_{2i}) \frac{\hat{p}_i \hat{q}_i}{(n_{2i} - 1)},$$

which reduces to the usual sample estimate of variance (cf. (1.9)) when f_1 is ignored. As

$$
\begin{aligned}
E[\hat{\bar{p}} \mid \{n_{1i}\}] &= \underset{k}{E}\, E[\hat{\bar{p}} \mid \{n_{1i}\}_k] \\
&= \underset{k}{E} \left[\sum_{i=1}^{k} p_i/k \right] \\
&= \left(\sum_{i=1}^{K} p_i \right)/K \\
&= \bar{p}, \text{ say,}
\end{aligned}
$$

then $\hat{\bar{p}}$ is unbiased when either H ($p_i = p$ for all i) is true or when $\bar{p} = p$. If H is true, $\hat{\bar{p}}$ seems preferable to $\hat{p}$ because of the general robustness of a mean with regard to normality and because $v[\hat{\bar{p}}]$ is more robust than $v_1[\hat{p}]$ with regard to departures from H. On the other hand, if H is rejected by the goodness-of-fit test and the sampling effort is uniform, then $\hat{p}$ can be used with $v_2[\hat{p}]$ of (3.34).

In some experiments the numbers n_{1i} of marked animals in the individual subareas are known and N can be estimated by (cf. **3.1.1**)

$$N' = K \sum_{i=1}^{k} N_i^*/k$$

$$= K \sum_{i=1}^{k} \left\{ \frac{(n_{1i} + 1)(n_{2i} + 1)}{(m_{2i} + 1)} - 1 \right\}/k .$$

Then

$$E[N'] = \mathop{E}_{k} E[N_i' \mid k]$$

$$\approx K \mathop{E}_{k} \left[\sum_{i=1}^{k} N_i/k \right]$$

$$= K\bar{N}$$

$$= N,$$

and using the general theory of Cochran [1963: 272–4, setting $y_i' = N_i^*/k$, $Y_i' = N_i/k$, $\pi_i = k/K$, etc.] it can be shown that an approximately unbiased estimate of the variance of N' is

$$v[N'] = \frac{K(K-k)}{k(k-1)} \sum_{i=1}^{k} (N_i^* - N')^2 + \frac{K}{k} \sum_{i=1}^{k} v[N_i^*],$$

where

$$v[N_i^*] = \frac{(n_{1i}+1)(n_{2i}+1)(n_{1i}-m_{2i})(n_{2i}-m_{2i})}{(m_{2i}+1)^2(m_{2i}+2)} .$$

When $n_{1i} + n_{2i} \geqslant N_i$ for each i, then N' and $v[N']$ are *exactly* unbiased.

In comparing N' with the Petersen estimate $\hat{N}$, we note from **3.2.5** that when $k = K$ (i.e. stratified sampling) the pooled estimate $\hat{N}$ will have smaller variance than N', the sum of the individual estimates. We would also expect this to be true when $k < K$. However, N' has several advantages. First of all, it is based on a mean, so that for large k it is approximately normally distributed by the Central Limit theorem. Secondly, if the assumptions underlying the Petersen method hold within each subarea then, in general, N' and $v[N']$ are almost unbiased estimates, irrespective of whether H is true or whether the same sampling effort is used in each subarea. This means that confidence intervals based on N' will often be more reliable than those based on $\hat{p}$ or $\hat{\bar{p}}$, particularly when these estimates of p are biased.

Example 3.10   Adult Sunn Pest (*Eurygaster integriceps* Put.): Banks and Brown [1962]

This pest seriously damages wheat and other cereals in various countries of the Middle East. In attempting to investigate chemical and biological methods of control, accurate estimates of the insect population density are required for assessing the effects of insecticides on the mortality of the insects; while quick, but less accurate, estimates are needed for deciding the best time to apply the insecticides to the overwintered insects newly arrived on the wheat fields. Banks and Brown considered three methods of estimation: sweeping with a hand net, quadrat sampling, and the Petersen recapture method.

A wheat plot of about 1000 square metres was studied within each of two fields (called A and B). Each plot was staked out into 36 equal sectors (of 28 sq.m. for A and 23 sq.m. for B) and each sector in turn was swept systematically with a hand net by the same person. The insects caught were counted and marked with a spot of quick-drying paint. As soon as the paint was dry, the marked insects were scattered evenly over the sector from which they were caught and the team then moved into the next sector. Sixteen hours were allowed for the insects to return to the plants and emerge from cracks in the soil after being disturbed, and for the marked to mix with the unmarked. Assuming uniform mixing of marked and unmarked, systematic sampling could then be used for taking the second sample (cf. **3.1.2**). The two systematic methods used here were the sweepnet method, again over the whole plot, and quadrat sampling. In the latter case fifty quadrats of one square metre were marked out systematically over the whole plot and each quadrat was carefully examined. All bugs visible on the plants were first picked off and then the vegetation and ground were thoroughly searched.

Applying the quadrat method to plot A, it was found that $K = 28(36) = 1008$, $k = 50$, $n_1 = 3538$, $n_2 = 980$, $m_2 = 106$, $\Sigma n_{2i}^2 = 30\,380$, $\Sigma m_{2i}^2 = 450$ and $\Sigma m_{2i} n_{2i} = 2700$. The authors effectively used the estimates $\hat{p}$ and $v_3[\hat{p}]$, and these are calculated as follows:

$$\hat{p} = 106/980 = 0 \cdot 108\,163,$$

$$\Sigma(m_i - n_i \hat{p})^2 = 450 - 2(2700)(0 \cdot 108\,163) + 30\,380(0 \cdot 108\,163)^2$$
$$= 221 \cdot 34,$$

and
$$v_3[\hat{p}] = \frac{50(221 \cdot 34)}{49(980)^2} = 0 \cdot 000\,235\,16.$$

Assuming $\hat{p}$ to be asymptotically normal, the 95 per cent confidence interval for p is $(0 \cdot 078\,45,\ 0 \cdot 1380)$ and the corresponding interval for N is $(25\,640,\ 45\,100)$. Also $\hat{N} = 3538(980)/106 = 32\,710$ and, from (3.35),

$$\{v[\hat{N}]\}^{\frac{1}{2}} = \frac{32\,710}{106} \left\{ \frac{50}{49}(221 \cdot 34) \right\}^{\frac{1}{2}}$$

$$= 4638.$$

We shall now briefly consider the assumptions underlying the above use of the Petersen method. To begin with, the population was expected to be fairly constant from one day to the next, as the deaths overnight were considered to be few and the dispersal was negligible owing to lack of horizontal movement shown by the insects. We can presume that there were no tag losses or unreported tags, and since no mention is made of mortality through handling it was no doubt negligible in both samples. It was assumed that a period of 16 hours was sufficient for the marked insects to get over the effects of handling and that the paint spot would not affect their movements. However, as far as the validity of the Petersen estimate is concerned, the main assumption that was difficult to satisfy was the assumption of uniform mixing of marked and unmarked which is needed when systematic rather than random sampling is carried out. The method used to achieve this mixing was to scatter the marked insects evenly over the sector from which they were caught; this would be satisfactory if the density of the population was uniform over the sector to start with and the marked insects dispersed in a random manner. To check on the mixing, we can calculate (cf. **3.1.2**)

$$\sqrt{v_1} \; = \; \frac{n_1}{m_2 + 1} \left\{ \frac{(n_2 + 1)(n_2 - m_2)}{(m_2 + 2)} \right\}^{\frac{1}{2}} \; = \; 2946.$$

Since this is much less than $\{v[\hat{N}]\}^{\frac{1}{2}}$ ($= 4638$) it would seem that the binomial model is not valid and that the dispersal of the marked bugs was not complete at the time of recapture. This lack of dispersal, and in fact the lack of mobility, indicates that the Petersen method is not suitable as a standard method of studying sunn pest populations. Another disadvantage of the Petersen method is that it is laborious, requiring a team of several people to obtain satisfactory results.

The two other methods of census suggested, namely quadrat counting and sweeping, are easier to carry out and would therefore be used more often. The main disadvantage of direct counting using quadrats is that a certain proportion of the population, though small in this case, would hide in soil crevices and would need to be smoked out for more accurate counts. On the credit side, however, the accuracy can be increased by increasing the number of quadrats: the same is true with the sweeping method, though in this case the maximum obtainable accuracy depends on the sweeping efficiency.

The quickest method of the three, namely sweeping, relies heavily on the assumption of uniform sweeping efficiency, so that this assumption needs to be checked from time to time. The authors conclude that the Petersen recapture method "may be useful as an occasional check on other sampling methods and, in particular, to provide a measure of the efficiency of routine sampling by sweeping".

3.4.3   Interpenetrating subsamples

Chapman and Johnson [1968] suggest one other estimate of N based on interpenetrating subsamples (cf. Cochran [1963: 383]). Here the total sample of size n_2 is subdivided randomly into r subsamples and an estimate N_i'' of N is obtained from each subsample, using one of the above methods. Then

N is estimated by the subsample mean $N'' = \sum\limits_{i=1}^{r} N_i''/r$, and $V[N'']$ is estimated by

$$v[N''] = \sum_{i=1}^{r} (N_i'' - N'')^2 / [r(r-1)].$$

Chapman and Johnson apply this method to a fur seal pup population.

If $N_{(1)}''$ and $N_{(r)}''$ are the smallest and largest of the N_i'' then (Overton and Davis [1969]) $\Pr[N_{(1)}'' < N < N_{(r)}''] = 1 - (\frac{1}{2})^{r-1}$ provides exact limits for the median of the distribution of N_i'': these are quite acceptable as limits for N.

3.5 INVERSE SAMPLING METHODS

We now consider an inverse sampling method for the second sample which, in contrast to the "direct" Petersen method considered so far, provides an unbiased estimate of N with an *exact* (rather than a large-sample) expression for the variance, and a coefficient of variation which is almost independent of N. The method is to tag or mark n_1 animals as before and then continue taking the second sample until a prescribed number m_2 of marked animals have been recovered. This means that n_1 and m_2 are now considered as fixed parameters and n_2 is the random variable. As in the direct Petersen method, the second sample can be taken with or without replacement, and we shall now consider these two cases separately.

3.5.1 Sampling without replacement

When the assumptions (a), (d), (e) and (f) of **3.1.1** are satisfied, Bailey [1951] shows that the probability function of n_2, conditional on n_1 and m_2, is the negative hypergeometric

$$f(n_2 \mid n_1, m_2) = \frac{\binom{n_1}{m_2-1}\binom{N-n_1}{n_2-m_2}}{\binom{N}{n_2-1}} \cdot \frac{n_1 - m_2 + 1}{N - n_2 + 1}, \tag{3.36}$$

where $n_2 = m_2, m_2 + 1, \ldots, N + m_2 - n_1$, and suggests the modified maximum-likelihood estimate of N

$$\hat{N}_2 = \{n_2(n_1 + 1)/m_2\} - 1 \tag{3.37}$$

This estimate is unbiased and has exact variance

$$V[\hat{N}_2 \mid n_1, m_2] = \{(n_1 - m_2 + 1)(N+1)(N-n_1)\}/\{m_2(n_1+2)\} \approx N^2/m_2.$$

Assuming $N + 1$ and $N - n_1$ to be approximately equal to N, the coefficient of variation of $\hat{N}_2$ is close to

$$C[\hat{N}_2] = [(n_1 - m_2 + 1)/\{m_2(n_1 + 2)\}]^{\frac{1}{2}},$$

and since n_1 is known, m_2 can be chosen beforehand, so that this coefficient has a prescribed value. For example, if $n_1 = 100$ and $C = 0\cdot1$, then sampling must continue until $m_2 = 50$ marked individuals have been caught.

Chapman [1952] mentions several useful properties of $\hat{N}_2$. As m_2 tends to infinity, n_2 is asymptotically normally distributed, so that for large m_2, $\hat{N}_2$ is approximately normal. Also, on the average, the inverse method is slightly more efficient than the direct Petersen method. By this we mean that for a given coefficient of variation, the inverse method provides an estimate of N with an expected sample size $E[n_2 \mid n_1, m_2] (= (N+1)m_2/(n_1+1))$ which is smaller than the sample size needed for the direct method though usually the difference is small (Robson and Regier [1964: 218]). The inverse method also shares the same advantage as the direct method in that it is still applicable when mortality is taking place, provided that ϕ, the probability of survival between the two samples, is the same for marked and unmarked (cf. page 71). In this case $\hat{N}_2$ remains approximately unbiased, and its variance is now (Chapman [1952: 301])

$$\frac{N^2}{m_2} + \frac{N(1-\phi)}{\phi}\left(1 + \frac{1}{m_2}\right),$$

which is still approximately N^2/m_2 unless ϕ is very small.

However, an undesirable feature of the inverse method is that if the experimenter knows absolutely nothing about N, the expected sample size may be very large with an improper choice of n_1 and m_2. Also the variance of n_2 (approximately $m_2 N^2/n_1^2$) is large, and this increases the unpredictability of what n_2 will actually turn out to be. These difficulties may be partly overcome by considering an inverse sampling scheme of Chapman's in which $u_2 (= n_2 - m_2)$, the number of unmarked individuals caught in n_2, is fixed instead of m_2; m_2 and n_2 are now both random variables, though completely dependent. For this scheme

$$E[n_2 \mid n_1, u_2] = u_2(N+1)/(N-n_1+1)$$

and

$$V[n_2 \mid n_1, u_2] = \frac{u_2 n_1 (N+1)(N-u_2-n_1+1)}{(N-n_1+1)^2(N-n_1+2)}$$

$$\approx u_2 n_1/(N-n_1) \quad \text{(for large } N\text{)},$$

so that in contrast with the method above, the variance of n_2 is small and the expected sample size does not depend so critically on N. We now find that an approximately unbiased estimate of N is given by

$$\hat{N}_3 = \frac{n_2(n_1+1)}{(m_2+1)} - 1 \tag{3.38}$$

with approximate variance

$$V[\hat{N}_3 \mid n_1, u_2] = (n_1+1)^2[q_{22} + q_{23} + 2q_{24} - q_{12} - q_{13} - 2q_{14}] - N^2 - 2N - 2, \tag{3.39}$$

where

$$q_{ij} = \frac{(N+i)_i(N-n_1)_k}{(n_1+j)_j(u_2-1)_k} \qquad (i \leqslant j, \quad k = j - i)$$

and $(a)_i = a(a-1) \ldots (a-i+1)$, etc.

Writing $p_0 = n_1/(N+1)$, Chapman [1952] shows that when n_2 is large,

$$z = \frac{n_2 - u_2/(1-p_0)}{\{u_2 p_0/(1-p_0)\}^{\frac{1}{2}}}$$

is approximately unit normal, so that approximate $100(1-\alpha)$ per cent confidence limits for p_0 are the roots of the quadratic

$$\{n_2(1-p_0) - u_2\}^2 = z_{\alpha/2}^2 u_2 p_0(1-p_0).$$

Inverting these limits leads to a confidence interval for N. In conclusion, we note that this method can be applied to the situation where n_1 is also unknown and an estimate of n_1/N ($\approx p_0$) is required, e.g. the estimation of sex or age ratios.

3.5.2   Sampling with replacement

We shall now consider the less common situation in which members of the second sample are caught one at a time and returned immediately to the population. For example, this model would apply when the animals are merely observed and not actually captured. In the inverse method, sampling is continued until a prescribed number m_2 of marked animals have been caught and released, so that the probability function of n is now the negative binomial

$$f(n_2 \mid n_1, m_2) = \binom{n_2 - 1}{m_2 - 1} p^{m_2} (1-p)^{n_2 - m_2}$$

where $n_2 = m_2, m_2 + 1, \ldots$, and $p = n_1/N$. As $E[n_2 \mid n_1, m_2] = Nm_2/n_1$, the obvious estimate for N is the Petersen estimate $\hat{N} = n_1 n_2/m_2$. This estimate is unbiased with variance

$$V[\hat{N} \mid n_1, m_2] = (N^2 - Nn_1)/m_2,$$

which is estimated unbiasedly by (Chapman [1952: 287])

$$v[\hat{N}] = \frac{n_2 n_1^2(n_2 - m_2)}{m_2^2(m_2 + 1)}.$$

The coefficient of variation of $\hat{N}$ is close to $1/\sqrt{m_2}$, so that for C to be no greater than 20 per cent, for example, sampling must continue until $m_2 = 25$ marked individuals have been captured. We note that, as in sampling without replacement, this inverse method is more efficient than Bailey's direct method of **3.1.2** (Goodman [1953: 67]). However, as the mean and variance of n_2 are almost the same as for sampling without replacement, the inverse method above suffers from the same disadvantages, namely that the variance of n_2 is large and the expected sample size may also be large with an improper choice of n_1 and m_2.

When m_2 is large, Chapman [1952] shows that

$$z = \frac{\hat{N} - N}{\{N(N - n_1)/m_2\}^{\frac{1}{2}}} \qquad (3.40)$$

is approximately distributed as the unit normal distribution, so that $100(1 - \alpha)$ per cent confidence limits for N are the roots of

$$(\hat{N} - N)^2 = z_{\alpha/2}^2 N(N - n_1)/m_2 .$$

When m_2 and p are both small, Chapman shows that $2n_2 p$ is approximately distributed as chi-squared with $2m_2$ degrees of freedom. Therefore an approximate $100(1 - \alpha)$ per cent confidence interval for N is $(2n_1 n_2/c_2, 2n_1 n_2/c_1)$ where c_1 and c_2 are the lower and upper $\alpha/2$ significance points respectively for $\chi_{2m_2}^2$.

If n_1 is also unknown, confidence intervals for p can be obtained by the above methods. For example, using the normal approximation (3.40), $100(1 - \alpha)$ per cent limits are the roots of

$$(n_2 p - m_2)^2 = z_{\alpha/2}^2 m_2 (1 - p),$$

while the chi-squared approximation leads to $(c_1/(2n_2), c_2/(2n_2))$. Haldane [1945] gives an unbiased estimate of p, namely

$$\hat{p} = (m_2 - 1)/(n_2 - 1)$$

which has a variance estimated by $\hat{p}(1 - \hat{p})^2/(m_2 - 2)$ when p is small.

In conclusion we note that several authors have recently considered the problem of finding a minimax estimate of N, given that N is known to be greater than a certain integer (Czen Pin [1962], Zubrzycki [1963, 1966]).

3.6   COMPARING TWO POPULATIONS

3.6.1   Goodness-of-fit method

Suppose we have two closed populations of unknown sizes N_a and N_b, and on the basis of a single Petersen experiment in each population we wish to test the null hypothesis H_0 that $N_a = N_b$. Such a situation could arise, for example, where a control area and an experimental area are under observation and one wishes to test for any difference in population size due to experimental management practice. Alternatively, N_a and N_b could refer to the same population area but at different times. A third possibility is when N_a and N_b refer to the same population at the same time but two different sampling methods are used, e.g. seine and gill-net fishing. A test of H_0 would then indirectly provide a test for the hypothesis that marked animals are equally vulnerable to both methods of sampling.

In the notation of **3.1.1**, we can test H_0 by assuming that

$$z = \frac{N_a^* - N_b^*}{\sqrt{(v_a^* + v_b^*)}}$$

is approximately unit normal when H_0 is true. A more sensitive test is given by Chapman [1951] as follows. Using suffixes a and b to denote the two populations, let

$$\widetilde{N} = \frac{\lambda_a^3 m_{2b} u_{1b} u_{2b} + \lambda_b^3 m_{2a} u_{1a} u_{2a}}{m_{2a} m_{2b} \left[\lambda_a^2 u_{1b} u_{2b} + \lambda_b^2 u_{1a} u_{2a} \right]} ,$$

where $\lambda_a = n_{1a} n_{2a}$, $u_{1a} = n_{1a} - m_{2a}$, $u_{2a} = n_{2a} - m_{2a}$, etc., and define

$$T_1 = \sum \frac{(m_{2c} - \lambda_c / \widetilde{N})^2}{\dfrac{\lambda_c}{\widetilde{N}} \left(1 - \dfrac{n_{1c}}{\widetilde{N}} \right) \left(1 - \dfrac{n_{2c}}{\widetilde{N}} \right)} ,$$

$$T_2 = \sum \frac{\lambda_c (m_{2c} - \lambda_c / \widetilde{N})^2}{m_{2c} u_{1c} u_{2c}} ,$$

and

$$T_3 = \sum \frac{2(m_{2c} - \lambda_c / \widetilde{N})^2}{\dfrac{\lambda_c}{\widetilde{N}} \left(1 - \dfrac{n_{1c}}{\widetilde{N}} \right) \left(1 - \dfrac{n_{2c}}{\widetilde{N}} \right) + \dfrac{m_{2c} u_{1c} u_{2c}}{\lambda_c}} ,$$

where $\sum$ denotes summation over the two values $c = a, b$. Then, when H_0 is true, T_1, T_2 and T_3 are each approximately distributed as chi-squared with one degree of freedom when N is large, and all three statistics are candidates for testing H_0. Chapman suggests using T_2 when $\lambda_a = \lambda_b$, T_3 when these quantities differ moderately, and T_1 when λ_a is widely different from λ_b. To test H_0 against the two-sided alternative $N_a \neq N_b$, the criterion is to reject H_0 at the 100α per cent level of significance when T_i is greater than the α critical value of χ_1^2.

3.6.2   Using Poisson approximations

When experimental circumstances are such that the hypergeometric distributions of m_{2a} and m_{2b} can be approximated by Poisson distributions (cf. **3.1.4**), the following technique of Chapman and Overton [1966] can be used for testing H_0. Let $m_2 = m_{2a} + m_{2b}$ and $N_b = kN_a$; then from **1.3.7**(1) the conditional probability function of m_{2a} given m_2 is

$$f(m_{2a} \mid m_2) = \binom{m_2}{m_{2a}} p^{m_{2a}} q^{m_{2b}},$$

where $p = 1 - q = k\lambda_a / (k\lambda_a + \lambda_b)$. Setting

$$p_0 = \lambda_a / (\lambda_a + \lambda_b),$$

we note that p is greater or less than p_0 if and only if k is greater or less than unity, so that testing H_0 against the two-sided alternative $N_a \neq N_b$ is equivalent to testing $p = p_0$ against $p \neq p_0$. Therefore, given m_2, m_{2a} and p_0, we can test H_0 by evaluating the exact tail probabilities, using such tables as those of the Harvard Computation Laboratory [1955], or we can obtain a confidence interval for p and reject H_0 if p_0 lies outside this interval. Confidence intervals for p can be determined from the Clopper–Pearson charts of Pearson and Hartley [1966: 228–9], using $\hat{p} = m_{2a}/m_2$ as the entering variable, or from tables such as Owen [1962: 273–85]. A bibliography of "equitail" confidence intervals is given by Owen [1962: 273], and for $m_2 \leqslant 30$, shorter intervals are given by Crow [1956]. When $\hat{p} < 0\cdot1$, and for any m_2 (Raff [1956]), the Poisson approximation to the binomial can be used to find a confidence interval for $m_2 p$, using m_{2a} as the entering variable in Poisson tables (e.g. Crow and Gardner [1959], Pearson and Hartley [1966: 227]). When $0\cdot1 \leqslant \hat{p} \leqslant 0\cdot9$, $m_2\hat{p}$ and $m_2\hat{q}$ are both greater than 5, and a correction for continuity is used, the normal approximation is applicable (Raff [1956: 296]) and we can use the statistic

$$z = \frac{|m_{2a} - m_2 p_0| - \frac{1}{2}}{\sqrt{(m_2 p_0 q_0)}}$$

for testing H_0.

We note that the above methods can be used to obtain a confidence interval for k. In particular, when N_a and N_b refer to the same population but at different times, and the population is closed except for mortality, k can be interpreted as the proportion of N_a surviving, i.e. as a survival probability. If we are interested in just a one-sided alternative, say $N_b > N_a$, then we would reject H_0 if m_{2a} lay in the upper tail of the binomial distribution with $p = p_0$; the appropriate tail probability could be evaluated using binomial tables or the normal approximation.

To illustrate the above theory we shall now consider a number of examples which have been adapted from Chapman and Overton.

Example 3.11

Suppose we wish to test H_0 against the two-sided alternative $N_a \neq N_b$ for the following data:

$$\lambda_a = 430, \ m_{2a} = 11; \quad \lambda_b = 1040, \ m_{2b} = 7.$$

Then $m_2 = 11 + 7 = 18$ and $p_2 = 430/1470 = 0\cdot2925$. For illustrative purposes we shall consider four methods of testing H_0.

(i) From Pearson and Hartley [1966], the 99 per cent confidence interval for p, namely $(0\cdot29, 0\cdot87)$ does not contain p_0, so we reject H_0 at the 1 per cent level of significance.

(ii) Working with m_{2b} instead of m_{2a}, we can use Owen's tables with $x = 7$, $n - x = 11$ to obtain a 99 per cent confidence interval $(0\cdot1284, 0\cdot7068)$ for $1 - p$. As this interval does not contain $1 - p_0$, we reject H_0 at the 1 per cent level of significance.

(iii) Using the normal approximation, we have

$$z = \frac{11 - 18(0\cdot2925) - \frac{1}{2}}{\sqrt{[18(0\cdot2925)(0\cdot7075)]}} = 2\cdot71,$$

123

which is greater than 2·58, the two-tailed 1 per cent significant point.

(iv) If we wish to carry out a more accurate test of H_0, the exact probabilities can be calculated with appropriate binomial tables. In this case the required probability for a two-tailed test is the sum of all terms $\Pr[m_{2a} = d \,|\, p = p_0]$ in both tails, such that

$$\Pr[m_{2a} = d \,|\, p = p_0] \leqslant \Pr[m_{2a} = 11 \,|\, p = p_0].$$

From binomial tables (interpolating between 0·29 and 0·30 for p_0) we obtain

$$\Pr[m_{2a} = 0 \,|\, p_0] = 0\cdot0020$$
$$\Pr[m_{2a} = 11 \,|\, p_0] = 0\cdot0038$$
$$\Pr[m_{2a} \geqslant 12 \,|\, p_0] = 0\cdot0011$$
$$\mathrm{Sum} = \overline{0\cdot0069}$$

Since Sum $< 0\cdot01$ we reject H_0 at the 1 per cent level of significance.

Example 3.12

Suppose that for a given fish population we wish to test the hypothesis that marked fish are more vulnerable to seining than to gill-net fishing. If the suffix a represents seining then this is equivalent to testing the hypothesis $p = p_0$ against the one-sided alternative $p > p_0$. For the following data

$$\lambda_a = 85\ 042, \quad m_{2a} = 73, \quad \lambda_b = 43\ 818, \quad m_{2b} = 20$$

we have $m_2 = 93$ and $p_0 = 0\cdot660$. The normal approximation is justified here, so that the test statistic is

$$z = \frac{73 - 93(0\cdot660) - \frac{1}{2}}{\sqrt{[93(0\cdot660)(0\cdot340)]}} = 2\cdot43.$$

For a one-sided test this corresponds to a significance level of $0\cdot0075$, so that we reject the null hypothesis at the 1 per cent level of significance. There is therefore evidence that the seine samples or gill-net samples are not random with respect to marked fish.

It should be noted that if m_{2a} and m_{2b} are too small, the test of H_0 will have low power and be very insensitive. To illustrate this point, Chapman and Overton [1966] discuss the following example.

Example 3.13

Suppose we wish to test H_0 against the one-sided alternative $N_b > N_a$ with $\lambda_a = \lambda_b = 1000$ (i.e. $p_0 = \frac{1}{2}$). If in actual fact $N_a = 100$, $N_b = 200$, then from the theory of the Petersen method (cf. **3.1.1**) the expected recaptures in the two populations (λ/N) will be 10 and 5, respectively. Hence from the binomial tables we have

$$\Pr[m_{2a} \geqslant 12 \,|\, m_2 = 15, \ p_0 = \tfrac{1}{2}] = 0\cdot018$$
$$\Pr[m_{2a} \geqslant 11 \,|\, m_2 = 15, \ p_0 = \tfrac{1}{2}] = 0\cdot059$$

so that H_0 will be rejected at the 0·05 level of significance for $m_{2a} \geqslant 12$. The power of this test for the alternative $N_b = 2N_a$ (i.e. $p = \frac{2}{3}$) is then given approximately by

$$\Pr[m_{2a} \geqslant 12 \mid m_2 = 15, \, p = \tfrac{2}{3}] \;=\; 1 - \Pr[m_{2a} \geqslant 4 \mid m_2 = 15, \, p = \tfrac{1}{3}]$$
$$= \; 0\text{·}209.$$

This means that the above program has only about 1 chance in 5 of detect-ing an effect as great as doubling the population. The main reason for this insensitivity is the small numbers of recaptures. We recall from **3.1.1** that the Petersen estimate may not even give the right order of magnitude when there are less than 10 recaptures. For example, if $\lambda_a = \lambda_b = 4000$ in the above experiment then the expected recaptures are 40 and 20 respectively, and the approximate power for $p = \frac{2}{3}$ is now 0·83, a considerable improve-ment over 0·209. We note that although such power calculations are only approximate, m_2 being strictly a random variable, they will be useful in providing some guidance on the choice of λ_a and λ_b.

In concluding this section we mention briefly two other problems con-sidered in the literature. Chapman and Overton [1966] describe a paired comparison technique, based on a number of pairs of areas, for testing whether a certain management technique has any effect on population size. Chapman [1951: 156] briefly considers the problem of comparing ratios of populations, say N_a/N_b and N_c/N_d, using four separate Petersen type experiments and a chi-squared goodness-of-fit test similar to those discussed in **3.6.1**. Such an experimental situation could arise, for example, in com-paring two populations before and after immigration or in detecting changes in sex or age ratios for populations which tend to segregate.

3.7   ESTIMATION BY LEAST SQUARES

In commercially exploited populations the second sample in the Petersen method may consist of a sequence of samples, each sample being permanently removed from the population. For this situation N can be estimated by the following least-squares method due to Paloheimo [1963].

It is convenient to have a change in notation. Let

$$
\begin{aligned}
N_0 &= \text{initial size of the total population,} \\
M_0 &= \text{initial size of the marked population,} \\
U_0 &= N_0 - M_0, \\
n_i &= \text{size of the ith sample removed from the population } (i = 1, 2, \ldots s), \\
m_i &= \text{number of marked individuals in the ith sample,} \\
u_i &= n_i - m_i, \\
y_i &= m_i/n_i, \\
M_i &= M_0 - \sum_{j=1}^{i-1} m_j, \text{ and} \\
N_i &= N_0 - \sum_{j=1}^{i-1} n_j.
\end{aligned}
$$

Then, if the assumptions underlying the Petersen method (**3.1.1**) hold for each sample,

$$E[y_i \mid M_i, N_i] = M_i/N_i \quad (i = 1, 2, \dots, s).$$

Paloheimo suggests estimating N_0 by minimising $\Sigma \, w_i(y_i - M_i/N_i)^2$ with respect to N_0, where the w_i's are appropriate weights, customarily taken to be proportional to the inverse of the variances of the y_i. When the sampling is random, or the marked and unmarked are randomly mixed, these variances may be calculated by assuming Poisson or binomial sampling. For example, assuming Poisson sampling, the variance of y_i equals its expected value M_i/N_i and the weights would have to be estimated iteratively as they contain the unknown N_0. Not only are such weights awkward to compute, but very often, in practice, the y_i vary more than expected on the assumption of random fluctuations. Under these circumstances De Lury [1958] argues that one should preferably choose weights equal to the sample sizes. Also, if the marked and unmarked are removed at the same rate, we have approximately $M_i/N_i = M_0/N_0$, so that

$$E[y_i \mid M_i, N_i] = M_0/N_0 = \beta_0, \quad \text{say.}$$

Therefore, assuming the y_i to be approximately independently and normally distributed with variances σ^2/n_i, and setting $w_i = n_i$, we can use the general theory of **1.3.5** (*1*) to obtain an estimate and $100(1 - \alpha)$ per cent confidence interval for β_0, namely

$$\tilde{\beta}_0 = \Sigma \, w_i y_i / \Sigma \, w_i$$
$$= \Sigma \, m_i / \Sigma \, n_i$$

and

$$\tilde{\beta}_0 \pm t_{s-1}[\alpha/2] \, (\tilde{\sigma}_0^2 / \Sigma \, n_i)^{\frac{1}{2}},$$

where $\quad (s-1)\tilde{\sigma}_0^2 = \sum_i \dfrac{m_i^2}{n_i} - \dfrac{(\Sigma \, m_i)^2}{\Sigma \, n_i}.$

The least-squares estimate of N_0 is then

$$\tilde{N}_0 = M_0/\tilde{\beta}_0 = M_0 \Sigma \, n_i / \Sigma \, m_i,$$

and Paloheimo notes that this is simply the usual Petersen estimate based on pooling the data from all the catches. Inverting the above interval for $\tilde{\beta}_0$ gives the following $100(1 - \alpha)$ per cent confidence interval for N_0, namely

$$\frac{M_0 \Sigma \, n_i}{\Sigma \, m_i \pm t_{s-1}[\alpha/2](\tilde{\sigma}_0^2 \Sigma \, n_i)^{\frac{1}{2}}}.$$

In the same way we can obtain an estimate and confidence interval for $\Sigma \, n_i/N_0$, the rate of exploitation.

As a first step in examining the underlying assumptions of the above least-squares method we can plot y_i against i as a visual check on the constancy of M_i/N_i. If necessary, a test of $\beta = 0$ for the model $E[y_i] = \beta_0 + \beta i$ could be carried out using the theory of **1.3.5** (*1*). Also, by drawing the line

$y = \tilde{\beta}_0$, an examination of the deviation of each y_i from this line would provide a rough check on the reliability of the weights $w_i = n_i$. When there is
mortality taking place the above method can still be used, provided that the
mortality rates for marked and unmarked are the same, so that M_i/N_i remains
approximately constant. However, if recruitment and immigration into the
population are appreciable then the more complex models of Chapter 6 are
required. Sometimes the effect of recruitment can be eliminated, and Ricker
[1958: 86] gives several methods for fish populations based on growth data.
For example, if the population can be divided into age-groups which overlap
only slightly in length (or some other suitable measurement), then by
choosing the minimum length of fish to be marked (L, say) at the trough
between two age-groups, a boundary can be established whose position will
advance as the season progresses and the fish grow larger. In this way
recruitment into the marked length-range can be eliminated, and M_i/N_i will
remain constant for this particular section of the population, provided the
marked grow as much as the unmarked and suffer the same mortality rates.
Thus N_0, the initial size of the population of all fish longer than L, can be
estimated as above. Alternatively, if suitable length boundaries are not
available, the recruits can be eliminated from each sample by the method of
Robson and Flick [1965] described on pp. 74–78.

Example 3.14   Lobsters: Paloheimo [1963]

A full discussion of the lobster fishery is given in Paloheimo's paper
and the reader is referred there for details. The data exhibited in Table 3.12
represent just one season's tagging experiments (1953–4) based on a
release of 1000 tagged lobsters in an area centred at Port Maitland on the
Atlantic coast of Canada. As lobsters are relatively non-migratory, the
population of lobsters in this study area can be treated as an isolated unit.

TABLE 3.12

**Recapture data for lobsters at Port Maitland (1953–4):
from Paloheimo [1963: extracts from Tables 2 and 3].**

i	m_i	n_i ('00)	y_i ($\times 10^2$)	$\dfrac{m_i^2}{n_i}$	i	m_i	n_i ('00)	y_i ($\times 10^2$)	$\dfrac{m_i^2}{n_i}$
1	95	323	0·29	0·279 412	8	30	77	0·39	0·116 883
2	28	173	0·16	0·045 318	9	36	193	0·19	0·067 150
3	53	202	0·26	0·139 059	10	42	253	0·16	0·069 723
4	17	155	0·11	0·018 645	11	34	336	0·10	0·034 405
5	7	61	0·11	0·008 033	12	53	286	0·18	0·098 217
6	3	58	0·05	0·001 552	13	45	135	0·33	0·150 000
7	12	92	0·13	0·015 652	Sum	455	2344		1·044 049

Since $M_0 = 1000$, $s = 13$, we have from Table 3.12:

$$\tilde{N}_0 = 1000(234\ 400)/455 = 515\ 200,$$
$$12\hat{\sigma}_0^2 = 1 \cdot 044\ 049 - (455)^2/234\ 400 = 0 \cdot 160\ 837,$$

and a 95 per cent confidence interval for N_0 is given by

$$\frac{1000\ (234\ 400)}{455 \pm 2 \cdot 179(3141 \cdot 682)^{1/2}} \quad \text{or } (406\ 100,\ 704\ 200).$$

A plot of y_i versus i shows little trend, so that the assumption of con-stant M_i/N_i seems reasonable. However, for other study areas considered by Paloheimo there is often a strong trend in the y_i, particularly at the beginning of an experiment. This trend is put down to the lack of mixing of tagged and untagged rather than to the effects of tagging on the lobsters. The tagged lobsters when released are distributed more or less uniformly over the study area, while general observations indicate that the resident population on the other hand would be more concentrated on ledges and on rocky parts of the sea-bed close to their potential hiding places. This concentration of resident seems to be greater when the water is cold, and less when the water is warm and the lobsters more active. The differences, therefore, between the uniform distribution of tagged lobsters and the non-uniform distribution of the residen population at the start of the season would be more pronounced in the cold-than in warm-water areas. These differences would presumably disappear as the tagged lobsters gradually establish themselves on the bottom. Other sources of trend in the y_i would be immigration into the study area through the depletion of stock by the fishery, and recruitment to legal size during the season by the moulting and growth of undersized lobsters.

MARKED AND UNMARKED EQUICATCHABLE. Let p_i be the average prob-ability of catching an unmarked individual in the ith sample: then

$$E[u_i \mid M_i, N_i] = (N_i - M_i)p_i$$
$$= \left(U_0 - \sum_{j=1}^{i-1} u_j\right)p_i$$
$$= (U_0 - x_i)p_i, \quad \text{say.}$$

If we assume that the average probability of catching a marked individual in the ith sample is p_i/k (where k is constant from sample to sample), then, estimating this probability by m_i/M_i, we are led to consider the regression model (Marten [1970a])

$$E\left[\frac{u_i M_i}{m_i} \,\middle|\, x_i\right] \approx k(U_0 - x_i)$$

or, adjusting for bias,

$$E\left[\frac{u_i(M_i+1)}{m_i+1}\mid x_i\right] \approx k(U_0-x_i).$$

The constancy of k can be checked visually by plotting $u_i(M_i+1)/(m_i+1)$ against x_i, and both k and U_0 can be estimated using the method of **1.3.5**(3). We can also test the hypothesis $k = 1$ using the usual t-test for the slope of a linear regression.

CHAPTER 4

CLOSED POPULATION : MULTIPLE MARKING

4.1 SCHNABEL CENSUS

4.1.1 Notation

A simple extension of the Petersen method to a series of s samples of sizes $n_1, n_2, \ldots, n_s$ is the so-called Schnabel census (Schnabel [1938]). In this method each sample captured (except the first) is examined for marked members and then every member of the sample is given another mark before the sample is returned to the population. If different marks or tags are used for different samples, then the capture—recapture history of any animal caught during the experiment is known. In particular, if individual numbered tags are used then the same information is available if just the untagged members in each sample are tagged.

For the closed population (i.e. one in which immigration, death, etc., are negligible) a variety of theoretical models have been suggested, but before we discuss these we shall need some notation. Let

N = total population size,
s = number of samples,
n_i = size of the ith sample $(i = 1, 2, \ldots, s)$,
m_i = number of marked individuals in n_i,
u_i = $n_i - m_i$,

$$M_i = \sum_{j=1}^{i-1} u_j \quad (i = 1, 2, \ldots, s + 1)$$

= number of marked individuals in the population just before the ith sample is taken.

Since there are no marked animals in the first sample, we have $m_1 = 0$, $M_1 = 0$, $M_2 = u_1 = n_1$ and we define M_{s+1} ($= r$ say) as the total number of marked animals in the population at the end of the experiment, i.e. the total number of *different* animals caught throughout the experiment.

4.1.2 Fixed sample sizes

1   The generalised hypergeometric model

Let a_w be the number of animals with a particular capture history w, where w is a non-empty subset of the integers $\{1, 2, \ldots, s\}$: thus a_{124} represents those animals caught in the first, second and fourth samples only, also $r = \sum_w a_w$. If P_w, the probability that an animal chosen at random from the population has history w, is the same for each animal, and animals act

130

independently, the animals may be regarded as N independent "trials" from a multinomial experiment. Therefore the joint probability function of the random variables $\{a_w\}$ is

$$f(\{a_w\}) = \frac{N!}{\prod\limits_w a_w! \, (N-r)!} \, Q^{N-r} \prod_w P_w^{a_w} \, , \qquad (4.1)$$

where $Q = 1 - \sum\limits_w P_w$. We shall assume that:

(i) all individuals, irrespective of recapture history (if any), have the same probability p_i ($= 1 - q_i$) of being caught in the ith sample, and

(ii) for any individual the events "caught in the ith sample ($i = 1, 2, \ldots,$ s)" are independent. Then

$$Q = \prod_{i=1}^{s} q_i, \quad P_{124} = p_1 p_2 q_3 p_4 \ldots q_s = \frac{p_1 p_2 p_4 Q}{q_1 q_2 q_4} \, , \quad \text{etc.,}$$

and Darroch [1958] shows that (4.1) reduces to

$$f(\{a_w\}) = \frac{N!}{\prod\limits_w a_w! \, (N-r)!} \prod_{i=1}^{s} p_i^{n_i} \, q_i^{N-n_i} \, . \qquad (4.2)$$

From the above assumptions the $\{n_i\}$ are independent binomial variables, so that

$$f(\{n_i\}) = \prod_{i=1}^{s} \binom{N}{n_i} p_i^{n_i} \, q_i^{N-n_i}$$

and the joint probability function of the $\{a_w\}$, conditional on fixed sample sizes $\{n_i\}$ (i.e. the sample sizes are chosen in advance), is

$$f(\{a_w\} \mid \{n_i\}) = \frac{N!}{\prod\limits_w a_w! \, (N-r)!} \prod_{i=1}^{s} \binom{N}{n_i}^{-1} \, . \qquad (4.3)$$

By setting $\Delta \log f(\{a_w\} \mid \{n_i\}) = 0$ and using the fact that $\Delta \log N! = \log N$, etc., Darroch shows that the maximum-likelihood estimate $\hat{N}$ of N for the model (4.3) is the *unique* root (cf. **A7**), greater than r, of the $(s-1)$th degree polynomial given by

$$\left(1 - \frac{r}{N}\right) = \prod_{i=1}^{s} \left(1 - \frac{n_i}{N}\right). \qquad (4.4)$$

This equation has a very simple interpretation when it is noted that the left-hand side is equal to $\prod\limits_i (1 - u_i/(N - M_i))$, the product of the probabilities that an unmarked individual is not caught in the ith sample. The right-hand side represents the same product of probabilities, but now with respect to all individuals rather than just the unmarked.

The above equation (4.4) was first obtained by Chapman [1952], using a slightly different model; summing on the $\{a_w\}$ in each sample, the joint probability function of the $\{m_i\}$ is

$$f(m_2, \ldots, m_s \mid \{n_i\}) = \prod_{i=2}^{s} \binom{M_i}{m_i} \binom{N-M_i}{u_i} \bigg/ \binom{N}{n_i}$$

$$= \frac{\prod_{i=2}^{s} \binom{M}{m_i}}{\prod_{i=1}^{s} u_i!} \cdot \frac{N!}{(N-r)!} \prod_{i=1}^{s} \binom{N}{n_i}^{-1}. \qquad (4.5)$$

This product of hypergeometric distributions, derived directly by Chapman on the assumption that each sample is a simple random sample, leads to the same maximum-likelihood equation (4.4).

When $s = 2$, (4.4) is of first degree and we find that $\hat{N} = n_1 n_2 / m_2$, the Petersen estimate. For $s = 3$ we have the quadratic

$$N^2 (m_2 + m_3) - N(n_1 n_2 + n_1 n_3 + n_2 n_3) + n_1 n_2 n_3 = 0,$$

which can be readily solved for the larger root $\hat{N}$. When $s > 3$ we require some iterative method of solution, and three useful techniques are described in **1.3.8**(1). However, since we are only interested in finding $\hat{N}$ to the nearest integer, Robson and Regier's technique will be used as it is the easiest method for a desk calculator. Let

$$g(N) = \prod_{i=1}^{s} (1 - n_i/N)$$

and let $h(N) = N - r - Ng(N)$; then the ith step of the iteration is given by

$$N_{(i+1)} = N_{(i)} - h(N_{(i)})/\nabla h(N_{(i)}),$$

where $\quad \nabla h(N_{(i)}) = h(N_{(i)}) - h(N_{(i)} - 1)$

$$= 1 - N_{(i)} g(N_{(i)}) + (N_{(i)} - 1) g(N_{(i)} - 1).$$

To begin the iterations we require first of all a trial solution $N_{(1)}$ where $N_{(1)} > r$. If N is large, then expanding $g(N)$ in powers of $1/N$, neglecting powers greater than the second and using

$$\sum_{i=1}^{s} n_i - r = \sum_{i=2}^{s} m_i,$$

we find that (4.4) yields the approximate solution

$$N_B = \left(\sum_{i=1}^{s} \sum_{j=i+1}^{s} n_i n_j \right) \bigg/ \left(\sum_{i=2}^{s} m_i \right)$$

$$= R_2/m, \text{ say,}$$

where R_2 can be expressed in the form

$$R_2 = \frac{1}{2} \left[\left(\sum_{i=1}^{s} n_i \right)^2 - \sum_{i=1}^{s} n_i^2 \right].$$

However, if the cubic terms are retained, (4.4) reduces to the quadratic

$$N^2 m - N R_2 + R_3 = 0, \qquad (4.6)$$

where $R_3 = \sum_{i=1}^{s} \sum_{j=i+1}^{s} \sum_{k=j+1}^{s} n_i n_j n_k$

$$= \frac{1}{3}\left[\left(\sum_{i=1}^{s} n_i\right)\left(R_2 - \sum_{i=1}^{s} n_i^2\right) + \sum_{i=1}^{s} n_i^3\right]$$

and the desired solution of the quadratic is the larger root, N_A say. Chapman [1952] shows that under certain conditions, which are generally satisfied when N is much greater than $\Sigma\, n_i$, $N_A < \hat{N} < N_B$. In fact, if N_A and N_B are both greater than r, $\hat{N}$ will lie between these limits if $h(N_A) < 0$ and $h(N_B) > 0$. Chapman also gives another pair of numbers

$$N_C = \text{maximum } \{r,\; \underset{2 \leqslant i \leqslant s}{\text{minimum }} (n_i M_i/m_i)\},$$

$$N_D = \underset{2 \leqslant i \leqslant s}{\text{maximum }} (n_i M_i/m_i)$$

and in general this pair will be satisfactory, provided that no m_i (except m_1) is zero. We note that the methods discussed later in this chapter can also be used for providing a first approximation (cf. Example 4.7 on p. 144).

In solving (4.4) we see that the only recapture information required is r, the number of different animals caught during the experiment. This follows from the fact that r is a sufficient statistic for N and means that as far as the estimation of N is concerned, distinguishing marks are not needed for each sample. In fact, at each stage, we need only mark the unmarked members of the sample. However, as pointed out in **4.1.1**, if the tags have sufficient information (e.g. are numbered) then we can record all the recapture histories: this information is useful for testing some of the underlying assumptions (cf. **4.1.5**).

Using the model (4.3), Darroch [1958] proves that asymptotically (i.e. $N \to \infty$, $n_i \to \infty$ such that n_i/N remains constant)

$$E[\hat{N}] = N + b,$$

where b, the bias, is estimated by

$$\hat{b} = \frac{\left[\frac{s-1}{\hat{N}} - \Sigma\left(\frac{1}{\hat{N}-n_i}\right)\right]^2 + \left[\frac{s-1}{\hat{N}^2} - \Sigma\left(\frac{1}{\hat{N}-n_i}\right)^2\right]}{2\left[\frac{1}{\hat{N}-r} + \frac{s-1}{\hat{N}} - \Sigma\left(\frac{1}{\hat{N}-n_i}\right)\right]^2},$$

and the asymptotic variance of $\hat{N}$ is estimated by

$$v[\hat{N}] = (\hat{N}-r)/h'(\hat{N})$$

$$= \left[\frac{1}{\hat{N}-r} + \frac{s-1}{\hat{N}} - \Sigma\left(\frac{1}{\hat{N}-n_i}\right)\right]^{-1},$$

where all summations are for $i = 1, 2, \ldots, s$. Obviously the last step in the Newton–Raphson method (which requires $h'(\hat{N})$, cf. **1.3.8**(1)) can be used for evaluating $v[\hat{N}]$. Also $\nabla h(\hat{N}) \approx h'(\hat{N})$, so that the last step of Robson and Regier's method will provide a reasonable approximation for $v[\hat{N}]$. It can be shown that

$$\hat{b} = -\tfrac{1}{2}(\hat{N}-r)h''(\hat{N})/[h'(\hat{N})]^2$$

so that $\hat{b}$ can be approximated by

$$-\tfrac{1}{2}(\hat{N}-r)\nabla^2 h(\hat{N})/[\nabla h(\hat{N})]^2.$$

Although this approximation may not be very accurate it does at least indicate the order of magnitude of $\hat{b}$.

Assuming $\hat{N}$ to be asymptotically normal, we have the approximate 95 per cent confidence interval for N, namely

$$\hat{N} - \hat{b} \pm 1\cdot96\sqrt{v[\hat{N}]}, \tag{4.7}$$

where $\hat{b}$ can be neglected if it is less than one-tenth of $\sqrt{v}$ (Cochran [1963: 12]). However, the statistic r is more nearly normally distributed than $\hat{N}$, and Darroch [1958: 348] shows that we can use r as a basis for a confidence interval as follows.

The expected value of r, regarded as a function of N, is

$$\rho(N) = N - \left(\prod_{i=1}^{s}(N-n_i)\right)/N^{s-1},$$

and equation (4.4) is simply $\rho(\hat{N}) = r$. The variance of r, expressed as a function of N, is approximately

$$\sigma^2(N) = (N-\rho)^2\left[\frac{1}{N-\rho} + \frac{S-1}{N} - \sum_{i=1}^{s}\left(\frac{1}{N-n_i}\right)\right]$$

and we have $\begin{aligned}0\cdot95 &\approx \Pr[r - 1\cdot96\,\sigma(N) < \rho(N) < r + 1\cdot96\,\sigma(N)] & (4.8)\\ &\approx \Pr[r - 1\cdot96\,\sigma(\hat{N}) < \rho(N) < r + 1\cdot96\,\sigma(\hat{N})]\\ &= \Pr[r_1 < \rho(N) < r_2]\\ &= \Pr[\rho^{-1}(r_1) < N < \rho^{-1}(r_2)]\\ &= \Pr[N_1 < N < N_2], \quad\text{say,} & (4.9)\end{aligned}$

since $\rho(N)$ is a monotonic increasing function of N. The confidence limits N_1 and N_2 can be calculated by setting $h(N) = \rho(N) - r_i$ and solving $h(N) = 0$ iteratively as above. Alternatively we could deal with the interval (4.8) directly by solving the equations $r \pm 1\cdot96\,\sigma(N) = \rho(N)$ iteratively on a computer.

RANDOM SAMPLE SIZES. We now mention briefly the more realistic situation in which the sample sizes n_i are random variables rather than fixed parameters. Darroch [1958] has investigated this model (namely 4.2) in some detail and shows that, as far as the point and interval estimation of N is concerned, there is no difference (asymptotically) between the two cases of fixed and random sample size. The reason for this is that the maximum-likelihood estimate $\hat{N}$ is the same in both cases, and in estimating the variance of $\hat{N}$, one effectively replaces $E[n_i]$ by n_i when n_i is random (cf. **3.1.3**).

Example 4.1   Cricket frog (*Acris gryllus*): Turner [1960a]

From the data in Table 4.1 we have $r = 87 + (41 - 36) = 92$ and $(N_C, N_D) = (92, 99)$. We note in passing that $N_B = 150$, but N_A is not applicable as (4.6) does not have real roots. Choosing $N_{(1)} = 97$ as our first approximation

TABLE 4.1
Capture–recapture data for a population of cricket frogs: from Turner [1960a: Table 3].

Sample	n_i	m_i	M_i	$n_i M_i / m_i$
1	32	–	–	–
2	54	18	32	96
3	37	31	68	81
4	60	47	74	94
5	41	36	87	99

and using Robson and Regier's iterative method, we have

$$N_{(1)} - r = 5$$
$$N_{(1)}g(N_{(1)}) = (97-32)(97-54)(97-37)(97-60)(97-41)/97^4$$
$$= 3 \cdot 924\ 965\ 0,$$
$$(N_{(1)} - 1)g(N_{(1)} - 1) = 3 \cdot 697\ 103\ 0,$$
$$\nabla h(N_{(1)}) = 1 + 3 \cdot 697\ 103 - 3 \cdot 924\ 965\ 0 = 0 \cdot 772\ 138\ 0$$
$$h(N_{(1)}) = 5 - 3 \cdot 924\ 965\ 0 = 1 \cdot 075\ 035\ 0$$

and the correction is

$$-h(N_{(1)})/\nabla h(N_{(1)}) = -1 \cdot 075\ 035\ 0/0 \cdot 772\ 138\ 0 = -1 \cdot 4.$$

Therefore our next approximation is

$$N_{(2)} = N_{(1)} - 1 = 96$$

from which we calculate

$$N_{(2)} - r = 4,$$
$$N_{(2)}g(N_{(2)}) = 3 \cdot 697\ 103\ 0,$$
$$(N_{(2)} - 1)g(N_{(2)} - 1) = 3 \cdot 476\ 320\ 4,$$
$$\nabla h(N_{(2)}) = 1 + 3 \cdot 476\ 320\ 4 - 3 \cdot 697\ 103\ 0 = 0 \cdot 779\ 217\ 4,$$
$$h(N_{(2)}) = 4 - 3 \cdot 697\ 103\ 0 = 0 \cdot 302\ 897$$

and the correction is

$$-0 \cdot 302\ 897/0 \cdot 779\ 217\ 4 = -0 \cdot 39.$$

Therefore, to the nearest integer, $N_{(3)} = N_{(2)}$ and $\hat{N} = 96$. Also

$$(N_{(2)} - r)/\nabla h(N_{(2)}) = 4/0 \cdot 779\ 217\ 4 = 5 \cdot 13$$

which is close to

$$v[\hat{N}] = 1/0 \cdot 189\ 32 = 5 \cdot 282.$$

Since

$$\nabla^2 h(\hat{N}) \approx \nabla^2 h(\hat{N} + 1)$$
$$= \nabla h(N_{(1)}) - \nabla h(N_{(2)})$$
$$= 0 \cdot 772\ 138\ 0 - 0 \cdot 779\ 217\ 4$$
$$= -0 \cdot 007\ 079,$$
$$\hat{b} \approx \tfrac{1}{2} \times 4 \times 0 \cdot 007\ 079/(0 \cdot 779\ 217\ 4)^2 = 0 \cdot 02,$$

indicating that the bias of $\hat{N}$ is negligible. The asymptotic confidence interval (4.7) is then found to be (91·5, 100·5).

We note that

$$1\cdot96 \; \sigma(\hat{N}) \;=\; 1\cdot96 \times 4(0\cdot18932)^{\frac{1}{2}} \;=\; 3\cdot412,$$

so that $r_1 = 88\cdot41$ and $r_2 = 95\cdot41$. Since $\rho(91) = 89\cdot3$, $\rho(90) = 88\cdot5$, $\rho(102) = 96\cdot29$, $\rho(101) = 95\cdot8$ and $\rho(100) = 95\cdot35$, we have, from (4.8), $(N_1, N_2) = (90, 100)$.

2 Samples of size one

The special case of sampling one at a time was considered by Craig [1953b] for the study of butterfly populations. Putting $n_i = 1$ in equation (4.4), we find that $\hat{N}$ is the solution of

$$\left(1 - \frac{r}{N}\right) \;=\; \left(1 - \frac{1}{N}\right)^s, \tag{4.10}$$

and this equation can be solved in much the same way as (4.4). However, since s is generally large, Craig suggests taking logarithms and solving

$$H(N) \;=\; (s-1) \log_{10} N + \log_{10} (N-r) - s \log_{10} (N-1) \;=\; 0,$$

using a good table of logarithms (e.g. Spenceley $et\;al.$ [1952]) and a suitable first approximation N_0 such as the following. Let f_x be the frequency of cases in which the same individual is caught x times ($x = 1, 2, \ldots$) and let $s_2 = \Sigma\; x^2 f_x$; then Craig suggests

$$N_0 \;=\; s^2/(s_2 - s).$$

Alternatively, following Darroch [1958: 349] and letting $N \to \infty$, $s \to \infty$, subject to s/N ($= D$, say) remaining constant, (4.10) becomes

$$1 - (r/N) \;=\; e^{-D}.$$

A first approximation to $\hat{N}$ (in fact an upper bound) is then $N_0' = s/D_0'$ where D_0', the solution of

$$(1 - e^{-D})/D \;=\; r/s \quad (= a \text{ say}) \tag{4.11}$$

is obtained by linear interpolation in Table **A3**. Samuel [1969] suggests a further approximation $N_0'' = s/D_0''$, where D_0'' is the solution of (4.11) with $a = r/(s + D_0')$.

For the limiting process mentioned above, Darroch [1958: 34] shows that asymptotically

$$E[\hat{N}] \;=\; N + b,$$

where $\quad b \;=\; \frac{1}{2}D^2(e^D - 1 - D)^{-2},$

and

$$V[\hat{N}] \;=\; N(e^D - 1 - D)^{-1}.$$

Both b and $V[\hat{N}]$ can be estimated by replacing D by $\hat{D} = s/\hat{N}$: extensive tables of e^D are given, for example, in Becker and Van Orstrand [1924], and Comrie [1959].

Confidence limits for N based on r can be calculated as in (4.8) above, using

$$\rho(N) = N(1-e^{-D})$$

and

$$\sigma^2(N) = Ne^{-2D}(e^D - 1 - D).$$

If the interval (4.9) is used, we have to solve two equations of the form $\rho(N) = r_i$ or

$$(1-e^{-D})/D = r_i/s$$

which, as for the case $r_i = r$ above (equation 4.11), can be solved for D by interpolating linearly in Table **A3**.

Example 4.2   Butterflies (*Colias eurytheme*): Craig [1953b]

From the data in Table 4.2 we have $r = 341$, $s = 435$, $s_2 = 645$ and $N_0 = (435)^2/(645-435) = 901$. Using $a = r/s = 0{\cdot}7839$ and interpolating linearly in

TABLE 4.2
Capture—recapture data for samples of size one from a population of butterflies: from Craig [1953b].

x	1	2	3	$\geqslant 4$	Total
f_x	258	72	11	0	341
xf_x	258	144	33	0	435
x^2f_x	258	288	99	0	645

Table **A3**, we find that $D_0' = 0{\cdot}5084$ and $\hat{N} < N_0' = 435/(0{\cdot}5084) = 856$. Also $r/(s+D_0') = 0{\cdot}7830$ so that $D_0'' = 0{\cdot}5110$ and $N_0'' = 851$. Since

$$H(851) = -0{\cdot}000\ 222 < 0,$$

where $H(N) = 434 \log_{10} N + \log_{10} (N-341) - 435 \log_{10} (N-1)$,

we have that $851 < \hat{N} < 856$.

To solve $H(N) = 0$ for $\hat{N}$ we shall use Robson and Regier's iterative method (cf. **1.3.8**(*1*)), starting with the trial solution $N_{(1)} = 851$. To get the next approximation we calculate

$$H(850) = -0{\cdot}000\ 329,$$
$$\nabla H(851) = H(851) - H(850) = 0{\cdot}000\ 107,$$

and the correction is

$$-H(851)/\nabla H(851) = 0{\cdot}000\ 222/0{\cdot}000\ 107 \approx 2,$$

so that $N_{(2)} = 851 + 2 = 853$. Successive iterations yield $N_{(3)} = N_{(4)} = 854$, so that to the nearest integer $\hat{N} = 854$. If we use Craig's first approximation $N_{(1)} = 901$, the successive approximations are $N_{(2)} = 867$, $N_{(3)} = 853$, $N_{(4)} = N_{(5)} = 854$.

Now $$\hat{D} = 435/854 = 0{\cdot}509\ 36,$$
$$\exp(\hat{D}) - 1 - \hat{D} = 0{\cdot}1549,$$

$$\hat{b} = \frac{1}{2}\left(\frac{0.509\ 36}{0.1549}\right)^2 = 5,$$

$$\hat{V}[\hat{N}] = 854/0.1549 = 5513,$$

and the approximate 95 per cent confidence interval (4.7) is 849 ± 146 or (703, 995). If we use (4.9) then we require

$$\sigma^2(\hat{N}) = 854(0.600\ 88)^2(0.1549) = 47.76,$$

$$r_1 = r - 1.96\sigma(\hat{N}) = 327.5,$$

$$r_2 = r + 1.96\sigma(\hat{N}) = 354.5,$$

$$r_1/s = 0.7529, \quad r_2/s = 0.8149,$$

and $(D_1, D_2) = (0.597, 0.424)$ where D_i is the solution of (4.11) when $r = r_i$. Hence the confidence interval (4.9) is

$$\left(\frac{435}{0.597}, \frac{435}{0.424}\right) \quad \text{or} \quad (729, 1026).$$

3 Mean Petersen estimate

At each stage of the sampling a modified Petersen estimate of N can be calculated, namely (cf. **3.1.1**)

$$N_i^* = \frac{(M_i+1)(n_i+1)}{(m_i+1)} - 1 \quad (i = 2, 3, \ldots, s)$$

with variance estimate

$$v_i^* = \frac{(M_i+1)(n_i+1)(M_i-m_i)(n_i-m_i)}{(m_i+1)^2(m_i+2)}.$$

Therefore a natural estimate of N, suggested by Chapman [1952: 293], is the average

$$\bar{N} = \Sigma N_i^*/(s-1).$$

(All summations throughout this section are for $i = 2, 3, \ldots, s$). Since the covariances of the N_i^* are asymptotically negligible compared with their variances, we have approximately

$$V[\bar{N} \mid \{n_i, M_i\}] = \Sigma V[N_i^* \mid n_i, M_i]/(s-1)^2.$$

This can be estimated by either

$$v^* = \Sigma v_i^*/(s-1)^2$$

which is approximately unbiased if and only if each v_i^* is almost unbiased, or by (cf. 1.9)

$$v[\bar{N}] = \Sigma (N_i^* - \bar{N})^2/(s-1)(s-2),$$

which is almost unbiased when the N_i^* have the same mean. When these conditions for unbiasedness are not satisfied, both estimates are conservative in that they tend to overestimate the true variance. For a numerical example using the above approach the reader is referred to p. 144 (Example 4.7).

A more efficient way of averaging the N_i^* is to take a weighted average with weights inversely proportional to the variances (cf. **1.3.2**). This produces an estimate with the same variance as the maximum-likelihood estimate $\hat{N}$ (Darroch [1958: 354] proves this for the case when the n_i are random variables). However, as the weights themselves have to be estimated, bias would be introduced, which $\bar{N}$ aims to avoid. Therefore, although $\bar{N}$ is less efficient than $\hat{N}$, it is almost unbiased (under certain reasonable conditions, cf. **3.1.1**), and we would expect $\bar{N}$ to have some degree of robustness with regard to departures from the assumptions underlying the Schnabel census.

4 Schnabel's binomial model

An alternative approach to the Schnabel census can be made by assuming that the M_i are fixed parameters (which they are, conditionally, at each sampling) and then using the binomial approximation of **3.1.2**. This leads to Schnabel's [1938] model (cf. (4.5))

$$f(m_2, \ldots, m_s \mid \{n_i, M_i\}) = \prod_{i=2}^{s} \binom{n_i}{m_i} \left(\frac{M_i}{N}\right)^{m_i} \left(1 - \frac{M_i}{N}\right)^{n_i - m_i}, \qquad (4.12)$$

and the maximum-likelihood estimate of N is now the appropriate root of

$$\sum_{i=2}^{s} \left\{\frac{(n_i - m_i)M_i}{N - M_i}\right\} = \sum_{i=2}^{s} m_i. \qquad (4.13)$$

This model assumes that at each stage, n_i/N is sufficiently small (say less than $0 \cdot 1$) for one to ignore the complications of sampling without replacement. If each M_i/N is also small, a first approximation to the solution of (4.13) is

$$N' = \left(\sum_{i=2}^{s} n_i M_i\right) \bigg/ \left(\sum_{i=2}^{s} m_i\right) = \lambda/m, \quad \text{say.}$$

We note that, irrespective of any assumptions concerning the probability function of the m_i or the magnitudes of the various parameters, N' has a certain intuitive appeal, being simply a weighted average of the Petersen estimates $n_i M_i/m_i$.

When n_i/N and M_i/N are both less than, say, $0 \cdot 1$ for each i, a modification of N' which is almost unbiased is

$$N'' = \lambda/(m + 1).$$

In fact, from Chapman [1952: 293] we have

$$E[N'' \mid \{n_i, M_i\}] = N(1 - \exp(-\lambda/N))$$

and (correcting a misprint in Chapman's equation)

$$V[N'' \mid \{n_i, M_i\}] \approx N^2 \left(\frac{N}{\lambda} + 2\frac{N^2}{\lambda^2} + 6\frac{N^3}{\lambda^3}\right).$$

These formulae were derived by Chapman on the basis that, when the above conditions hold, m_i is approximately distributed as Poisson with parameter $M_i n_i/N$. Hence the sum, m, of independent Poisson variables is also Poisson with parameter λ/N. Actually a study of Raff [1956: Table 4] would suggest that the Poisson approximation still applies, even if $0\cdot1 < M_i/N < 0\cdot2$, provided that n_i/N is much less than $0\cdot1$, so that the hypergeometric distribution of m_i is well approximated by the binomial. But the errors in these approximations have an accumulative effect on the sum m, so that for m to be approximately Poisson we could not have more than one or two samples with M_i/N greater than $0\cdot1$.

Assuming N'' to be asymptotically normal, we can calculate a confidence interval for N in the usual manner. However, as in the Petersen method (cf. **3.1.4**), it is recommended to base confidence intervals on the distribution of m. For $m \leqslant 50$ we can use Chapman's Poisson table (Table **A1**) to obtain the shortest interval for N/λ, and hence for N. When $m > 50$ we can use the normal approximation to the Poisson, and a 95 per cent confidence interval for N is given by the roots of the quadratic

$$\frac{(m - \lambda/N)^2}{\lambda/N} = 1\cdot96^2,$$

namely

$$(N/\lambda) = \frac{2m + 1\cdot96^2 \pm 1\cdot96\sqrt{(4m + 1\cdot96^2)}}{2m^2}. \tag{4.14}$$

If the Poisson approximation to the binomial is not satisfactory, the variance of m will be less than λ/N, and the above normal confidence interval will be too wide. In this case λ/N is replaced by a sum of binomial variances, and (4.14) becomes

$$(N/\lambda) = \frac{2m + 1\cdot96^2(1-\delta) \pm 1\cdot96\sqrt{(1-\delta)[4m + 1\cdot96^2(1-\delta)]}}{2m^2} \tag{4.15}$$

where $\delta = \sum_{i=2}^{s} n_i M_i^2/(\lambda N')$.

Example 4.3

Consider the data in the first half of Table 4.3, opposite. Here $m = 15$, $\lambda = 22\,000$ and $N'' = 22\,000/16 = 1375$. Entering Table **A1** with $m = 15$, a 95 per cent confidence interval for N/λ is $(0\cdot0365, 0\cdot111)$ and the corresponding interval for N is $(803, 2442)$. This interval is very wide because of the small number of recaptures.

COMPARING TWO POPULATIONS. Chapman and Overton [1966] consider the problem of comparing two populations of sizes N_a and N_b respectively. Suppose a Schnabel census is carried out in each population, yielding $\lambda_a, m_a, \lambda_b$ and m_b as the respective values of λ and m. Then, provided m_a and m_b can be regarded as Poisson random variables, the methods described in **3.6.2** can be used here to test the hypothesis H_0 that $N_a = N_b$.

Example 4.4

Artificial data for two hypothetical populations are set out in Table 4.3. We have $m_a = 15$, $m_b = 14$, $m_a + m_b = 29$, $\lambda_a = 22\,000$, $\lambda_b = 27\,765$ and $p_0 = 22\,000/49\,765 = 0\cdot4421$. To test H_0 against the two-sided alternative

$N_a \neq N_b$ we can use the unit normal statistic

$$z = \frac{|15 - 29(0 \cdot 4421)| - \frac{1}{2}}{\sqrt{29(0 \cdot 4421)(0 \cdot 5579)}} = 0 \cdot 63 \, ,$$

which is not significant at the 5 per cent level of significance.

4.1.3   Regression methods

The maximum-likelihood method described in **4.1.2**(*1*) will give the most efficient estimate $\hat{N}$ of N, provided the assumptions underlying the model are

TABLE 4.3
Artificial capture—recapture data for two hypothetical populations.

n_i	m_i	N_a M_i	$n_i M_i$	n_i	m_i	N_b M_i	$n_i M_i$
70	—	—	—	204	—	—	—
43	3	70	3 010	90	9	204	18 360
75	6	110	8 250	33	5	285	9 405
60	6	179	10 740				
Total	15		22 000	Total	14		27 765

satisfied. $\hat{N}$ will, however, tend to be sensitive to departures from the underlying assumptions, particularly those relating to constant N and the random behaviour of marked animals. Therefore, in practice, $\hat{N}$ may sometimes be an inefficient estimate and $v[\hat{N}]$ may be unreliable. For this reason less efficient but more robust estimates of N, like $\bar{N}$ of **4.1.2**(*3*), are desirable. In particular a useful regression method has been suggested by Schumacher and Eschmeyer [1943] and we now discuss this technique in detail.

1   Schumacher and Eschmeyer's method

In Schnabel's model (4.12) each m_i is assumed to be binomially distributed, so that $y_i = m_i/n_i$ has mean M_i/N and variance

$$\sigma_i^2 = \frac{M_i}{N} \left(1 - \frac{M_i}{N} \right) \cdot \frac{1}{n_i} .$$

We may therefore write

$$y_i = \beta M_i + e_i \quad (i = 2, 3, \ldots, s) ,$$

where $\beta = 1/N$ and the "error" e_i has mean zero and variance σ_i^2. If we plot y_i against M_i, the plotted points should lie approximately on a straight line of slope β passing through the origin. Since the variance of e_i is not constant, the least-squares fitting of a straight line should be done using weights w_i, say, as in **1.3.5**(*1*). Thus $\tilde{N}$, the least-squares estimate of N, is given by

$$1/\tilde{N} = \tilde{\beta} = (\Sigma \, w_i y_i M_i)/(\Sigma \, w_i M_i^2) ,$$

where all summations throughout this section are for $i = 2, 3, \ldots, s$. If the weights are chosen in the usual manner, namely proportional to the reciprocal of the variances, then the above equation becomes the maximum-likelihood equation (4.13). However, although these weights will give the most efficient estimate of N when sampling is truly random, we are computationally no better off than before, as these unknown weights have to be estimated iteratively. De Lury [1958] also points out that "owing to the tendency of fishes to stratify, and for other reasons that lead to similar effects, the proportion of marked individuals available to the sampling at any one time is likely to differ widely from the 'true' proportion, and the weights are therefore likely to be seriously wrong. In these circumstances, weighting by sample size alone is preferable to weighting according to proportions tagged." In support of this last statement we note that $(M_i/N)[1-(M_i/N)]$ does not vary much as M_i/N varies from $0\cdot2$ to $0\cdot8$. Therefore, putting $w_i = n_i$, $\tilde{\beta}$ is now given by

$$\tilde{\beta} = (\Sigma\, m_i M_i)/(\Sigma\, n_i M_i^2), \tag{4.16}$$

which is equivalent to the formula given by Schumacher and Eschmeyer [1943]. This formula was also given by Hayne [1949b], so that the method is sometime called Hayne's method. The mean and variance of $\tilde{N}$ could be calculated using the delta method (**1.3.3**) and an approximate confidence interval for N obtained in the usual manner. However, following De Lury [1958] it seems preferable to assume that the e_i's are independently normally distributed with variances σ^2/n_i, and to invert the confidence interval for β given in **1.3.5** (*1*). Hence a $100(1-\alpha)$ per cent confidence interval for N is given by

$$\frac{\Sigma\, n_i M_i^2}{\Sigma\, m_i M_i \pm t_{s-2}[\alpha/2](\tilde{\sigma}^2\Sigma\, n_i M_i^2)^{\frac{1}{2}}}, \tag{4.17}$$

where $(s-2)\tilde{\sigma}^2 = \Sigma\,\dfrac{m_i^2}{n_i} - \dfrac{(\Sigma\, m_i M_i)^2}{\Sigma\, n_i M_i^2}.$

From bead sampling experiments, De Lury showed that the above confiden interval compared favourably with the confidence interval based on the more efficient binomial weights. We would also expect (4.17) to be robust with regard to departures from the underlying assumptions, and this model should therefore be used in conjunction with the other methods mentioned so far in this chapter. In particular a graph is always a useful indicator of any marked departures from the assumptions underlying the model.

In conclusion we mention a small point concerning the number of degrees of freedom used in the above theory. Some authors (e.g. Hayne [1949b], Ricker [1958]) include the first sample in the theory, so that the point $(0, 0)$ is used in the regression analysis. In this case the number of degrees of freedom should be $s - 1$ rather than $s - 2$. However, as y_1 is always zero when $M_1 = 0$, y_1 is not strictly a random observation and for this reason is not included in the above theory.

Example 4.5   Red-ear sunfish (*Lepomis microlophus*): Ricker [1958: 103]

The data and calculations are set out in Table 4.4. We have

$$\tilde{N} = 970\ 296/2294 = 423\ ,$$

$$\tilde{\sigma}^2 = \tfrac{1}{12}\{7 \cdot 7452 - (2294)^2/970\ 296\} = 0 \cdot 1935\ ,$$

and (4.17) becomes

$$\frac{970\ 296}{2294 \pm 2 \cdot 179 [0 \cdot 1935(970\ 296)]^{\frac{1}{2}}}$$

TABLE 4.4

Capture—recapture data for a population of sunfish: from Ricker [1958: 103].

i	n_i	m_i	y_i	M_i	$n_i M_i$	$m_i M_i$	$n_i M_i^2$	m_i^2/n_i
1	10	—	—	—	—	—	—	—
2	27	0	0	10	270	0	2 700	0
3	17	0	0	37	629	0	23 273	0
4	7	0	0	54	378	0	20 412	0
5	1	0	0	61	61	0	3 721	0
6	5	0	0	62	310	0	19 220	0
7	6	2	0·33	67	402	134	26 934	0·6667
8	15	1	0·07	71	1 065	71	75 615	0·0667
9	9	5	0·56	85	765	425	65 025	2·7778
10	18	5	0·28	89	1 602	445	142 578	1·3889
11	16[2]	4	0·25	102	1 632	408	166 464	1·0000
12	5	2	0·40	112	560	224	62 720	0·8000
13	7[1]	2	0·29	115	805	230	92 575	0·5714
14	19	3	0·16	119	2 261	357	269 059	0·4737
Total		24			10 740	2294	970 296	7·7452

[2,1] Number of deaths in this sample.

or (300, 719). A cursory glance at Table 4.4 shows that there is no obvious linear relation between y_i and M_i, but any trend could have been masked by the wide fluctuations of the values of y_i due to the small numbers of recaptures m_i.

For comparison, the Schnabel estimates are given by (cf. **4.1.2**(4))

$$N' = 10\ 740/24 = 448\ ,$$

and

$$N'' = 10\ 740/25 = 430\ .$$

Although $m < 50$, $M_i/\tilde{N} > 0 \cdot 1$ for most samples, so that m will not have a Poisson distribution. It would seem, therefore, that the normal approximation (4.15) is appropriate here, and this yields a 95 per cent confidence interval for N of

(314, 639). We note that the confidence intervals are wide as the number of recaptures in each sample is small.

Example 4.6   Cricket frog (*Acris gryllus*): Turner [1960a]

Using the data of Example 4.1 (p. 135) we find that the plot of y_i versus M_i is approximately linear and the regression estimate is $\tilde{N} = 93$ with 95 per cent confidence interval (82, 108). This may be compared with the Schnabel method: $N' = 93$, (85, 101), and, from Example 4.1, the maximum-likelihood method: $\hat{N} = 96$, (90, 100).

Example 4.7   Cricket frog (*Acris crepitans*): Pyburn [1958]

A population of cricket frogs was studied to determine the population si and to obtain information about the movements of the frogs. Cricket frogs commonly occur about the margins of permanent and semi-permanent areas of water and are usually found in large numbers where conditions are favourabl The study site was centred round a pond $112' \times 79'$ which was staked at in- tervals of 10 feet around its circumference so that the points of capture coul be located. The sampling method was to proceed around the pond from a give point, attempting to take frogs at random. Each frog was marked by toe clip- ping, located according to the nearest stake, and released at the point of capture. Six times round the pond constituted a "sample", and the data from six consecutive "samples" are given in Table 4.5.

TABLE 4.5

Capture—recapture data for a population of cricket frogs: from Pyburn [1958: Table 1].

Date of capture	n_i	m_i	M_i	N_i^*	v_i^*
Sept. 19–Sept. 21	109	—	—	—	—
Sept. 22–Sept. 25	133	15	109	920·3	37 568
Sept. 28–Oct. 2	138	30	227	1021·3	21 927
Oct. 7–Oct. 9	72	23	335	1021·0	26 041
Oct. 11–Oct. 14	134	47	384	1081·8	13 498
Oct. 16–Oct. 21	72	33	471	1012·4	14 548
Total		148		5056·8	113 582

The assumptions underlying the Schnabel census seemed to be satisfied throughout the experiment. For example, the pond was completely isolated fr any other source of water, and the fact that only 15 of the 510 caught were 15 mm or less indicated that the recruitment of young frogs was small. The method of marking by toe clipping did not seem to affect the behaviour of the frogs, and the good fit of the data to a straight line (Fig. 4.1) would suggest that marking did not affect their catchability.

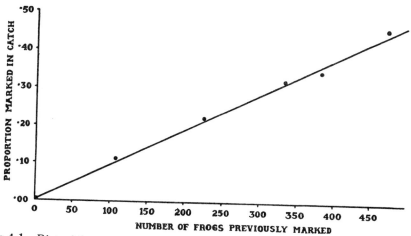

Fig 4.1 Plot of the proportion marked in the sample (y_i) versus the number of frogs previously marked (M_i): from Pyburn [1958].

From Table 4.5 we find that $r = 471 + (72-33) = 510$, $\widetilde{N} = 1056$ and the 95 per cent regression confidence interval for N is (1012, 1104). The Schnabel estimate is $N' = 1049$, and (4.15) leads to the confidence interval (957, 1149). Using $\widetilde{N}$ as a first approximation, (4.4) can be solved iteratively using the method of Robson and Regier to give $\hat{N} = 1052$ and a confidence interval (cf. 4.7) of (919, 1181). From **4.1.2**(3) we have the mean estimate

$$\overline{N} = 5056 \cdot 8/5 = 1011 ,$$

with variance estimate

$$v^* = 113\ 582/25 = 4543 ,$$

and $\overline{N} \pm 1 \cdot 96\sqrt{v^*}$ gives the interval (879, 1143). Also

$$\sum_{i=2}^{s} (N_i^* - \overline{N})^2 = \Sigma N_i^{*2} - (\Sigma N_i^*)^2/(s-1) = 13\ 347 ,$$

so that

$$v[\overline{N}] = 13\ 347/20 = 6673$$

and $\overline{N} \pm 1 \cdot 96 \sqrt{(v[\overline{N}])}$ leads to (851, 1171).

The general agreement of the above confidence intervals indicates that the assumptions underlying the various models are approximately satisfied.

2   Tanaka's model

Sometimes a plot of y_i versus M_i, as in the previous section, yields a graph which is definitely curved. For this situation Tanaka [1951, 1952] has proposed a non-linear relationship of the form $y = (M/N)^\gamma$ or, taking logarithms, the linear regression model

$$E[-\log_{10} y_i] \approx \gamma(\log_{10} N - \log_{10} M_i), \quad (i = 2, 3, \ldots, s) .$$

Least-squares estimates and confidence intervals for γ and $\theta = \log_{10} N$ can be obtained by setting $Y_i = -\log_{10} y_i = \log_{10} (n_i/m_i)$, $x_i = \log_{10} M_i$ and using the methods of **1.3.5**(3). A visual estimate of θ can also be obtained by drawing the regression line by eye and extending this line to meet the x-axis. However, before actually looking up the logarithms it is simpler to plot y_i versus M_i on log–log paper first.

The parameter γ can be interpreted as an index of trap response (e.g. Tanaka and Teramura [1953]). For example, if $\gamma < 1$, $E[m_i/n_i] > M_i/N$ or, rearranging, $E[m_i/M_i] > E[n_i/N]$, and the marked individuals have a higher probability of capture than the unmarked. However, care should be exercised in interpreting the graph of y_i versus M_i as several interpretations are possible (e.g. Tanaka [1951: 452], Hayne [1949b: 407] and Davis [1963: 111]). If, for example, the graph curves downwards, Hayne argues that the fall-off in the proportion of marked in the sample could be due to the immigration of unmarked animals into the trapping area. But if the graph is interpreted in the light of Tanaka's model we have $\gamma < 1$ and the curvature is due to the marked animals having a higher probability of capture than the unmarked. In this case the fall-off is simply due to the curve settling down to its "correct" position instead of dropping away from its "correct" position as suggested by Hayne. Obviously both interpretations are possible, and one could perhaps distinguish between the two by an analysis of the recaptures to see whether any individuals were being recaptured more often than expected (cf. **4.1.5**). Alternatively Marten's regression model discussed below may be applicable.

Example 4.8   Red-backed vole (*Clethrionomys smithi*): Tanaka [1951]

Tanaka studied populations and home ranges of voles and mice in a bush area near the summit of Mt Ishizuchi in central Shikoku. Thirty-two small live-traps were set in a grid pattern spaced 15 metres apart, and the traps were baited with peanuts and sweet potatoes. Unfortunately some individuals escaped because of a deficiency in the door apparatus, and a number of animals were found dead in the traps. The voles were numbered using toe clipping, and in Table 4.6 we have the capture–recapture data for a red-backed vole population.

TABLE 4.6

Capture–recapture data for a population of red-backed voles: from Tanaka [1951: Table 2].

Day (i)	$n_i - m_i$	m_i	n_i	M_i	y_i (m_i/n_i)	Y_i $\log_{10}(n_i/m_i)$	x_i $\log_{10} M_i$
1	5[1]	—	5	—	—	—	—
2	5[1]	2	7	4	0·29	0·544 07	0·6021
3	7[1]	3	10	8	0·30	0·522 84	0·9031
4	5	7	12	14	0·58	0·234 01	1·1461
5	5[1]	11	16	19	0·69	0·162 86	1·2788
6	5[1]	12	17	23	0·71	0·151 37	1·3617
7	4	14	18	27	0·78	0·109 24	1·4314

[1]One death.

From (4.16) we find that the Schumacher–Eschmeyer regression estimate of N is $\tilde{N} = 21$ with a 95 per cent confidence interval $\left(\text{cf. } (4.17)\right)$ of $(27, 37)$. This method is obviously unsatisfactory both graphically (Fig. 4.2) and

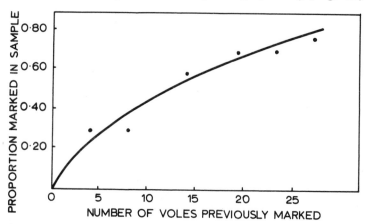

Fig 4.2   Plot of the proportion marked in the sample (y_i) versus the number of voles previously marked (M_i): from Tanaka [1951].

mathematically, as the initial population size is at least $r = 36$. However, a plot of Y_i versus x_i (Fig. 4.3) is fairly linear, thus indicating that Tanaka's method can be used. The steps in the calculations are as follows:

$$\bar{Y} = 0{\cdot}287\ 40, \qquad \bar{x} = 1{\cdot}120\ 52,$$
$$\Sigma\,(Y_i - \bar{Y})^2 = 0{\cdot}189\ 888, \qquad \Sigma\,(x_i - \bar{x})^2 = 0{\cdot}496\ 467,$$
$$\Sigma\,(Y_i - \bar{Y})(x_i - \bar{x}) = \Sigma\,Y_i\,(x_i - \bar{x}) = -0{\cdot}293\ 474.$$

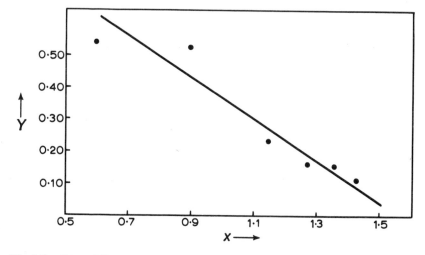

Fig 4.3   Plot of Y_i $(= \log_{10}(n_i/m_i))$ versus x_i $(= \log_{10} M_i)$ for a population of red-backed voles: redrawn from Tanaka [1951].

Setting $w_i = 1$ in **1.3.5** (3), we have

$$\hat{\gamma} = 0\cdot293\ 474/0\cdot496\ 467 = 0\cdot591\ 125,$$
$$\hat{\theta} = \log_{10} \hat{N} = \bar{x} + (\bar{Y}/\hat{\gamma}) = 1\cdot606\ 71,$$
$$\hat{N} = 40,$$
$$\hat{\sigma}^2 = \tfrac{1}{4}\{\Sigma\ (Y_i - \bar{Y})^2 - [\Sigma\ Y_i(x_i - \bar{x})]^2/\Sigma\ (x_i - \bar{x})^2\}$$
$$= 0\cdot004\ 101\ 9,$$
$$t_4[0\cdot025] = 2\cdot776,$$
$$\hat{\gamma}^2 - A = \hat{\gamma}^2 - \hat{\sigma}^2 (2\cdot776)^2/\Sigma\ (x_i - \bar{x})^2 = 0\cdot285\ 769,$$
$$2\bar{Y}\hat{\gamma} = 0\cdot339\ 776,$$
$$\bar{Y}^{2\cdot} - B = \bar{Y}^2 - \hat{\sigma}^2 (2\cdot776)^2/6 = 0\cdot077\ 329\ 4,$$

and the quadratic to be solved, namely (1.17), is

$$0\cdot285\ 769\ d^2 - 0\cdot339\ 776\ d + 0\cdot077\ 329\ 4 = 0.$$

This has roots $0\cdot8823$, $0\cdot3067$, and adding $\bar{x}$ to each root gives ($1\cdot4272$, $2\cdot0028$) as the 95 per cent confidence interval for θ, or (27, 100) for N. We note from (1.18) that

$$\hat{V}[\hat{\theta}] \approx \frac{\hat{\sigma}^2}{\hat{\gamma}^2}\left\{\frac{1}{s} + \frac{(\hat{\theta} - \bar{x})^2}{\Sigma\ (x_i - \bar{x})^2}\right\}$$

$$= 0\cdot007\ 546,$$

and $\hat{\theta} \pm 1\cdot96\sqrt{\hat{V}[\hat{\theta}]}$ is ($1\cdot436$, $1\cdot777$), which leads to the confidence interval (27, 60) for N. However, as s is small, this confidence interval is of doubtfu validity and the wider interval based on t_4 [$0\cdot025$] = $2\cdot776$ is probably more reliable. One should expect a wide interval as the recaptures m_i are small.

The 95 per cent confidence interval for γ, namely

$$\hat{\gamma} \pm t_{s-2}[\alpha/2]\ \sqrt{\hat{\sigma}^2/\Sigma\ (x_i - \bar{x})^2},$$

is ($0\cdot3388$, $0\cdot8434$). As this interval does not contain the value 1, γ is signi cantly less than unity, which, interpreted in the light of Tanaka's model, suggests the presence of trap addiction among the marked animals.

There were five deaths during the experiment through trapping, which w be expected to have some effect on the Schumacher–Eschmeyer regression model. In particular, with decreasing N the expected value of y_i would incre from M_i/N to $M_i/(N - D_i)$, where D_i is the total number of deaths before the i sample. This increase, ranging from about 3 per cent for $i = 2$ to about 14 pe cent for $i = 7$, would have the effect of curving the regression line upwards. As the reverse is true, it would seem that the degree of trap addiction is eve greater than that indicated by the above analysis.

Example 4.9   Meadow vole (*Microtus pennsylvanicus*): Hayne [1949b]

Table 4.7 gives the capture–recapture data for a population of adult fe- male meadow voles trapped in a field in East Lansing, Michigan. It is found

TABLE 4.7

Capture—recapture data for a population of meadow voles: from Hayne [1949b: Table 2].

Date of capture	n_i	m_i	M_i	y_i (m_i/n_i)	Y_i $\log_{10}(n_i/m_i)$	x_i $\log_{10} M_i$
July 19 p.m.	8	—	—	—	—	—
20 a.m.	19	0	8	0	—	—
20 p.m.	10	2[1]	27	0·20	0·698 97	1·431 36
21 a.m.	23	8	34	0·35	0·458 64	1·531 48
21 p.m.	9	0	49	0·00	—	—
22 a.m.	14	9	58	0·64	0·192 01	1·763 43
22 p.m.	9	7[1]	63	0·78	0·109 24	1·799 34
23 a.m.	21	13	64	0·62	0·208 17	1·806 18

[1]One death

that the Schumacher—Eschmeyer regression estimate of N is $\tilde{N} = 106$ with a 95 per cent confidence interval of (78, 165). However, the graph of y_i versus M_i (Fig. 4.4) seems to curve upwards. Therefore, ignoring the points for which $y_i = 0$ and replotting the remaining five points on log—log paper, we obtain approximate linearity (Fig. 4.5), thus indicating that Tanaka's model can be

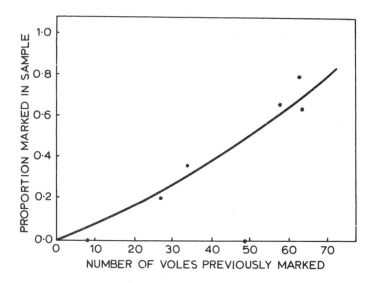

Fig 4.4   Plot of the proportion marked in the sample (y_i) versus the number of voles previously marked (M_i): from Hayne [1949b]. The above freehand curve is drawn to demonstrate the upward trend in the data.

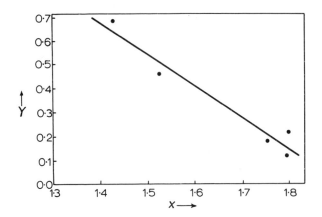

Fig 4.5   Plot of Y_i (= $\log_{10}(n_i/m_i)$) versus x_i(= $\log_{10}M_i$) for a population of meadow
voles: data from Hayne [1949b].

used. Using the same technique as outlined in the previous example, we find
that N is now estimated as 81 with a confidence interval of (66, 126). Also
$\hat{\gamma} = 1 \cdot 36$ with a confidence interval for γ of ($0 \cdot 79$, $1 \cdot 94$). This indicates that
γ is not significantly different from unity.

A further application of Tanaka's method is given in Pearson [1955: 251
spadefoot toad].

3   Marten's model

One of the difficulties in using Tanaka's regression model above is the
interpretation of the parameter γ. Although one may obtain a good straight-li
fit to the graph of Y_i versus x_i, the model lacks a simple "physical" interpre
tation and other regression curves may give just as good a fit. For example,
suppose that the average catchability of the m_i marked individuals in each
sample bears a constant ratio to the average catchability of the u_i unmarked
members, then we have approximately

$$\frac{m_i}{M_i} = \frac{u_i}{k(N-M_i)} \quad \left(= \frac{n_i}{k(N-M_i) + M_i} \right)$$

and

$$E[y_i|M_i] \approx \frac{M_i}{k(N-M_i) + M_i}.$$

This means that the plot of y_i versus M_i will be curved upwards or downward
depending on whether k is less than or greater than unity. Instead of fitting
Tanaka's model we can rearrange the above equation, apply a bias correction
and obtain the linear regression model

$$E[Y_i|M_i, n_i] \approx k(N-M_i), \tag{4.1}$$

where $Y_i = u_i(M_i + 1)/(m_i + 1)$. This model, first suggested by Marten [1970a]
can be analysed using the methods of **1.3.5**(3) (with $\theta = N$, $\gamma = k$, $x_i = M_i$) t

150

obtain point and interval estimates of N, and to test $k = 1$. Marten points out that although the plot may not be linear, there may be a subclass M'_i, say, whose average catchability relative to the unmarked animals remains constant. If m'_i members of this subclass are caught in the ith sample and

$$Y'_i = u_i(M'_i + 1)/(m'_i + 1),$$

we can plot Y'_i versus M_i (not M'_i): an example of this is given below.

Seber [1970a] derives (4.18) from the general theory of **3.2.2** (2), and such a derivation throws some light on the nature of k. Using the notation of **3.2.2**(2), we define x_j as the probability that the jth member of the population ($j = 1, 2, \ldots, N$) is caught at least once before the ith sample, and y_j, z_j now refer to the conditional probabilities of the jth member being caught in the ith sample, given that it is caught or not caught, respectively, before the ith sample, i.e., in terms of **3.2.2**(2), the first $i - 1$ samples are regarded as the "first" sample and the ith sample as the "second". Therefore, setting $n_1 = M_i$, $u_2 = u_i$, $m_2 = m_i$, and defining Y_i as above, we have from (3.15)

$$E[Y_i \,|\, M_i, \, n_i] \approx k_i(N - M_i),$$

which reduces to (4.18) if $k_i = k$. The plot of Y_i versus M_i will indicate whether k_i can be regarded as approximately constant and will also show up any marked heterogeneity of variance. In many cases the variances of the Y_i will not be too different. For example, since the variance of Y_i depends on the product $M_i n_i$ (cf. (3.2)), the variance of Y_i will be constant if, at each stage, n_i is chosen so that $M_i n_i$ has some predetermined constant value; methods for doing this are discussed in **3.1.5**.

If $k \neq 1$ we see from the discussion in **3.2.2**(2) that we cannot be sure whether this departure is due to marking affecting future catchability, or to variation in the inherent catchability of individuals, or both. When there is a variable catchability the more catchable individuals are caught first, resulting in marked animals becoming, on the average, more catchable than the unmarked. If marking does not affect catchability, then excluding the unlikely case when the catchability of an individual before every sample is uncorrelated with its catchability after the sample, a test of $k = 1$ is a test of constant catchability. On the other hand, if the inherent catchability is constant over the population, it will not remain constant (except for the unmarked) if marking affects catchability. In this case one way of separating the two effects is to consider a regression using the subclass $M'_i = u_{i-1}$ of individuals first caught in the $(i-1)$th sample. Defining x_j to be the probability that the jth member of the population is first caught in the $(i-1)$th sample, then, if the unmarked are equicatchable, we would expect x_j to be constant over the population. This implies from p. 87 that a test of $k = 1$ is then a test that, for each sample, the average catchability of the marked members from u_{i-1} in the ith sample is the same as the (constant) catchability of the unmarked in the ith sample.

Example 4.10   Tide-pool snail (*Polinices duplicatus*): Marten [1970a]

Marten used data from Hunter and Grant [1966] on tide-pool snails to illustrate the above regression technique. From Table 4.8 a plot of Y_i versus

TABLE 4.8

Capture—recapture data for tide-pool snails: from Marten [1970a: Table 1].

i	u_i	m_i	M_i	Y_i	m'_i	M'_i	Y'_i
1	142	—	—	—	—	—	—
2	129	3	142	4611	—	—	—
3	122	23	271	1383	19	139	854
4	99	58	393	661	35	248	685
5	94	67	492	682	63	335	494
6	147	273	586	315	219	425	285

M_i showed considerable scatter, thus indicating that the entire class of marked snails was not suitable for applying the regression method. However, Hunter and Grant found that after handling, the snails tended to burrow into the ground and remain immobile for about 24 hours. This meant that in any particular sample there were fewer individuals than expected from those caught in the previous sample. Therefore, choosing the subgroup M'_i consisting of all marked individuals except those caught in the $(i-1)$th sample, Marten found the regression of Y'_i on M_i to be closely linear (Fig. 4.6). Hence, using Fieller's technique of **1.3.5**(*3*), N is estimated to be 756 with a 95 per cent confidence interval of (670, 923). Also $\hat{k} = 1\cdot81$ with a 95 per cent confidence interval of ($1\cdot21$, $2\cdot41$). As this interval does not contain unity, we reject the hypothesis that $k = 1$ and conclude that marked snails are under-represented in the sample.

4.1.4   Allowing for known removals

1   The hypergeometric model

In many population experiments there are accidental deaths due to trapping and handling, and some animals may be deliberately removed for further study. If the percentage of such removals is appreciable, some allowance must be made for them in the particular model used. For example, the removal could form a major part of the sample as in commercially exploited populations, with the remainder of the sample being tagged (or retagged) and returned to the population. Suppose, then, that d_i members of the ith sample are not returned to the population and let M_i be the number of marked animals alive in the population before the ith sample is taken. Then, assuming the n_i to be fixed parameters, Chapman's model (4.5) now becomes

$$f(m_2,\dots, m_s \mid \{n_i\}) = \prod_{i=2}^{s} \binom{M_i}{m_i}\binom{N-M_i-D_i}{n_i-m_i} \Big/ \binom{N-D_i}{n_i},$$

where $D_i = \sum_{j=1}^{i-1} d_j$ (the total removal up to but not including the ith sample), and N is now the *initial* population size. It is readily shown that N_D, the maximum-likelihood estimate of N, is the unique root greater than r of the polynomial

$$\frac{N-r}{N} = \prod_{i=1}^{s} \left\{ \frac{N - D_i - n_i}{N - D_i} \right\} \qquad (4.19)$$

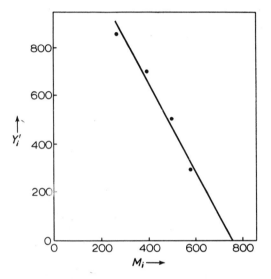

Fig 4.6 Application of Marten's regression model to a population of tide-pool snails: modified from Marten [1970a].

where r is the total number of different animals caught during the whole experiment, *including* the ones not returned. By setting $g(N)$ equal to the right-hand side of (4.19) and defining $h(N) = N - r - Ng(N)$, (4.19) can be solved in exactly the same way as (4.4). If N is large, so that each n_i/N is less than about 0.1, a reasonable first approximation of (4.19) is (Robson and Regier [1968: 138])

$$N_{(1)} = \frac{[\Sigma (n_i + D_i)]^2 - \Sigma (n_i + D_i)^2 - (r + \Sigma D_i)^2 + r^2 + \Sigma D_i^2}{2 \Sigma m_i}, \qquad (4.20)$$

where all summations are for $i = 1, 2, \ldots, s$ ($m_1 = 0$, $D_1 = 0$). This approximation is obtained by cross-multiplying in (4.19), dividing both sides by N^s, expanding the products, and neglecting powers of $1/N$ greater than the second. Another possible first approximation is the mean estimate

$$\bar{N}_D = \sum_{i=2}^{s} (N_i^* + D_i)/(s-1), \qquad (4.21)$$

153

where N_i^* is the modified Petersen estimate $[(n_i + 1)(M_i + 1)/(m_i + 1)] - 1$.

The mean and variance of $\hat{N}_D$ can be evaluated using the method outlined in Darroch [1958: 346–7]. It transpires that, asymptotically,

$$E[\hat{N}_D] = N + b,$$

where b, the bias, is estimated by

$$\hat{b} = \frac{\left[\frac{1}{\hat{N}_D} + \sum_{i=1}^{s}\left(\frac{1}{\hat{N}_D - D_i - n_i} - \frac{1}{\hat{N}_D - D_i}\right)\right]^2 - \left[\frac{1}{\hat{N}_D^2} + \sum_{i=1}^{s}\left(\frac{1}{(\hat{N}_D - D_i - n_i)^2} - \frac{1}{(\hat{N}_D - D_i)^2}\right)\right]}{2\left[\frac{1}{\hat{N}_D - r} - \frac{1}{\hat{N}_D} - \sum_{i=1}^{s}\left(\frac{n_i}{(\hat{N}_D - D_i - n_i)(\hat{N}_D - D_i)}\right)\right]^2}$$

and the asymptotic variance of $\hat{N}_D$ is estimated by

$$v[\hat{N}_D] = (\hat{N}_D - r)/h'(\hat{N}_D)$$

$$= \left[\frac{1}{\hat{N}_D - r} - \frac{1}{\hat{N}_D} - \sum_{i=1}^{s}\frac{n_i}{(\hat{N}_D - D_i - n_i)(\hat{N}_D - D_i)}\right]^{-1}.$$

Since

$$\hat{b} = -\tfrac{1}{2}(\hat{N}_D - r)h''(\hat{N}_D)/[h'(\hat{N}_D)]^2,$$

the bias and the variance of $\hat{N}_D$ can be estimated from the last steps in the Robson–Regier iterative procedure as for the case of no removals (cf. **4.1.2** (1)); a numerical example is given in Robson and Regier [1968: 134–138].

2   Overton's method

Overton [1965] has given the following method for modifying Schnabel's estimate N' (p. 139) to allow for known removals. Since

$$E[m_i \mid \{M_i, n_i, D_i\}] = M_i n_i/(N - D_i), \quad (i = 2, 3, \ldots, s),$$

then summing this equation for $i = 2, 3, \ldots, s$ and setting $m = \sum_{i=2}^{s} m_i$ leads to

$$E[m \mid \{M_i, n_i, D_i\}] = \frac{1}{N} \frac{\sum n_i M_i (N - D_i + D_i)}{N - D_i}$$

$$= \frac{\sum n_i M_i}{N} + \frac{\sum n_i M_i D_i}{N(N - D_i)}.$$

Equating m to its expected value leads to an estimate N'_D of N given by

$$N'_D = \frac{\sum n_i M_i}{m} + \frac{\sum n_i M_i D_i}{(N'_D - D_i)m} \tag{4.22}$$

$$= N' + A, \quad \text{say},$$

where A is to be added to the usual Schnabel estimate N'. Equation (4.22) must be solved iteratively for N'_D, and Overton suggests the following first approximation which will usually be close unless the removal is heavy, namel

$$N_{(1)} = N' + A_{(1)},$$

where $\qquad A_{(1)} = \Sigma\, n_i M_i D_i / (mN')$

$\qquad\qquad\qquad = (\Sigma\, n_i M_i D_i)/(\Sigma\, n_i M_i)\,,$

so that $N_{(1)} < N_D'$. Another first approximation, suggested by Robson and Regier [1968: 143, $D_s < 0\cdot 1\, N$], is obtained from (4.22) directly by neglecting D_i in the denominator of the right-hand side and solving for N_D', namely

$$\tfrac{1}{2}\{N' + \sqrt{N'^2 + 4\, \Sigma\, n_i M_i D_i / m}\}\,.$$

Whichever first approximation is used, however, subsequent approximations are

$$N_{(j)} = N' + A_{(j)}\,,$$

where $\qquad A_{(j)} = \Sigma\, n_i M_i D_i / (N_{(j)}^{*} - D_i)m \qquad\qquad\qquad (4.23)$

and $N_{(j)}^{*}$ is to be determined by $N_{(j-1)}$. In determining a suitable method for choosing $N_{(j)}^{*}$, Overton points out that the iterative process is not necessarily convergent if we set $N_{(j)}^{*} = N_{(j-1)}$. But if $N_{(j)}^{*} < N_D'$, then $N_{(j)} > N_D'$ and vice versa, so that N_D' will be between any pair $N_{(j)}^{*}$, $N_{(j)}$. Overton therefore sugges the reasonable procedure of choosing

$$N_{(j+1)}^{*} = \tfrac{1}{2}(N_{(j)}^{*} + N_{(j)})$$

and taking $N_{(2)}^{*}$, as the integer nearest to $N_{(1)}$.

Confidence limits for N can be obtained as in **4.1.2** (4) on p. 140, using $\lambda = \Sigma\, n_i M_i + \Sigma\, n_i M_i D_i / (N_D' - D_i)$ (which can be obtained from the last step of (4.23)) and either Chapman's table (Table **A1**, $m \leqslant 50$) or the normal approximation (4.14); (4.15) can also be used with

$$\delta = \frac{N_D'}{\lambda} \sum_{i=2}^{s} \left\{ \frac{n_i M_i^2}{(N_D' - D_i)^2} \right\}. \qquad\qquad (4.24)$$

Example 4.11 Hypothetical data: Overton [1965]

In Table 4.9, d_i consists of the members of n_i permanently removed from the population, $a_i (= n_i - m_i - d_i)$ is the *net* increase in the marked population after the ith sample,

$$D_i = \sum_{j=1}^{i-1} d_j \quad \text{and} \quad M_i = \sum_{j=1}^{i-1} a_j\,.$$

Now

and $\qquad\qquad N' = 6231/16 = 389\cdot 4$

$$N_{(1)} = N' + A_{(1)}$$

$$\qquad = \frac{6231}{16} + \frac{362\,781}{6231} = 447\cdot 6.$$

Taking the nearest integer $N_{(2)}^{*} = 448$, and substituting in (4.23), leads to

TABLE 4.9

Hypothetical data for demonstrating Overton's iterative procedure for estimating population size when there are known removals: from Overton [1966: Table 1].

i	n_i	m_i	d_i	a_i	M_i	D_i	n_iM_i	$n_iM_iD_i$	$N^*_{(2)'}-D_i$	$\dfrac{n_iM_iD_i}{N^*_{(2)}-D_i}$
1	20	–	10	10	–	0	–	–	448	0
2	22	1	11	10	10	10	220	2 200	438	5
3	18	0	10	8	20	21	360	7 560	427	18
4	21	1	11	9	28	31	588	18 228	417	44
5	16	2	7	7	37	42	592	24 864	406	61
6	18	1	11	6	44	49	792	38 808	399	97
7	17	3	9	5	50	60	850	51 000	388	131
8	14	2	8	4	55	69	770	53 130	379	140
9	17	2	8	7	59	77	1003	77 231	371	208
10	16	4	7	5	66	85	1056	89 760	363	247
					(71)	(92)				
Total		16	92				6231	362 781		951

$$N_{(2)} = \frac{6231}{16} + \frac{951}{16} = 448 \cdot 875.$$

From the discussion above we now have $448 < N'_D < 448 \cdot 875$ and normally it would be sufficient to choose N'_D equal to 448 or 449. Overton, however, takes the calculations a step further to determine which integer N'_D is nearer. Thus, choosing $N^*_{(3)} = 448 \cdot 5$, it transpires that $N_{(3)} = 448 \cdot 86$ and $448 \cdot 5 < N'_D < 448 \cdot 86$, so that, to the nearest integer, $N'_D = 449$.

To calculate an approximate confidence interval for N we require

$$\lambda = \Sigma\, n_iM_i + \Sigma\, n_iM_iD_i/(N'_D - D_i)$$
$$\approx 6231 + 951 = 7182.$$

Entering $m = 16$ in Chapman's table (Table A1), we obtain 95 per cent limits $(0 \cdot 0350,\ 0 \cdot 1020)$ for N/λ, or (251, 733) for N. However, as five of the values $M_i/(N^*_{(2)}-D_i)$ are greater than $0 \cdot 1$ (cf. p. 140), m will no longer be strictly Poisson. Therefore calculating $\delta = 0 \cdot 1271$ from (4.24), (4.15) leads to the approximate 95 per cent confidence interval for N (285, 707): if δ is neglected (4.14) leads to the interval (276, 729).

3 Regression methods

All the regression methods discussed in 4.1.3 depend heavily on the assumption of N remaining constant, and these methods cannot be used unless the removal is negligible. However, the mean estimate $\bar{N}_D$ given above by (4.21) will provide a robust estimate of N, provided the number of recaptures in each sample is not too small (cf. **3.1.1**). The variance of $\bar{N}_D$ can be estimated by (cf. (1.9))

$$v[\bar{N}_D] \;=\; \sum_{i=2}^{s} (N_i^* + D_i - \bar{N}_D)/(s-1)(s-2),$$

or, when the assumptions underlying the Petersen method hold for each sample, more efficiently by

$$v_D^* \;=\; \sum_{i=2}^{s} \left\{ \frac{(M_i+1)(n_i+1)(M_i-m_i)(n_i-m_i)}{(m_i+1)^2\,(m_i+2)(s-1)^2} \right\}.$$

4.1.5 Testing the underlying assumptions

1 Validity of the models

MULTINOMIAL MODEL. Using the notation of **4.1.2**(*1*) we wish to test the hypothesis H that the probabilities P_w in (4.1) can be written as products of the $\{p_i\}$ and $\{q_i\}$ so that (4.1) reduces to Darroch's multinomial model (4.2). From the theory of **1.3.6**(*3*), an appropriate goodness-of-fit test statistic for H is therefore given by

$$T \;=\; \sum_{w} (a_w - \hat{N}\hat{P}_w)^2/(\hat{N}\hat{P}_w),$$

where, for example, $\hat{P}_{124} = \hat{p}_1\hat{p}_2\hat{p}_4\hat{Q}/(\hat{q}_1\hat{q}_2\hat{q}_4)$ and $\hat{p}_i = 1 - \hat{q}_i = n_i/\hat{N}$. When H is true, T is asymptotically distributed as chi-squared with $d - s - 1$ degrees of freedom, where d is the number of different recapture histories w. If any of the groups a_w are too small (some of them may be zero) they can be pooled in the usual manner.

HYPERGEOMETRIC MODEL. Chapman [1952: 295] has suggested a non-parametric test for the validity of the model (4.3) using the random variables $\{b_{ij}\}$, where b_{ij} $(i < j)$ is the number of marked in the jth sample which were first caught and marked in the ith. When the sampling is random we have

$$E[b_{ij}/n_j \mid n_j, u_i] \;=\; u_i/N \;(= \theta_i, \text{ say}), \tag{4.25}$$

and an array

$$\frac{b_{12}}{u_1 n_2}, \quad \frac{b_{13}}{u_1 n_3}, \quad \frac{b_{14}}{u_1 n_4}, \dots, \frac{b_{1s}}{u_1 n_s}$$

$$\frac{b_{23}}{u_2 n_3}, \quad \frac{b_{24}}{u_2 n_4}, \dots, \frac{b_{2s}}{u_2 n_s}$$

$$\cdots \qquad \cdots$$

$$\frac{b_{s-1,s}}{u_{s-1} n_s}$$

may be formed, in which each element is a random variable with expectation $1/N$. These random variables are independent within each row, but are dependent between rows as $b_{i_1 j}$, $b_{i_2 j}$ belong to the same sample and are therefore correlated. For large N, however, the correlation is small and Chapman suggests testing for the validity of the underlying model by testing whether the $t = \frac{1}{2} s(s-1)$ elements formed by putting the rows one after another is a sequenc of random observations from a common distribution. The test suggested is th sign test of Moore and Wallis [1943] based on D, the number of negative sig in the sequence of successive differences of observations (i.e. first observation minus the second, etc.). When the hypothesis of a common distributio is true, $E[D] = \frac{1}{2}(t-1)$, $\sigma^2[D] = \frac{1}{12}(t+1)$, and $(D-E[D])/\sigma[D]$ is approximate unit normal for $t \geqslant 12$; Moore and Wallis have tabled the exact probability distribution of D for small values of t.

Chapman points out that in many cases the alternatives to randomness a essentially one-sided. For example, possible alternatives are:

(a) some marked individuals die off more rapidly or disappear, so that they a not available for sampling;
(b) the marked individuals disperse from the tagging location slowly and are more likely to be recaptured in the samples taken soon after marking tha later; and
(c) the population size N is increasing through recruitment.

If any of these alternatives is true, the numbers in each row of the array wil tend to decrease from left to right. In this case a test based on the whole ar as a single sequence has the following defect: in each row the probability o a negative difference between successive elements is less than $\frac{1}{2}$, but the probability of a negative difference between the last element of any row and the first element of the next row will be much greater than $\frac{1}{2}$. Also another disadvantage of considering the whole array as a single sequence is that th variances of the elements will vary from row to row. However, if the sample sizes n_i are approximately the same, the elements within a given row will h approximately the same distribution when the underlying model is valid. The fore, to overcome the above objections, Chapman recommends treating each row separately so that the array may be considered as $(s-1)$ sequences of observations decreasing in length from $(s-1)$ to 1. A test of randomness ma then be made using the statistic

$$X = D_1 + D_2 + \ldots + D_{s-2},$$

where D_i is the number of negative differences in row i (no difference is obtained from the last row). Then

$$E[X] = \tfrac{1}{4}(s-1)(s-2), \quad \sigma^2[X] = \tfrac{1}{4}(s+3)(s-2),$$

and X is asymptotically normal. A partial tabulation of the distribution of X (reproduced from Chapman [1952]) for $s = 5, 6, 7, 8$ is given in Table 4.10.

An alternative test can be carried out using Table 4.11. Neglecting the complications of sampling without replacement, the columns of Table 4.11 represent independent multinomial distributions. Let p_{ij} be the probability c being in the class containing b_{ij} individuals, then from (4.25), $p_{ij} = \theta_i$ ($j = i+1, \ldots, s$) and the likelihood function for Table 4.11 is proportional

158

$$\prod_{j=2}^{s}\left\{\left(\prod_{i=1}^{j-1}\theta_i^{\,b_{ij}}\right)(1-\theta_1-\theta_2-\ldots-\theta_{j-1})^{\,u_j}\right\}$$
$$=\theta_1^{\,b_1\cdot}\,\theta_2^{\,b_2\cdot}\ldots(1-\theta_1)^{u_2}(1-\theta_1-\theta_2)^{u_3}\ldots\,.$$

The maximum-likelihood estimates $\hat{\theta}_i$ of θ_i are then solutions of the equations

$$\frac{b_{1\cdot}}{\theta_1}-\frac{u_2}{(1-\theta_1)}-\frac{u_3}{(1-\theta_1-\theta_2)}-\cdots-\frac{u_s}{(1-\theta_1-\theta_2-\ldots-\theta_{s-1})}=0$$

$$\frac{b_{2\cdot}}{\theta_2}-\frac{u_3}{(1-\theta_1-\theta_2)}-\cdots-\frac{u_s}{(1-\theta_1-\theta_2-\ldots-\theta_{s-1})}=0$$

$$\cdot\quad\cdot\quad\cdot\quad\cdot\quad\cdot\quad\cdot\quad\cdot\quad\cdot\quad\cdot\quad\cdot\quad\cdot\quad\cdot$$

$$\frac{b_{s-1\cdot}}{\theta_{s-1}}-\frac{u_s}{(1-\theta_1-\theta_2-\ldots-\theta_{s-1})}=0$$

TABLE 4.10
Cumulative distribution of X: from Chapman [1952: Table 2].

				$\Pr[X \leqslant x]$				
s \ x	0	1	2	3	4	5	6	7
5	0·0035	0·0590	0·3056	0·6944	0·9410	0·9965	1·0000	1·0000
6	–	0·0012	0·0172	0·1052	0·3392	0·6608	0·8948	0·9828
7	–	–	0·0001	0·0020	0·0166	0·0627	0·2010	–
8	–	–	–	–	0·0001	0·0009	0·0323	0·1103

TABLE 4.11
Contingency table for carrying out Leslie's test for dilution.

	b_{12}	b_{13}	b_{14}	$\ldots$	b_{1s}	Total $b_{1\cdot}$
		b_{23}	b_{24}	$\ldots$	b_{2s}	$b_{2\cdot}$
			b_{34}	$\ldots$	b_{3s}	$b_{3\cdot}$
				$\ldots$		$\ldots$
					$b_{s-1,s}$	b_{s-1}
	u_2	u_3	u_4	$\ldots$	u_s	$u_\cdot$
Total	n_2	n_3	n_4	$\ldots$	n_s	$n_\cdot$

and the expected frequencies corresponding to the observed frequencies b_{ij} are $n_j\hat{\theta}_i$; the expected frequencies for the u_j are obtained by subtraction. The goodness-of-fit statistic based on comparing the observed frequencies with the expected frequencies in Table 4.11 is chi-squared with $(s-1)(s-2)/2$ degrees of freedom.

Unfortunately the above maximum-likelihood equations do not seem to have explicit solutions, and alternative estimates of the θ_i are desirable. For example, if there were no blanks in Table 4.11 the estimate of θ_i would be $b_{i\cdot}/n_\cdot$; Leslie

[1952: 385] uses this estimate in his so-called test for "dilution" (immigratio and recruitment). Another problem that arises in the use of the above test is t the expected frequencies are often small and pooling may be needed. Leslie *et al.* [1953] suggest pooling the b_{ij} (and their expected frequencies) in each column, thus reducing the table to $s - 1$ pairs of frequencies (m_j, u_j).

The above method can still be used when mortality is taking place, provi that all subclasses of marked and unmarked have the same mortality rates between successive samples. In this case the proportion of the population first marked at the ith sample will remain constant and equal to u_i/N, so that (4.25) is still satisfied.

Examples of Leslie's test are given in Leslie *et al.* [1953], Turner [1960a], Dunnet [1963: 92], Krebs [1966: 243] and Delong [1966].

REGRESSION MODEL. As far as the Schumacher–Eschmeyer regression method (cf. **4.1.3(1)**) is concerned, the best evidence for its validity is obviously the linearity of the graph. Any change in N through recruitment, mortality, etc., or any variation in catchability will affect the basic equation $E[m_i/n_i] = M_i/N$, and this in turn will show up in the graph, provided the m_i are not too small (say greater than 10). In practice, point and interval estimates of N should be obtained using as many different methods as possible, as any departures from underlying assumptions will usually affect different models in different ways. A substantial agreement among the estimates would then give qualitative support for the validity of the models concerned. If mortality, recruitment, etc. are definitely affecting the population then the recapture data should be analysed using the general methods of Chapter 5.

2   Tests for random sampling

Apart from poor experimentation and inadequate experimental design, there are three basic sources of non-randomness:

(a) There may be subcategories in the population due to size, sex, species, etc. for which the sampling is random within each subcategory but not between the subcategories (e.g. Takahasi [1961]). In this case, if there is no mortality, the chi-squared goodness-of-fit test based on the contingency table 3.3 on page 71 can be applied to each sample (except the first) using the pairs m_{ix}, $M_{ix} - m_{ix}$. As pointed out by Robson and Regier [1968], these $(s-1)$ chi-squares are independent and can be added together to give a total chi-square.

(b) Catching and handling may affect catchability, so that marked and unmarked have different probabilities of capture in a given sample. However, in some populations, once an animal has been caught its catchability remains fairly constant irrespective of future recaptures (e.g. Young *et al.* [1952]). In this case the ratio k of the average probability of capture of an unmarked animal to the average probability of capture of a marked may remain approximately constant from sample to sample, so that Marten's regression method

$(\mathbf{4.1.3}\,(3))$ can be used for testing $k = 1$.

(c) If the catching and handling go on affecting the catchability of marked individuals after their first capture, then the sampling will not be random within the marked population. Such non-randomness can be detected using the following technique of Leslie [1958], based on the frequency of recapture of individuals. This technique has also been examined by Carothers [1971].

Suppose a multiple-recapture experiment consisting of t samples is carried out in a closed population containing an identifiable group of animals, and let G denote both the group and the number in the group. If g_j of this group are caught in the jth sample $(j = 1, 2, \ldots, t)$, then, on the assumption of simple random sampling, the probability P_j $(= 1 - Q_j)$ that an individual member of G bears the recovery mark j is g_j/G. Suppose a particular member of G is caught x times, then from Kendall and Stuart [1969: 126–7] we have

$$E[x \mid \{g_j\}] = \sum_{i=1}^{t} P_j = \mu, \quad \text{say,}$$

$$V[x \mid \{g_j\}] = \sum_{i=1}^{t} P_j Q_j$$

$$= \mu - \sum_{j=1}^{t} g_i^2/G^2$$

$$= \sigma^2, \quad \text{say.}$$

If f_x members of G are caught x times, then

$$\mu = \Sigma\, g_i/G$$

and

$$= \Sigma\, x f_x / \Sigma\, f_x = \bar{x},$$

$$T = \sum_{x=0}^{t} f_x(x-\mu)^2/\sigma^2$$

is approximately distributed as chi-squared with $G - 1$ degrees of freedom when the sampling is random. Leslie suggests that the approximation is satisfactory when $G > 20$ and $t \geqslant 3$. We note that any samples for which $g_j = 0$ are ignored in the above analysis.

To apply this test to a Schnabel census of s samples we first of all define $G = n_1$ $(= u_1)$, the animals tagged in the first sample. In this case g_1, g_2, etc. are the members of this group caught in the second and third samples, etc., and $t = s - 1$. We can then apply this procedure to the newly tagged individuals in each sample, so that G successively represents u_2, $u_3, \ldots, u_{s-3}$ with corresponding t values $s - 2, s - 3, \ldots, 3$ respectively. If there are accidental deaths through catching and handling, then G refers to the members of u_i which are still alive at the end of the experiment. Since the test statistics thus obtained are based on different individuals they are independent and can be combined to give a total chi-square. In practice, G will often be greater than the degrees of freedom tabled, so that one must use the usual normal approximation

$$z = \sqrt{(2T)} - \sqrt{(2G-3)},$$

which is approximately distributed as the unit normal distribution when sampling is random. (The above test has been generalised by Carothers [1971]

We note that any group of identifiable animals can be used for G. In particular, if the population size is known (e.g. Crowcroft and Jeffers [1961], Huber [1962]) we can put $G = N$, and T is now a test that sampling is random with respect to the whole population and not just the marked population. One or two authors (e.g. Geis [1955]) have put G equal to the group of animals caught at least once throughout the experiment and then applied one of the approximate methods described below, with x now representing the number of *recaptures* rather than captures (i.e. $x = 0, 1, 2, \ldots, s-1$). However, I cannot see how such a model can be justified in terms of the above theory.

One advantage of the above method is that it can be adapted to "open" populations in which there is natural death and recruitment. In this case the group G consists of the members of u_i known to be alive over a certain sequence of samples, say samples $i + 1, i + 2, \ldots, i + t$, through having been caught after sample $i + t$. This method, however, does not apply if there is migration, as some of the marked animals may be out of the sampling area for several sampling occasions.

Example 4.12   Hypothetical data

In a Schnabel census of 5 samples the following recapture histories were recorded:

a_1	a_{12}	a_{13}	a_{14}	a_{15}	a_{123}	a_{124}	a_{125}	a_{134}	a_{135}	a_{145}	a_{1235}	other	n_1
84	20	14	15	12	~2	1	3	3	1	4	1	0	160

a_2	a_{23}	a_{24}	a_{25}	a_{234}	a_{235}	other	u_2
83	8	7	10	1	1	0	110

From this set of data Table 4.12 was constructed. For the group n_1 we have

TABLE 4.12

Data for demonstrating Leslie's test for simple random sampling.

Sample	g_i	g_i^2	x	f_x	$x^2 f_x$	Sample	g_i	g_i^2	x	f_x	$x^2 f_x$
	$G = 160$						$G = 110$				
2	27	729	0	84	0	3	10	100	0	83	0
3	21	441	1	61	61	4	8	64	1	25	25
4	23	529	2	14	56	5	11	121	2	2	8
5	21	441	3	1	9		29	285	3	0	0
	92	2140	4	0	0					110	33
				160	126						

$$G = 160$$
$$g_1 = a_{12} + a_{123} + a_{124} + a_{125} + a_{1235} = 27$$
$$g_2 = a_{13} + a_{123} + a_{134} + a_{135} + a_{1235} = 21, \text{ etc.}$$
$$\mu = 92/160 = 0\cdot575$$
$$\sigma^2 = 0\cdot575 - 2140/160^2 = 0\cdot4914$$

$$\Sigma\, f_x(x-\mu)^2 \;=\; 126 - 92^2/160 \;=\; 73\!\cdot\!1$$

$$T \;=\; 73\!\cdot\!1/(0\!\cdot\!4914) \;=\; 148\!\cdot\!7$$

and

$$z \;=\; \sqrt{297\!\cdot\!4} - \sqrt{317} \;=\; -0\!\cdot\!6,$$

which is not significant. Similarly for the group u_2 we find that $z = -0\!\cdot\!5$, which again is not significant. Therefore, on the basis of these two tests we do not reject the hypothesis of simple random sampling as far as the marked population is concerned.

APPROXIMATE TESTS. If $g_j = g$, say $(j = 1, 2, \ldots, t)$, we have

$$P_j \;=\; \Sigma\, P_j/t \;=\; \bar{P}\; (= \bar{x}/t)$$

and x is the outcome of t binomial trials. The test for randomness is then a test that the G values of x constitute a random sample of size G from a binomial distribution with parameters t and $\bar{P}$. In this case $\sigma^2 = t\bar{P}\bar{Q}$ and T reduces to the standard Binomial Index of Dispersion $(\mathbf{1.3.6}\,(1))$

$$T' \;=\; \frac{\Sigma\, f_x(x-\bar{x})^2}{\bar{x}\,(1-\bar{x}/t)}$$

which may be regarded as an approximation for T when the g_j are not too different. In fact

$$t\bar{P}\bar{Q} - \Sigma\, P_j Q_j \;=\; -\Sigma\,(P_i - \bar{P})(Q_i - \bar{Q})$$
$$=\; \Sigma\,(P_i - \bar{P})^2$$
$$\geqslant\; 0,$$

so that $T' \leqslant T$ with equality only in the unlikely event of the g_j being equal. Therefore, if T' is significant then T will be significant, and T' is a "conservative" approximation.

When P is small $(< 0\!\cdot\!05$ say$)$, we can use the Poisson approximation to the binomial with $\sigma^2 = \mu = t\bar{P}$. The statistic T then becomes the Poisson Index of Dispersion, $\Sigma\, f_x(x - \bar{x})^2/\bar{x}$, and since $\bar{P} > \bar{P}\bar{Q}$ this statistic will be smaller than T'. The Poisson approximation is particularly relevant to the situation where the sampling is a continuous process and the animals are caught one at a time (i.e. for the $\Sigma\, xf_x$ samples in which a member of G is caught we have $P_j = 1/G$).

COMPARING OBSERVED AND EXPECTED FREQUENCIES. We note that the statistic T is based on comparing the observed variance of x with the theoretical variance, calculated on the assumption of random sampling. In general this test will be more sensitive than a goodness-of-fit test based on comparing the observed frequencies f_x with the expected frequencies (Cochran [1954]). However, if T is significant, a comparison of the observed and expected frequencies can be helpful in detecting where departures from random sampling occur. Unfortunately, using Leslie's method, the expected frequencies require lengthy calculations, particularly for large values of x. For example,

$$\Pr[x = 0] \;=\; Q_1 Q_2 \ldots Q_t,$$

$$\Pr[x = 1] = Q_1 Q_2 \ldots Q_t \sum_i (P_i/Q_i),$$

$$\Pr[x = 2] = Q_1 Q_2 \ldots Q_t \sum_{i < j} \sum (P_i P_j /Q_i Q_j), \text{ etc.}$$

Using the binomial approximation, however, the expected frequencies are more readily calculated, namely

$$E_x = G \binom{t}{x} \bar{P}^x \bar{Q}^{t-x}.$$

4.1.6 Models based on constant probability of capture

1 General theory

Suppose that the trapping effort is the same for each sample, so that p_i, the probability of capture in the ith sample ($i = 1, 2, \ldots, s$), is constant ($= p$, say). Then (4.2) reduces to

$$f(\{a_w\}) = \frac{N!}{\prod\limits_{w} a_w! \, (N-r)!} \, p^{\Sigma n_i} \, q^{sN - \Sigma n_i}$$

and $\hat{N}_p$, the maximum-likelihood estimate of N, is now the unique root greater than r of

$$\left(1 - \frac{r}{N}\right) = \left(1 - \frac{\Sigma n_i}{sN}\right)^s. \tag{4.26}$$

Darroch [1958: 355] shows that as $N \to \infty$ the asymptotic variance of $\hat{N}_p$ is

$$V[\hat{N}_p] = N \left[\frac{1}{q^s} + s - 1 - \frac{s}{q}\right]^{-1}$$

which may be compared with

$$V[\hat{N}] = N \left[\frac{1}{q_1 q_2 \ldots q_s} + s - 1 - \Sigma \frac{1}{q_i}\right]^{-1},$$

the corresponding expression when the p_i are unequal. Since $V[\hat{N}_p]$ follows from $V[\hat{N}]$ by simply putting $p_i = p$, Darroch concludes that, asymptotically, no information is gained by using the knowledge that p_i is constant. It is therefore recommended that the methods of **4.1.2** be used irrespective of whether one suspects p_i to be constant or not.

2 Frequency of capture

Several models based on the frequency of capture have been developed recently, and these have been used mainly for detecting any variation in trap response. For example, if p is constant, the probability that an animal is caught x times ($x = 0, 1, 2, \ldots, s$) is given by the binomial probability function

$$f(x) = \binom{s}{x} p^x q^{s-x}. \tag{4.27}$$

If N is known, we can regard the animals as representing N independent observations from (4.27) and carry out a standard goodness-of-fit test (**1.3.6** (*1*)) to test for constant p. When p is small (< 0.05), the Poisson approximation to (4.27) can be used. Examples of such tests are given in Crowcroft and Jeffers [1961] and in Example 4.14 below.

Some authors have also used these tests for populations in which N has been estimated by independent methods not affected by trap response. For example, Geis [1955], in his study of a rabbit population, estimated N by the Petersen method using data from a hunting kill. He assumed that in spite of a variable trap response, shooting would provide an unbiased sample from the population. Keith and Meslow [1968] also used the Petersen estimate for a population of snowshoe hares. Three independent methods of estimation were used, based on the ratios of (i) colour-marked to unmarked hares observed after winter and spring trapping periods, (ii) marked to unmarked hares taken in late summer live-snaring, and (iii) radioactive to non-radioactive young in summer resulting from implantation of adult females with calcium-45 in spring. Edwards and Eberhardt [1967] carried out two studies: one based on a known population of rabbits which had been released in an enclosure and the other using an unconfined population. In the second study, N was estimated using the Petersen method: the tag ratio was obtained from a drive census, which would not be biased by any differential trap response.

It should be noted that the above theory is not just a repetition of the approximate method given on p. 163. In the above theory, p_i is the probability of catching *any* individual in the ith sample, while P_i in Leslie's method is the conditional probability that an individual from an identifiable group is caught in the ith sample, *given* that at least one member of this group is caught in the sample.

Example 4.13 House mice (*Mus musculus* L.): Young *et al.* [1952]

A program of live-trapping was carried out in two populations of house mice in an endeavour to detect any heterogeneity in trap response. Both populations inhabited essentially the same kind of environment, namely heated buildings with an ample supply of food (stored grain), water and cover. Since there appeared to be no differences in the trap reactions of the two populations, they were treated as one population and their records pooled.

Mice were caught in a fixed number of live-traps which operated for three consecutive nights each week followed by a four-day rest period. Each mouse, when first caught, was marked with an individual pattern of toe-clipping, and the termination of any mouse's exposure to trapping was indicated either by its being found dead in the course of the study, or by its capture during an extensive snap-trapping programme at the conclusion. Mice not finally recovered were treated as residents of the population only up to the dates on which they last appeared in live-traps. By interpreting the recapture records in terms of exposure to trapping, the authors eliminated some of the effects of mortality and migration. An examination of local movements in the two populations indicated that "there were essentially no irregular wandering movements and shifts of base which might have affected the calculations of exposure to trapping (Young *et al.* [1950])".

In order to detect any tendencies for the pattern of trap reaction to change with time, the mice were separated into groups known to have survived through successive periods of ten nights of trapping following the date of their original capture. The frequencies of capture within each period were then compared with the frequencies expected under a Poisson model. In each case the agreement was poor as the observed frequencies were too large at either end of the distribution, thus indicating a heterogeneity of trap response. From the histograms, however, there did not appear to be any substantial change in the form of the capture distribution with respect to time, indicating perhaps that the tendency towards trap addiction or avoidance is a relatively permanent characteristic of any one mouse, i.e. once supplied with an attitude towards traps, a mouse "maintains" this attitude in spite of an accumulation of experiences with traps. In support of this statement the authors presented Table 4.13 in which the trapping history of a cohort of 56

TABLE 4.13

Frequencies of captures as a function of time for 56 mice surviving 50 nights of trapping: from Young *et al.* [1952: Table 2].

| Period | \multicolumn{6}{c}{Number of captures (x)} | Total |
	0	1	2	3	4	5	
Nights 1–10	41	12	2	0	1	0	56
Nights 11–20	44	9	3	0	0	0	56
Nights 21–30	44	9	3	0	0	0	56
Nights 31–40	43	11	1	0	1	0	56
Nights 41–50	48	6	1	0	1	0	56

mice that survived through at least 50 nights of trapping was broken down into five periods of ten nights. The capture distributions for each of these periods are remarkably similar.

The large observed frequencies at both ends of the distributions can be explained by the presence of trap-shy individuals boosting the "zero" class and by trap-addicted animals affecting the higher frequencies. However, an alternative explanation is available. In a previous study (Young *et al.* [1950]) there were indications that mice had a tendency to be recaptured in traps from which they were most recently released. This is an important factor when the normal foraging distances of the mice are small in relation to trap distances, for then some of the marked mice might be expected to have moved temporarily out of range of any trap, while others may tend to establish themselves close to a trap in which they would then be caught frequently. But the aim of the study was not to determine whether the departures from randomness were due to trap distribution or to innate trap vulnerability, but rather to show that such departures actually occur and to demonstrate how they may be detected.

Example 4.14   Cottontail rabbit (*Sylvilagus floridanus mearnsii*): Geis [1955]
The trap response of rabbits was evaluated by considering live-trapping data collected during Nov.–Dec. 1951 and Oct.–Nov. 1952 from a population of cottontails at the Kellogg Station of Michigan State College. Fifty wooden traps were used, and to ensure complete coverage, the main study area of 160 acres, largely surrounded by open, cultivated land, was divided into halves which were trapped separately. An irregular spacing of traps was used because an irregularly shaped lake in the centre of the study area made the operation of a grid or a straight trap-line impractical.

Rabbits were marked with numbered metal tags inserted near the centre of each ear. There was no evidence of tags being lost except that an· occasional one was torn out by shot. The trap location, age, sex, weight and other data were recorded each time a rabbit was handled. After each of the trapping periods, closely supervised hunting took place and the location at which each rabbit was shot was recorded.

The capture frequencies for one particular trap-line (called trap-line A) are given in Table 4.14, and f_0, the number not captured, is obtained from

TABLE 4.14

Expected distribution of captures for a population of rabbits made up of subpopulations with three different probabilities of capture, trap-line A: from Geis [1955: Table 4].

Times captured x	Observed frequency f_x	Expected frequency E_x	Expected frequencies			
			(1) (Poisson) $\hat{p}=0.027$	(2) (Binomial) $\hat{p}=0.314$	(3) (Binomial) $\hat{p}=0.571$	Total (4) $=(1)+(2)+(3)$
0	121	65·8	118·2	0·2	–	118·4
1	35	77·2	41·5	1·1	–	42·6
2	12	40·8	7·3	3·0	0·0	10·3
3	6	12·8	0·8	5·1	0·1	6·0
4	11	3·0	0·2	5·8	0·2	6·2
5	8	0·4	0·0	4·8	0·6	5·4
6	3	0·0	–	2·9	1·1	4·0
7	0	–	–	1·4	1·5	2·9
8	3	–	–	0·6	1·5	2·1
9	0	–	–	0·1	1·1	1·2
10	1	–	–	0·0	0·6	0·6
11	0	–	–	–	0·2	0·2
12	0	–	–	–	0·1	0·1
13	0	–	–	–	0·0	0·0
Total	200	200·0	168·0	25·0	7·0	200·0

a Petersen estimate of total population size using the marked–unmarked ratio in the hunting kill. From this table we have that p, the probability of capture in a sample, is estimated by

$$\hat{p} = \frac{\Sigma\, xf_x}{\Sigma\, f_x s} = \frac{213}{200(13)} = 0\cdot081\ 92$$

and the expected frequencies E_x are given by

$$E_x = 200 \binom{13}{x} \hat{p}^x \hat{q}^{\,13-x}.$$

Pooling for $x > 3$, the chi-squared statistic $\Sigma\,(f_x - E_x)^2/E_x = 244$ which at 3 degrees of freedom is highly significant, thus indicating that the assumption of constant p is not true. In fact, from Table 4.14 we see that there are too many animals in the no-capture and many-capture categories. This could be due to the fact that the data for adults and juveniles, males and females are pooled, whereas the four groupings of sex and age may have different probabilities of capture. To investigate this possibility we can calculate the expected frequencies for each class separately as in Table 4.15, add the

TABLE 4.15

Comparison to show the effect of combining data from four age and sex combinations of rabbits: adapted from Geis [1955: Table 3].

Sex and Age	Number captured	$\hat{p}$	$x=$ 0	1	2	3	4	5	6	Total
Adult—Male	5	0·084	1·60	1·91	1·05	0·35	0·08	0·02	0·00	5·01
Adult—Female	9	0·131	1·45	2·84	2·57	1·42	0·54	0·14	0·05	9·01
Juv.—Male	28	0·084	8·96	10·67	5·88	1·96	0·45	0·08	0·00	27·99
Juv.—Female	27	0·071	10·37	10·29	4·73	1·32	0·24	0·03	0·00	26·98
Total	69	0·082	22·38	25·71	14·23	5·05	1·31	0·27	0·05	69·00
E_x	69	0·082	22·70	26·63	14·08	4·42	1·04	0·14	0·00	69·01

expected frequencies together, and compare these with the expected frequencies based on the pooled data (Geis uses 69 of the 79 animals captured). As these two groups of expected frequencies are close together, it is clear that the differences observed in Table 4.14 cannot be accounted for by the pooling of the four categories. Geis suggests, however, that the observed frequencies may still be the product of pooling segments of the population with different probabilities of capture, and to demonstrate this possibility he divides the data into three groups: 168 caught 0—2 times, 25 caught 3—5 times and 7 caught 6—10 times. The corresponding values of $\hat{p}$ for each of these groups are 0·027, 0·314 and 0·571 respectively, and three separate expected distributions are computed as in Table 4.14 (columns (1), (2) and (3)). A combined distribution (column (4), formed by adding the corresponding frequencies from the separate distributions, is compared with the distribution of observed frequencies f_x, and the difference is found to be not significant. But there is a highly significant difference between the compound distribution and the distribution of expected frequencies E_x.

This analysis shows that the observed frequencies of capture can be explained in terms of a compound distribution which reflects a variable probability of capture. According to this view, most animals had a low probability of capture and relatively few had higher probabilities. Geis offers three explanations as to why some rabbits entered traps more readily than others: first, some animals may have a small home range within which a trap was favourably located and therefore frequently encountered; second, rabbits vary in an innate tendency to enter traps; and, third, once a rabbit has had some experience with traps its behaviour is altered so that it is more likely to be recaptured. The first explanation is ruled out by the fact that frequently captured rabbits turned up in many trap locations. The third explanation is supported by the fact that the mean time-interval between first and second captures was significantly longer than the interval between later captures. Additional evidence for trap conditioning comes from the fact that rabbit tracks were observed in the snow around sprung, unbaited traps after the trapping period had been completed.

3   Truncated models: constant probability of capture

BINOMIAL. When N is unknown we can test whether sampling is random with respect to just the marked population by using Leslie's method of **4.1.5** (2). When p_i is constant, an alternative approach is to truncate (4.27) by ignoring the group of $N - r$ animals not captured during the experiment. Thus x, the number of times an animal is captured *given* that it is captured at least once, has probability function

$$f(x) = \binom{s}{x} p^x q^{s-x}/(1-q^s), \quad x = 1, 2, \ldots, s.$$

For this model the maximum-likelihood estimate $\hat{q}$ of q is the unique root of

$$0 = h_s(q) = \frac{1-q^s}{1-q} - \frac{s}{x} = 1 + q + \ldots + q^{s-1} - (s/\bar{x}),$$

where $\bar{x}$ ($= \Sigma\, x_i/r$) is now the mean number of captures per animal for the r animals actually captured. For $s > 3$ this equation can be solved iteratively using the Newton–Raphson method; the ith step is given by

$$q_{i+1} = q_i - h_s(q_i)/h_s'(q_i),$$

and a possible first approximation q_1 is given by the positive root of the quadratic $h_3(q) = 0$. Alternatively we can use the following general technique given by Hartley [1958] for handling truncated distributions. Beginning with a first approximation $N_{(1)}$ of N, we carry out the chain of iterations

$$p_{(i)} = \frac{\bar{x}}{s} \cdot \frac{r}{N_{(i)}}$$

and

$$N_{(i+1)} = \frac{r}{(1 - q_{(i)}^s)}.$$

This procedure not only gives us $\hat{p}$ but, as a bonus, we also get $\hat{N}_p$, the solution of (4.26): this follows from the fact that $r\bar{x} = \Sigma\, n_i$. Once $\hat{p}$ is calculated, a standard goodness-of-fit test for the above truncated binomial model can be carried out.

POISSON. When p is small we can use the Poisson approximation to the binomial and consider the truncated distribution (Craig [1953b])

$$f(x) = \frac{e^{-\lambda}}{(1-e^{-\lambda})} \cdot \frac{\lambda^x}{x!}, \quad x = 1, 2, \ldots. \tag{4.28}$$

This distribution has been investigated by David and Johnson [1952], who show that the maximum-likelihood estimate $\hat{\lambda}$ of λ is the solution of

$$(1-e^{-\lambda})/\lambda = 1/\bar{x},$$

which can be solved by interpolating in Table **A3**. One can then carry out a chi-squared goodness-of-fit test by comparing observed and expected frequencies in the usual manner (examples of this are given in Keith and Meslow [1968]). David and Johnson suggest the alternative procedure of using the usual Poisson Index of Dispersion, but with the class of zero captures left out, i.e.

$$T = \sum_{x=1}^{X} f_x(x-\bar{x})^2/\bar{x},$$

where $\quad \bar{x} = \sum_{x=1}^{X} x f_x / \sum_{x=1}^{X} f_x$

and X is the largest observed value of x. They show that treating T as chi-squared with $r-1$ degrees of freedom leads to a conservative test, for if T is significant then the Poisson Index of Dispersion derived from the complete data is also significant. Rao and Chakravarti [1956] and Chakravarti and Rao [1959] suggest using $\bar{x}(1+\hat{\lambda}-\bar{x})$ in the denominator of T: they also give an exact small-sample test based on $\Sigma\, x_i^2$.

We note that, strictly speaking, (4.28) should be truncated on the right at $x = s$ as no more than s recaptures are possible. However, if s is sufficiently large for $\Pr[X \leqslant s]$ to be almost 1, the effect of truncation on the data will be negligible.

An estimate of N, the population size, is given by $r/[1 - \exp(-\hat{\lambda})]$ or $r\bar{x}/\hat{\lambda}$.

4 Truncated models: allowing for trap response

GEOMETRIC. Eberhardt *et al.* [1963] found that the capture frequencies for a rabbit population are well fitted by the geometric distribution[*]

$$f(x) = PQ^x, \quad x = 0, 1, 2, \ldots, (0 < P < 1; Q = 1-P),$$

and we now outline one of the two derivations that they give for this model. Suppose that conditional on λ, the "average capture rate", x has a Poisson distribution

---

[*] Overton and Davis [1969: 445–7] question this: they also give a useful non-parametric method.

$$f(x \mid \lambda) = e^{-\lambda} \lambda^x / x!$$

Then, assuming a circular home range of radius R, we would expect the average capture rate to be proportional to the area of the home range, i.e. $\lambda = d\pi R^2$, where d is a constant depending on such factors as the density of traps and the probability of capture, given that there is "contact" with one or more traps. Following Calhoun and Casby [1958], it is assumed that R has density function

$$f_1(R) = \frac{R}{\sigma^2} e^{-R^2/(2\sigma^2)}$$

so that if $c = 2d\pi\sigma^2$,

$$f_2(\lambda) = c^{-1} e^{-\lambda/c}, \quad (\lambda \geqslant 0).$$

Hence

$$f(x) = \int_0^\infty f(x \mid \lambda) f_2(\lambda) \, d\lambda = PQ^x,$$

where $P = 1/(1+c)$.

When the size of the zero class is unknown, this distribution can be truncated at the origin as in the previous models. If trapping is carried out on s occasions then the distribution should also be truncated on the right, so that we are led to consider

$$f(x) = PQ^{x-1}/(1 - Q^s), \quad x = 1, 2, \ldots, s. \tag{4.29}$$

For a sample of r observations from this distribution, the maximum-likelihood estimate $\hat{P}$ (= $1 - \hat{Q}$) is the unique solution of

$$\bar{x} = \frac{sQ^{s+1} - (s+1)Q^s + 1}{Q^{s+1} - Q^s - Q + 1}$$

which can be solved by interpolating linearly in Table **A4** (reproduced from Thomasson and Kapadia [1968]). For example, if $s = 10$ and $\bar{x} = 3 \cdot 644$,

$$\hat{P} = 0 \cdot 20 + \frac{3 \cdot 797 - 3 \cdot 644}{3 \cdot 797 - 3 \cdot 043} (0 \cdot 10) = 0 \cdot 2203.$$

When s is large, the effect of truncation on the right is negligible and $\hat{P} = 1/\bar{x}$; or allowing for bias (Eberhardt [1969a]), $\hat{P} = (r-1)/(r\bar{x} - 1)$ which, from Chapman and Robson [1960], is the minimum variance unbiased estimate of P. In this case the total population size can be estimated by $\hat{N} = r/\hat{Q}$ (Edwards and Eberhardt [1967]). To determine when the truncation can be neglected, we enter $1/\bar{x}$ at the top of Table **A4** and in the nearest column we note when the entry becomes independent of s.

If f_x animals are caught x times then, truncating the distribution of x on the right only,

$$E[f_x] = NPQ^x/(1 - Q^{s+1}), \quad x = 0, 1, 2, \ldots, s,$$

and taking logarithms we have

$$E[\log f_x] \approx \log [NP/(1 - Q^{s+1})] + x \log Q$$
$$= \beta_0 + \beta x, \text{ say}, \quad (x = 1, 2, \ldots, s),$$

which, neglecting Q^{s+1}, is the regression model suggested by Edwards and Eberhardt [1967] (see also Kelley and Barker [1963]). We can then estimate N and P from the usual least-squares estimates of β_0 and β. One method of obtaining a confidence interval for N is to calculate the confidence limits for $NP/(1-Q^{s+1})$, the expected number of animals not caught, and add r to both limits.

In applying these methods to squirrel populations Edwards and Eberhardt make the empirical suggestion that reliable estimates of N are obtained when about 50 per cent of the population is captured and $\bar{x}$ is about $1\frac{1}{2}$ to 2.

Example 4.15   Squirrel (*Sciurus carolinensis*, etc.): Nixon *et al.* [1967]

The aim of the study was to investigate the accuracy of the Schnabel (**4.1.2**(*4*)) and Schumacher–Eschmeyer (**4.1.3**(*1*)) estimates of population size for squirrel populations, and to consider the possible application of the above geometric model. The study area occupied 237 acres of continuous forest habitat in the 1250 acre Waterloo Wildlife Experiment Station, Athens County, Ohio. Both fox and grey squirrels occurred in the area, with the grey squirrels comprising about 85 per cent of the squirrel population. The area was gridded on a 3×3 chain interval, with a trap placed at the discretion of the trapper within a $\frac{1}{5}$-acre plot surrounding each point of intersection. This yielded a trap density of about one (0·96) trap per acre.

Prebaiting was used for 10 days before the experiment and trapping was carried out for 11 consecutive days just before the hunting season. All squirrels captured were ear-tagged and released at their points of capture. Squirrels killed by the hunters on the study area during the hunting season provided an estimate of the proportion tagged, from which N could be estimated using the Petersen method.

Recapture data for the year 1962 are given in Table 4.16, and a plot of m_i/n_i versus M_i is not linear (cf. p. 141), thus suggesting that the Schnabel estimate and its modifications will not give reliable estimates of N. This was borne out from a more detailed analysis by the authors, who felt that the Schnabel and Schumacher–Eschmeyer methods led to an underestimation of population size. This is in agreement with Flyger [1959: 221] who reported considerably lower estimates using Schnabel's method than those derived from sight records of colour-marked squirrels, and it also agrees with the data obtained on cottontail rabbits by Edwards and Eberhardt [1967].

Using the capture frequencies f_x as given in Table 4.17, a goodness-of-fit test for (4.29) can be carried out. We have $1/\bar{x} = 72/223 = 0·323$, and from Table **A4** we find that the truncation of the distribution at $s = 11$ must be taken into account. Therefore, entering the table with $s = 11$ and $\bar{x} = 3·097$ we find that $\hat{P} = 0·300$. The expected frequencies are then given by

$$E_x = 72\,\hat{P}\hat{Q}^{\,x-1}/(1-\hat{Q}^{\,11})$$

and

$$\Sigma\,(f_x - E_x)^2/E_x = 2·3,$$

TABLE 4.16
Capture–recapture data from a Schnabel census: from Nixon *et al.*
[1967: Table 1, 1963].

Trap day i	Sample size n_i	Marked m_i	M_i	m_i/n_i
1	38	–	–	–
2	29	19	33	0·66
3	31	23	48	0·74
4	16	13	56	0·81
5	20	19	59	0·95
6	18	17	60	0·94
7	17	14	61	0·82
8	19	13	64	0·68
9	16	14	70	0·88
10	14	14	72	1·00
11	5	5	72	1·00
Total	223	151		

TABLE 4.17
Observed capture frequencies fitted to zero-truncated geometric and
Poisson distributions: data from Nixon *et al.* [1967: Table 3].

Number of captures x	Observed frequencies f_x	$x f_x$	Expected frequencies Geometric $(s=11)$ E_x	Geometric $(s=\infty)$	Poisson $(s=\infty)$
1	23	23	22·0	23·3	11·9
2	14	28	15·4	15·8	17·4
3	9	27	10·8	10·7	17·2
4	6	24	7·6	7·2	12·6
5	8	40	5·3	4·9	7·4
6	7	42			
7	3	21			
8	0	0	10·9	10·1	5·5
9	2	18			
10	0	0			
11	0	0			
Total	72	223	72·0	72·0	72·0

which at 4 degrees of freedom, indicates a close fit. The authors actually
ignored the effect of truncation in their analysis ($s=\infty$) and used $\hat{P} = 0.323$.
However, we see from Table 4.17 that this makes little difference to the
analysis. They also fitted a truncated Poisson which gave a very poor fit
to the observed frequencies ($\chi_4^2 = 26\cdot1$).

The authors concluded that the probability of capture did not seem to be the same for all individuals and that the geometric model gave a reasonable fit to the observed frequencies. However, as pointed out by Eberhardt *et al.* [1963], a good fit to this model provides no conclusive evidence that the assumptions used in deriving the model actually occur in Nature: different sets of assumptions can give rise to the same model. For example, suppose that radio-tracking data indicate that the number of visits y of an animal to some small regular area around the trap follows a geometric distribution (e.g. Tester and Siniff [1965])

$$f_1(y) = \theta(1-\theta)^y, \quad y = 0, 1, 2, \dots .$$

If x, the number of captures, is conditionally binomial with parameters (y, p), then we find that the unconditional distribution is once again geometric (Eberhardt [1969a]), namely

$$f(x) = w(1-w)^x, \quad x = 0, 1, 2, \dots , \tag{4.30}$$

where $w = \theta/[\theta + (1-\theta)p]$.

Further examples comparing the merits of the above Poisson and geometric models are given in Eberhardt [1969a].

NEGATIVE BINOMIAL. The above derivation of (4.30) applies to the situation where a single trap is randomly located within a given animal's home range. If, however, the traps are closer together, so that k traps fall within the home range, and assuming that traps act independently, then it is readily shown (Eberhardt [1969a]) that the sum of k random variables independently sampled from the geometric distribution (4.30) has a negative binomial distribution

$$\frac{k(k+1) \dots (k+x-1)}{x!} w^k(1-w)^x, \quad x = 0, 1, 2, \dots \tag{4.31}$$

(which reduces to (4.30) when $k = 1$). Alternatively this distribution can also be derived by assuming that the Poisson model with parameter λ is appropriate, but with λ varying according to a Pearson Type III distribution (Kendall and Stuart [1969: 129]). In this case x has probability function

$$\frac{k(k+1) \dots (k+x-1)}{x!} \cdot \frac{a^x}{(1+a)^{k+x}}, \quad x = 0, 1, 2, \dots , \tag{4.32}$$

which reduces to (4.31) by putting $1 - w = a/(1+a)$. Thus, whichever method of derivation is used, the distribution of x, truncated at $x = 0$, is given by

$$f(x) = \frac{k(k+1) \dots (k+x-1)}{x!} \cdot \frac{w^k(1-w)^x}{(1-w^k)} \quad x = 1, 2, \dots , \tag{4.33}$$

where k may or may not be an integer.

This model seems to have been first applied to recapture data by Tanton [1965: 14]: it has also been used by Taylor [1966: 125] and by Tanton [1969: 515]. From Sampford [1955] the maximum-likelihood estimates of w and k

for a sample of r observations from this distribution are the solutions of

$$\frac{rk}{w(1-w^k)} - \frac{r\bar{x}}{1-w} = 0$$

and

$$\frac{r \log w}{(1-w^k)} + \sum_{x=1}^{X} \left(\frac{1}{k} + \frac{1}{k+1} + \dots + \frac{1}{(k+x-1)} \right) f_x = 0,$$

where X is the maximum observed value of x; and Sampford describes several methods for solving these equations iteratively. Alternatively we can use the iterative technique of Hartley [1958], which consists basically of estimating $N - r$, the size of the "zero" class, and then using Bliss and Fisher's [1953] technique for estimating k and a from (4.32) and the "completed" data. However, a simpler method of obtaining estimates for k and w has been proposed by Brass [1958: method A] as follows.

Let $\pi_1 = \Pr[x=1] = kw^k(1-w)/(1-w^k)$; then if μ and σ^2 are the mean and variance of x for (4.33), Brass shows that

$$w = \mu(1-\pi_1)/\sigma^2$$

and

$$k = (w\mu - \pi_1)/(1-w).$$

Therefore, replacing μ, σ^2 and π_1 by their sample estimates

$$\bar{x} = \sum_{x=1}^{X} xf_x/r,$$

$$s^2 = \sum_{x=1}^{X} f_x(x-\bar{x})^2/(r-1),$$

and

$$\hat{\pi}_1 = f_1/r,$$

respectively, we have the simple estimates

$$\tilde{w} = \bar{x}(1-\hat{\pi}_1)/s^2$$

and

$$\tilde{k} = (\tilde{w}\bar{x} - \hat{\pi}_1)/(1-\tilde{w}).$$

Table 4.18, taken from Brass, gives the efficiency (based on the determinant of the variance–covariance matrix) of the above estimation procedure as

TABLE 4.18
Percentage efficiency of estimation by Method A: from Brass [1958: Table 1].

M \ k	0·5	1	2	3	4	5	10	∞
0·5	93·4	97·1	98·9	99·3	99·4	99·5	99·3	98·6
1	88·0	93·6	97·1	98·1	98·4	98·6	98·0	97·4
2	81·0	88·2	93·7	95·7	96·6	97·0	97·4	96·1
5	70·9	78·8	86·5	90·4	92·7	94·2	97·2	98·5
10	63·6	70·9	78·9	83·7	87·0	89·3	95·1	100·0
∞	22·7	38·8	57·5	67·6	73·9	78·2	88·0	100·0

compared to the maximum-likelihood method for different values of k and $M = k(1-w)/w$ (the mean of the complete distribution (4.32)). We see that for small M or large k, Brass's procedure is remarkably efficient.

Example 4.16 Wood mouse (*Apodemus sylvaticus* (L.)): Tanton [1965]

Using Brass's method, Tanton estimates w and k for a population of wood mice and carries out a goodness-of-fit test for (4.33) as set out in

TABLE 4.19

Capture frequencies of a male population of wood mice fitted to a zero-truncated negative-binomial distribution: from Tanton [1965: Table 5].

x	f_x	$x\,f_x$	$x^2\,f_x$	E_x
1	71	71	71	68·10
2	59	118	236	56·82
3	41	123	369	45·90
4	39	156	624	36·46
5	20	100	500	28·68
6	26	156	936	22·40
7	19	133	931	17·41
8	12	96	768	13·45
9	9	81	729	10·38
10	5	50	500	8·00
11	8	88	968	6·15
12	4	48	576	4·72
13	9 ⎫ 11	117	1 521	3·62 ⎫ 6·39
14	2 ⎭	28	392	2·77 ⎭
15	1 ⎫	15	225	2·12 ⎫
16	3 ⎪ 10	48	768	1·62 ⎪
17	3 ⎬	51	867	1·24 ⎪
18	3 ⎭	54	972	0·94 ⎬ 9·13
19	0	0	0	0·72 ⎪
20	0	0	0	0·55 ⎪
21	0	0	0	0·42 ⎪
> 21	0	0	0	1·52 ⎭
Total	334	1 533	11 953	333·99

Table 4.19. From this table we have $r = 334$,

$$\bar{x} = 1533/334 = 4\cdot5898,$$

$$s^2 = [11\ 953 - 1533^2/334]/333 = 14\cdot765,$$

and $\hat{\pi}_1 = 71/334 = 0\cdot212\ 57$
so that

$$\tilde{w} = 4\cdot5898\,(0\cdot787\ 43)/14\cdot765 = 0\cdot2448.$$

Actually Tanton uses a divisor of 334 in calculating s^2, and his estimates $\tilde{w} = 0\cdot2455$ and $\tilde{k} = 1\cdot2118$ are used in calculating the column of expected frequencies E_x. He obtains a value of $1\cdot10$ for his chi-squared goodness-of-fit statistic which, at 11 degrees of freedom, indicates a good fit of the observed to the expected frequencies. He also suggests estimating the population size by

$$\tilde{N} = r/(1-\tilde{w}^k) = 408\cdot5 ,$$

though no variance formula for this estimate is given.

Finally, since $\tilde{M} = \tilde{k}(1-\tilde{w})/\tilde{w} = 3\cdot7$, we see from Table 4.18 that Brass's method has an efficiency relative to the maximum-likelihood method of approximately 80 per cent.

SKELLAM'S MODEL. Suppose that for a given animal the frequency of capture follows the binomial distribution (4.27) with parameters s and p. In many experimental situations p may not be the same for each animal but will vary according to some distribution $f_1(p)$ (cf. 3.2.2(2)). For example, Skellam [1948] used the beta-distribution

$$f_1(p) = \frac{1}{B[\alpha,\beta]} p^{\alpha-1}(1-p)^{\beta-1} \! ; \quad 0 \leqslant p \leqslant 1$$

where $B[\alpha,\beta] = \Gamma(\alpha)\Gamma(\beta)/\Gamma(\alpha+\beta) ,$

and hence showed that

$$f(x) = \int_0^1 f(x|p)f_1(p)dp$$

$$= \binom{s}{x} \frac{B[\alpha+x,\, \beta+s-x]}{B[\alpha,\, \beta]} , \quad x = 0, 1, 2, \ldots, s. \qquad (4.34)$$

If $s \to \infty$, $\beta \to \infty$, $\beta/s \to c$ $(c \neq 0)$ and $p \to 0$ in such a way that sp remains finite, then Skellam showed that sp tends to a Pearson Type III distribution, and the limit of $f(x)$ is the negative binomial (4.32) with $k = \alpha$, $a = 1/c$.

The truncated version of (4.34), which is appropriate when the zero class is not observed, is given by

$$\binom{s}{x} \frac{B[\alpha+x,\, \beta+s-x]}{B[\alpha,\beta] - B[\alpha,\beta+s]} , \quad x = 1, 2, \ldots, s .$$

Unfortunately estimation for this distribution is not easy, which rather precludes its use in practice. However, when $s = 2$, (4.34) can be used for investigating the Petersen estimate. For example, if n_i is the size of the ith sample $(i = 1, 2)$, m_2 the number of recaptures in the second sample, and $\hat{N} = n_1 n_2/m_2$, then

$$E[n_i] = NE[p] = N\alpha/(\alpha+\beta), \qquad (4.35)$$

$$E[m_2] = NE[p^2] = N\alpha(\alpha+1)/[(\alpha+\beta)(\alpha+\beta+1)], \qquad (4.36)$$

and asymptotically

$$E[\hat{N}] = E[n_1]E[n_2]/E[m_2]$$

$$= \frac{N\alpha^2}{(\alpha+\beta)^2} \cdot \frac{(\alpha+\beta)(\alpha+\beta+1)}{\alpha(\alpha+1)}$$

$$= \frac{N\alpha(\alpha+\beta+1)}{(\alpha+1)(\alpha+\beta)}$$

$$= NB, \text{ say},$$

where B is tabulated in Table 4.20 for selected values of α and β. We note that the value of B could have also been derived directly from the general theory of **3.2.2**(2) by setting $x = y = z = p$, so that

$$B = p_1 p_2 / p_{12} = (E[p])^2 / E[p^2].$$

TABLE 4.20

Table of $B = \alpha(\alpha+\beta+1)/[(\alpha+1)(\alpha+\beta)]$ for selected values of α and β.

β \ α	1	2	3	5	10	∞
1	0·75	0·89	0·94	0·97	0·99	1·00
2	0·67	0·83	0·90	0·95	0·98	1
3	0·63	0·80	0·88	0·94	0·98	1
5	0·58	0·76	0·84	0·92	0·97	1
10	0·55	0·72	0·81	0·89	0·94	1
∞	0·50	0·67	0·75	0·83	0·91	1

It was also mentioned there that B is small when a large proportion of the population has a low probability of capture (i.e. α small).

For the special case $\alpha = 1$ the limiting form of (4.34) is geometric rather than negative-binomial, and Eberhardt [1969a] uses this special case to derive a new estimate of N when $s = 2$. Thus, setting $\alpha = 1$ in (4.35) and (4.36), we have

$$E[n_1 + n_2] = 2N/(\beta+1)$$

$$E[m_2] = 2N/[(\beta+1)(\beta+2)],$$

and solving, we have the moment estimates

$$\hat{\beta} + 2 = (n_1 + n_2)/m_2$$

and

$$\hat{N}_\beta = (n_1 + n_2)(\hat{\beta}+1)/2$$

$$= (n_1 + n_2)(n_1 + n_2 - m_2)/(2m_2).$$

If in fact p is actually constant, then $\hat{N}$ is asymptotically unbiased (since $B = 1$) and, asymptotically,

$$E[\hat{N}_\beta] = E[n_1 + n_2]E[n_1 + n_2 - m_2]/E[2m_2]$$

$$= 2Np(2Np - Np^2)/(2Np^2)$$

$$= N(2-p).$$

which lies between N and $2N$.

To find the asymptotic variance of $\hat{N}_\beta$, let y_i be the number of animals caught i times, i.e. $y_1 = n_1 + n_2 - 2m_2$ and $y_2 = m_2$. Then the joint distribution of y_1 and y_2 is multinomial, namely

$$f(y_1, y_2)$$

$$= \frac{N!}{y_1! \, y_2! \, (N - y_1 - y_2)!} \; P_1^{y_1} \, P_2^{y_2} \, P_0^{N - y_1 - y_2}$$

$$= \frac{N!}{y_1! \, y_2! \, (N - y_1 - y_2)!} \left[\frac{2\beta}{(\beta + 1)(\beta + 2)} \right]^{y_1} \left[\frac{2}{(\beta + 1)(\beta + 2)} \right]^{y_2} \left[\frac{\beta}{\beta + 2} \right]^{N - y_1 - y_2}$$

since, from (4.34) with $s = 2$, $P_i = \Pr[x = i] = f(i)$. Then the maximum-likelihood estimate of N is once again $\hat{N}_\beta$ which now takes the form

$$\hat{N}_\beta, = (y_1 + 2y_2)(y_1 + y_2)/(2y_2).$$

Hence, using the delta method, we find, after some algebra, that

$$V[\hat{N}_\beta] \approx \frac{N\beta}{2(\beta + 1)^2 (\beta + 2)^2} (\beta^5 + 7\beta^4 + 20\beta^3 + 29\beta^2 + 21\beta + 6).$$

5 Models based on waiting times between captures

TIME TO FIRST RECAPTURE. The probability that an animal is caught for the first time in the yth sample, *given* that it is caught at least once in s samples, is given by the truncated geometric probability function

$$f(y) = q^{y-1}p/(1 - q^s), \quad y = 1, 2, \ldots, s,$$

where p is the probability of capture in a sample. This model has already been discussed above (p. 171), though in a different context. However, a slightly different model has been suggested by Young *et al.* [1952] which can still be used when there is migration and mortality.

Suppose that an animal is captured for the second time in sample number $y + z$ ($z = 1, 2, \ldots, s - y$). Then, given y, and given that an animal is caught at least twice, z has probability function

$$f(z \mid y) = q^{z-1}p/(1 - q^{s-y}), \quad z = 1, 2, \ldots, s - y.$$

If we consider only those animals for which $s - y$ is large, then the truncation at $z = s - y$ can be neglected, and we are led to consider the model

$$f(z) = q^{z-1}p, \quad z = 1, 2, \ldots .$$

This model has the simple maximum-likelihood estimate $\hat{p} = 1/\bar{z}$. Young *et al.* [1952] point out that once an animal has been recaptured, we are not interested in its subsequent fate, so that we do not need to "correct" the data for those dying or disappearing before the end of the experiment. Also, if the tendency to die or emigrate is not related to trap vulnerability, then those animals which die or emigrate before being recaptured at all will be distributed randomly over the groups that would have been recaptured after

1, 2, 3,... samples, and the disappearance of such animals will therefore not bias $\hat{p}$ and the associated goodness-of-fit test.

The above model has been used by several authors in bird banding studies (e.g. Young [1958], Swinebroad [1964]). Young also utilises 3rd, 4th, etc. captures in the estimation of p. This means that one animal can give rise to several values of z, representing the intervals between 1st and 2nd, 2nd and 3rd captures, etc. Obviously this estimate of p is more sensitive to the presence of any trap addiction among the animals, as the "repeaters" are counted several times.

TIME OF RESIDENCE. Suppose that the population under study is such that animals move into the population area, stay for a random length of time, and move out and stay out of the area for the remainder of the investigation. If the trapping is carried out at equally spaced intervals of time with constant trapping effort (i.e. p constant), then it is not unreasonable to assume that θ, the probability that an animal does not leave the trapping area some time between two successive trappings, is the same for all animals in the area and for all successive pairs of trappings. On the basis of these assumptions, Holgate [1964b] gives the following method for estimating θ and p from the observed values of Y, the recorded period of residence, i.e. the interval between the first and last occasions when it is actually captured.

Let Z denote the true period of residence of an individual in the study area, i.e. the interval between the first and last occasions when it is exposed to capture; then (ignoring truncation on the right)

$$\Pr[Z = z] = (1-\theta)\theta^z, \quad z = 0, 1, 2, \ldots. \tag{4.37}$$

Now an animal that remains in the area for z complete intervals is exposed to capture on $z + 1$ occasions, so that the probability of its not being caught at all (i.e. Y undefined) is

$$\Pr[Y \text{ undefined} \mid Z = z] = q^{z+1}.$$

Also

$$\Pr[Y = 0 \mid Z = z] = (z+1)pq^z$$

and, noting that as far as Y is concerned, it does not matter how often an animal is recaptured between its first and last capture,

$$\Pr[Y = y \mid Z = z] = (z-y+1)p^2q^{z-y}, \quad y = 1, 2, \ldots, z.$$

Now

$$\Pr[Y \text{ undefined}] = \sum_{z=0}^{\infty} \Pr[Y \text{ undefined} \mid Z = z]\Pr[Z = z]$$

$$= \frac{(1-\theta)q}{1 - q\theta}, \tag{4.38}$$

and in a similar fashion it is readily shown that

$$\Pr[Y = 0] = \frac{(1-\theta)p}{(1-q\theta)^2} \tag{4.39}$$

and

$$\Pr[Y = y] = \frac{p^2(1-\theta)\theta^y}{(1-q\theta)^2} \qquad y = 1, 2, \ldots. \qquad (4.40)$$

Finally we have (dividing (4.39) and (4.40) by one minus the probability (4.38)) the zero modified geometric distribution

$$\Pr[Y = 0 \,|\, Y \text{ defined}] = \frac{1-\theta}{1-q\theta} \qquad (4.41)$$

and

$$\Pr[Y = y \,|\, Y \text{ defined}] = \frac{(1-q)(1-\theta)\theta^y}{(1-q\theta)}, \qquad y = 1, 2, \ldots. \qquad (4.42)$$

If a sample of r observations is taken from the above distribution then the maximum-likelihood estimates of θ and q are

$$\hat{\theta} = 1 - (u/\bar{y})$$

and

$$\hat{q} = \frac{\hat{\theta} - u}{\hat{\theta}(1-u)}$$

where u is the proportion of individuals caught more than once (i.e. with $y > 0$). Holgate shows that as $r \to \infty$ the asymptotic variances and covariances of these estimates are given by

$$V[\hat{\theta} \,|\, r] = \frac{(1-\theta)^2(1-q\theta)}{r(1-q)},$$

$$V[\hat{q} \,|\, r] = \frac{(1-q)(1-q\theta^2)(1-q\theta)}{\cdot r\theta^2(1-\theta)},$$

and

$$\text{cov}[\hat{\theta}, \hat{q} \,|\, r] = \frac{(1-\theta)(1-q\theta)}{r\theta}.$$

    It is noted that, strictly speaking, (4.37) should be truncated on the right since the number of trappings is finite. Otherwise, at the end of the trapping series, the animals still in the area will be ascribed a duration of residence which is too short. Unfortunately, since the time of the first capture varies, each animal will have a different truncation point, thus leading to a complicated likelihood function for the estimation of θ and q. However, if the study period is long compared with the average time of residence, this effect will be negligible and the truncation can be ignored.

    Andrzejewski and Wierzbowska [1961] and Wierzbowska and Petrusewicz [1963] consider using an exponential distribution for the true residence period Z. They find that the values of y (treated as grouped data since the observed period of residence is measured in integral units) are not well fitted by an exponential distribution as there are too many observations in the zero class. However, if this class of animals, which they call "ephemeral", is omitted, it is found that the remaining data are well fitted by an exponential distribution with its zero class omitted (which is again an exponential distribution). From this the authors conclude that "residents" have a constant disappearance rate, while the greater rate for the ephemeral group is caused by the capture of "migrants" passing through the area. Therefore by extrapolating the successfully fitted exponential distribution the number of residents can be estimated; subtracting this number from the ephemerals gives an estimate of the migrants. But the authors effectively assume that $Y = Z$, i.e. the observed period of residence is the actual period of residence, which is not true as not all the animals in the trapping area are captured on every trapping occasion ($p < 1$). Holgate [1964b] points out that this feature of the model could account for the excess of animals in the zero class. In carrying out a goodness-of-fit test for Holgate's model the estimates of p and θ are such that the zero class is fitted exactly.

Holgate [1966] has recently considered two generalisations of the above theory. In the first generalisation he assumes that θ varies from animal to animal according to a beta-distribution, and he derives a new distribution for Y which he calls the "zero modified Appell distribution". However, this distribution involves hypergeometric functions and is therefore rather difficult to fit to data. In the second generalisation θ is constant but p is assumed to vary according to a beta-distribution. This time the distribution of Y is again a zero modified geometric distribution like (4.41) and (4.42), though unfortunately α, β and θ cannot be estimated uniquely.

BIVARIATE DISTRIBUTION. Holgate [1966] has utilised the joint distribution of Y (the recorded period of residence) and W (the number of captures during the *intervening* period, i.e. between the first and last capture) to obtain more efficient estimates of q and θ as follows (in Holgate's notation $Y = X$ and $W = Y$).

Since $Y = y$ implies $y + 1$ possible captures, the range of W is 0 to $y - 1$ and

$$\Pr[W = w \,|\, Y = y] = \binom{y-1}{w}(1-q)^{w} q^{\,y-w-1} \quad (y > 0).$$

Hence, from (4.41) and (4.42), the joint probability function is given by

$$\Pr[Y = 0 \,|\, Y \text{ defined}] = (1-\theta)/(1-q\theta)$$

and

$$\Pr[Y = y,\, W = w \,|\, Y \text{ defined}] = \binom{y-1}{w}(1-q)^{w+1} q^{\,y-w-1} \frac{(1-\theta)\theta^{y}}{(1-q\theta)},$$

where $w = 0, 1, \ldots, y - 1$ and $y = 1, 2, \ldots$. (For $Y = 0$, W is not defined.) Let n_0 and n_{yw} denote the corresponding sample frequencies and let r be the total number in the sample ($= n_0 + \sum_y \sum_w n_{yw}$). Then the likelihood function for the sample is proportional to

$$\left(\frac{1-\theta}{1-q\theta}\right)^{n_0} \prod_{y,w} \left\{ \frac{(1-q)^{w+1} q^{\,y-w-1}(1-\theta)\theta^{y}}{(1-q\theta)} \right\}^{n_{yw}}$$

and it transpires that the maximum-likelihood estimates $\tilde{q}$ and $\tilde{\theta}$ are given by

$$\tilde{q} = (\bar{y}-\bar{w}-u)(1+\bar{w}+u)/[\bar{y}-(\bar{w}+u)(1-\bar{y})]$$

and
$$\tilde{\theta} = [\bar{y}-(\bar{w}+u)(1-\bar{y})]/[\bar{y}(1+\bar{w}+u)],$$

where $\bar{w} = \underset{y}{\Sigma}\underset{w}{\Sigma} wn_{yw}/r = \Sigma wn_{.w}/r, \quad$ say

$\bar{y} = \underset{y}{\Sigma}\underset{w}{\Sigma} yn_{yw}/r = \Sigma yn_{y.}/r, \quad$ say

and

$u = 1 - (n_0/r).$

Holgate shows that the asymptotic variance–covariance matrix of $\tilde{q}$ and $\tilde{\theta}$ is given by

$$\frac{1}{r}\begin{bmatrix} \dfrac{q(1-q)(1-q\theta^2)(1-\theta)}{\theta^2(1-q\theta)}, & \dfrac{q(1-\theta)^3}{\theta(1-q\theta)} \\[3mm] \dfrac{q(1-\theta)^3}{\theta(1-q\theta)}, & \dfrac{(q+\theta+q^2\theta^2-3q\theta)(1-\theta)^2}{(1-q)(1-q\theta)} \end{bmatrix}$$

where the (1, 1) element is the asymptotic variance of $\tilde{q}$.

Using the determinant of the variance–covariance matrix as a measure of asymptotic efficiency, Holgate shows that the relative efficiency of the above method with respect to the previous method based on the marginal distribution of Y only is

$$e = q(1-\theta)^2/(1-q\theta)^2 .$$

Example 4.17 Wood mouse (*Apodemus sylvaticus* (L.)): Holgate [1966]

Holgate applied the above theory to the data in Table 4.21. From this table we have

$$n_0 = 65, \qquad r = 65 + 47 = 112,$$
$$u = 47/112, \quad \bar{y} = 255/112 \quad \text{and} \quad \bar{w} = 26/112.$$

Multiplying through by r^2, we have

$$\tilde{q} = \frac{(255-36-47)(112+36+47)}{[255(112) - (36+47)(112-225)]} = 0{\cdot}8298,$$

and similarly

$$\hat{\theta} = 0{\cdot}8131.$$

Using these estimates in the variance formulae, we find that

$$\tilde{\sigma}[\tilde{q}] = 0{\cdot}0215 \quad \text{and} \quad \tilde{\sigma}[\hat{\theta}] = 0{\cdot}0203.$$

If only the marginal distribution of Y is used, we have

$$u = 47/112, \qquad \bar{y} = 255/112,$$
$$\hat{\theta} = 1 - (u/\bar{y}) = 1 - (47/255) = 0{\cdot}8157,$$

and

$$\hat{q} = \frac{\bar{y} -u(\bar{y}+1)}{(\bar{y}-u)(1-u)} = \frac{255(112) - 47(255+112)}{(255-47)(112-47)} = 0{\cdot}8366.$$

TABLE 4.21

Bivariate frequency distribution (n_{yw}) of length of residence and number of times caught in the intervening period for a population of wood mice: from Holgate [1966: Table 3].

Length of residence in weeks, y	Frequency of capture, w							$n_y.$	$yn_y.$
	0	1	2	3	4	5	6		
0	65	–	–	–	–	–	–	65	
1	6	0	0	0	0	0	0	6	6
2	5	1	0	0	0	0	0	6	12
3	6	0	0	0	0	0	0	6	18
4	9	0	0	0	0	0	0	9	36
5	1	2	0	0	0	0	0	3	15
6	0	1	1	0	0	0	0	2	12
7	0	2	0	0	0	0	0	2	14
8	0	0	0	1	1	0	0	2	16
9	3	0	1	0	0	0	0	4	36
10	0	1	0	0	0	0	0	1	10
11	0	0	1	1	0	0	0	2	22
13	0	0	1	0	1	0	0	2	26
16	0	1	0	0	0	0	1	2	32
$n._w$	30	8	4	2	2	0	1	47	255
$wn._w$	0	8	8	6	8	0	6	36	

Also $\hat{e} = 0·28$, indicating that the method using the joint distribution is much more efficient.

Unfortunately the numbers are too small for an adequate test of fit to the bivariate distribution. However, using the estimates $\hat{q}$ and $\hat{\theta}$, Holgate carried out a goodness-of-fit test of the marginal distribution of Y to the modified geometric distribution given by (4.41) and (4.42). Here the observed frequencies are given by $f_0 = n_0$ and $f_y = n_y.$; the expected frequencies are

$$E_0 = r(1-\hat{\theta})/(1-\hat{q}\hat{\theta})$$

and

$$E_y = r\frac{(1-\hat{q})(1-\hat{\theta})\hat{\theta}^y}{(1-\hat{q}\hat{\theta})} \quad (y > 0).$$

These are set out in Table 4.22, and (pooling as indicated)

$$\Sigma (f_y - E_y)^2/E_y = 7·65,$$

which, at 6 degrees of freedom, indicates a good fit. Further examples of this goodness-of-fit test are given in Tanton [1969: 519].

6 *A runs test for randomness*

When the probability of capture is the same for each sample, Young [1961] gives a runs test for testing the randomness of the captures. He

TABLE 4.22

Observed length of residence compared with expected length of residence, fitted by Holgate's modified geometric distribution for a population of wood mice: from Holgate [1966: Table 1].

Length of residence in weeks, y	Observed frequency f_y	Expected frequency E_y	$\dfrac{(f_y - E_y)^2}{E_y}$
0	65	65·00	0
1	6	8·66	0·82
2	6	7·07	0·16
3	6	5·76	0·01
4	9	4·70	3·93
5	3	3·83	0·18
6	2	3·13	0·41
7	2 ⎫	2·55 ⎫	0·09
8	2 ⎭	2·08 ⎭	
9	4 ⎫	1·70 ⎫	1·85
10	1 ⎥	1·38 ⎥	
11	2 ⎭	1·13 ⎭	
12	0 ⎫	0·92 ⎫	
13	2 ⎥	0·75 ⎥	
14	0 ⎥	0·61 ⎥	0·20
15	0 ⎥	0·50 ⎥	
16	2 ⎥	0·41 ⎥	
⩾17	0 ⎭	1·82 ⎭	
Total	112	112·00	7·65

demonstrates the test with the following example.

From January 29 to April 21, 1958, a ten-trap banding station was in operation in a local cemetery. The Slate-coloured Junco (*Junco hyemalis*) was one of the species captured, and the data were first organised by designating as A any day on which at least one junco was caught, and as B any day on which no juncos were captured. These were then arranged in their natural chronological sequence, giving the pattern shown below:

BB A B AA B AA B AAA B A BBB A BB A B A BBBBBB AAAA BBB A BB A BBBB A BBBBB A

From this pattern we have $n_A = 20$ A's, $n_B = 32$ B's and $x = 26$ runs of either A or B (a single letter counts as a run). If the sampling is random, then (Mood [1940])

$$E[x \mid n_A, n_B] = 2np_A q_A + 1 = \mu, \quad \text{say,}$$

and
$$V[x \mid n_A, n_B] = 2np_A q_A (2np_A q_A - 1)/(n-1) = \sigma^2, \quad \text{say,}$$

where $n = n_A + n_B$ and $p_A = 1 - q_A = n_A/n$.

Provided n_A and n_B are both greater than 10, x is approximately normal, so that we can test for randomness using the statistic $z = (x - \mu)/\sigma$. Therefore, from the above data we have $z = (26 - 25 \cdot 62)/11 \cdot 33$ which, for the unit normal, is not significant, so that the hypothesis of randomness is not rejected.

Strictly speaking, the above conditional test is not correct, as n_A and n_B are random variables rather than fixed parameters chosen in advance (cf. Brunk [1965: 354] for an excellent discussion). From the delta method and **1.3.4**, equation (1.12), we have approximately
$$E[x] = 2npq + 1$$
and
$$V[x] = 2npq (2npq - 1)/(n-1) + V[2np_A q_A],$$

where p is the probability of capture on a given day. Since p can be estimated by p_A, we see that $V[x]$ exceeds σ^2 by approximately $V[2np_A q_A]$. This means that the above conditional test may reject the hypothesis of randomness when in fact the trapping is random. For a proper application of the conditional test one chooses the numbers n_A and n_B in advance, and if, for example, $n_A + k$ A's actually turn up, k of the A's are chosen at random (e.g. using tables of random numbers) and changed to B's. A correction for continuity and tables for carrying out an exact test are given in Swed and Eisenhart [1943].

In conclusion we note that the above test can be applied to the trapping record of a single marked individual.

4.2   INVERSE SCHNABEL CENSUS

Consider a Schnabel census in which, for each sample n_i, the sampling is now continued until a *predetermined* number of marked animals m_i are captured. This modification is a generalisation of **3.5** and for obvious reasons can be suitably called the "inverse" Schnabel census. Using the same notation as in **4.1.1**, we have fixed parameters N, s, n_1 $(= M_2)$, m_2, $m_3, \dots, m_s$, random variables $M_3, M_4, \dots, M_s, r, n_2, \dots, n_s$, and the joint probability function of the random variables is a straightforward generalisation of (3.36), namely

$$\prod_{i=2}^{s} \left\{ \frac{\dbinom{M_i}{m_i - 1} \dbinom{N - M_i}{u_i}}{\dbinom{N}{n_i - 1}} \cdot \frac{M_i - m_i + 1}{N - n_i + 1} \right\}.$$

We find that the maximum-likelihood estimate $\hat{N}$ is still given by the appropriate root of (4.4), but the asymptotic bias and variance of $\hat{N}$ are not so readily found as in the "direct" Schnabel census. Alternative methods are therefore desirable, and by analogy with (3.37) Chapman [1952: 297] suggests the mean

$$\bar{N}_2 = \sum_{i=2}^{s} \left\{ \frac{n_i(M_i + 1)}{m_i} - 1 \right\} \bigg/ (s-1),$$

which is unbiased with approximate variance

$$V[\bar{N}_2] = \frac{N^2}{(s-1)^2} \sum_{i=2}^{s} \frac{1}{m_i}.$$

Here the coefficient of variation

$$C[\bar{N}_2] = \left\{ \sum_{i=2}^{s} \frac{1}{m_i} \right\}^{\frac{1}{2}} \bigg/ (s-1)$$

can be used for choosing the fixed parameters to give a predetermined precision. However, the correct choice of the m_i is also important from another point of view, for in **3.5.1** it was pointed out that a wrong choice of m_i, coupled with an unfavourable M_i, could give rise to a large n_i. Therefore a reasonable criterion, suggested by Chapman, for choosing these parameters is to minimise $E[\Sigma \, n_i]$, subject to $C[\bar{N}_2]$ being held constant; unfortunately this does not have a simple solution (a few values are tabulated in Chapman [1952: 306]). As pointed out by Cormack [1968: 448], an increasing sequence of the m_i spreads the sampling effort most evenly, and an examination of Chapman's table seems to indicate that a constant m_i provides the maximum precision for approximately the same expected sample size $E[\Sigma \, n_i]$.

To avoid the possibility of large n_i we can modify the above model as in **3.5.1** and continue the sampling until a predetermined number u_i of unmarked individuals is taken in the ith sample ($i = 2, 3, \ldots, s$). This means that our fixed parameters are now N, s, n_1, $u_2, \ldots, u_s$, $M_3, \ldots, M_s$, r and the random variables $n_2, n_3, \ldots, n_s, m_2, \ldots, m_s$. By analogy with (3.38) an approximately unbiased estimate of N is the mean

$$\bar{N}_3 = \sum_{i=2}^{s} \hat{N}_{3i}/(s-1),$$

where $\hat{N}_{3i} = \frac{n_i(M_i + 1)}{(m_i + 1)} - 1,$

and, asymptotically,

$$V[\bar{N}_3] = \sum_{i=2}^{s} V[\hat{N}_{3i}]/(s-1)^2,$$

where $V[\hat{N}_{3i}]$ can be evaluated using (3.39). A simpler estimate of variance is given by (cf. (1.9))

$$v[\bar{N}_3] = \sum_{i=2}^{s} (\hat{N}_{3i} - \bar{N}_3)^2/(s-1)(s-2).$$

4.3 SEQUENTIAL SCHNABEL CENSUS

4.3.1 General methods

In the inverse Schnabel census we fixed the m_i or u_i in advance and took a *fixed* number of samples. A more flexible, sequential-type method suitable for small recaptures m_i would be to fix the sequence $n_1, n_2, \ldots$ in advance and continue this sequence until a predetermined number of re-captures $\Sigma\, m_i$ have been taken, s being now a random variable. Thus we have the fixed parameters $N, n_1, n_2, \ldots, L$, and the random variables $m_2, m_3, \ldots, M_3, M_4, \ldots, r, s, n\ (= \sum_{i=1}^{s} n_i)$, and the procedure is to stop sampling at the completion of the sth sample, s being defined by

$$\sum_{i=2}^{s-1} m_i < L, \qquad \sum_{i=2}^{s} m_i \geqslant L.$$

1 Goodman's model

Goodman [1953] showed that r is sufficient for N and he deduced the existence of a minimum variance unbiased estimate (MVUE) as the ratio of two determinants of order $n - n_1 + 1$. In general, these determinants will be of high order, so that a fair amount of calculation is involved. However, an asymptotic MVUE is available in the form of

$$\hat{N}_4 = n^2/(2L)$$

with asymptotic variance

$$V[\hat{N}_4] = N^2/L.$$

Confidence intervals for $1/N$ (and hence for N) can then be obtained from n^2/N, which is asymptotically distributed as chi-squared with $2L$ degrees of freedom. The coefficient of variation of $\hat{N}_4$ takes the simple form $1/\sqrt{L}$, which can be used for choosing L to give a predetermined precision.

In looking up the chi-squared table we find that for $2L > 30$, which is generally the case for a reasonable coefficient of variation, we use

$$z = \sqrt{(2n^2/N)} - \sqrt{(4L-1)},$$

which is asymptotically distributed as $\mathcal{N}(0, 1)$. Thus an approximate 95 per cent confidence interval for N is given by

$$\frac{2n^2}{[1\cdot 96 + \sqrt{(4L-1)}]^2} < N < \frac{2n^2}{[-1\cdot 96 + \sqrt{(4L-1)}]^2}. \qquad (4.43)$$

When $2L > 30$ we have $E[n] \approx \sqrt{(2NL)}$ which, for the special case $n_1 = n_2 = \ldots (= n_0$ say), leads to $E[s] \approx \sqrt{(2NL)}/n_0$. Therefore, if L has been chosen and a rough estimate of N is available before the experiment, we can either calculate the expected duration $E[s]$ of the experiment for a given n_0, or determine n_0 for a given expected duration.

Example 4.18 Hypothetical data: Goodman [1953: Table 2]

Seven samples of 100 $(n_1 = n_2 = \ldots = n_7 = 100)$, followed by samples of size 1 $(n_8 = n_9 = \ldots = 1)$, were drawn from a population of $N = 10\,000$

random numbers. Sampling ceased when a total of $L = 25$ "recaptures" were made, giving the following data:

Sample	1	2	3	4	5	6	7	8, 9,..., 72	Total
n_i	100	100	100	100	100	100	100	65	765
m_i	—	4	4	1	3	2	7	4	25

From the above table we have $n = 765$, $\hat{N}_4 = (765)^2/50 = 11\ 704$ and the confidence interval (4.43) is (8251, 18 334).

2   Chapman's model

Assuming the M_i to be fixed parameters as in (4.12) and approximating the distribution of m_i by the Poisson distribution with mean $n_i M_i/N$ (which is satisfactory if n_i/N and M_i/N are both less than $0\cdot 1$), Chapman [1954: 6] derives an alternative (MVUE) for N, namely

$$\hat{N}_5 = \sum_{i=2}^{s} n_i M_i/L = \lambda/L, \quad \text{say},$$

with variance N^2/L. Confidence intervals for N can be based on $2\lambda/N$, which is approximately distributed as chi-squared with $2L$ degrees of freedom. For $2L > 30$ an approximate 95 per cent confidence interval for N is given by

$$\frac{4\lambda}{[1\cdot 96 + \sqrt{(4L-1)}]^2} < N < \frac{4\lambda}{[-1\cdot 96 + \sqrt{(4L-1)}]^2} .$$

We note that Chapman's model would still be applicable even when there is a small natural mortality, provided that marked and unmarked have the same mortality rate, so that M_i/N remains constant between the $(i-1)$th and ith samples. An additional advantage of Chapman's model is that the marking can be done independently of the sampling.

4.3.2   Samples of size one

1   Fixed $s - r$

If we put $n_i = 1$ in Goodman's model above, the sampling procedure amounts to sampling one at a time with replacement until $s - r = L$ marked animals have been caught. The general theory of **4.3.1**(1) still applies, with $n = s$, but now the exact MVUE mentioned there has a simpler form, namely (Goodman [1953: 61])

$$\hat{N}_6 = K(r, L)/K(r, L-1),$$

where $K(r, 0) = r$ and $K(r, L) = r \sum_{i=1}^{r} K(i, L-1)$.

Darroch [1958: 351] has also shown that

$$\hat{N}_6 = \sigma_{r+L}^r/\sigma_{r+L-1}^r,$$

where $\sigma_y^x = \Delta^x(0^y)/x!$, a Stirling number of the second kind. As far as maximum-likelihood estimation is concerned, it transpires that (cf. Darroch [1958: 350], Samuel [1968]) the maximum-likelihood estimate $\hat{N}$ of N is the same as for the direct census and is therefore given by (4.10).

A different approach to the problem is given by Boguslavsky [1955] as follows. He shows that

$$g(N, j) = \Pr[S \leqslant j \,|\, N]$$
$$= \binom{N}{N-j+L}^{-1} \sum_{i=0}^{j-L} \left\{ (-1)^i \binom{j-L}{i} \left(\frac{j-L-i}{N} \right)^j \binom{N-j+L}{N-j+L+i} \right\}, \qquad (4.44)$$

where S is the random variable taking values s, and uses the above function as a basis for constructing uniformly most powerful (UMP) one-sided tests. Thus a UMP test of size α for the null hypothesis $H_0 : N \geqslant N_0$ against the alternative $H_1 : N < N_0$ consists of rejecting H_0 at the 100α per cent level of significance if $s \leqslant a$, where s is the observed value of S and a is a function of N_0 determined by $g(N_0, a) = \alpha$. Similarly the rejection rule for a UMP test of $H_0 : N \leqslant N_0$ against $H_1 : N > N_0$ is given by $s \geqslant b$, where $g(N_0, b-1) = 1 - \alpha$. However, for testing $H_0 : N = N_0$ against the two-sided alternative $H_1 : N \neq N_0$ we find that a UMP test does not exist. Therefore an approximation to the best unbiased test is to accept H_0 if $c + 1 \leqslant s \leqslant d - 1$, where c and d are functions of N_0 such that

$$g(N_0, c) = \alpha/2 \quad \text{and} \quad g(N_0, d-1) = 1 - (\alpha/2).$$

Such a method can also be used to obtain a $100(1-\alpha)$ per cent two-sided confidence interval for N as the set of all N_0 such that the interval $[c+1, d-1]$ contains the observed value s. Therefore, if this interval is $N_c \leqslant N \leqslant N_d$ then it can be shown that N_c and N_d are given by

$$g(N_c, s) = \alpha/2 \quad \text{and} \quad g(N_d, s-1) = 1 - (\alpha/2).$$

As N_c and N_d are integers, these equations, in general, will not have exact solutions, and we would choose the nearest integers so that

$$g(N_c, s) \leqslant \alpha/2 \quad \text{and} \quad g(N_d, s) \geqslant 1 - (\alpha/2).$$

Boguslavsky gives a chart for reading off N_c and N_d for $\alpha = 0 \cdot 2$, $L = 1(1)5$ and $s = 1$ to 80 (in his notation $s = n_k$, $L = k$, $N_c = N_L$ and $N_d = N_U$). However, it should be noted that, as in **4.3.1** (1), $1/\sqrt{L}$ is still a rough measure of the coefficient of variation and, in particular, of half the proportional width of the confidence interval. This means that the interval (N_c, N_d) will be wide if L is small. Boguslavsky also gives a method for testing the adequacy of the above model (4.44), but unfortunately the calculations involved are considerable unless L is small.

2   *Fixed r*

A variation on the above theme is to continue sampling until a fixed number r of unmarked individuals is caught. Darroch [1958: 350] shows that the maximum-likelihood estimate of N is still $\hat{N}$ and points out that (4.10) has no solution $\hat{N} \geqslant r$ when

$$s > r \left(1 + \frac{1}{2} + \ldots + \frac{1}{r} \right).$$

He adds that this situation is unlikely to happen in practice and proves that asymptotically ($N \to \infty$, $r \to \infty$, r/N constant)

$$E[\hat{N}] = N\left\{1 + \sum_{k=1}^{r-1} \frac{k}{(N-k)^3}\left[\sum_{k=1}^{r-1} \frac{k}{(N-k)^2}\right]^{-2}\right\}$$

and

$$V[\hat{N}] = N\left[\sum_{k=1}^{r-1} \frac{k}{(N-k)^2}\right]^{-1}$$

The above sums can be simplified using the following integral approximations:

$$\sum_{k=1}^{r-1} \frac{k}{(N-k)^2} \approx \int_0^1 \frac{x}{\left(\frac{N}{r} - x\right)^2}\, dx$$

and

$$\sum_{k=1}^{r-1} \frac{k}{(N-k)^3} \approx \frac{1}{r}\int_0^1 \frac{x}{\left(\frac{N}{r} - x\right)^3}\, dx\,.$$

Darroch also mentions how s, which is sufficient for N, can be used for constructing a confidence interval for N in much the same way as r was used on p. 134.

3 Variable stopping rules

Boguslavsky [1956], Darling and Robbins [1967] and Samuel [1968] have given a number of methods in which the stopping rule at any particular instant depends on the number of unmarked individuals already caught. The method given in the first two articles amounts to sampling until the absence of further unmarked individuals in the samples leads the experimenter to believe that he has marked the whole population. Mathematically this means that we require a stopping rule for s such that if r is the number of different (unmarked) individuals caught throughout the entire experiment, then

$$\Pr[r = N] \geqslant 1 - \alpha, \tag{4.45}$$

i.e. we have at least $100(1-\alpha)$ per cent confidence that the whole population has been marked. Since $r \leqslant N$, (4.45) can be rewritten as

$$\Pr[r = 1] + \Pr[r = 2] + \ldots + \Pr[r = N-1] \leqslant \alpha,$$

which was used by Boguslavsky to give the following stepwise procedure for calculating a suitable stopping rule.

The simplest situation that one can have is when the same individual keeps turning up, i.e. $r = 1$, so that at the end of $s = s_1$ samples the observer may wish to terminate the experiment with the statement that $N = 1$. In making this decision, the observer rejects the most likely alternative hypothesis that $N = 2$ with a Type II error of $2(\frac{1}{2})^s$ or $(\frac{1}{2})^{s-1}$. If we set this probability equal to α (say $0 \cdot 10$) and solve for s, we find that $s = 4 \cdot 3$. Therefore a suitable stopping rule would be to stop sampling if there are $s_1 = 5$ consecutive drawings of the same individual and accept $N = 1$ with, using an obvious notation,

$$\Pr[r = 1 \,|\, N = 2] = \Pr[s = s_1 \,|\, N = 2]$$

$$= \left(\frac{1}{2}\right)^4$$

$$< 0{\cdot}10.$$

However, if a second individual turns up within the 5 samples, our stopping rule changes and we now look for an $s_2 > s_1$ such that we accept $N = 2$ if no new individual turns up by sample number s_2. This amounts to finding s_2 such that

$$\Pr[r = 1 \,|\, N = 3] + \Pr[r = 2 \,|\, N = 3] = \Pr[s = s_1 \,|\, N = 3] + \Pr[s = s_2 \,|\, N = 3]$$

$$\leqslant 0{\cdot}10.$$

Thus, proceeding stepwise in this manner it is possible to calculate the following sequence $s_1, s_2, \ldots$, which we reproduce from Boguslavsky:

i	1	2	3	4	5	6	7	8	9	10	11	12	13	14	15
s_i	5	9	14	18	23	28	33	39	44	50	55	61	67	73	79

In general one would use this method for small populations only, as s is much greater than N. This means that the above table ($N \leqslant 15$, $\alpha = 0{\cdot}10$) is probably satisfactory for most practical purposes. To illustrate the use of the table, suppose that unmarked animals are caught at samples 1, 3, 6, 11, 17 and no new individuals are captured from samples 18 to 23 inclusive. Then our sampling would stop at $s = s_5$, and we would accept $N = 5$ with 90 per cent confidence.

    Darling and Robbins [1967] also derive a stopping rule for which (4.45) is true, but use a different sequence $b_1, b_2, \ldots$, where b_i is the longest "waiting time" allowed between the appearances of the ith and $(i + 1)$th unmarked individuals. Suppose the ith unmarked animal appears on the (Y_i)th sample ($Y_1 = 1$, $Y_{N+1} = \infty$) and let $X_i = Y_{i+1} - Y_i$ ($i = 1, 2, \ldots, N$) be the ith "waiting time". Then, if the stopping rule is such that one stops sampling the first time the event $X_i > b_i$ occurs for some i, it transpires that

$$\Pr[r = N] = \prod_{i=1}^{N-1} \{1 - (i/N)^{b_i}\}. \tag{4.46}$$

Darling and Robbins suggest choosing the sequence (b_i) in such a way that the right-hand side of (4.46) is greater than or equal to α for $N = 2, 3, \ldots$. (i.e. essentially Boguslavsky's stepwise approach). Their solution is the sequence (c_i), where $c_1, c_2, c_3, \ldots$ are the smallest integers satisfying

$$1 - \left(\frac{1}{2}\right)^{c_1} \geqslant 1 - \alpha, \quad \left[1 - \left(\frac{1}{3}\right)^{c_1}\right]\left[1 - \left(\frac{2}{3}\right)^{c_2}\right] \geqslant 1 - \alpha,$$

$$\left[1 - \left(\frac{1}{4}\right)^{c_1}\right]\left[1 - \left(\frac{2}{4}\right)^{c_2}\right]\left[1 - \left(\frac{3}{4}\right)^{c_3}\right] \geqslant 1 - \alpha, \text{ etc.}$$

Therefore, if $\alpha = 0 \cdot 10$, we have, using logarithms, $c_1 = 4$, $c_2 = 9$, etc., so that if, for example, the second unmarked animal does not turn up after $1 + 4 = 5$ samples, then $X_1 > 4$ and we stop sampling. If, however, the second unmarked animal turns up at sample 3, then we stop sampling if a third unmarked animal has not been captured by the $3 + 9 = 12$th sample.

Darling and Robbins also consider a stopping rule based on the "cumulative waiting times" Y_i, but this turns out to be the same as that given by Boguslavsky above. They also suggest an asymptotic solution to the problem of finding a suitable sequence (s_i), namely (cf. Samuel [1968: rule D with $c = D + 1$])

$$s_i \geqslant \max \, (i+1, \, i \, \log i + ci), \quad -\infty < c < \infty,$$

where $\quad \lim \Pr[r = N] = \exp \, (-\exp(-c)) \geqslant 1 - \alpha, \quad N \to \infty.$

But this stopping rule needs further investigation for small and moderate N as s may be impracticably large for large N: an upper bound for s is $N \log N + cN$.

One further stopping rule is considered in some detail by Samuel [1969: rule C]. Let $C > 0$ be a fixed constant, then one samples until the ratio of the number of samples which are marked $(s-r)$ to the number (r) which are unmarked is at least C, i.e. until $(s-r)/s \geqslant C$. This rule is investigated numerically by Samuel for different values of C, and the reader is referred to her article for further details.

Samuel also shows that for all the stopping rules considered in this section, $\hat{N}$, the solution of (4.10), is still the maximum-likelihood estimate of N.

4.4 THE MULTI-SAMPLE SINGLE-RECAPTURE CENSUS

We shall now consider a census technique which is specially suited to commercially exploited populations such as fisheries, where the samples are permanently removed from the population. The method is as follows. The experimenter, using differentiated marking, releases batches of marked individuals containing R_1, R_2, ..., R_s members respectively into the population, and after each batch R_i is released, a sample of size n_i is removed permanently from the population, thus giving the sequence R_1 added, n_1 removed, R_2 added, n_2 removed, etc. Ideally the marked individuals should either be caught before the experiment and stored, or perhaps taken from a similar population not connected with the one under investigation. In actual practice, however, the experimenter will usually take each batch from the population during the experiment because the R_i, although large, will generally be much smaller than the n_i, so that the recaptures in the successive batches will be negligible. Also the overall reduction in the number of unmarked individuals due to the marking of $\Sigma \, R_i$ individuals will be small compared with the total population size. Let

N = initial population size,

m_j = number of marked in n_j $(j = 1, 2, \ldots, s)$,

m_{ij} = the number of individuals from R_i caught in n_j $(j = i, i + 1,$
$\ldots, s)$,

u_j = $n_j - m_j$,

and $r = \sum\limits_{j=1}^{s} u_j$.

Then, regarding the $\{n_j\}$ as fixed parameters and assuming that each sample is a simple random sample, the joint probability function of $\{m_{ij}, u_j\}$ is the multi-hypergeometric distribution (cf. Seber [1962: 347])

$f(\{m_{ij}, u_j\} | \{R_i, n_j\})$

$$= \prod_{i=1}^{s} \left\{ \frac{\binom{R_1 - \sum\limits_{j=1}^{i-1} m_{1j}}{m_{1i}}\binom{R_2 - \sum\limits_{j=2}^{i-1} m_{2j}}{m_{2i}} \cdots \binom{R_i}{m_{ii}}\binom{N - \sum\limits_{j=1}^{i-1} u_j}{u_i}}{\binom{N + \sum\limits_{j=1}^{i} R_j - \sum\limits_{j=1}^{i-1} n_j}{n_i}} \right\}$$

Setting $\Delta \log f = 0$, the maximum-likelihood estimate $\hat{N}$ of N is the unique root, greater than r, of $h(N) = 0$, where

$$h(N) = N - r - N \prod_{i=1}^{s} \left(1 - \frac{n_i}{N + \sum\limits_{j=1}^{i} R_j - \sum\limits_{j=1}^{i-1} n_j} \right).$$

Seber shows that $\hat{N}$ is asymptotically unbiased and that the asymptotic variance of $\hat{N}$ is estimated by

$v[\hat{N}] = (\hat{N} - r)/h'(\hat{N})$

$$= \left[\frac{r}{\hat{N}(\hat{N} - r)} - \sum_{i=1}^{s} \frac{n_i}{\left(\hat{N} + \sum\limits_{j=1}^{i} R_j - \sum\limits_{j=1}^{i} n_j \right)\left(\hat{N} + \sum\limits_{j=1}^{i} R_j - \sum\limits_{j=1}^{i-1} n_j \right)} \right]^{-1}$$

As in **4.1.2 (1)**, $h(N) = 0$ can be solved using Robson and Regier's iterative method, and once again the last iteration provides an approximation for $v[\hat{N}]$. By setting $D_i = \sum\limits_{j=1}^{i-1} n_j - \sum\limits_{j=1}^{i} R_j$, a first approximation to $\hat{N}$ is given by (4.20) on page 154: alternatively one can use the mean estimate (4.21) with $M_i = \sum\limits_{j=1}^{i} R_j - \sum\limits_{j=1}^{i-1} m_j$ in N_i^*.

For the more realistic situation where the mark releases R_i are obtained from the population during the course of the experiment, we have

$f(\{m_{ij}, u_j\} | \{R_i, n_j\})$

$$= \prod_{i=1}^{s} \left\{ \frac{\binom{R_1 - \sum\limits_{j=1}^{i-1} m_{1j}}{m_{1i}} \cdots \binom{R_i}{m_{ii}}\binom{N - \sum\limits_{j=1}^{i} R_j - \sum\limits_{j=1}^{i-1} u_j}{u_i}}{\binom{N - \sum\limits_{j=1}^{i-1} n_j}{n_i}} \right\}$$

In this case R_i refers to the *newly* marked individuals released; any recaptures are not retagged but simply returned to the population. The maximum-likelihood estimate $\hat{N}$ is the largest root of

$$\prod_{i=1}^{s} \left\{ \frac{N - \sum_{j=1}^{i} R_j - \sum_{j=1}^{i} u_j}{N - \sum_{j=1}^{i} R_j - \sum_{j=1}^{i-1} u_j} \right\} = \prod_{i=1}^{s} \left\{ \frac{N - \sum_{j=1}^{i} n_j}{N - \sum_{j=1}^{i-1} n_j} \right\}$$

and we note that this equation has the same simple interpretation as (4.4).

In actual practice the $\{n_j\}$ will be random variables, and to set up a probability model for this situation we make the usual assumption that p_i ($= 1 - q_i$), the probability of capture of a given animal in the ith sample, given that the animal is in the population at the time of the sample, is the same for all animals, irrespective of whether they are marked or unmarked. Under this assumption the random variables $\{m_{ij}\}$ and $\{u_j\}$ are independent and

$$f(\{m_{ij}\}|\{R_i\})$$

$$= \prod_{i=1}^{s} \left\{ \frac{R_i!}{\prod_{j=i}^{s} m_{ij}! \ (R_i - r_i)!} p_i^{m_{ii}} \ (q_i p_{i+1})^{m_{i, i+1}} \dots (q_i q_{i+1} \dots q_s)^{R_i - r_i} \right\},$$

$$(4.47)$$

where $r_i = \sum_{j=1}^{s} m_{ij}$. As in the Schnabel census of **4.1.2(1)**, $\hat{N}$ and $v[\hat{N}]$ are the same, irrespective of whether the n_i are regarded as fixed parameters or random variables. However, (4.47) can be used to provide a goodness-of-fit test to the hypothesis H that all *marked* individuals have the same probability p_i of being caught in the ith sample, as follows. Let p_{ij} be the probability that a member of R_i is caught in n_j, then H is the hypothesis that $p_{ij} = q_i \dots q_{j-1} p_j$ ($j > i$) and $p_{ii} = p_i$. This can be tested using (cf. Mitra [1958])

$$T = \sum_{i=1}^{s} \sum_{j=i}^{s+1} \frac{(m_{ij} - R_i \hat{p}_{ij})^2}{R_i \hat{p}_{ij}},$$

where $\hat{p}_{ij} = \hat{q}_i \dots \hat{q}_{j-1} \hat{p}_j$, $\hat{p}_{i, s+1} = \hat{q}_i \hat{q}_{i+1} \dots \hat{q}_s$,

$$\hat{p}_i = m_i \bigg/ \left(\sum_{j=1}^{i} R_j - \sum_{j=1}^{i-1} m_j \right) \quad \text{and} \quad m_{i, s+1} = R_i - r_i.$$

When H is true and the R_i are large, T is approximately distributed as chi-squared with $\frac{1}{2}s(s+1) - s$ degrees of freedom.

CHAPTER 5

OPEN POPULATION:
MARK RELEASES DURING SAMPLING PERIOD

5.1 THE JOLLY–SEBER METHOD

5.1.1 The Schnabel census

In this chapter we shall apply the Schnabel method of **4.1.1** to an open population in which there is possibly death, recruitment, immigration and *permanent* emigration (i.e. animals enter and leave the population only once). We shall again consider a sequence of s samples of sizes $n_1, n_2, ..., n_s$ but with the added generalisation that, for each sample, only R_i of the n_i are marked and returned to the population. This more general model allows for accidental deaths due to marking and handling, and also includes the case when some of the R_i are zero, as in commercial exploitation where the sample is permanently removed from the population. The following discussion of this model is based mainly on Jolly [1965] and Seber [1965].

1 Assumptions and notation

We shall assume the following:

(a) Every animal in the population, whether marked or unmarked, has the same probability p_i ($i = 1 - q_i$) of being caught in the ith sample, given that it is alive and in the population when the sample is taken.

(b) Every marked animal has the same probability ϕ_i of surviving from the ith to the $(i+1)$th sample and of being in the population at the time of the ith sample, given that it is alive and in the population immediately after the ith release ($i = 1, 2, ..., s-1$).

(c) Every animal caught in the ith sample has the same probability ν_i of being returned to the population: in many experiments $1 - \nu_i$ can be regarded as the probability of accidental death through handling, etc.

(d) Marked animals do not lose their marks and all marks are reported on recovery.

(e) All samples are instantaneous, i.e. sampling time is negligible.

Let

t_i = time when the ith sample is taken,
N_i = total number in the population just before time t_i,
M_i = total number of marked animals in the population just before time t_i,
U_i = $N_i - M_i$,
n_i = number caught in the ith sample,

m_i = number of marked animals caught in the ith sample,

u_i = $n_i - m_i$,

m_{hi} = number caught in the ith sample last captured in the hth $(1 \leqslant h \leqslant i - 1)$,

R_i = number of marked animals released after the ith sample,

r_i = number of marked animals released after the ith sample which are subsequently recaptured,

z_i = number of animals caught before the ith sample which are not caught in the ith sample but are caught subsequently,

B_i = number of *new* animals joining the population in the interval from time t_i to time t_{i+1} which are still alive and in the population at time t_{i+1},

$N_i(h)$ = number in the population at time t_i which first joined the population between times t_h and t_{h+1}, that is, which are members of B_h $(1 \leqslant h \leqslant i - 1)$,

ρ_i = M_i/N_i.

(The above notation is different from Jolly's in that his symbols N_{i0}, n_{i0}, n_{ji}, R_i, s_i, Z_i, α_i and η_i are replaced by U_i, u_i, m_{ij}, r_i, R_i, z_i, ρ_i and ν_i respectively.) Our symbols can be categorised as follows:

Unknown fixed parameters: p_i, ϕ_i, ν_i; for mathematical convenience we also define U_i to be a fixed parameter. We shall make use of the intermediary parameters

$$\alpha_i = \phi_i q_{i+1}, \quad \beta_i = \phi_i p_{i+1} \quad \text{and}$$
$$\chi_i = (1 - \phi_i) + \phi_i q_{i+1}(1 - \phi_{i+1}) + \dots + \phi_i q_{i+1} \dots \phi_{s-1} q_s$$
$$= 1 - \phi_i p_{i+1} - \phi_i q_{i+1} \phi_{i+1} p_{i+1} - \dots - \phi_i q_{i+1} \dots \phi_{s-2} q_{s-1} \phi_s p_s$$
$$= 1 - \beta_i - \alpha_i \beta_{i+1} - \dots - \alpha_i \alpha_{i+1} \dots \alpha_{s-1} \beta_s, \tag{5.1}$$

the last expression being the conditional probability that a marked animal released in the ith release of R_i animals is not caught again.

Known random variables: n_i, m_i, u_i $(u_1 = n_1)$, m_{hi}, R_i, r_i and z_i; we define $m_1 = r_s = z_1 = z_s = 0$. The random variables m_i, r_i and z_i are all functions of the m_{hi} and their calculation is discussed later.

Unknown random variables: N_i, M_i, B_i and ρ_i; we define $M_1 = 0$ and $B_0 = N_1$.

2   Estimation

PROBABILITY MODEL. To demonstrate how the probability model for the Schnabel census is set up, we shall consider first of all the simple case of $s = 3$. Using the notation of **4.1.1**, let w be a non-empty subset of the set of integers $\{1, 2, \dots, s\}$ and let a_w be the number of marked animals with capture history w. Suppose that on the last occasion when the group of a_w animals is captured, d_w are not returned to the population. Let $b_w = a_w - d_w$, then the conditional distribution of the random variables $\{b_w, d_w\}$ conditional on the $\{u_i\}$ is given by

$f(\{b_w, d_w\} \mid \{u_i\})$

$$= \left\{ \frac{u_1!}{\underset{w \supset 1}{\prod} \{b_w! \, d_w!\}} (\nu_1 \chi_1)^{b_1} (\nu_1 \phi_1 p_2 \nu_2 \chi_2)^{b_{12}} (\nu_1 \phi_1 q_2 \phi_2 p_3 \nu_3 \chi_3)^{b_{13}} \right.$$

$$\times (\nu_1 \phi_1 p_2 \nu_2 \phi_2 p_3 \nu_3 \chi_3)^{b_{123}} (1 - \nu_1)^{d_1} (\nu_1 \phi_1 p_2 (1 - \nu_2))^{d_{12}}$$

$$\left. \times (\nu_1 \phi_1 q_2 \phi_2 p_3 (1 - \nu_3))^{d_{13}} (\nu_1 \phi_1 p_2 \nu_2 \phi_2 p_3 (1 - \nu_3))^{d_{123}} \right\}$$

$$\times \left\{ \frac{u_2!}{b_2! b_{23}! d_2! d_{23}!} (\nu_2 \chi_2)^{b_2} (\nu_2 \phi_2 p_3 \nu_3 \chi_3)^{b_{23}} (1 - \nu_2)^{d_2} (\nu_2 \phi_2 p_3 (1 - \nu_3))^{d_{23}} \right\}$$

$$\times \left\{ \frac{u_3!}{b_3! d_3!} \nu_3^{b_3} (1 - \nu_3)^{d_3} \right\}. \tag{5.2}$$

For the general case of s samples it can be shown by induction (e.g. Seber [1965: 252]) that

$$f(\{b_w, d_w\} \mid \{u_i\}) = \frac{\overset{s}{\underset{i=1}{\prod}} u_i!}{\underset{w}{\prod} \{b_w! \, d_w!\}} \overset{s-1}{\underset{i=1}{\prod}} \left\{ \chi_i^{R_i - r_i} \, \alpha_i^{z_i + 1} \, \beta_i^{m_i + 1} \right\}$$

$$\times \overset{s}{\underset{i=1}{\prod}} \left\{ \nu_i^{R_i} (1 - \nu_i)^{n_i - R_i} \right\} \tag{5.3}$$

$$= L_1 \times L_2, \text{ say.}$$

Since we have binomial sampling,

$$f(\{u_i\}) = \overset{s}{\underset{i=1}{\prod}} \left\{ \binom{U_i}{u_i} p_i^{u_i} q_i^{U_i - u_i} \right\} \tag{5.4}$$

and we finally have the unconditional distribution

$$f(\{b_w, d_w\}) = f(\{b_w, d_w\} \mid \{u_i\}) \, f(\{u_i\}). \tag{5.5}$$

(This distribution was obtained indirectly by Jolly [1965: 234] using a different method.)

As mentioned above, m_i, r_i and z_i are functions of the m_{hi}, so that all the information about the fixed parameters is contained in the set of sufficient statistics $\{u_i, m_{hi}, R_i\}$. We also note that the marked population, represented by (5.3), supplies information on the $\{\phi_i, p_i\}$ via the parameters $\{\alpha_i, \beta_i\}$, and assuming that p_i is the same for marked and unmarked, the $\{u_i\}$ supply information on the parameters $\{U_i\}$.

Since (5.3) factorises into two components L_1 and L_2, where L_1 does not contain the parameters $\{\nu_i\}$, the maximum-likelihood estimates of the α_i and β_i will be unchanged if some of the ν_i are put equal to unity. On the other hand, if $\nu_j = 0$, so that there is a 100 per cent removal on the jth sample ($R_j = r_j = 0$), we find that not all the parameters can be estimated.

In particular, we find that α_{j-1}, α_j and β_j are no longer identifiable, only the products $\alpha'_j = \alpha_{j-1}\alpha_j$ and $\beta'_j = \alpha_{j-1}\beta_j$ can be estimated. To see this we note the following:

$$1 - \chi_k = 1 - \beta_k - \alpha_k\beta_{k+1} - \ldots - \alpha_k \ldots \alpha_{j-2}(\alpha_{j-1}\beta_j) - $$
$$- \alpha_k \ldots \alpha_{j-2}(\alpha_{j-1}\alpha_j)\beta_{j+1} - \ldots$$

for $k < j$, χ_j does not appear in (5.3), and using the relationship $z_{j+1} + m_{j+1} = z_j + r_j$ with $r_j = 0$ we have

$$\alpha_{j-1}^{z_j}\ \beta_{j-1}^{m_j}\ \alpha_j^{z_{j+1}}\ \beta_j^{m_{j+1}} = (\alpha_{j-1}\alpha_j)^{z_{j+1}}\ (\alpha_{j-1}\beta_j)^{m_{j+1}}\ \beta_{j-1}^{m_j}.$$

From the estimates of α'_j and β'_j we can then estimate p_{j+1} ($= \beta'_j/(\alpha'_j + \beta'_j)$) and $\phi_{j-1}q_j\phi_j$ ($= \alpha'_j + \beta'_j$), the probability of survival from immediately after time t_{j-1} to time t_{j+1}.

If two releases are zero, say $R_j = R_{j+1} = 0$, then it transpires that the parameters α_{j-1}, α_j, α_{j+1}, β_j a β_{j+1} are no longer identifiable, only the products β'_j, $\alpha'_{j+1} = \alpha_{j-1}\alpha_j\alpha_{j+}$ nd $\beta'_{j+1} = \alpha_{j-1}\alpha_j\beta_{j+1}$ can be estimated. From the estimates of α'_{j+1} an β'_{j+1} we can then obtain estimates of p_{j+2} and $\phi_{j-1}q_j\phi_j q_{j+1}\phi_{j+1}$, the prob bility of survival from immediately after time t_{j-1} to time t_{j+2}. The que...ion of zero releases is discussed further on p. 210.

Finally it should be stressed that (5.2) can only be applied to a population in which there is no emigration, or the emigration is "permanent" in the sense that if a marked animal moves outside the population it remains outside for the remainder of the experiment. This requirement is clearly seen by examining the probabilities associated with the b_w in (5.2). For example, the class of b_{13} individuals has a probability $\nu_1\phi_1 q_2\phi_2 p_3\nu_3\chi_3$ which only applies if the members of this class are actually in the population at time t_2 and are not temporarily outside the sampling area during the second sample.

MAXIMUM-LIKELIHOOD ESTIMATES. By differentiating the logarithm of (5.5), maximum-likelihood estimates of the unknown parameters can be derived in the usual fashion (cf. Seber [1965] for the special case of no losses on capture, i.e. $\nu_i = 1$). Also, simple moment type estimates of the unknown random variables N_i, M_i and B_i can then be obtained from the relations:

$$E[N_{i+1} \mid N_i, B_i] = B_i + \phi_i(N_i - n_i + R_i),$$

and

$$E[n_i \mid N_i] = N_i p_i,$$
$$E[m_i \mid M_i] = M_i p_i.$$

However, the same set of estimates for the unknown parameters and random variables can also be obtained by an intuitive argument which we reproduce from Jolly [1965].

The number of marked animals M_i can be thought of as consisting of the number captured, m_i, plus the number $M_i - m_i$ not captured in the ith sample. Immediately after the ith sample there are two groups of marked animals, the

$M_i - m_i$, of which z_i are subsequently caught, and the R_i just released, of which r_i are subsequently caught. Since the chances of recapture are assumed to be the same for both groups, we would expect

$$\frac{z_i}{M_i - m_i} \approx \frac{r_i}{R_i},$$ (5.6)

which leads to the estimate

$$\hat{M}_i = \frac{R_i z_i}{r_i} + m_i, \quad (i = 2, 3, \dots, s-1).$$ (5.7)

Also the proportion of marked in the sample will represent the proportion of marked in the population, i.e.

$$\frac{m_i}{n_i} = \frac{\hat{M}_i}{\hat{N}_i}$$

or $\hat{N}_i = \hat{M}_i n_i / m_i, \quad (i = 2, 3, \dots, s-1).$ (5.8)

A natural estimate of ϕ_i would be the ratio of the marked animals alive at time t_{i+1} to the marked animals in the population just after the ith release, namely

$$\hat{\phi}_i = \frac{\hat{M}_{i+1}}{\hat{M}_i - m_i + R_i} \quad (i = 2, 3, \dots, s-2)$$ (5.9)

and

$$\hat{\phi}_1 = \frac{\hat{M}_2}{R_1}.$$ (5.10)

Intuitive estimates of p_i, ν_i, ρ_i, B_i and U_i are then given by

$$\hat{p}_i = \frac{n_i}{\hat{N}_i} = \frac{m_i}{\hat{M}_i}, \quad (i = 2, 3, \dots, s),$$ (5.11)

$$\hat{\nu}_i = \frac{R_i}{n_i}, \quad (i = 1, 2, \dots, s),$$

$$\hat{\rho}_i = \frac{\hat{M}_i}{\hat{N}_i} = \frac{m_i}{n_i}, \quad (i = 2, 3, \dots, s),$$

$$\hat{B}_i = \hat{N}_{i+1} - \hat{\phi}_i(\hat{N}_i - n_i + R_i), \quad (i = 2, 3, \dots, s-2),$$ (5.12)

and $\hat{U}_i = \hat{N}_i - \hat{M}_i.$

It should be noted that $\hat{M}_i$ and $\hat{N}_i$ are not maximum-likelihood estimates but are simply used as intermediate steps in the calculation of the maximum-likelihood estimates $\hat{\phi}_i$, $\hat{p}_i$ and $\hat{U}_i$. Also the estimate of B_i is only valid if we replace assumption (b) by the stronger assumption that ϕ_i is the same for all individuals and not just for the marked.

To calculate the above estimates we require n_i, m_i, R_i, r_i and z_i. In particular, if

$$c_{ij} = \sum_{h=1}^{i} m_{hj} \quad (i < j),$$

the number in the jth sample last caught in the ith or before, then

$$m_i = c_{i-1, i},$$

$$r_i = \sum_{j=i+1}^{s} m_{ij},$$

and

$$z_i = \sum_{j=i+1}^{s} c_{i-1, j}$$

can be calculated from arrays of the m_{hi} and c_{hi} (cf. Example 5.1 below). The z_i can also be calculated iteratively from the relations

$$z_{i+1} = z_i + r_i - m_{i+1},$$

starting with $z_1 = 0$.

This method of grouping the recaptures according to when they were last caught (m_{hi}) is often known as Method B (Leslie and Chitty [1951]). In setting out the array of m_{hi} (cf. Table 5.1, p. 206), we shall use Leslie and Chitty's method of display rather than that of Jolly who uses an array $\{m_{ji}\}$, i.e. the "transpose" of Table 5.1.

It is of interest to compare Leslie's estimates (cf. Leslie et al. [1953]) with those given above. For example, Leslie's estimate of M_i ($= \hat{\psi}_i$ in his notation) is obtained from the equation

$$\frac{c_{i-1, i+1}}{M_i - m_i} = \frac{m_{i, i+1}}{R_i},$$

i.e. by equating the proportions of the $M_i - m_i$ marked animals and the release R_i which are caught in just the $(i+1)$th sample (and not in all future samples as in equation (5.6)). Leslie's estimates of N_i and ϕ_i then follow from the above estimate of M_i in the same way that $\hat{N}_i$ and $\hat{\phi}_i$ follow from the $\hat{M}_i$.

One or two comments on the range of the suffix i in the above estimates are appropriate at this point. Since $M_1 = 0$ (i.e. no marked animals in the population before the first sample), N_1, p_1 and B_1 cannot be estimated from (5.8), (5.11) and (5.12). However, from the trend in $\hat{N}_2, \ldots, \hat{N}_{s-1}$ or the lack of trend (e.g. $N_1' = \hat{N}_2$) we can estimate p_1 and B_1 by

$$p_1' = n_1/N_1'$$

and

$$B_1' = \hat{N}_2 - \hat{\phi}_1(N_1' - n_1 + R_1).$$

Also z_s and r_s are both zero, so that M_s, and consequently N_s, p_s, ϕ_{s-1} and B_{s-1}, cannot be estimated. The product $\beta_{s-1} = \phi_{s-1} p_s$ can be estimated by

$$\hat{\beta}_{s-1} = m_s/\hat{M}_{s-1} - m_{s-1} + R_{s-1}),$$

and if an estimate of p_s is available (e.g. $p_s' = \hat{p}_i$, where the samples numbered s and i are taken under similar conditions and with the same effort), then ϕ_{s-1}' can be estimated by

$$\phi_{s-1}' = \hat{\beta}_{s-1}/p_s'. \tag{5.13}$$

This leads to the estimates

$$N_s' = n_s/p_s'$$

and

$$B_{s-1}' = N_s' - \phi_{s-1}'(\hat{N}_{s-1} - n_{s-1} + R_{s-1}).$$

Alternatively, if ϕ_{s-1} can be estimated from the trend (or lack of trend) in $\hat{\phi}_1, \hat{\phi}_2, \ldots, \hat{\phi}_{s-2}$, then p_s can be estimated from (5.13). Finally we note that ϕ_s and B_s are undefined.

VARIANCES AND COVARIANCES. Every animal in the population belongs to just one of the mutually exclusive groups of B_i new animals (including $B_0 = N_1$). Also each of these groups can be split up into various multinomial classes which contribute to the random variables used in the above estimates. Therefore, treating the $\{B_i\}$ as fixed parameters, and utilising the fact that the multinomial distributions arising from different B_i are mutually independent, Jolly uses the delta method to derive asymptotic expressions for the variances and covariances of the estimates. His formulae reduce to the following when the expectations (conditional on $\{B_i\}$) of the random variables N_i, M_i, R_i, etc. are replaced by their "observed" values:

$$V[\hat{N}_i] = N_i(N_i - n_i)\left\{\frac{M_i - m_i + R_i}{M_i}\left(\frac{1}{r_i} - \frac{1}{R_i}\right) + \frac{1 - \rho_i}{m_i}\right\} +$$

$$+ N_i - \sum_{h=0}^{i-1} \frac{N_i^2(h)}{B_h}, \quad (i = 2, 3, \ldots, s-1), \tag{5.14}$$

$$\mathrm{cov}\,[\hat{N}_i, \hat{N}_j] = \sum_{h=0}^{i-1}\left\{N_j(h) - \frac{N_i(h)N_j(h)}{B_h}\right\} \quad (i < j),$$

$$V[\hat{\phi}_i] = \phi_i^2\left\{\frac{(M_{i+1} - m_{i+1})(M_{i+1} - m_{i+1} + R_{i+1})}{M_{i+1}^2}\left(\frac{1}{r_{i+1}} - \frac{1}{R_{i+1}}\right) + \right.$$

$$\left. + \frac{M_i - m_i}{M_i - m_i + R_i}\left(\frac{1}{r_i} - \frac{1}{R_i}\right) + \frac{1 - \phi_i}{M_{i+1}}\right\}, \quad \begin{array}{l}(i = 1, 2, \ldots, s-2;\\ M_1 = m_1 = 0),\end{array} \tag{5.15}$$

$$\mathrm{cov}\,[\hat{\phi}_i, \hat{\phi}_{i+1}] = -\frac{\phi_i\phi_{i+1}(M_{i+1} - m_{i+1})}{M_{i+1}}\left(\frac{1}{r_{i+1}} - \frac{1}{R_{i+1}}\right), \tag{5.16}$$

$$\mathrm{cov}\,[\hat{\phi}_i, \hat{\phi}_j] = 0 \quad (j > i + 1), \tag{5.17}$$

$$V[\hat{B}_i] = \frac{B_i^2(M_{i+1} - m_{i+1})(M_{i+1} - m_{i+1} + R_{i+1})}{M_{i+1}^2} \left(\frac{1}{r_{i+1}} - \frac{1}{R_{i+1}} \right) +$$

$$+ \frac{M_i - m_i}{M_i - m_i + R_i} \left[\frac{\phi_i R_i (1 - \rho_i)}{\rho_i} \right]^2 \left(\frac{1}{r_i} - \frac{1}{R_i} \right) +$$

$$+ \frac{(N_i - n_i)(N_{i+1} - B_i)(1 - \rho_i)(1 - \phi_i)}{M_i - m_i + R_i} + N_{i+1}(N_{i+1} - n_{i+1}) \left(\frac{1 - \rho_{i+1}}{m_{i+1}} \right) +$$

$$+ \phi_i^2 N_i (N_i - n_i) \left(\frac{1 - \rho_i}{m_i} \right), \quad (i = 2, 3, \dots, s-2),$$

$$\text{cov}\,[\hat{B}_i, \hat{B}_{i+1}] = -\phi_{i+1}(N_{i+1} - n_{i+1})(1 - \rho_{i+1})$$

$$\times \left\{ \frac{B_i R_{i+1}}{M_{i+1}} \left(\frac{1}{r_{i+1}} - \frac{1}{R_{i+1}} \right) + \frac{N_{i+1}}{m_{i+1}} \right\},$$

and

$$\text{cov}\,[\hat{B}_i, \hat{B}_j] = 0, \quad (j > i + 1).$$

The above formulae are asymptotic and are only valid for large expectations of the random variables n_i, m_i, R_i, r_i and z_i; the small sample properties of the formulae need further investigation. Estimates of the variances and covariances are obtained by replacing each unknown by its estimate.

The first expression in (5.14) represents the error of estimation $V[\hat{N}_i \mid N_i]$, while the final terms represent an approximation for

$$V[N_i] = E[N_i] - \sum_{h=0}^{i-1} \frac{\{E[N_i(h)]\}^2}{B_h}, \tag{5.18}$$

where the expectations are conditional on the $\{B_i\}$. Jolly shows that the terms $E[N_i(h)]$ in (5.18) are most readily obtained as successive products of $(E[N_{k+1}] - B_k)/E[N_k]$ using the relations

and

$$N_{h+1}(h) = B_h \tag{5.19}$$

$$E[N_{i+1}(h)] = \frac{E[N_{i+1}] - B_i}{E[N_i]} \cdot E[N_i(h)], \quad (i > h). \tag{5.20}$$

These relations give all the $E[N_i(h)]$ for $h > 1$, but we run into difficulty with $N_i(0)$ and $N_i(1)$ as neither B_0 nor B_1 are estimable. However, since N_2 and ϕ_1 can be estimated, $B_0 \, (= N_1)$ and $B_1 \, (\approx N_2 - \phi_1(N_1 - n_1 + R_1))$ can also be estimated if an estimate of N_1 is available from the trend (or lack in trend) in the $\hat{N}_i$. Estimates of the $N_i(h)$ then follow from (5.19) and (5.20) using $\hat{B}_i$ and $\hat{N}_i$ to estimate $\hat{B}_i$ and $E[N_i]$ respectively. However, in general, $V[N_i]$ will usually be much smaller than $V[\hat{N}_i \mid N_i]$ and can therefore be ignored in most cases (except possibly when p_i is large); this avoids the awkward computations just outlined.

In conclusion we note from $V[\hat{\phi}_i]$ and $V[\hat{N}_i]$ that the accuracies of

$\hat{\phi}_i$ and $\hat{N}_i$ depend both on the numbers of recaptures (r_i and m_i) and on the sampling intensities p_i. The dependence on p_i is reflected in the variance terms $N_i - n_i$ and $M_i - m_i$; as n_i approaches N_i, $V[\hat{N}_i | N_i]$ tends to zero. The above variance formulae have been studied by Manly [1971a] using computer simulation techniques.

BIAS. In deriving $\hat{M}_i$ and $\hat{N}_i$ (equations (5.7) and (5.8)), we used arguments which are reminiscent of the Petersen method of **3.1.1**. For this reason it is natural to consider whether we can "patch up" our estimates to reduce their bias. To determine how this can be done we note that the Petersen estimate $\hat{N}$ for a two-sample experiment satisfies

$$\frac{u_2}{\hat{N} - n_1} = \frac{m_2}{n_1},$$

while the almost unbiased modification N^* satisfies

$$\frac{u_2}{N^* - n_1} = \frac{m_2 + 1}{n_1 + 1}. \tag{5.21}$$

Therefore, on comparison with (5.6) we are led to consider

$$M_i^* = \frac{R_i + 1}{r_i + 1} \cdot z_i + m_i \tag{5.22}$$

which, using Jolly's formula for the asymptotic bias of $\hat{M}_i$ (cf. Jolly [1965: 238]) is readily shown to have a bias of smaller order than that of $\hat{M}_i$. It can also be shown that the modified estimates

$$N_i^* = \frac{n_i + 1}{m_i + 1} \cdot M_i^*,$$

$$\phi_i^* = \frac{M_{i+1}^*}{\hat{M}_i - m_i + R_i}, \quad (i = 2, 3, \dots, s-2),$$

$$\phi_1^* = \frac{M_2^*}{R_1},$$

and

$$B_i^* = N_{i+1}^* - \phi_i^*(N_i^* - n_i + R_i)$$

are all approximately unbiased. The corrections for bias will not affect the asymptotic variances, so that asymptotically $V[N_i^*] = V[\hat{N}_i]$, etc.

It was pointed out in **3.1.1** that the Petersen estimate and its modification may be unreliable for $m_2 < 10$. Therefore it is recommended that m_i and r_i should be greater than 10 for a satisfactory application of the above method.

ALTERNATIVE PROBABILITY MODELS. A number of deterministic formulations of the above capture—recapture model have been developed since the 1930's and these are reviewed in Cormack [1968] and compared in Parr [1965].

As these approaches have been largely superseded by the simpler and more general stochastic models of this chapter they will not be discussed. However, it should be mentioned that the most general stochastic model developed so far is a hypergeometric type model set up by Robson [1969]. This model, although complex, makes few assumptions and therefore represents a major step forward in the field of capture—recapture analysis. Robson obtains maximum-likelihood estimates for some specific situations and derives $\hat{N}_i$ above as a special case. But the question of finding asymptotic variances and confidence intervals for these estimates is left open. An excellent discussion of the various models that can arise has been given by Cormack [1972] and should be consulted by the reader.

The above Jolly—Seber method has not yet had much time to be used in practice. However, applications are given in Jolly [1965: capsids], Parr [1965: damselflies], Sadleir [1965: deer mice], Lidicker [1966: house mice], Delong [1966: house mice], Parker [1968: pink salmon; cf. Appendix, where the method is applied to a population of dead fish in the spawning area], Parr et al. [1968: grasshoppers, butterflies and damselflies], and White [1971: grasshoppers]. The method can also be applied to previous recapture experiments where the m_{hi} are tabulated, that is, where recaptures are grouped according to Leslie and Chitty's [1951] method B (e.g. Orians [1958: Table 3], Turner [1960a: Table 1], Sonleitner and Bateman [1963: Table 2], Kikkawa [1964: 296]). An extensive Fortran computer program for the method has been written by White [1971].

Example 5.1   Black-kneed capsid (*Blepharidopterus angulatus*): Jolly [1965]

Thirteen successive samples were taken at alternatively 3- and 4-day intervals from an apple orchard population of female black-kneed capsid, and the values of m_{hi} are recorded in Table 5.1. The steps in calculating the estimates are as follows:

(i) Sum the rows in Table 5.1 to give the values r_h (the number from the release R_h subsequently recaptured), e.g. $r_7 = 108$.

(ii) Sum each column in Table 5.1 cumulatively from top to bottom, entering the accumulated totals in the columns of Table 5.2. For example, the fourth column of Table 5.2 is 5, 5 + 18, 5 + 18 + 33.

(iii) In Table 5.2 sum each of the rows, excluding the first entry, to obtain the z_i; the first entries are the m_i, e.g. $m_7 = 112$, $z_7 = 110$. The z_i's can be checked from the relation $z_{i+1} - z_i = r_i - m_{i+1}$.

(iv) Calculate $\hat{M}_i$, $\hat{\rho}_i$, $\hat{N}_i$, $\hat{\phi}_i$ and $\hat{B}_i$ as in Table 5.3. For example:

$$\hat{M}_7 = \frac{R_7 z_7}{r_7} + m_7 = \frac{243(110)}{108} + 112 = 359 \cdot 50,$$

$$\hat{\rho}_7 = \frac{m_7}{n_7} = \frac{112}{250} = 0 \cdot 4480,$$

205

TABLE 5.1

Tabulation of m_{hi}, the number caught in the ith sample last captured in the hth sample, for a black-kneed capsid population: data from Jolly [1965: Table 2].

i	1	2	3	4	5	6	7	8	9	10	11	12	13	
n_i	54	146	169	209	220	209	250	176	172	127	123	120	142	
R_i	54	143	164	202	214	207	243	175	169	126	120	120	–	Total
h														r_h
1		10	3	5	2	2	1	0	0	0	1	0	0	24
2			34	18	8	4	6	4	2	0	2	1	1	80
3				33	13	8	5	0	4	1	3	3	0	70
4					30	20	10	3	2	2	1	1	2	71
5						43	34	14	11	3	0	1	3	109
6							56	19	12	5	4	2	3	101
7								46	28	17	8	7	2	108
8									51	22	12	4	10	99
9										34	16	11	9	70
10											30	16	12	58
11												26	18	44
12													35	35
Total	m_i 0	10	37	56	53	77	112	86	110	84	77	72	95	

TABLE 5.2

Tabulation of c_{hi}, the number caught in the ith sample last caught in or before the hth sample, for a black-kneed capsid population: from Jolly [1965: Table 3].

i / h	1	2	3	4	5	6	7	8	9	10	11	12	13	Total	
1	10	3	5	2	2	1	0	0	0	1	0	0		14	z_2
2		37	23	10	6	7	4	2	0	3	1	1		57	z_3
3			56	23	14	12	4	6	1	6	4	1		71	z_4
4				53	34	22	7	8	3	7	5	3		89	z_5
5					77	56	21	19	6	7	6	6		121	z_6
6						112	40	31	11	11	8	9		110	z_7
7							86	59	28	19	15	11		132	z_8
8								110	50	31	19	21		121	z_9
9									84	47	30	30		107	z_{10}
10										77	46	42		88	z_{11}
11											72	60		60	z_{12}
12												95			

TABLE 5.3
Population estimates for a black-kneed capsid population: from Jolly [1965: Table 4].

i	$\hat{\rho}_i$	$\hat{M}_i$	$\hat{N}_i$	$\hat{\phi}_i$	$\hat{B}_i$	$\hat{\sigma}[\hat{N}_i]$	$\hat{\sigma}[\hat{\phi}_i]$	$\hat{\sigma}[\hat{B}_i]$	$\hat{\sigma}[\hat{N}_i \mid N_i]$
1	–	0	–	0·649	–	–	0·114	–	–
2	0·0685	35·02	511·2	1·015	263·2	151·2	0·110	179·2	150·8
3	0·2189	170·54	779·1	0·867	291·8	129·3	0·107	137·7	128·9
4	0·2679	258·00	963·0	0·564	406·4	140·9	0·064	120·2	140·3
5	0·2409	227·73	945·3	0·836	96·9	125·5	0·075	111·4	124·3
6	0·3684	324·99	882·2	0·790	107·0	96·1	0·070	74·8	94·4
7	0·4480	359·50	802·5	0·651	135·7	74·8	0·056	55·6	72·4
8	0·4886	319·33	653·6	0·985	– 13·8	61·7	0·093	52·5	58·9
9	0·6395	402·13	628·8	0·686	49·0	61·9	0·080	34·2	59·1
10	0·6614	316·45	478·5	0·884	84·1	51·8	0·120	40·2	48·9
11	0·6260	317·00	506·4	0·771	74·5	65·8	0·128	41·1	63·7
12	0·6000	277·71	462·8	–	–	70·2	–	–	68·4
13	0·6690	–	–	–	–	–	–	–	–

$$\hat{N}_7 = \hat{M}_7/\hat{\rho}_7 = 802\cdot5,$$

$$\hat{\phi}_7 = \frac{\hat{M}_8}{\hat{M}_7 - m_7 + R_7} = \frac{319\cdot33}{359\cdot50 - 112 + 243} = 0\cdot651,$$

and

$$\hat{B}_7 = \hat{N}_8 - \hat{\phi}_7(\hat{N}_7 - n_7 + R_7) = 135\cdot7.$$

(v) The calculation of the variances is straightforward, though tedious if not computerised. For example,

$$\hat{V}[\hat{N}_7 \mid N_7] = \hat{N}_7(\hat{N}_7 - n_7)\left\{\frac{\hat{M}_7 - m_7 + R_7}{\hat{M}_7}\left(\frac{1}{r_7} - \frac{1}{R_7}\right) + \frac{1 - \hat{\rho}_7}{m_7}\right\}$$

$$= 5241\cdot6$$

$$\hat{V}[N_7] = \hat{N}_7 - \sum_{h=0}^{6} \hat{N}_7^2(h)/\hat{B}_h.$$

As mentioned in the general theory, $V[N_7]$ can only be estimated if an estimate of N_1 is available. Jolly sets $\hat{N}_1 = 500 (= \hat{B}_0)$, so that

$$\hat{B}_1 = \hat{N}_2 - \hat{\phi}_1(\hat{N}_1 - n_1 + R_1) = 186\cdot7,$$

and the $\hat{N}_7(h)$ can be obtained from the recurrence formulae (5.19) and (5.20). Thus,

$$\hat{V}[N_7] = 353\cdot28$$

and

$$\hat{V}[\hat{N}_7] = 5241\cdot76 + 353\cdot28 = 5595\cdot04.$$

Comparing $\hat{\sigma}[\hat{N}_i]$ with $\hat{\sigma}[\hat{N}_i \mid N_i]$ in Table 5.3, we see that it was not worth calculating the terms $V[\hat{N}_i]$.

In conclusion we note that, apart from $\hat{N}_2$, the coefficients of variation of the $\hat{N}_i$ vary from about 10 to 16 per cent, while those for the $\hat{\phi}_i$ are approximately 10 per cent. This accuracy was achieved with a sampling intensity ($100p_i$) of approximately 30 per cent. Cormack [1968] estimates that by the end of the experiment about 60 per cent of the living population was marked, despite the influx of an estimated 1500 new individuals.

Example 5.2   Damselfly (*Ischnura elegans* (Van der Linden)): Parr [1965]

A colony of the damselfly (*Ischnura elegans*) centred on a chain of four ponds at Maryborough Farm, Pembrokeshire, was studied from 24 June to 7 July, 1964. Butterfly nets having six-foot handles were used to catch the daily sample. The sampling was random to the extent that every individual that could be caught was taken, irrespective of age (teneral or adult) and sex; the whole area of the colony being covered as far as possible. Those individuals flying over the water or resting on low emergent vegetation were more difficult to capture unharmed than those over land. This is because dragonflies tend to fly very low over the water surface and, in attempting capture with a wet heavy net, there is danger of causing damage to the delicate insect. The wings and head suspension are most likely to be damaged in this way, particularly in teneral insects. However, the insects were successfully marked if handled carefully, and observations on freshly marked and released teneral insects failed to reveal wing damage or any degree of abnormal behaviour.

The insects were marked with small spots of quick-drying cellulose paint, and the method of handling and marking the insects is described in some detail by the author. In practice it was found most convenient to capture, mark, and release the damselflies from a specific area occupied by the colony before continuing the sampling in another part. The insects were thus held captive for only a relatively short time, and the number of individuals caught more than once on a particular day was small. Individuals incapable of normal flight after capture were not released, though this seldom happened. Marking, recording and releasing were mostly carried out as soon as a batch of approximately 30 insects had been captured: this would generally be done twice or three times during each day's collecting period.

Although the sampling was not strictly random, the marked insects dispersed rapidly, particularly in fine weather, so that there appeared to be a uniform mixing of marked and unmarked. Evidence for this dispersal came from the recapture of adult specimens within minutes of being marked at distances up to 150 yards away.

The population estimates for the male population only are set out in Table 5.4.

5.1.2   General applications

Jolly has pointed out the wide applicability of the general theory, and we shall now consider some variations in the above model. Much of the

TABLE 5.4
Estimates of population parameters for a male damselfly population:
from Parr [1965: Table 15].

Day	$\hat{\rho}_i$	$\hat{M}_i$	$\hat{N}_i$	$\hat{\phi}_i$	$\hat{B}_i$	$\hat{\sigma}[\hat{N}_i \mid N_i]$
1	–	0	–	1·0710	–	–
2	0·0545	82·5	1514·0	0·4319	– 52·4	992·5
3	–	–	–	–	–	–
4	0·0953	57·2	600·7	1·1210	– 181·3	257·3
5	0·2586	126·9	490·8	0·5296	+ 305·7	161·0
6	0·1600	90·5	565·6	1·1990	+ 36·5	208·4
7	0·2222	158·8	714·6	0·6417	– 8·8	263·0
8	0·2771	124·4	448·6	0·8481	+ 171·1	104·5
9	0·2836	156·4	551·4	1·1800	– 87·9	149·4
10	0·4287	241·2	562·8	0·4960	+ 166·5	182·2
11	0·3085	137·5	445·6	0·9184	+ 226·1	138·0
12	0·2927	186·0	635·4	–	–	484·2
13	0·1935	–	–	–	–	–

following discussion (except for the examples) is based on Jolly [1965: 239–41].

1   Release and future recapture operated independently

From (5.14) we see that, given M_i, m_i and R_i, $V[\hat{N}_i \mid N_i]$ will be a minimum when the number of recaptures r_i is as large as possible. To achieve this it could be advantageous (and cheaper) to have a separate organisation for recording future recaptures from that of marking and releasing animals. Since for each time t_i it is necessary to distinguish only two classes of marked animals in the future recaptures, namely those marked before t_i and not at t_i, and those released at t_i, a very simple code of marks might be used in specific situations, thus enabling untrained persons to recapture marked animals over a wide area. Such a recapture system could proceed continuously, since the time at which an animal is recaptured or even the number of times it is recaptured after time t_i is of no importance as far as time t_i is concerned; only the fact that it *is* recaptured after t_i is relevant. Releases, on the other hand, would only be made at the particular times for which the estimates $\hat{M}_i$ were required, the catching, marking, and releasing being done by more experienced staff.

Example 5.3

Suppose $s = 10$ with samples 1, 4, 7, giving rise to marked releases by experienced staff; individual marks or tags are used. In the remaining samples 2, 3, etc. the sample is taken from the population (or just observed) by untrained staff, the marked members are noted, and the sample returned *without* any additional marking. For this situation $R_i \leqslant m_i$ for $i \neq 1, 4, 7$ with equality if there are no casualties among the marked individuals caught. The two tables of m_{hi} and c_{hi}, respectively, can then be set up and all the parameters estimated as before.

In practice, the sampling between "official" releases may be an almost continuous process, so that the marked releases on these sampling occasions are small; in fact, R_i is zero if no marked individuals are caught. This means that for these samples the estimates of N_i, ϕ_i, etc. would either have large variances (because of the term $1/r_i$ in the variance formulae) or would not be applicable ($R_i = m_i = 0$). However, with such an experimental set-up one would normally be concerned with just the periods of time t_1 to t_4, t_4 to t_7 and t_7 to t_{10}, and the recaptures between official releases would simply be used to augment r_1, r_4, r_7, z_4 and z_7. Thus, if $\phi_{(k)}$, $B_{(k)}$ ($k = 1, 2, 3$) refer to these three intermediate periods of recapture time, then the estimates of $\phi_{(1)}$, $\phi_{(2)}$ and $B_{(2)}$ are

$$\hat{\phi}_{(1)} = \hat{M}_{(2)}/R_1,$$
$$\hat{\phi}_{(2)} = \hat{M}_{(3)}/(\hat{M}_{(2)} - m_{(2)} + R_{(2)}),$$

and

$$\hat{B}_{(2)} = \hat{N}_{(3)} - \hat{\phi}_{(2)}(\hat{N}_{(2)} - n_{(2)} + R_{(2)}),$$

where $\hat{M}_{(2)} = \hat{M}_4$, $\hat{M}_{(3)} = \hat{M}_7$, $\hat{M}_{(2)} - m_{(2)} + R_{(2)} = \hat{M}_4 - m_4 + R_4$, etc. If an estimate $\hat{N}_1$ is available from the trend, or lack of trend, in $\hat{N}_4$, $\hat{N}_7$, then $B_{(1)}$ can be estimated by

$$\hat{B}_{(1)} = \hat{N}_4 - \hat{\phi}_{(1)}(\hat{N}_1 - n_1 + R_1).$$

The variances and covariances given on p. 202 still apply here by simply replacing i by (i) in the formulae and using the above notation.

The parameter $\phi_{(2)}$ represents the probability of a marked animal surviving from immediately after time t_4 to t_7, so that

$$\phi_{(2)} \leqslant \phi_4 \phi_5 \phi_6$$

with equality if there are no permanent removals through samples 5 and 6. In this case of no permanent removals we have $R_5 = m_5$, $R_6 = m_6$ and

$$\hat{\phi}_{(2)} = \frac{\hat{M}_7}{\hat{M}_4 - m_4 + R_4}$$
$$= \frac{\hat{M}_5}{\hat{M}_4 - m_4 + R_4} \cdot \frac{\hat{M}_6}{\hat{M}_5} \cdot \frac{\hat{M}_7}{\hat{M}_6}$$
$$= \hat{\phi}_4 \hat{\phi}_5 \hat{\phi}_6.$$

If the population is commercially exploited so that the removal is 100 per cent ($R_i = 0$ for $i \neq 1, 4, 7$), then a simple tagging code representing the three releases 1, 4, 7 is sufficient. We then have

$$\phi_{(2)} = \phi_4 q_5 \phi_5 q_6 \phi_6.$$

(This case was discussed on p. 199; see also p. 212). However, if the removal is not 100 per cent, individual tags should be used, so that an animal recaptured more than once is counted only once in the r_i and z_i.

2   Non-random sampling

All the information from the Schnabel census comes essentially from three types of samples; the first of size n_i, giving information on $\rho_i = M_i/N_i$, the second, of size $z_i + r_i$, concerned with future recaptures, and the third of size R_i released into the population. Although up till now we have assumed that all sampling is strictly random, there may be particular situations in which complete random sampling is not necessary or even possible. For example, as long as the probability of capturing a member of M_i is the same as that of capturing one of N_i, then $\hat{\rho}_i = m_i/n_i$ will be a satisfactory estimate of ρ_i. The same argument also applies to z_i/r_i as an estimate of $(M_i - m_i)/R_i$ when members of $M_i - m_i$ and R_i have the same probability of being caught after the ith sample. We shall now consider four types of non-randomness.

NON-RANDOM n_i. If the released animals mix freely and randomly with the population, and if the probability of selection is independent of "mark" status, then non-random samples will still be random with respect to mark status, and $\hat{\rho}_i$ will be a satisfactory estimate of ρ_i (cf. **3.1.2**). Therefore, if the n_i are likely to be non-random then the releases R_i should be arranged so that mixing can take place as much as possible.

POPULATION NOT ALL CATCHABLE. Suppose that the animals have been marked and released from random samples, and that recapturing is to be done during the breeding season when the animals are most easily observed. Supposing that a substantial proportion are either not breeding at all or are breeding in inaccessible places away from the main breeding areas, then a sample observed in the main breeding areas may differ in many respects from a random sample. The experimenter then faces the problem of whether the locality of an animal in the breeding season could in any way be associated with its mark status. For example, if the breeding members of the population have increased their numbers since the date of marking, either by immigration or by young members not born at the time of marking having reached breeding age, then the proportion of marked animals is likely to be higher among the non-breeders than the population as a whole. The estimate $\hat{\rho}_i$ would then be biased by the exclusion of non-breeders from the sample. If, on the other hand, the marking has been sufficiently recent for no further members to have reached breeding age and if immigration was considered to affect breeders and non-breeders alike, ρ_i could be validly estimated from the non-random sample. An example of this problem of accessibility is given in Phillips and Campbell [1970].

Whether or not additions to the breeding members had taken place since marking, the ratio z_i/r_i would not be biased by the exclusion of inaccessible breeders unless place of breeding was associated with age, the r_i being in general younger than the z_i (only r_i can include animals first marked as recently as time t_i). However, since non-breeders may well be older on the average than breeders, the exclusion of non-breeders would almost certainly bias z_i/r_i. Under these circumstances Jolly suggests that one way of removing bias would be to exclude from both R_i and r_i all individuals first marked at time t_i; larger numbers would have to be captured to compensate for the resulting loss in information. The R_i would then be a random selection of the *marked* animals in the population just after time t_i, and assuming (as has been done throughout) that capture does not affect behaviour, r_i would be a random sample from R_i, and z_i a random sample from $M_i - m_i$, even if both non-breeders and distant breeders were excluded. Thus r_i/R_i would estimate $z_i/(M_i - m_i)$ and $\hat{M}_i$ would be a valid estimate of M_i.

RELEASE INDEPENDENT OF SAMPLE. So far it has been assumed that the R_i released will consist of part or all of the n_i, and generally this will be the most convenient way of obtaining R_i. However, as Jolly points out, R_i could be *any* group of animals which is known to behave similarly to the remainder of the population. In fact, the R_i could be introduced from outside the population and have nothing to do with the n_i. For example, in commercial hunting and fishing the samples or catches, n_i, are completely removed from the population by the hunters, and the releases could be made independently by the scientists immediately after each commercial catch (except of course for the first release). Although losses on capture in this case are 100 per cent, this represents no loss of information, for once an animal released at time t_i is recaptured, it has yielded all its information about time t_i.

This particular situation in which release and capture are operated separately with a 100 per cent loss on capture has been considered by Seber [1962] under the title of the "multi-sample single-recapture" census (cf. **4.4** for the case of a closed population). If we count the sample after the first release as sample number 2, then, in terms of the above notation, Seber's model for the marked population is given by the distribution

$$f(\{m_{ij}\} \mid \{R_i\}) = \prod_{i=1}^{s-1} \left\{ \frac{R_i!}{[\prod_{j=i+1}^{s} m_{ij}!][(R_i - r_i)!]} \beta_i^{m_{i,i+1}} (\alpha_i \beta_{i+1})^{m_{i,i+2}} \cdots \right. $$
$$\left. (\alpha_i \alpha_{i+1} \cdots \alpha_{s-2} \beta_{s-1})^{m_{is}} \chi_i^{R_i - r_i} \right\}. \qquad (5.23)$$

Since animals can only be recaptured once, m_{ij} is now the number from the ith release caught in the jth sample.

It can be shown that the above probability model simplifies to

$$\frac{\prod_{i=1}^{s-1} R_i!}{\prod_{i=1}^{s-1} \{[\prod_{j=i+1}^{s} m_{ij}!][(R_i - r_i)!]\}} \prod_{i=1}^{s-1} \{\chi_i^{R_i - r_i} \alpha_i^{z_{i+1}} \beta_i^{m_{i+1}}\} \qquad (5.24)$$

which, apart from a constant term, is the same as L_1 of (5.3). Therefore, since (5.24) contains the same information about $\{p_i, \phi_i\}$ as (5.5), the maximum-likelihood estimates of p_i and ϕ_i are unchanged and are still given by (5.9), (5.10) and (5.11); this was demonstrated by Jolly [1965: 243]. In addition, if we equate Seber's symbols $\{\phi_i, p_{i-1}, a_i, \beta_i, \theta_i, x_{1i}, x_{2i}, x_{3i}, x_{4i}, x_{5i}\}$ with $\{\phi_i, p_i, \beta_i, a_i, 1 - X_i, z_{i+1}, r_{i+1}, m_{i+1}, r_i, z_i + r_i\}$, it can be shown that Seber's estimate of $V[\hat{\phi}_i]$ is equivalent to (5.15). Intuitively this is to be expected since in estimating variances the expected values of random variables are estimated by the observed values, so that, asymptotically, there is no difference between treating $\{R_i\}$ as fixed or random (cf. **1.3.4**). Therefore we would also expect the asymptotic variances of the other estimates $\hat{N}_i$ and $\hat{B}_i$ to have the same estimates as before when the R_i are fixed parameters.

The above model can be extended to the case when the sampling or exploitation is continuous and is carried on for some time after the last release. Suppose that for s years a release of R_i individuals ($i = 1, 2, \ldots, s$) is made at the beginning of each year, and for t years ($t \geq s$) the population is continuously exploited. Let

a_i = probability of an individual surviving the ith year, given that it is alive at the beginning of the year;

β_i = probability of an individual being caught and removed from the population in the ith year, given that it is alive at the beginning of the year;

and m_{ij} = number from the ith release which are removed from the population in the jth year ($i = 1, 2, \ldots, s; \; j = i, i + 1, \ldots, t$).

Then it transpires that, apart from a difference in the interpretation of the parameters a_i and β_i, the probability model for the above experimental situation is identical with one discussed by Seber [1970b] and analysed in **5.4.1**. Hence, from (5.36), the joint distribution of the $\{m_{ij}\}$ is proportional to

$$\left\{ \prod_{i=1}^{s-1} a_i^{z_{i+1}} \, \beta_i^{m_i} (1 - \theta_i)^{R_i - r_i} \right\} \left\{ \beta_s^{m_s} (1 - \theta_s)^{R_s - r_s} \right\} \left\{ \gamma_{s+1}^{m_{s+1}} \, \gamma_{s+2}^{m_{s+2}} \cdots \gamma_t^{m_t} \right\},$$

where $\gamma_j = a_s a_{s+1} \cdots a_{j-1} \beta_j, \qquad (j = s + 1, s + 2, \ldots, t),$

and $\theta_i = \beta_i + a_i \beta_{i+1} + \cdots + a_i a_{i+1} \cdots a_{t-1} \beta_t.$

Equating $\hat{\phi}_i$ in **5.4.1** with $\hat{a}_i$, we find that the maximum-likelihood estimate a_i is

$$\hat{a}_i = \frac{z_{i+1}}{r_i + z_i} \cdot \frac{r_i}{R_i} \cdot \frac{R_{i+1}}{r_{i+1}} \qquad (i = 1, 2, \ldots, s - 1)$$

and from (5.37),

$$V[\hat{a}_i] \approx a_i^2 \left\{ \frac{1}{E[r_i]} + \frac{1}{E[r_{i+1}]} + \frac{1}{E[z_{i+1}]} - \frac{1}{R_i} - \frac{1}{R_{i+1}} - \frac{1}{E[r_i + z_i]} \right\},$$

$$\text{cov} \, [\hat{a}_i, \hat{a}_j] \approx 0, \quad j > i + 1,$$

and

$$\text{cov} \, [a_i, a_{i+1}] \approx - a_i a_{i+1} \left\{ \frac{1}{E[r_{i+1}]} - \frac{1}{R_{i+1}} \right\}.$$

213

Here z_i is once again the number of tags recovered after the ith release from releases prior to the ith, r_i is the number of tags recovered from the ith release and m_j is the number of tags recovered in year j. Although r_i and z_i now have slightly different algebraic representations in terms of sums of the m_{ij} (e.g. $r_i = \sum_{j=i}^{t} m_{ij}$ instead of $r_i = \sum_{j=i+1}^{s} m_{ij}$, as the year's catch after the first release is treated as sample number 1 and not sample number 2 as in the general theory of **5.1.1**), the tabular methods described in Example 5.1 (p. 205) for calculating r_i and z_i still apply (cf. Example 5.11 on p. 248).

Robson [1963] has also considered the above model for the case $s = t$, and although he uses Poisson approximations to the multinomial distribution, his model contains the same "information". Therefore, equating Robson's symbols $\{M_i, R_i, T_i\}$ with our symbols $\{R_i, r_i, z_i + r_i\}$, we are not surprised to find that Robson's estimate of survival, $\hat{S}$, is the same as $\hat{a}_i$. Also, neglecting R_i^{-1} and R_{i+1}^{-1}, we find that $V[\hat{a}_i]$ and cov $[\hat{a}_i, \hat{a}_j]$ reduce to Robson's formulae, as expected.

RELEASE IS A NON-RANDOM SUB-SAMPLE OF SAMPLE CAPTURED. In some situations (e.g. Cormack [1964]) it is not possible or practicable to obtain a random selection R_i from the sample n_i, and the question arises as to how much of the above general theory can still be used. Jolly points out that since M_i is estimated solely from counts of marked animals, we can imagine the existence of a subpopulation of which the R_i are a random sample and for which the $\hat{M}_i$ and $\hat{\phi}_i$ are satisfactory estimates. Whether $\hat{\phi}_i$ can now be applied to the parent population is a matter of conjecture and depends on further knowledge of the population. As far as further estimation is concerned, if members of the subpopulation have a different probability of being captured from those of the parent population, then we can go no further. However, if the probabilities are the same then we can estimate ρ_i and hence N_i, the size of the parent population — irrespective of whether or not the survival probabilities are the same for the two populations. Finally, Jolly notes that B_i cannot be estimated unless both populations have the same probability of capture and the same survival probability ϕ_i.

Example 5.4 Fulmar petrel (*Fulmarus glacialis*): Cormack [1964]

Cormack applied the above method to the study of a colony of over 100 breeding pairs of fulmars on a small island in Orkney. Each bird captured there was marked *individually* by a set of coloured leg bands which were clearly visible in flight. Sampling was carried out in successive years, and each sample, except for the first, consisted of two parts: the banded birds (m_i) which were simply observed and not recaptured, and the u_i unbanded birds which were caught on their nests, banded and released.

In carrying out an analysis of the data it was assumed that:

(i) All banded birds alive at the time of the ith sample have the same probability p_i of being "captured" (that is, seen) in the ith sample.
(ii) The capture and banding of a bird does not alter its expectation of life.

(iii) The probability ϕ_i of a banded bird surviving from the ith sample to the $(i+1)$th is independent of the age of the bird.

Since the banded members of the sample were simply observed in flight and the unbanded members captured in their nests, the usual assumption that p_i is the same for banded and unbanded did not hold. However, in support of assumption (i) there was a strong tendency for birds to use the same nest site each year, and even the inaccessibility of a nest did not affect the probability of a banded bird being sighted and identified in flight around the island. Assumption (ii) is not unreasonable as, in contrast to the usual multiple-recapture experiments, the birds were handled only once. Although a goodness-of-fit test seemed to indicate that the probability model based on the above assumptions was adequate, assumption (iii) was open to question as there were insufficient data to test this assumption.

Using the above assumptions, Cormack considered just the banded population and assumed that the numbers u_i of newly banded birds released into the population were not random variables but fixed parameters. Theoretically this amounts to dealing with just the conditional density function (5.3) which, since $R_i = n_i$, is given by L_1 (with $d_w = 0$). In fact, equating $\{r_i, R_i, r_i + z_i, m_i, R_i - m_i, R_i - r_i\}$ with Cormack's parameters $\{t_i, s_i, v_i, a_i, b_i, c_i\}$, we find that L_1 is proportional to Cormack's likelihood function [1964: 431] as expected. Now L_1 contains the same information about $\{p_i, \phi_i\}$ as the whole likelihood (5.5), so that the maximum-likelihood estimates of ϕ_i and p_i are unchanged. Hence

$$\hat{M}_i = \frac{R_i z_i}{r_i} + m_i, \quad (i = 2, 3, \dots, s - 1),$$

$$\hat{\phi}_i = \hat{M}_{i+1}/(\hat{M}_i - m_i + R_i), \quad (i = 1, 2, \dots, s - 2; \; m_1 = \hat{M}_1 = 0),$$

and $\quad \hat{p}_i = m_i/\hat{M}_i, \quad (i = 2, 3, \dots, s - 1).$

But the variance formulae for the $\{\hat{\phi}_i\}$ will not be the same as those given by Jolly, since we want variances conditional on the $\{u_i\}$. Using the delta method, Cormack showed that asymptotically

$$V[\hat{\phi}_i \mid \{u_i\}]$$

$$= \phi_i^2 \left\{ \frac{\chi_{i+1}(1-p_{i+1})^2}{(1-\chi_{i+1})E[R_{i+1}]} + \frac{\chi_{i+1}^2 p_{i+1}^2(1-p_{i+1})\phi_i}{(1-\chi_i)(1-\chi_{i+1})E[m_{i+1}]} + \frac{\chi_i}{(1-\chi_i)E[R_i]} \right\}, \quad (5.25)$$

where χ_i, defined in (5.1), satisfies the recurrence relation

$$(1-\chi_i) = \phi_i(1-\chi_{i+1}+\chi_{i+1}p_{i+1}). \quad (5.26)$$

However, using (5.26), replacing $E[R_i]$, $E[m_i]$ by R_i and m_i respectively, and replacing ϕ_i, p_i, χ_i by their estimates $(\hat{\chi}_i = (R_i - r_i)/R_i)$, we find that the estimates of (5.25) and cov $[\hat{\phi}_i, \hat{\phi}_j \mid \{u_i\}]$ are the same as Jolly's estimates. This means that equations (5.15), (5.16) and (5.17) can also be used here. This is not surprising since we would expect variance estimates to be asymptotically the same, irrespective of whether we regard the u_i as random variables or fixed parameters.

Another parameter of interest is E_L, the expected life-span, which we now discuss in some detail. Let T_i be the time from the ith release to the $(i+1)$th sample and suppose that the mortality process in this interval is Poisson with parameter μ_i. Then, if there is no immigration, we have from (1.1)

$$\phi_i = e^{-\mu_i T_i},$$

and

$$\hat{\mu}_i = -\frac{1}{T_i} \log \hat{\phi}_i, \quad (i = 1, 2, \dots, s-2),$$

is the maximum-likelihood estimate of μ_i. The average instantaneous mortality rate, μ, can be estimated by

$$\hat{\mu} = \sum_{i=1}^{s-2} \hat{\mu}_i / (s-2) = -\log \hat{\phi}, \quad \text{say,}$$

where

$$\hat{\phi} = \{ \prod_{i=1}^{s-2} \hat{\phi}_i^{1/T_i} \}^{1/(s-2)}$$

(which reduces to the geometric mean of the $\{\hat{\phi}_i\}$ when $T_i = 1$). If $\mu_i = \mu$ $(i = 1, 2, \dots, s-2)$ then from (1.3)

$$E_L = 1/\mu,$$

which can be estimated by

$$\hat{E}_L = 1/\hat{\mu} \tag{5.27}$$

$$= -1/\log \hat{\phi}.$$

Using the delta method we find that

$$V[\hat{E}_L] \approx V[\hat{\phi}] \, \phi^{-2} (\log \phi)^{-4},$$

where

$$V[\hat{\phi}] = \frac{\phi^2}{(s-2)^2} \left\{ \sum_{i=1}^{s-2} \frac{V[\hat{\phi}_i]}{\phi_i^2 T_i^2} + 2 \sum_{i=1}^{s-3} \frac{\operatorname{cov}[\hat{\phi}_i, \hat{\phi}_{i+1}]}{\phi_i \phi_{i+1} T_i T_{i+1}} \right\}.$$

An alternative estimate of $V[\hat{E}_L]$ is available if we use (5.27) and (1.10), namely

$$v[\hat{E}_L] = v[\hat{\mu}]/\hat{\mu}^4$$

where (defining $\hat{\mu}_{s-1} = \hat{\mu}_1$)

$$v[\hat{\mu}] = \frac{1}{(s-2)(s-5)} \{3 \sum_{i=1}^{s-2} (\hat{\mu}_i - \hat{\mu})^2 - \sum_{i=1}^{s-2} (\hat{\mu}_{i+1} - \hat{\mu}_i)^2 \}.$$

Table 5.5 was derived from Cormack's table, and we see that a coefficient of variation of about 4 per cent was obtained for the estimates $\hat{\phi}_i$. This accuracy was achieved by a sampling intensity ($100\,p_i$) of about 60 per cent; this may be compared with 30 per cent in Jolly's study (Example 5.1). Also $\hat{\phi}_i$ appears to be fairly constant, so that it is appropriate to consider ϕ and E_L. Cormack shows that for a bird just starting to breed, $\hat{\phi}$ and $\hat{E}_L$, together with their standard deviations, are given by 0.9420 ± 0.01 and

216

TABLE 5.5
Recapture data and population estimates for a fulmar population:
data from Cormack [1964: Table 1].

Year i	Captured u_i	Seen m_i	R_i	r_i	z_i	$\hat{M}_i$	$\hat{p}_i$	$\hat{\phi}_i$	$\hat{\sigma}[\hat{\phi}_i]$
1950	11	0	11	10	0	–	–	0·9697	0·097
1951	66	4	70	63	6	10·667	0·38	0·9287	0·040
1952	28	36	64	60	33	71·200	0·51	0·9735	0·039
1953	2	43	45	42	50	96·571	0·45	0·9619	0·041
1954	4	54	58	54	38	94·815	0·57	0·9593	0·036
1955	51	63	104	104	29	94·788	0·66	0·9664	0·035
1956	13	69	73	73	64	140·890	0·49	0·9419	0·040
1957	5	99	86	86	38	144·953	0·68	0·8546	0·040
1958	19	85	94	94	39	128·149	0·66	0·9444	0·038
1959	8	51	55	55	82	138·964	0·37	0·9662	0·036
1960	26	102	128	112	35	142·000	0·72	0·9028	0·032
1961	3	133	136	102	14	151·667	0·88	–	–
1962	18	116	134	–	–	–	–	–	–

16·7 ± 3·0 years respectively. He points out that the rather tentative start to the experiment with only 11 individuals marked in the first year caused $V[\hat{\phi}_1]$ to be much larger than any subsequent variance. Omitting all references to these 11 birds, the data were re-analysed, giving 0·9378 ± 0·0075 and 15·58 ± 1·93 years, a considerable increase in precision.

In conclusion, we raise the question of whether $\hat{\phi}_i$ applies to the whole population rather than to just the banded population. For example, some birds of the colony nest on sites totally inaccessible to the experimenter and therefore could not enter the banded population. Our estimate of $\hat{\phi}_i$ would then not apply to these birds if the choice of nesting site affected the survival of the adult bird. Although it is true that the choice of nesting site would probably affect the chance of successfully rearing young, it was only the adult breeding population that was considered in this investigation. Cormack also mentioned the fact that the pattern of attendance at and near the nest was different for the two sexes, so that the probability of capture and recapture was different for males and females. However, Cormack felt that this would not greatly affect the validity of the above analysis, provided the survival probabilities were the same for males and females. Unfortunately this assumption could not be tested as there were insufficient data.

5.1.3 Special cases

1 Enclosed populations

Suppose that the population area is bounded or enclosed, so that there is no migration. Then losses are due solely to deaths, and the influx of new individuals represents recruitment only. If, however, there is no recruitment, as for example in a non-breeding season, or if the new recruits can be distinguished from the others, for example by their size, then the unrecruited population can be analysed as one in which only mortality is operating (i.e. each $B_i = 0$, except $B_0 = N_1$). Jolly [1965] shows that the estimates are then given by

$$\hat{N}_i = \frac{R_i Z_i}{r_i} + n_i, \quad (i = 1, 2, \ldots, s - 1),$$

$$\hat{\phi}_i = \frac{\hat{N}_{i+1}}{\hat{N}_i - n_i + R_i}, \quad (i = 1, 2, \ldots, s - 2),$$

and

$$\hat{p}_i = \frac{n_i}{\hat{N}_i}, \quad (i = 1, 2, \ldots, s - 1),$$

where $Z_i = z_i + \sum_{j=i+1}^{s} u_j$, the number of animals not caught in the ith sample but caught subsequently; $Z_i - z_i$ is the number of animals caught for the *first* time after the ith sample. These estimates were originally obtained by Darroch [1959] for the case of no losses on handling ($R_i = n_i$).

By analogy with (5.22), approximately unbiased estimates are given by

$$N_i^* = \frac{(R_i + 1)}{(r_i + 1)} Z_i + n_i$$

$$= \frac{(R_i + 1)(Z_i + r_i + 1)}{r_i + 1} - 1 + (n_i - R_i)$$

and when $R_i = n_i$ this reduces to the unbiased estimate suggested by Darroch (called N_i'' in his notation).

From Jolly we have (replacing expected values of random variables by "observed" values)

$$V[\hat{N}_i] = (N_i - n_i)(N_i - n_i + R_i)\left(\frac{1}{r_i} - \frac{1}{R_i}\right) + N_i - N_i^2/N_1 ,$$

$$\text{cov}\,[\hat{N}_i, \hat{N}_j] = N_j - \frac{N_i N_j}{N_1}, \quad (i < j),$$

$$V[\hat{\phi}_i] = \phi_i^2 \left\{ \frac{(N_{i+1} - n_{i+1})(N_{i+1} - n_{i+1} + R_{i+1})}{N_{i+1}^2}\left(\frac{1}{r_{i+1}} - \frac{1}{R_{i+1}}\right) + \right.$$

$$\left. + \frac{N_i - n_i}{N_i - n_i + R_i}\left(\frac{1}{r_i} - \frac{1}{R_i}\right) + \frac{1 - \phi_i}{N_{i+1}} \right\},$$

$$\text{cov}\,[\hat{\phi}_i, \hat{\phi}_{i+1}] = -\phi_i \phi_{i+1} \frac{(N_{i+1} - n_{i+1})}{N_{i+1}}\left(\frac{1}{r_{i+1}} - \frac{1}{R_{i+1}}\right),$$

and

$$\text{cov}\,[\hat{\phi}_i, \hat{\phi}_j] = 0, \quad (j > i + 1).$$

We note that it is now possible to estimate N_1 as $Z_1 \neq 0$.

2   No death or emigration

For completeness we shall consider the rather uncommon situation in which there is no death or emigration, so that the only changes in population size are due to recruitment and possibly immigration. In this case $\phi_i = 1$ ($i = 1, 2, \ldots, s-1$) and from Jolly [1965: 242] we have

$$M_i = \sum_{h=1}^{i-1} (R_h - m_h), \qquad \hat{\rho}_i = m_i/n_i,$$

leading to

$$\hat{N}_i = M_i/\hat{\rho}_i, \qquad (i = 2, 3, \ldots, s)$$

and

$$\hat{B}_i = \hat{N}_{i+1} - (\hat{N}_i - n_i + R_i), \qquad (i = 2, 3, \ldots, s-1).$$

We note that N_s and B_{s-1} are estimable but not N_1. Expressions for the asymptotic variances and covariances are given by Jolly, namely

$$V[\hat{N}_i] = N_i(N_i - n_i)\left(\frac{1 - \rho_i}{m_i}\right) + N_i - \sum_{h=0}^{i-1} \frac{N_i^2(h)}{B_h},$$

$$\text{cov }[\hat{N}_i, \hat{N}_j] = \sum_{h=0}^{i-1} \left\{N_j(h) - \frac{N_i(h)N_j(h)}{B_h}\right\},$$

$$V[\hat{B}_i] = N_{i+1}(N_{i+1} - n_{i+1})\left(\frac{1 - \rho_{i+1}}{m_{i+1}}\right) + N_i(N_i - n_i)\left(\frac{1 - \rho_i}{m_i}\right),$$

$$\text{cov }[\hat{B}_i, \hat{B}_{i+1}] = -N_{i+1}(N_{i+1} - n_{i+1})\left(\frac{1 - \rho_{i+1}}{m_{i+1}}\right),$$

$$\text{cov }[\hat{B}_i, \hat{B}_j] = 0, \qquad (j > i + 1)$$

and

$$V[\hat{N}_i \mid N_i] = N_i(N_i - n_i)\left(\frac{1 - \rho_i}{m_i}\right).$$

The difference between $V[\hat{N}_i]$ and $V[\hat{N}_i \mid N_i]$, namely $V[N_i]$, arises entirely from stochastic death due to loss on capture, there being no death from other causes. Therefore, when $R_i = n_i$, $N_i(h) = B_h$ (for $h \leq i - 1$) and $V[N_i] = \text{cov }[\hat{N}_i, \hat{N}_j] = 0$; this case was first considered by Darroch [1959] who also obtained $\hat{N}_i$, $\hat{B}_i$ and $V[\hat{N}_i]$.

3   Three-point census: triple-catch method

When $s = 3$ we have from the general theory the estimates

$$\hat{M}_2 = \frac{R_2 z_2}{r_2} + m_2, \qquad (5.28)$$

$$\hat{N}_2 = n_2 \hat{M}_2/m_2,$$

$$\hat{\phi}_1 = \hat{M}_2/R_1,$$

$$\hat{p}_2 = m_2/\hat{M}_2$$

and

$$\hat{\phi}_2 \hat{p}_3 = m_3/(\hat{M}_2 - m_2 + R_2).$$

If an estimate $\hat{N}_1$ is available we can also add

$$\hat{B}_1 = \hat{N}_2 - \hat{\phi}_1(\hat{N}_1 - n_1 + R_1).$$

The asymptotic variances of $\hat{N}_2$ and $\hat{\phi}_1$ are given by

$$V[\hat{N}_2] = N_2(N_2 - n_2)\left\{\frac{M_2 - m_2 + R_2}{M_2}\left(\frac{1}{r_2} - \frac{1}{R_2}\right) + \frac{1 - p_2}{m_2}\right\} +$$
$$+ (N_2 - B_1)\{1 - (N_2 - B_1)/N_1\}$$

and

$$V[\hat{\phi}_1] = \phi_1^2\left\{\frac{(M_2 - m_2)(M_2 - m_2 + R_2)}{M_2^2}\left(\frac{1}{r_2} - \frac{1}{R_2}\right) + \frac{1 - \phi_1}{M_2}\right\}.$$

As pointed out in the general theory, the expression in $V[\hat{N}_2]$ involving $N_2 - B_1$ will usually be negligible.

If there is no emigration, and mortality is a Poisson process with parameter μ, then from (1.1)

$$\phi_i = e^{-\mu T_i}, \quad (i = 1, 2),$$

where T_i is the time from the ith release to the $(i+1)$th sample. We can now estimate the remaining parameters using the following chain of estimates:

$$\hat{\mu} = -\frac{1}{T_1}\log\hat{\phi}_1$$

$$\hat{\phi}_2 = e^{-\hat{\mu}T_2} = (\hat{\phi}_1)^{T_2/T_1},$$

$$\hat{M}_3 = (\hat{M}_2 - m_2 + R_2)\hat{\phi}_2,$$

$$\hat{p}_3 = m_3/\hat{M}_3, \quad \hat{N}_3 = n_3\hat{M}_3/m_3$$

and

$$\hat{B}_2 = \hat{N}_3 - \hat{\phi}_2(\hat{N}_2 - n_2 + R_2).$$

From the above discussion we see that if $T_1 = T_2$ then, on the assumption that $\phi_1 = \phi_2$, three samples is the minimum number needed to estimate all the unknowns (except B_1). Although such estimates may not be very accurate with such a short sample sequence, it may be all the samples that an experimenter can obtain.

Bailey [1951, 1952] was the first to introduce the so-called triple-catch method and considered the case of no losses on capture. His estimates were based on a deterministic model, and we shall briefly compare his estimate of M_2 with ours. Putting $R_2 = n_2$ and rearranging (5.28), we have

$$\frac{r_2 + z_2}{\hat{M}_2 + u_2} = \frac{r_2}{n_2},$$

where (denoting a_w as the number with capture history w)

$$\frac{r_2}{n_2} = \frac{a_{23} + a_{123}}{u_2 + m_2}.$$

In words, $\hat{M}_2$ is obtained by equating the proportions of the marked population $M_2 + u_2$ and the release n_2 which are caught in the third sample. It can be shown that Bailey's estimate M_2' of M_2 is given by

$$\frac{r_2 + z_2}{M_2' + u_2} = \frac{a_{23}}{u_2},$$

which ignores the recapture information a_{123} given by the release of m_2 previously marked individuals.

Example 5.5   Meadow grasshopper (*Chorthippus parallelus*): Parr *et al.* [1968]

A colony of the meadow grasshopper was studied, using the mark-recapture method, on 9, 10, 11 September, 1966. The small field supporting the colony contained meadow grasses cut short for hay and was bounded by a road on one side and thick scrubby vegetation on the remaining sides. As *C. parallelus* is unable to fly, it may be caught by sweeping on warm sunny days when it is very active; on dull days it may be more efficient to capture the insects by hand. An attempt to mark the grasshoppers with cellulose paint failed as in many cases it was seen to peel off the insect's cuticle in a matter of minutes. The use of black Indian ink proved more satisfactory, although this was liable to flake off partially, and was difficult to apply because of the greasy nature of the cuticle. (Oil paint would have been a better marking agent but was not available.) However, it was found that once the insect was marked, some particles of ink remained attached to the insect, and even if most of the ink flaked off, these particles could easily be seen using a hand-lens. The insects were marked in batches from limited parts of the colony area, and after marking were released as near to the point of capture as possible.

The recapture data for just the male population are summarised below:

$$(n_1, n_2, n_3) = (52, 41, 39), \quad (R_1, R_2) = (52, 40), \quad m_{12} = 5,$$

$$m_{13} = 7, \quad m_{23} = 8,$$

where m_{hi} is the number caught in the ith sample last captured in the hth. Then $r_1 = m_{12} + m_{13} = 12$, $r_2 = m_{23} = 8$, $z_2 = m_{13} = 7$, $m_2 = m_{12} = 5$, $m_3 = m_{13} + m_{23} = 15$,

$$\hat{M}_2 = \frac{40(7)}{8} + 5 = 40,$$

$$\hat{N}_2 = 40(41)/5 = 328,$$

$$\hat{\phi}_1 = 40/52 = 0 \cdot 769,$$

$$\sqrt{\{\hat{V}[\hat{N}_2 \mid N_2]\}} = 184 \cdot 0$$

and

$$\sqrt{\{\hat{V}[\hat{\phi}_1]\}} = 0 \cdot 317.$$

We note that both $\hat{N}_2$ and $\hat{\phi}_1$ have large coefficients of variation, and this is due to the small values of m_2 and r_2. It was recommended on p. 204 that these values should be at least as great as 10.

Since $T_1 = T_2 = 1$ day, then, assuming $\phi_1 = \phi_2$, we have

$$\hat{\phi}_2 = \hat{\phi}_1,$$
$$\hat{M}_3 = 75(40)/52 = 57 \cdot 7,$$
$$\hat{p}_3 = 0 \cdot 26, \quad \hat{N}_3 = 150 \quad \text{and} \quad \hat{B}_2 = -96.$$

From p. 204, approximately unbiased estimates are given by

$$M_2^* = \frac{41(7)}{9} + 5 = 36 \cdot 9,$$

and

$$N_2^* = 36 \cdot 9(42)/6 = 258 \cdot 3$$
$$\phi_1^* = 36 \cdot 9/52 = 0 \cdot 71.$$

4   Ricker's two-release method

Ricker [1958: 128] has suggested a useful method for estimating the probability of survival for a given period of time by making a release at the beginning and end of the period followed by a sample. For $i = 1, 2$ let R_i be the size of the ith release and let m_{i3} be the number from the ith release caught in the sample. Let a_1 be the probability of survival between releases, and assume that every marked individual alive just after the second release has the same probability β_2 of surviving to the time of the sample and being caught in the sample. Then, neglecting the complications of sampling without replacement, m_{13} and m_{23} will be independent binomial variables, so that

$$f(m_{13}, m_{23}) = \binom{R_1}{m_{13}}(a_1\beta_2)^{m_{13}}(1 - a_1\beta_2)^{R_1 - m_{13}}\binom{R_2}{m_{23}}\beta_2^{m_{23}}(1 - \beta_2)^{R_2 - m_{23}}.$$

It is readily shown that the maximum-likelihood estimates of a_1 and β_2 (which are also the moment estimates) are

$$\hat{a}_1 = \left(\frac{m_{13}}{R_1}\right)\bigg/\left(\frac{m_{23}}{R_2}\right)$$

and

$$\hat{\beta}_2 = m_{23}/R_2.$$

However, a slight modification of $\hat{a}_1$, namely

$$\tilde{a}_1 = \frac{m_{13}(R_2 + 1)}{R_1(m_{23} + 1)}$$

is almost unbiased, as

$$E[\tilde{a}_1] = E\left[\frac{m_{13}}{R_1}\right]E\left[\frac{R_2 + 1}{m_{23} + 1}\right]$$

$$= a_1\beta_2 \cdot \frac{1}{\beta_2}(1 - (1 - \beta_2)^{R_2 + 1})$$

$$\approx a_1.$$

Also

$$E\left[\frac{m_{13}(m_{13}-1)}{R_1(R_1-1)}\cdot\frac{(R_2+1)(R_2+2)}{(m_{23}+1)(m_{23}+2)}\right] = E\left[\frac{m_{13}(m_{13}-1)}{R_1(R_1-1)}\right]E\left[\frac{(R_2+1)(R_2+2)}{(m_{23}+1)(m_{23}+2)}\right]$$

$$= (\alpha_1\beta_2)^2\cdot\frac{1}{\beta_2^2}[1-(1-\beta_2)^{R_2+2}-(R_2+2)\beta_2(1-\beta_2)^{R_2+1}]$$

$$\approx \alpha_1^2,$$

so that

$$v[\tilde{\alpha}_1] = \tilde{\alpha}_1^2 - \frac{m_{13}(m_{13}-1)(R_2+1)(R_2+2)}{R_1(R_1-1)(m_{23}+1)(m_{23}+2)}$$

is an almost unbiased estimate of $V[\tilde{\alpha}_1]$.

Ricker's scheme is very flexible as it can be used when exploitation is taking place continuously between releases and when the sample is not instantaneous but extends over a period of time. If the number m_{12} of tagged removed from the population between releases is recorded, then β_1, the probability of being caught between releases, is estimated by

$$\hat{\beta}_1 = m_{12}/R_1.$$

In this case the joint distribution of m_{12}, m_{13}, m_{23} is given by (5.23) with $s = 3$.

The case when the instantaneous natural mortality rate and the instantaneous exploitation rate are both constant is discussed in **6.4.2**(2); Ricker's method is also mentioned again in **9.2.2**(1).

Example 5.6 Bluegills (*Lepomis macrochirus*): Ricker [1958]

The above method was applied to a population of bluegills in Muskellunge Lake, Indiana. Of $R_1 = 230$ bluegills marked before the start of the 1942 fishing season, $m_{13} = 13$ were captured in 1943. Of $R_2 = 93$ marked before the start of the 1943 season, $m_{23} = 13$ were captured in 1943. Thus α_1, the probability of survival in 1942, is estimated by

$$\tilde{\alpha}_1 = \frac{13(94)}{230(14)} = 0\cdot379\ 50$$

and

$$v[\tilde{\alpha}_1] = (0\cdot3795)^2 - \frac{13(12)(94)(95)}{230(229)(14)(15)}$$

$$= 0\cdot144\ 02 - 0\cdot125\ 95 = 0\cdot018\ 07.$$

The approximate 95 per cent confidence interval, $\tilde{\alpha}_1 \pm 1\cdot96\sqrt{v}$, for α_1 is $(0\cdot10, 0\cdot66)$.

5.1.4 Underlying assumptions*

1 Validity of underlying model

GOODNESS-OF-FIT TEST. Since (5.3) comes from a product of independent multinomial distributions (cf. 5.2), we can calculate a goodness-of-fit test for

---

* See also Cormack [1972].

the adequacy of this model with regard to the *marked* population. For example, when $s = 3$, the test statistic is

$$\frac{(b_1 - u_1 \hat{\nu}_1 \hat{\chi}_1)^2}{u_1 \hat{\nu}_1 \hat{\chi}_1} + \frac{(b_{12} - u_1 \hat{\nu}_1 \hat{\phi}_1 \hat{p}_2 \hat{\nu}_2 \hat{\chi}_2)^2}{u_1 \hat{\nu}_1 \hat{\phi}_1 \hat{p}_2 \hat{\nu}_2 \hat{\chi}_2} + \ldots + \frac{(d_1 - u_1(1 - \hat{\nu}_1))^2}{u_1(1 - \hat{\nu}_1)} + \ldots +$$

$$+ \frac{(b_2 - u_2 \hat{\nu}_2 \hat{\chi}_2)^2}{u_2 \hat{\nu}_2 \hat{\chi}_2} + \ldots + \frac{(d_2 - u_2(1 - \hat{\nu}_2))^2}{u_2(1 - \hat{\nu}_2)} + \ldots +$$

$$+ \frac{(b_3 - u_3 \hat{\nu}_3)^2}{u_3 \hat{\nu}_3} + \frac{(d_3 - u_3(1 - \hat{\nu}_3))^2}{u_3(1 - \hat{\nu}_3)} \,,$$

which is asymptotically distributed as chi-squared. For general s the goodness-of-fit statistic has $(f_1 - f_2)$ degrees of freedom, where f_1 is the number of squared terms (after any necessary pooling) and $f_2 = 4s - 3$ (since we have $3s - 3$ parameters $\nu_1, \ldots, \nu_s, \alpha_1, \ldots, \alpha_{s-2}, \beta_1, \ldots, \beta_{s-1}$ and s constraints — one for each multinomial distribution).

In calculating the goodness-of-fit statistic we note that

$$\hat{\chi}_i = \frac{R_i - r_i}{R_i}, \quad \hat{\nu}_i = \frac{R_i}{n_i}$$

and

$$\hat{\alpha}_i = \hat{\phi}_i \hat{q}_{i+1} = (\hat{M}_{i+1} - m_{i+1})/(\hat{M}_i - m_i + R_i)$$

$$\hat{\beta}_i = \hat{\phi}_i \hat{p}_{i+1} = m_{i+1}/(\hat{M}_i - m_i + R_i).$$

Since the d_w will generally be small it is recommended that they be pooled with the b_w. In fact in some situations many of the expected frequencies may be less than unity, so that the large-scale pooling required would lead to an approximate test only.

METHOD OF LESLIE, CHITTY AND CHITTY. Another method which indirectly tests for the validity of the underlying model with respect to the marked population has been proposed by Leslie *et al.* [1953]. This method, which we shall now discuss in detail, consists of comparing the increase in the marked population at time t_i due to the release of freshly marked individuals, with an estimate of this increase obtained from the marked population of animals which are caught at least twice.

Suppose we consider just the marked population; this is generated by the releases, v_i say, of individuals marked for the first time ($v_i \leqslant u_i$). We now regard these releases as constituting the "unmarked" population, and those captured from this population (i.e. those caught at least twice) constitute the "marked" population. To begin with, we have a population of R_1 "unmarked" individuals of which m_2 are caught in the second sample. These m_2 animals now represent the "marked" population, and v_2 "unmarked" animals are added to the population by "recruitment". In the third sample, animals previously caught twice are regarded as "recaptures", while those caught in only one of the previous samples are from the "unmarked" population and are regarded as having been caught for the "first" time: once again we have a "recruitment" of v_2 "unmarked". This new population of "marked" and "unmarked", consisting solely of marked animals, can be analysed using exactly the same method as demonstrated in Example 5.1 (p. 205). The only difference is that one excludes from the group m_{hi} all animals caught only once before. This new group we note by $m_{.hi}$, the number of animals caught in the ith sample which were last caught in the hth sample and at least once prior to the hth sample. From the array $\{m_{.hi}\}$ we can calculate the corresponding array $\{c_{.hi}\}$ and the quantities $r_{.i}$, $z_{.i}$. Using the "dot" notation to denote membership of the marked population, we find that $n_{.i} = m_i$; $m_{.i}$ is the number of marked caught at least twice before the ith sample and in the ith sample; $R_{.i} = m_i - d_i$, where d_i is the number of marked caught in the ith sample which are not returned to the population. The quantities $N_{.i}$, $\phi_{.i}$ and $\rho_{.i}$ now refer to the marked population while $M_{.i}$ refers to the "marked" population of those caught at least twice. We note that $N_{.1} = R_{.1}$, and define $M_{.1} = M_{.2} = 0$, $m_{.1i} = 0$ for $i = 2, 3, \ldots, s$.

Since
$$E[N_{.i+1} \mid N_{.i}] = \phi_{.i}(N_{.i} + v_i - d_i),$$

we can "estimate" v_i by (cf. equations (5.6) to (5.12))

$$\hat{v}_i = (\hat{N}_{.i+1}/\hat{\phi}_{.i}) - \hat{N}_{.i} + d_i$$

$$= \frac{\hat{M}_{.i} - m_{.i} + R_{.i}}{\hat{\rho}_{.i+1}} - \frac{\hat{M}_{.i}}{\hat{\rho}_{.i}} + d_i, \quad (i = 3, 4, \ldots, s-1).$$

To find the asymptotic variance of $\hat{v}_i$ we note from Jolly [1965: 237] that the component due to the errors of estimation in the covariance of any pair of $\hat{M}_{.i}$, $\hat{\rho}_{.i}$ and $\hat{\rho}_{.i+1}$ is zero. Therefore, using the delta method, treating d_i as a constant, and neglecting covariance terms, we have from Jolly

$$\sigma^2[\hat{v}_i] = V[\hat{M}_{.i} \mid M_{.i}]\left(\frac{1}{\rho_{.i+1}} - \frac{1}{\rho_{.i}}\right)^2 +$$

$$+ V[\hat{\rho}_{.i+1} \mid \rho_{.i+1}]\frac{(M_{.i} - m_{.i} + R_{.i})^2}{\rho_{.i+1}^4} + V[\hat{\rho}_{.i} \mid \rho_{.i}]\frac{M_{.i}^2}{\rho_{.i}^4}$$

$$= (M_{.i} - m_{.i})(M_{.i} - m_{.i} + R_{.i})\left(\frac{1}{r_{.i}} - \frac{1}{R_{.i}}\right)\left(\frac{1}{\rho_{.i+1}} - \frac{1}{\rho_{.i}}\right)^2 +$$

$$+ \rho_{.i+1}(1 - \rho_{.i+1})\left(\frac{1}{n_{.i+1}} - \frac{1}{N_{.i+1}}\right)\frac{(M_{.i} - m_{.i} + R_{.i})^2}{\rho_{.i+1}^4} +$$

$$+ \rho_{.i}(1 - \rho_{.i})\left(\frac{1}{n_{.i}} - \frac{1}{N_{.i}}\right)\frac{M_{.i}^2}{\rho_{.i}^4}.$$

It can be shown that this formula is equivalent to that given by Leslie et al. [1953: 168] if we ignore all terms involving $R_{.i}^{-1}$, $N_{.i}^{-1}$, $N_{.i+1}^{-1}$ (i.e. the finite population corrections) and use $c_{.i-1,i+1}/m_{.i,i+1}$ instead of $z_{.i}/r_{.i}$ in the estimation of $M_{.i} - m_{.i}$ (cf. p. 201).

Having calculated an estimate $\hat{v}_i$ we now see whether the observed value of v_i lies in the interval $\hat{v}_i \pm k \hat{\sigma}[\hat{v}_i]$, where k is to be chosen. We can also compare $\hat{N}_{\cdot i}$ and $\hat{M}_i$ as they are estimates of the marked population; $\hat{\phi}_{\cdot i}$ and $\hat{\phi}_i$ should be fairly similar as they are both estimates of ϕ_i.

Example 5.7   Six-spot Burnet moth (*Zygaena filipendula* L.): Manly and Parr [1968]

The above analysis was applied to mark-recapture data given by Manly and Parr. One sample was taken on each of five days, and sunny weather throughout the period of study ensured adequate mixing of the marked and unmarked between samples. The insects were marked with cellulose "dope" applied to the underside of the hindwings, and the following colour code was used:

Day 1: green   (*g*)

Day 2: white   (*w*)

Day 3: blue   (*b*)

Day 4: orange (*o*)

There were no losses on capture (i.e. $R_i = n_i$, $R_{\cdot i} = n_{\cdot i} = m_i$, $v_i = u_i$) and using the above colour abbreviations the recapture data were recorded as follows:

Day 1 (19 July): 57 captured, marked and released.
Day 2 (20 July): 52 captured; 25*g*, 27 unmarked.
Day 3 (21 July): 52 captured; 8*g*, 9*w*, 11*gw*, 24 unmarked.
Day 4 (22 July): 31 captured; 2*g*, 3*w*, 4*b*, 5*gb*, 1*wb*, 2*gwb*, 14 unmarked.
Day 5 (24 July): 54 captured; 1*g*, 2*w*, 7*b*, 5*o*, 4*gw*, 2*gb*, 2*go*, 4*wb*, 1*wo*, 1*bo*, 5*gbo*, 1*gwbo*, 19 unmarked.

In working with such data it is helpful to convert the colours to numbers.

The arrays $\{m_{hi}\}$ and $\{m_{\cdot hi}\}$ are recorded in Table 5.6: for example $m_{34} = 4b + 5gb + 1wb + 2gwb = 12$ and $m_{\cdot 34} = 5gb + 1wb + 2gwb = 8$. From these data the arrays $\{c_{hi}\}$, $\{c_{\cdot hi}\}$ were obtained, and the estimates in Table 5.7 were then calculated. As there is a reasonable agreement between $\hat{N}_{\cdot i}$, $\hat{\phi}_{\cdot i}$ and $\hat{v}_i$, and $\hat{M}_i$, $\hat{\phi}_i$ and v_i respectively, we have no reason to reject the underlying model for the marked population.

In order to estimate $\sigma[\hat{v}_4]$ we must either ignore terms involving $1/N_{\cdot 5}$ or else estimate $N_{\cdot 5}$ from the trend, or lack of trend, in $\hat{N}_{\cdot 3}$ and $\hat{N}_{\cdot 4}$; for this experiment both methods led to almost the same value of $\hat{\sigma}[\hat{v}_4]$.

If a goodness-of-fit test is required for the above data, the first step is to find the b_w (= a_w as d_w = 0). This can be done using, for example, a tree diagram for each release of u_i insects marked for the first time. An example of this is given in Fig. 5.1.

2. Equal probability of capture for marked individuals

METHOD OF LESLIE.   Assumption (a) of p. 196 will be satisfied if all sampling is random or if there is random mixing of marked and unmarked and

TABLE 5.6

Tabulation of m_{hi} and $(m_{.hi})$ for a moth population: data from **Manly and Parr** [1968: 86].

i	1	2	3	4	5	
$n_i\ (n_{.i})$	57	52(25)	52(28)	31(17)	54(35)	Total
h						$r_h\ (r_{.h})$
1		25	8	2	1	36
2			20(11)	3(0)	6(4)	29(15)
3				12(8)	13(6)	25(14)
4					15(10)	15(10)
u_i	57	27	24	14	19	

TABLE 5.7

Estimation of population parameters for a moth population from the data in Table 5.6.

i	$\hat{M}_i$	$\hat{\phi}_i$	$\hat{N}_{.i}$	$\hat{\phi}_{.i}$	v_i	$\hat{v}_i$	$\hat{\sigma}[\hat{v}_i]$
1	—	0·78	—	—	57	—	—
2	44·7	0·74	—	0·76	27	—	—
3	53·0	0·76	48·4	0·69	24	28	18
4	58·3	—	53·1	—	14	6	12
5	—	—	—	—	19	—	—

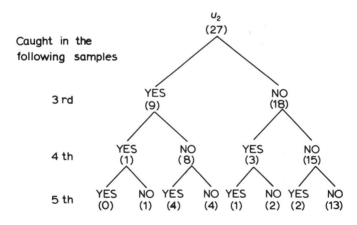

Fig 5.1 Tree diagram for calculating the b_w from the second release of newly marked individuals.

all individuals have the same catchability. Leslie's method described on p. 161 can be used for testing whether the sampling is random with respect to just the marked population by studying the capture frequencies of members of a group of animals known to be in the population over a given period of time. For example, consider the group of animals released after sample i

227

which had been marked for the first time, and suppose G of this group were captured for the last time in sample $i + t + 1$. Then members of G could have been recaptured up to t times during the intervening interval, and we can apply Leslie's chi-squared test to see if the sampling is random with respect to this group. By varying i and t subject to $i \geqslant 1$, $t \geqslant 3$ and $i + t + 1 \leqslant s$ we have a whole series of intervals each with its own chi-squared test. For example, if $s = 8$ we have the intervals defined by the pairs (1, 5), (1, 6), (1, 7), (1, 8); (2, 6), (2, 7), (2, 8); (3, 7), (3, 8); (4, 8). Since the groups of animals corresponding to these intervals are mutually exclusive, the various chi-squareds are independent and can be added. One way of doing this would be to add the chi-squareds which are based on the same release. However, if one uses the above intervals, the groups may be too small (say < 20 members) for a valid application of the chi-squared test. In this case it is preferable to pool the information from several samples. For example, if $s = 8$ we can use the pairs (1, 5), (1, 6) and (2, 6), where (1, 6) represents the group of individuals from the first release which are recaptured in at least one of the samples 6, 7 and 8.

Examples of the above method are given in Orians [1958] and Turner [1960b]. However, a more useful procedure has been given by Carothers [1971].

Example 5.8

Using the data of Example 5.7, we have 5 samples and therefore just one interval defined by (1, 5). Here $G = 15$ (the number of animals caught in the 5th sample bearing a green mark) and, using the notation of **4.1.5** (2), Table 5.8 was derived. For example, g_i, for sample 3, is the number caught

TABLE 5.8
Leslie's test for random sampling using the data from Example 5.6.

Sample	g_i	g_i^2	x	f_x	$x f_x$	$x^2 f_x$
2	5	25	0	1	0	0
3	8	64	1	8	8	8
4	8	64	2	5	10	20
	21	153	3	1	3	9
				15	21	37

in all three samples 1, 3 and 5, namely $2gb + 5gbo + 1gwbo = 8$. We find that $\mu = 1 \cdot 4$, $\sigma^2 = 1 \cdot 4 - 0 \cdot 68 = 0 \cdot 72$ and

$$T = \sum_{x=0}^{3} f_x (x - \mu)^2 / \sigma^2 = 10 \cdot 6,$$

which, at 14 degrees of freedom, indicates that the observed and expected variances are close. Although such a test should be viewed with caution as G is small, it does at least support the hypothesis of random sampling as far as just the first release in concerned.

UNIFORM MIXING. The question of random sampling is a crucial one when the animals are relatively immobile so that mixing of marked and unmarked is slow. The following example is a good illustration of this.

Example 5.9   Land snail (*Helix aspersa*): Parr *et al.* [1968]

In September 1966 an attempt was made to estimate the number in a colony of *H. aspersa* living on a wall. This species is known to have a low mortality rate and observations showed that there was very little migration. It was therefore expected that a mark-recapture study of this snail using the Jolly–Seber method could be used to demonstrate high survival probabilities ϕ_i, a series of similar daily estimates of population size, and a high overall proportion of the population caught by the end of the experiment. In fact, none of these expectations was fully realised, and although 935 different snails were marked in six days the average daily estimate of population size was only about 850. On the last day of observation, 102 out of a total of 230 snails had not been seen previously. This breakdown of the Jolly–Seber method was due to the fact that random mixing of marked and unmarked snails did not occur during the six-day sampling period owing to irregular spells of activity and aestivation (rest). Thus, if snails aestivated on the surface of the wall they were recaptured every day, but if they aestivated in holes in the wall they were inaccessible.

The experiment was then repeated on the same colony in September 1967, but for a period of ten days, in the hope that this might be long enough to allow mixing of marked and unmarked. A Petersen estimate (cf. **3.1.1**) of the population size based on the first two days of sampling gave a value of $\hat{N} = 743$ which was far too low. Similar estimates on subsequent days based on the numbers of recaptures, y_i, of the individuals originally marked and released on the first day are shown in Table 5.9. These estimates are valid if there is negligible mortality or the mortality rates are the same for marked and unmarked (cf. **3.2.1**(2)). The authors felt that the slow upward drift in the estimates reflected the slow mixing of marked and unmarked snails. The later estimates would therefore be more reliable as the mixing became more uniform. The day-to-day fluctuations in the estimates were due to the relatively large standard deviations.

The population was again sampled on 28 September 1967 (16 days after the last previous sampling) when 357 snails were caught of which 254 were marked. Since there were 1077 + (318 − 206) = 1189 marked snails in the population the Petersen estimate is

$$\hat{N} = 1189(357)/254 = 1672$$

with a standard deviation of (cf. **3.1.1**) $\sqrt{v^*} = 49 \cdot 6$. A similar overall estimate was obtained from the last entry in Table 5.9 using the total population of marked, namely

$$\hat{N} = 318(1077)/206 = 1662 \cdot 5.$$

This value is very close to the above overall estimate and supports the

229

TABLE 5.9
Recapture data for a population of snails, September 3–12, 1967:
from Parr *et al.* [1968: Table 10].

Day i	n_i	m_i	M_i	b_{1i}	$\hat{N} = \dfrac{n_1 n_i}{y_i}$
1	217	–	–	–	–
2	243	71	217	71	742·7
3	310	110	389	56	1201·3
4	326	198	590	71	996·4
5	371	248	717	69	1166·8
6	306	243	840	51	1302·0
7	274	233	903	40	1486·5
8	296	250	943	51	1259·5
9	258	170	989	32	1749·6
10	318	206	1077	48	1437·6

y_i = number captured on day i which were marked on day 1

authors' view that random mixing of the snails had occurred by the end of
the period of regular sampling.

The authors finally mentioned that the weather was generally dry between
3 and 10 September and many snails may have aestivated throughout this
period, but heavy rain caused considerable movement of snails on 10 September
and provided the necessary mixing of marked and unmarked.

3   Equal probability of survival for marked population

Possible departures from assumption (b) of p. 196, that ϕ_i is the same
for all marked individuals, are: (i) The method of catching (and tagging, if
individual tags are not used) has a deleterious effect, so that animals caught
many times have a higher mortality rate. (ii) If individual tags are used, so
that only the untagged in each sample are given a tag, newly tagged animals
may have a higher mortality rate than those tagged on a previous occasion.
(iii) Different age-classes have different mortality rates.

We shall now describe contingency table methods for detecting the above
three types of departures.

SURVIVAL INDEPENDENT OF MARK STATUS. Let v_{xi} be the number of
animals released from the ith sample which have been caught x times in the
first i samples ($x = 0, 1, 2, \ldots, i$) and let $v_{xi,i+1}$ be the number from this
group recaptured in the $(i+1)$th sample. Then, using an obvious notation,

$$E[v_{xi,i+1}/v_{xi}] = \phi_{xi}p_{xi}, \quad (x = 0, 1, \ldots, i),$$

and we can test the hypothesis $H : \phi_{xi}p_{xi} = \phi_i p_i$ using the pairs ($v_{xi,i+1}$,
$v_{xi} - v_{xi,i+1}$) in a 2-by-i contingency table; any groups which are too small
can be pooled. If the sampling is random so that $p_{xi} = p_i$ ($x = 0, 1, \ldots, i$)

then a test of H is a test that $\phi_{xi} = \phi_i$, that is, survival in the marked population is independent of mark status. Alternatively, if $\phi_{xi} = \phi_i$ then we are testing whether p_i is independent of mark status. A contingency table can be constructed for each $i = 2, 3, \ldots, s - 1$, and since the chi-squareds are independent they can be added together to give a total chi-square. Further contingency tables can be constructed if we consider v_{xij}, the number of v_{xi} which are not caught again until the jth sample $(j > i)$.

INITIAL TAGGING MORTALITY. In fishery investigations it is well known that the operation of attaching or inserting a tag places considerable stress on the fish. This stress may lead to an increase in mortality during the period immediately following release. To test for this initial mortality (called "type I losses" in the fisheries literature) Robson [1969] suggests using the 2-by-2 contingency table, Table 5.10; once again the tests are independent for $i = 2, 3, \ldots, s - 1$. An essentially equivalent test has been proposed by Manly [1971b: 184].

TABLE 5.10
Contingency table for detecting type I losses: from Robson [1969].

	Captured after ith sample	Not captured after ith sample
First captured in the ith sample and released	*	*
Recaptured in the ith sample and released	*	*

TABLE 5.11
Contingency table for detecting sustained type I losses: from Robson[1969].

	Captured in sample $i + 1$ and	
	captured after sample $i + 1$	not captured after sample $i + 1$
First caught in sample i	*	*
Caught before sample i	*	*

TABLE 5.12
Contingency table for detecting sustained type I losses: from Robson[1969].

	Captured after sample * and	
	captured in sample $i + 1$	not captured in sample $i + 1$
First caught in sample i	*	*
Caught before sample i	*	*

If it is suspected that the mortality effect for those first tagged in the ith sample carries over after the $(i+1)$th release, Robson [1969] suggests two contingency tables, Table 5.11 and 5.12, for detecting this. These tests are independent within pairs as well as between pairs.

SURVIVAL INDEPENDENT OF AGE. If we suspect that survival depends on age (or sex), we can split up the release of R_i individuals into age—sex classes and record the subsequent recaptures from these classes in the $(i+1)$th sample or in all subsequent samples. Once again, contingency tables can be constructed for $i = 1, 2, \ldots, s-1$ and tests of homogeneity carried out. However, if survival is independent of mark status and probability of capture independent of age, the Jolly—Seber method will not be greatly affected by age-dependent mortality. This follows from the fact that the two groups of $M_i - m_i$ and R_i marked individuals will than have much the same age-distribution, so that the basic equation (cf. (5.6))

$$z_i/(M_i - m_i) \approx r_i/R_i$$

will still hold.

4 Special cases

COMMERCIALLY EXPLOITED POPULATIONS. When the multi-sample single-recapture census is used, the marked population has probability function (5.23). A goodness-of-fit test for this model is given by

$$T = \sum_{h=1}^{s} \sum_{i=h+1}^{s} \frac{(m_{hi} - R_h \hat{p}_{hi})^2}{R_h \hat{p}_{hi}},$$

where $\hat{p}_{hi} = \hat{\phi}_h \hat{q}_{h+1} \hat{\phi}_{h+1} \hat{q}_{h+2} \cdots \hat{\phi}_{i-1} \hat{p}_i$

               $= \hat{\alpha}_h \hat{\alpha}_{h+1} \cdots \hat{\alpha}_{i-2} \hat{\beta}_{i-1}.$

T is asymptotically χ^2_k, where $k = \frac{1}{2}s(s-1) - (2s-3)$; $2s-3$ degrees of freedom are subtracted as there are $2s-3$ parameters estimated, namely $\alpha_1, \ldots, \alpha_{s-2}, \beta_1, \ldots, \beta_{s-1}$.

NO MIGRATION. In this situation each consecutive pair of samples may be regarded as a Petersen-type experiment and the discussion in **3.2.1** applies. For example, if there is no recruitment, and mortality is such that it removes a random portion of the population between samples, we can derive a chain of Petersen estimates. Procedures for eliminating recruits from samples are given in **3.2.1** (4), p. 72.

NO MORTALITY. When there is no mortality or emigration one can test for "dilution" by recruitment or immigration using Leslie's dilution test of p. 159.

5.2 THE MANLY–PARR METHOD

5.2.1 Population size

PETERSEN-TYPE ESTIMATE. Recently Manly and Parr [1968] suggested a simple method of estimating N_i without making any assumptions about the survival probabilities (cf. assumption (b) on p. 196). They point out that when all animals have the same probability of capture in the ith sample, we have

$$E[n_i \mid N_i] = N_i p_i.$$

Therefore, given an estimate $\tilde{p}_i$ of p_i, we have the obvious estimate

$$\tilde{N}_i = n_i / \tilde{p}_i.$$

Here p_i can be estimated by picking out a class of C_i animals known to be in the population at the time of the ith sample, and then observing the number c_i of this class caught in the ith sample, i.e.

and hence

$$\tilde{p}_i = c_i / C_i$$
$$\tilde{N}_i = n_i C_i / c_i.$$

We note that $\tilde{N}_i$ is basically a Petersen estimate in which the n_i individuals caught in the ith sample represent the "first" sample and the C_i animals represent the "second" sample. It seems that the above method was first used by Davis *et al.* [1964: 4–5] for the case of just three samples only (see also Meslow and Keith [1968: 814]).

MEAN AND VARIANCE. Manly [1969] has obtained asymptotic expressions for the mean and variance of $\tilde{N}_i$ by considering the four categories set out in Table 5.13. Suppose that an animal has probability θ_i of being included in

TABLE 5.13

Animals alive at the time of the ith sample: from Manly [1969].

	Captured in sample i	Not captured in sample i	Total
In class C_i	c_i	$C_i - c_i$	C_i
Not in class C_i	$n_i - c_i$	$N_i - C_i - n_i + c_i$	$N_i - C_i$
Total	n_i	$N_i - n_i$	N_i

the class of C_i animals, then if the probability of capture in the ith sample is assumed to be independent of class status, the four random variables in Table 5.13 will have a joint multinomial distribution proportional to

$$\frac{N_i!}{(N_i - C_i - n_i + c_i)!} (p_i \theta_i)^{c_i} ((1-p_i)\theta_i)^{C_i - c_i} (p_i(1-\theta_i))^{n_i - c_i} ((1-p_i)(1-\theta_i))^{N_i - C_i - n_i + c_i}$$

$$= \frac{N_i!}{(N_i - C_i - n_i + c_i)!} \, p_i^{n_i} (1-p_i)^{N_i - n_i} \theta_i^{C_i} (1-\theta_i)^{N_i - C_i}. \qquad (5.29)$$

Manly shows that the maximum-likelihood estimates of p_i, N_i and θ_i are

$$\tilde{p}_i = c_i/C_i,$$
$$\tilde{N}_i = n_i/\tilde{p}_i$$

and

$$\tilde{\theta}_i = C_i/\tilde{N}_i.$$

Using the delta method he proves that asymptotically (i.e. for large $N_i\theta_i$ and C_ip_i)

$$E[\tilde{N}_i \mid N_i] = N_i + (1-p_i)(1-\theta_i)/(p_i\theta_i) \tag{5.30}$$

and

$$E[(\tilde{N}_i - N_i)^2 \mid N_i] = N_i(1-p_i)(1-\theta_i)/(p_i\theta_i). \tag{5.31}$$

An estimate of this last expression is given by

$$\tilde{V}_i = \tilde{N}_i(\tilde{N}_i - n_i)(\tilde{N}_i - C_i)/(n_iC_i)$$
$$= \tilde{N}_i(C_i - c_i)(n_i - c_i)/c_i^2. \tag{5.32}$$

As $\tilde{N}_i$ is a Petersen-type estimate it is worth considering Chapman's modification (cf. **3.1.1**)

$$N_i^* = \frac{(n_i+1)(C_i+1)}{(c_i+1)} - 1$$

$$\approx \frac{n_iC_i}{c_i}\left(1+\frac{1}{n_i}\right)\left(1+\frac{1}{C_i}\right)\left(1-\frac{1}{c_i}\right) - 1$$

$$\approx \frac{n_iC_i}{c_i} - \left(1-\frac{C_i}{c_i}\right)\left(1-\frac{n_i}{c_i}\right).$$

Taking expectations and using the delta method we find that (cf. (5.30))

$$E[N_i^* \mid N_i] \approx E[\tilde{N}_i \mid N_i] - \left(1-\frac{1}{p_i}\right)\left(1-\frac{1}{\theta_i}\right)$$

$$\approx N_i,$$

so that N_i^* is approximately unbiased. On the basis of the discussion in **3.1.1** it is recommended that c_i should be greater than 10 if $\tilde{N}_i$ or N_i^* is to give even the order of magnitude of the true population size. It is also noted that (5.29) and the approximate unbiasedness of N_i^* follow directly from Darroch [1958: 352–3 with $p_1 = p_i$, $p_2 = \theta_i$].

In conclusion we observe that since $\tilde{N}_i$ is a Petersen estimate, the general theory of **3.2.2**(2) can be applied here. This means that if p_i and θ_i are not the same for all members of the population but rather follow a bivariate distribution, then $\tilde{N}_i$ and N_i^* will still be asymptotically unbiased, and (5.32) will be a satisfactory estimate of the *true* variance, provided p_i and θ_i are independent. In this case the only change in (5.30) and (5.31) is to replace p_i and θ_i by $E[p_i]$ and $E[\theta_i]$ where the expectations are with respect to the bivariate distribution.

SPECIFYING THE CLASS C_i. We shall now turn our attention to the problem of specifying a suitable class of C_i individuals for a population in which all emigration is permanent (the case considered in the Jolly–Seber method) and there are no losses on sampling ($R_i = n_i$). Let $r_{\bullet i}$ be the number of individuals from the group of m_i marked animals caught in the ith sample which are later recaptured. Then defining $C_i = r_{\bullet i} + z_i$, the number of marked animals caught both before and after the ith sample, p_i can be estimated by

$$\tilde{p}_i = r_{\bullet i}/(r_{\bullet i} + z_i).$$

This estimate can be compared with Jolly's estimate

$$\hat{p}_i = \frac{m_i}{\hat{M}_i}$$

$$= \frac{m_i}{\left(\dfrac{n_i z_i}{r_i} + m_i\right)}$$

$$= \frac{\left(r_i \dfrac{m_i}{n_i}\right)}{\left(r_i \dfrac{m_i}{n_i} + z_i\right)}.$$

Assuming random sampling, then, when assumption (b) of constant survival probability is true, $r_{\bullet i}/r_i$ will be approximately equal to m_i/n_i and the two estimates of p_i will give similar values. However, if assumption (b) is not true and survival depends on age, for example, $\hat{p}_i$ may be unsuitable if the age structure of the m_i marked individuals differs significantly from the age structure of the whole sample of n_i individuals. This could occur if sampling is not random with respect to age, so that mark status depends on age.

With this choice of C_i we find that θ_i, the probability of being captured both before and after the ith sample, is not the same for all the N_i animals. For example, new immigrants or recruits entering the population between the $(i-1)$th and ith samples will have zero probability of belonging to this class of C_i. However, as pointed out above, Manly's model can still be used if the events "caught in sample i" and "belonging to the class C_i" are independent for each individual. We note that if there are deaths through handling, then these two events are dependent, as a marked animal has a better chance of being recaptured after the ith sample, that is of belonging to C_i, if it is not caught in the ith sample.

Manly and Parr [1968] have given a method for setting out the recapture data and calculating $r_{\bullet i}$ and z_i. However, these quantities are more readily calculated using the methods of this chapter; z_i can be obtained from the table of m_{hi} (cf. Example 5.1 on p. 205) and $r_{\bullet i}$ is easily obtained from the table of $m_{\bullet hi}$ (cf. Example 5.7 on p. 226). For example, using the data of Example 5.7, we find that $(r_{\bullet 2}, r_{\bullet 3}, r_{\bullet 4}) = (15, 14, 10)$ and $(z_2, z_3, z_4) = (11, 12, 20)$. Hence the Manly–Parr estimates can be calculated as in Table

5.14, where they are compared with Jolly's estimates ($\tilde{\sigma}_i = \tilde{V}_i^{\frac{1}{2}}$). There is not a great deal of difference between the two sets of estimates.

TABLE 5.14

**Comparison of estimates for a moth population:
data from Manly and Parr [1968: 86].**

i	Manly–Parr method			Jolly–Seber method		
	$\tilde{p}_i$	$\tilde{N}_i$	$\tilde{\sigma}_i$	$\hat{p}_i$	$\hat{N}_i$	$\hat{\sigma}[\hat{N}_i \mid N_i]$
1	–	–	–	–	–	–
2	0·5769	90·13	12·8	0·5590	93·03	15·3
3	0·5385	96·56	15·0	0·5287	98·35	14·2
4	0·3333	93·01	19·8	0·2914	106·37	23·6
5	–	–	–	–	–	–

5.2.2 Survival estimates

Suppose that $S_i n_i$ of the n_i animals released from the ith sample survive and remain in the population until the $(i+1)$th sample, and let $m_{i,i+1}$ be the number of animals caught in both the ith and $(i+1)$th samples. Then, if sampling is random,

$$E[m_{i,i+1} \mid S_i n_i] = S_i n_i p_{i+1},$$

and S_i can be estimated by

$$\tilde{S}_i = m_{i,i+1}/(n_i \tilde{p}_{i+1}).$$

This estimate may be compared with Jolly's estimate

$$\hat{\phi}_i = \frac{\hat{M}_{i+1}}{\hat{M}_i - m_i + n_i}$$

$$= \frac{m_{i+1}}{(\hat{M}_i - m_i + n_i)\hat{p}_{i+1}}$$

$$= \frac{m_{i+1}\left(\dfrac{r_i}{r_i + z_i}\right)}{n_i \hat{p}_{i+1}}.$$

If survival is independent of mark status then we would expect $m_{i,i+1}/r_i$ to to be approximately equal to $m_{i+1}/(r_i + z_i)$ and the two estimates will give similar results. For example, using the data of Example 5.7 (p. 226) we find that $(\tilde{S}_1, \tilde{S}_2, \tilde{S}_3) = (0·76, 0·71, 0·69)$ which may be compared with $(\hat{\phi}_1, \hat{\phi}_2, \hat{\phi}_3) = (0·78, 0·74, 0·76)$.

Manly and Parr also give an estimate of B_i, the number of new entries into the population between the ith release and $(i+1)$th sample, namely

$$\tilde{B}_i = \tilde{N}_{i+1} - \tilde{S}_i \tilde{N}_i.$$

Thus, for Example 5.7 we have $(\tilde{B}_2, \tilde{B}_3) = (32, 26)$ which are similar to $(\hat{B}_2, \hat{B}_3) = (30, 32)$.

Asymptotic variances for $\tilde{S}_i$ and $\tilde{B}_i$ have not yet been given, though work in this direction is under way (Manly [personal communication]). Recently Manly [1970] has studied the small-sample properties of the Jolly–Seber and Manly–Parr estimates using simulation.

5.3   REGRESSION METHODS

5.3.1   Instantaneous samples

In addition to the notation given in **5.1.1** (*1*) we shall define:

v_i = number of newly marked individuals in the ith release,

R_{ij} = number of animals from the group of v_i which are available for capture in the jth sample,

b_{ij} = number of animals from the group of v_i animals which are caught in the jth sample,

ϕ_{ij} = probability that an animal from the ith release is alive and in the population at the time of the jth sample,

t_{ij} = time from the ith release to the jth sample.

Then if sampling is random, we have, using Bailey's modification of the Petersen estimate (**3.1.2**),

$$E\left[\frac{R_{ij}(n_j+1)}{(b_{ij}+1)} \mid R_{ij}, n_j\right] \approx N_j,$$

and

$$E[R_{ij} \mid v_i] = v_i\phi_{ij}.$$

Therefore, combining these equations and taking logarithms, we are led to consider the regression model

$$E[y_{ij}] \approx \log N_j - \log \phi_{ij}. \tag{5.33}$$

where

$$y_{ij} = \log\left\{\frac{v_i(n_j+1)}{(b_{ij}+1)}\right\}.$$

Now, for given j and random n_j, the b_{ij} and hence the y_{ij} are independent as they come from different mark releases. Also, for given i, the y_{ij} are virtually independent, provided there are no losses on capture ($v_i = u_i$). Even if there are some losses on capture any dependence effect will be small in comparison with sampling errors, so that in general we may assume that the y_{ij} are all independently distributed. There is also some evidence that the logarithmic transformation will, in general, stabilise the variance. Therefore, defining the functional form of ϕ_{ij} and assuming the y_{ij} to be independently normally distributed, (5.33) can be analysed using multiple regression methods. Two examples are considered below.

PERMANENT EMIGRATION. If all emigration is permanent (as was assumed for the Jolly–Seber method) and the probability of surviving and remaining in the population for unit time is constant throughout the experiment (= ϕ say), then, for no losses on sampling,

and
$$\phi_{ij} = \phi^{t_{ij}}$$
$$E[y_{ij}] \approx \log N_j + (\log \phi)t_{ij}.$$

This multiple linear-regression model was proposed by Chapman [1954: 7] and least-squares estimates of $\log \phi$ and $\{\log N_j\}$ can be obtained in the usual manner. As a check on the validity of the model we can plot y_{ij} against t_{ij} for each j. These plots should be approximately linear, approximately parallel, and have much the same scatter about their fitted lines if the assumption of constant variance is applicable.

NO MIGRATION OR RECRUITMENT. When there is no migration or recruitment and N_0 is the population size at time zero, then

and
$$N_j = N_0 \phi^{t_j}$$
$$E[y_{ij}] \approx \log N_0 + (\log \phi)(t_j - t_{ij}).$$

5.3.2 Continuous sampling

We shall now consider the case of an exploited population in which the sampling (exploitation) is continuous. When there is no migration or recruitment and both the instantaneous natural mortality rate μ and the instantaneous exploitation rate μ_E are constant, Paulik [1963a] gives the following general regression model for estimating μ and μ_E:

Since the sampling is continuous we shall divide up the total experimental period into equal time-intervals and consider the marked catch only in each interval. It is assumed that the releases are timed so that each release is made at the beginning of one of the time-intervals. Let

m_{ij} = number from the ith release caught in the jth time-interval, where j is now measured *relative* to the ith release ($i = 1, 2, \ldots, I$; $j = 1, 2, \ldots, J_i$), and

p_{ij} = probability that a marked individual from the ith release is caught in the jth time-interval after the release.

Then from (6.9),

$$p_{ij} = \text{(probability of survival for } j - 1 \text{ intervals)}$$
$$\times \text{(probability of capture in one interval)}$$
$$= e^{-Z(j-1)}P_1,$$

where $Z = \mu + \mu_E$ (the total instantaneous mortality rate) and, putting $J = 1$ in (6.6),

$$P_1 = \mu_E(1 - e^{-Z})/Z.$$

Since $E[m_{ij}] = R_i p_{ij}$, we are led to consider the regression model

$$E[Y_{ij}] \approx -(Z + \log P_1) + Zj, \qquad (5.34)$$

where $Y_{ij} = \log(R_i/m_{ij})$. As P_1 and Z are functions of μ and μ_E, estimates of Z and $-(Z + \log P_1)$ will provide estimates of μ and μ_E.

Several procedures for estimation are available. For example, (5.34) can be analysed as a series of independent regression lines (one for each release), using the methods of **6.2.2**. The estimates of μ and μ_E can be averaged and sample estimates of the variances calculated, using the methods of **1.3.2**. Alternatively (5.34) can be analysed as a single multiple linear-regression model, using the model of **1.3.5** (2) with

$$\mathbf{Y}' = (Y_{11}, Y_{12}, \ldots, Y_{1J_1}, \ldots, Y_{I1}, Y_{I2}, \ldots, Y_{IJ_I})$$

$$\mathbf{X} = \begin{bmatrix} 1 & 1 \\ 1 & 2 \\ \cdots & \\ 1 & J_1 \\ \hline \cdots & \\ \hline 1 & 1 \\ 1 & 2 \\ \cdots & \\ 1 & J_I \end{bmatrix}, \qquad \boldsymbol{\beta} = \begin{bmatrix} -Z & -\log P_1 \\ & Z \end{bmatrix},$$

and $\mathbf{B}^{-1}$ given by (6.37) and (6.38) (except that j ranges from 1 to J_i and not 1 to J in $\mathbf{B}_i^{-1}$).

The case when the releases are made at different times *before* exploitation commences is considered in **6.4**.

5.4   TAG RETURNS FROM DEAD ANIMALS

5.4.1   Survival independent of age: general model

We shall now discuss a variation of the multi-sample single-recapture census (p. 212) which is particularly useful for estimating survival probabilities for bird populations. Consider an experiment in which tagged animals are released at the beginning of each year for s consecutive years. The animals then die through natural mortality, predation, hunting, etc., and for t consecutive years ($t \geqslant s$) a record is kept of tags returned from dead animals.

NOTATION AND ASSUMPTIONS. We shall assume that:

(i) Every tagged animal (irrespective of age or time of release) has the same probability ϕ_i of survival for year i, given that it is alive at the beginning of the year ($i = 1, 2, \ldots, t$).

(ii) Every tagged animal which dies in year i has the same probability λ_i of being found and its tag reported in year i; $100\lambda_i$ per cent is usually called the *recovery rate* for year i.

Assumption (i) implies that survival is time-dependent rather than age-dependent. This is not an unreasonable assumption for, say, adult bird populations where the survival rate is fairly constant except for fluctuations due to environmental changes in weather, hunting pressure, etc. The definition of $1 - \lambda_i$ includes three types of non-return: first, there are tagged individuals which die and are not found; secondly, the tags may be found and not returned; and thirdly the animals may be found but their tags are missing or possibly unreadable.

Let

R_i = number of tagged animals released at the beginning of year i ($i = 1$, $2, \ldots, s$),

m_{ij} = number from the ith release which die in year j and are reported in the same year ($j = i, i+1, \ldots, t$),

m_j = number of tags recovered in year j,

D_j = number of tags returned from animals which die in their jth year after release, i.e. j is measured *relative* to the year of release,

r_i = $\displaystyle\sum_{j=i}^{t} m_{ij}$ (the number of tags recovered from the ith release),

z_i = $\displaystyle\sum_{j=i}^{t} m_{1j} + \sum_{j=i}^{t} m_{2j} + \ldots + \sum_{j=i}^{t} m_{i-1,j}$ (the total number of tags recovered after the ith release from releases prior to the ith; $i = 1, 2, \ldots, s$ and $z_1 = 0$),

T_i = $r_i + z_i$,

and define the "working" parameters

α_i = ϕ_i,
β_i = $(1 - \phi_i)\lambda_i$,
γ_j = $\alpha_s \alpha_{s+1} \ldots \alpha_{j-1}\beta_j$,   ($j = s+1, s+2, \ldots, t$), and
θ_i = $\beta_i + \alpha_i\beta_{i+1} + \ldots + \alpha_i\alpha_{i+1} \ldots \alpha_{t-1}\beta_t$.

Here β_i is the probability that a tag is recovered from an animal in year i, given that the animal is alive at the beginning of the year; θ_i is the probability that a tag is recovered from an animal after the ith release, given that the animal is alive at the time of the ith release; and γ_j is the probability that a tag is recovered from an animal in year j, given that it is alive at the time of the sth release.

ESTIMATION. The joint probability function of the $\{m_{ij}\}$ is

$$f(\{m_{ij}\}\,|\,\{R_i\})$$

$$= \prod_{i=1}^{s} \left\{ \frac{R_i!}{[\prod_{j=i}^{s} m_{ij}!][(R_i - r_i)!]} ((1-\phi_i)\lambda_i)^{m_{ii}} (\phi_i(1-\phi_{i+1})\lambda_{i+1})^{m_{i,i+1}} \cdots \right.$$

$$\left. \times (\phi_i\phi_{i+1} \cdots \phi_{t-1}(1-\phi_t)\lambda_t)^{m_{it}} (1-\theta_i)^{R_i - r_i} \right\}$$

$$\propto \prod_{i=1}^{s} \{\beta_i^{m_{ii}}(\alpha_i\beta_{i+1})^{m_{i,i+1}} \cdots (\alpha_i\alpha_{i+1} \cdots \alpha_{t-1}\beta_t)^{m_{it}} (1-\theta_i)^{R_i - r_i}\} \tag{5.35}$$

$$= \left\{ \prod_{i=1}^{s-1} \alpha_i^{z_{i+1}} \beta_i^{m_i} (1-\theta_i)^{R_i - r_i} \right\} \{\beta_s^{m_s} (1-\theta_s)^{R_s - r_s}\} \{\gamma_{s+1}^{m_{s+1}} \gamma_{s+2}^{m_{s+2}} \cdots \gamma_t^{m_t}\}$$

$$\tag{5.36}$$

and the maximum-likelihood estimates $\hat{\alpha}_i$, $\hat{\beta}_i$ and $\hat{\gamma}_i$ are obtained in the usual manner. However, Seber [1970b] has shown that these estimates can also be obtained by intuitive arguments. For example, the obvious estimate of θ_i is $\hat{\theta}_i = r_i/R_i$, so that from

$$\beta_i = \Pr[\text{recovered in year } i \mid \text{recovered after } i\text{th release}]$$
$$\times \Pr[\text{recovered after } i\text{th release}]$$

we have the intuitive estimate

$$\hat{\beta}_i = \frac{m_i}{T_i} \cdot \hat{\theta}_i, \quad (i = 1, 2, \dots, s).$$

As $\alpha_i = (\theta_i - \beta_i)/\theta_{i+1}$, α_i can be estimated by

$$\hat{\alpha}_i = \frac{\hat{\theta}_i - \hat{\beta}_i}{\hat{\theta}_{i+1}}$$

$$= \frac{T_i - m_i}{T_i} \cdot \frac{\hat{\theta}_i}{\hat{\theta}_{i+1}},$$

and from

$$\frac{E[m_s]}{E[m_j]} = \frac{\beta_s}{\gamma_j}, \quad (j > s),$$

we have the estimate

$$\hat{\gamma}_j = \hat{\beta}_s m_j/m_s, \quad (j = s+1, \dots, t).$$

Finally, as the transformation

$$\hat{\phi}_i = \hat{\alpha}_i \quad (i = 1, 2, \dots, s-1)$$
$$\hat{\lambda}_i = \hat{\beta}_i/(1-\hat{\alpha}_i) \quad (i = 1, 2, \dots, s-1)$$
$$\hat{\beta}_s = \hat{\beta}_s$$
$$\hat{\gamma}_j = \hat{\gamma}_j \quad (j = s+1, \dots, t)$$

is one-to-one, $\hat{\phi}_i$ and $\hat{\lambda}_i$ are the maximum-likelihood estimates of ϕ_i and λ_i respectively. We note that ϕ_s and λ_s cannot be separately estimated through lack of identifiability; only the product $\beta_s = (1-\phi_s)\lambda_s$ is estimable. However,

241

if $\hat{\phi}_1, \hat{\phi}_2, \ldots, \hat{\phi}_{s-1}$ are approximately the same, we could assume that ϕ_i is constant $(= \phi$ say), so that

$$\lambda_s = \beta_s/(1-\phi)$$

and

$$\lambda_{j+1} = \lambda_j \gamma_{j+1}/(\phi \gamma_j), \quad (j = s, \ldots, t-1: \gamma_s = \beta_s).$$

Estimating ϕ by ϕ^* (say the average of the $\hat{\phi}_i$), we have the estimates

$$\lambda_s^* = \hat{\beta}_s/(1-\phi^*)$$

and

$$\lambda_{j+1}^* = \lambda_j^* m_{j+1}/(\phi^* m_j), \quad (j = s, \ldots, t-1).$$

If we had assumed $\phi_i = \phi$ $(i = 1, 2, \ldots, t)$ right from the start, we would find that the maximum-likelihood equations cannot be solved explicitly, and iterative methods of solution are required.

CALCULATIONS. In performing the calculations we require r_i $(i = 1, 2, \ldots, s)$, m_j $(j = 1, 2, \ldots, t)$ and T_i $(i = 1, 2, \ldots, s)$; one method for obtaining these quantities is given by Seber (in his symbols they are called R_i, C_j and T_i respectively; also α_i and β_i are interchanged). However, it transpires that the tabular methods used for the Jolly–Seber method (Example 5.1, p. 206) can also be used here if we work with the random variables r_i, m_j and z_i $(z_1 = 0)$. We then have the sequence of estimates

$$\hat{\theta}_i = r_i/R_i,$$
$$\hat{\beta}_i = \hat{\theta}_i m_i/(r_i + z_i)$$

and, since $T_i - m_i = T_{i+1} - r_{i+1} = z_{i+1}$,

$$\hat{\phi}_i = \frac{z_{i+1}}{r_i + z_i} \cdot \frac{\hat{\theta}_i}{\hat{\theta}_{i+1}}, \quad (i = 1, 2, \ldots, s-1),$$

and

$$\hat{\lambda}_i = \hat{\beta}_i/(1-\hat{\phi}_i), \quad (i = 1, 2, \ldots, s-1).$$

It is noted that (5.35) is a slight extension of the model (5.23) with $\theta_i = 1 - \chi_i$. The main differences between the two models lie in the additional parameters γ_i and the ranges of the suffixes. In (5.23) the first sample after the first release (which is equivalent to the first recovery year in (5.35)) is called sample number 2, so that one has the term m_{i+1} in (5.24) but m_i in (5.36).

VARIANCES AND COVARIANCES. Using the delta method, Seber [1970b] shows that the asymptotic variance and covariance formulae for the estimates $\hat{\phi}_i$ and $\hat{\lambda}_i$ are

$$V[\hat{\phi}_i] = \phi_i^2 \left\{ \frac{1}{E[r_i]} + \frac{1}{E[r_{i+1}]} + \frac{1}{E[z_{i+1}]} - \frac{1}{R_i} - \frac{1}{R_{i+1}} - \frac{1}{E[r_i + z_i]} \right\}, \quad (5.37)$$

$$\text{cov}[\hat{\phi}_i, \hat{\phi}_j] = 0, \quad j > i+1,$$

$$\text{cov}[\hat{\phi}_i, \hat{\phi}_{i+1}] = -\phi_i \phi_{i+1} \left\{ \frac{1}{E[r_{i+1}]} - \frac{1}{R_{i+1}} \right\},$$

and

$$V[\hat{\lambda}_i] = \frac{\lambda_i^2}{(1-\phi_i)^2} \left\{ \frac{1}{E[r_i]} - \frac{1}{E[r_i+z_i]} - \frac{1}{R_i} + \frac{(1-\phi_i)^2}{E[m_i]} + \right.$$

$$\left. + \phi_i^2 \left(\frac{1}{E[r_{i+1}]} + \frac{1}{E[z_{i+1}]} - \frac{1}{R_{i+1}} \right) \right\}. \tag{5.38}$$

The above variances and covariances can be estimated by simply replacing the unknown parameters by their estimates and the expectations by the observed values of the random variables, e.g. r_i replaces $E[r_i]$, etc.

Since

$$\frac{1}{E[r_i]} - \frac{1}{E[r_i+z_i]} = \frac{1}{\theta_i} \left\{ \frac{1}{R_i} - \frac{1}{\sum\limits_{h=1}^{i} R_h \beta_h \cdots \beta_{i-1}} \right\},$$

and θ_i increases with t, $V[\hat{\phi}_i]$ and $V[\hat{\lambda}_i]$ both decrease as t increases. This means that by increasing t we can increase the number of recoveries and thereby increase the precision of our estimates.

It should be noted that the above formulae are asymptotic, and their usefulness will depend not only on the validity of the underlying model but also on the expected numbers of recaptures. For example, if the underlying model is appropriate, (5.37) is valid for large $E[r_i]$, $E[r_{i+1}]$, etc., while (5.38) generally holds for large $(1-\phi_i)E[r_i]$, $(1-\phi_i)E[r+z_i]$, etc. If $(1-\phi_i)$ is small (i.e. low mortality rates), there will be few dead animals available for recovery, so that $\hat{\lambda}_i$ will be based on small samples and $V[\hat{\lambda}_i]$ will be large. This sensitivity of the variance to $(1-\phi_i)$ is indicated by the presence of the term $(1-\phi_i)^2$ in the denominator of $V[\hat{\lambda}_i]$.

GOODNESS-OF-FIT TEST. Assuming the R_i to be large, a standard chi-squared goodness-of-fit test for the model can be carried out using the statistic (cf. Mitra [1958])

$$T = \sum_{i=1}^{s} \sum_{j=i}^{t} (m_{ij} - E_{ij})^2 / E_{ij},$$

which is approximately distributed as chi-squared with $(s-1)(t-1) - \frac{1}{2}s(s-1)$ degrees of freedom (i.e. $st - \frac{1}{2}s(s-1)$ square terms minus $s+t-1$, the number of parameters estimated). Here,

$$E_{ii} = R_i\hat{\beta}_i \qquad (i = 1, 2, \ldots, s)$$

$$E_{ij} = \begin{cases} R_i\hat{\phi}_i\hat{\phi}_{i+1} \cdots \hat{\phi}_{j-1}\hat{\beta}_j & (j = i+1, \ldots, s) \\ R_i\hat{\phi}_i\hat{\phi}_{i+1} \cdots \hat{\phi}_{s-1}\hat{\gamma}_j & (j = s+1, \ldots, t), \end{cases}$$

and the terms involving $R_i - r_i$ are not included in T as their contribution is zero $(R_i - r_i - R_i(1-\hat{\theta}_i) = 0)$.

Example 5.10 Pink-footed geese: Seber [1970b]

The above method was applied to banding data (Table 5.15) on adult

TABLE 5.15

Numbers of ringed, pink-footed geese m_{ij} recovered over a period of 4 years (from Boyd [1956: Table 1]); the expected recoveries E_{ij} are in brackets.

Release R_i	Recovery year (j) 1	2	3	4	Recoveries r_i
301	32 (32)	22 (23·79)	16 (15·35)	7 (5·85)	77
766		70 (68·21)	50 (44·01)	9 (16·78)	129
897			52 (58·64)	29 (22·36)	81
m_i	32	92	118	45	287

TABLE 5.16

Cumulated columns of Table 5.15 for the calculation of z_i.

Recovery year (j) 1	2	3	4	
<u>32</u>	22	16	7	$45 = z_2$
	<u>92</u>	66	16	$82 = z_3$
		<u>118</u>	45	$45 = z_4$

pink-footed geese from Boyd [1956: Table 1, adults only]. As a first step in the calculations, the r_i and z_i are obtained from Tables 5.15 and 5.16; the method for constructing Table 5.16 is described in Example 5.1 (p. 205). Then, for example,

$$\hat{\theta}_1 \ = \ 77/301, \quad \hat{\theta}_2 \ = \ 129/766,$$

$$\hat{\phi}_1 \ = \ \frac{45}{77 + 0} \cdot \frac{77}{301} \cdot \frac{766}{129} \ = \ 0\cdot887\ 74,$$

$$\hat{\beta}_1 \ = \ \frac{77}{301} \cdot \frac{32}{77 + 0} \ = \ 0\cdot106\ 31,$$

and

$$\hat{\lambda}_1 \ = \ 0\cdot106\ 31/(1 - 0\cdot887\ 74) \ = \ 0\cdot9470.$$

Estimates together with standard deviations are given in Table 5.17.

TABLE 5.17

Estimates of the survival and recovery probabilities for a population of pink-footed geese: from Seber [1970b: Table 5].

i	$\hat{\beta}_i$	$\hat{\phi}_i \pm \hat{\sigma}[\hat{\phi}_i]$	$\hat{\lambda}_i \pm \hat{\sigma}[\hat{\lambda}_i]$
1	0·106 31	0·8877 ± 0·141	0·9470 ± 1·18
2	0·089 04	0·8789 ± 0·137	0·7352 ± 0·83
3	0·065 37	—	—

Since $\hat{\phi}_1 \approx \hat{\phi}_2$, we can assume ϕ_i to be approximately constant and estimate λ_3 and λ_4 using the method given on p. 242. Let $\phi^* = (\hat{\phi}_1 + \hat{\phi}_2)/2 = 0\cdot8833$; then

$$\lambda_3^* = \hat{\beta}_3/(1-\phi^*) = 0\cdot5602$$

and

$$\lambda_4^* = \lambda_3^* m_4/(\phi^* m_3) = 0\cdot242.$$

To carry out the goodness-of-fit test we require

$$\hat{\gamma}_4 = m_4\hat{\beta}_3/m_3 = 0\cdot024\ 93.$$

The E_{ij} can then be calculated, e.g. $E_{13} = R_1\hat{\phi}_1\hat{\phi}_2\hat{\beta}_3$, $E_{14} = R_1\hat{\phi}_1\hat{\phi}_2\hat{\gamma}_4$, etc., and these are given in brackets in Table 5.15. It is found that $T = \Sigma\Sigma\ (m_{ij}-E_{ij})^2/E_{ij} = 7\cdot58$, which, for 3 degrees of freedom, is not significant at the 5 per cent level (though only just; $\chi_3^2[0\cdot05] = 7\cdot8$).

In conclusion we note from the values of $\hat{\sigma}[\hat{\lambda}_i]$ that the estimates $\hat{\lambda}_i$ are not reliable, in fact they seem far too high.

5.4.2 Survival independent of age: special cases

1   Constant recovery probability

It was noted briefly in the previous section that for the case $\phi_i = \phi$ $(i = 1, 2, \ldots, t)$ the maximum-likelihood equations for ϕ and λ_i must be solved iteratively. The same is true when $\lambda_i = \lambda$ $(i = 1, 2, \ldots, t)$ and ϕ_i varies; both these models were used by Boyd[1956], though details of the calculations are not given. In the latter case, however, if t is much greater than s, so that the probability of a banded bird dying by the end of the tth year is virtually unity, i.e.

$$(1-\phi_i) + \phi_i(1-\phi_{i+1}) + \ldots + \phi_i\phi_{i+1} \ldots (1-\phi_t) \approx 1,$$

then

$$\beta_i = (1-\phi_i)\lambda,$$
$$\theta_i = \lambda,$$

and ϕ_i can be estimated by ϕ_i^*,

where $\quad 1 - \phi_i^* = \hat{\beta}_i/\hat{\theta}_i = m_i/T_i.$

This estimate has been used by several authors e.g. Hickey [1952: 14, Table 3, where T_i is calculated for $i = s, s + 1, \ldots, t$ only] and Coulson [1960: 206]. For a further discussion of this method of estimation cf. Seber [1972].

2   Constant survival and recovery probabilities

If $\phi_i = \phi$ and $\lambda_i = \lambda$ $(i = 1, 2, \ldots, t)$, then from (5.36)

$$f(\{m_{ij}\}|\{R_i\}) \propto (1-\phi)^{m_{11}} (\phi(1-\phi))^{m_{12}} \ldots (\phi^{t-1}(1-\phi))^{m_{1t}} \lambda^{r_1}(1-\theta_1)^{R_1-r_1}$$
$$\times (1-\phi)^{m_{22}} (\phi(1-\phi))^{m_{23}} \ldots (\phi^{t-2}(1-\phi))^{m_{2t}} \lambda^{r_2}(1-\theta_2)^{R_2-r_2} \ldots ,$$

where $\theta_i = \lambda(1-\phi^{t-i+1})$. Unfortunately the maximum-likelihood equations are again complicated by the presence of λ and must be solved iteratively. However, since the r_i are independent binomial variables with parameters

245

R_i and θ_i, it transpires that the conditional distribution $f(\{m_{ij}\}|\{r_i\})$ does not depend on λ. Thus

$$f(\{m_{ij}\}|\{r_i\}) \propto (1-\phi)^{D_1} (\phi(1-\phi))^{D_2} \dots (\phi^{t-1}(1-\phi))^{D_t}$$

$$\times (1-\phi^t)^{-r_1} (1-\phi^{t-1})^{-r_2} \dots (1-\phi^{t-s+1})^{-r_s}$$

$$= (1-\phi)^{\sum_{j=1}^{t} D_j} \phi^{\sum_{j=1}^{t} (j-1)D_j} \prod_{i=1}^{s} \{(1-\phi^{t-i+1})^{-r_i}\}, \qquad (5.39)$$

where D_j is the number of tags returned from animals which die in their jth year after release (see Table 5.19 below).

The maximum-likelihood estimate $\hat{\phi}$ of ϕ for (5.39) is the solution of

$$\sum_{j=1}^{t} \frac{(j-1)D_j}{\phi} - \sum_{j=1}^{t} \frac{D_j}{1-\phi} + \sum_{i=1}^{s} \left\{ \frac{r_i(t-i+1)\phi^{t-i}}{1-\phi^{t-i+1}} \right\} = 0, \qquad (5.40)$$

which must be solved iteratively. For computational convenience (5.40) can be rewritten as

$$-\frac{\sum_{j=1}^{t} D_j}{1-\phi} + \sum_{i=1}^{s} \left\{ \frac{r_i(t-i+1)}{1-\phi^{t-i+1}} \right\} = \sum_{i=1}^{s} r_i(t-i+1) - \sum_{j=1}^{t} jD_j \qquad (5.41)$$

or

$$g(\phi) = c,$$

which can be solved using the methods of **1.3.8** *(1)*.

If f represents (5.39) then from maximum-likelihood theory (cf. (1.4))

$$V[\hat{\phi}] \approx -1/E\left[\frac{\partial^2 \log f}{\partial \phi^2}\right],$$

and $V[\hat{\phi}]$ can be estimated by

$$\hat{V}[\hat{\phi}] = -1 \Big/ \left[\frac{\partial^2 \log f}{\partial \phi^2}\right]_{\phi=\hat{\phi}}.$$

Therefore, carrying out the differentiation and using (5.40), we find that

$$\hat{V}[\hat{\phi}] = \left(\sum_{j=1}^{t} \frac{D_j}{\hat{\phi}(1-\hat{\phi})^2} - \sum_{i=1}^{s} \left\{ \frac{r_i(t-i+1)^2\hat{\phi}^{t-i-1}}{(1-\hat{\phi}^{t-i+1})^2} \right\} \right)^{-1}.$$

When $t = s$, the case considered by Haldane [1955], we note that

$$\sum_{i=1}^{s} \frac{r_i(s-i+1)}{1-\phi^{s-i+1}} = \frac{r_s}{1-\phi} + \frac{2r_{s-1}}{1-\phi^2} + \dots + \frac{sr_1}{1-\phi^s}$$

in (5.41).

A regression method for estimating both ϕ and λ by least squares is given in **5.4.3** (cf. equation (5.45)).

METHOD OF LACK. If t is much greater than s, so that $\phi^{t-s+1} \approx 0$, then from (5.39)

$$f(\{m_{ij}\} \mid \{r_i\}) \propto (1-\phi)^{\Sigma D_j} \phi^{\Sigma(j-1)D_j} .$$

Hence $\widetilde{\phi}$, the maximum-likelihood estimate of ϕ, is now given by

$$\widetilde{\phi} = 1 - (\sum_j D_j / \sum_j jD_j),$$

an estimate suggested by Lack [1943, 1951] and Farner [1945]. Using the same method as for $\hat{\phi}$, it is readily shown that

$$\widetilde{V}[\widetilde{\phi}] = (1-\widetilde{\phi})^2 \widetilde{\phi} / \sum_j D_j$$

is an estimate of the asymptotic variance of $\widetilde{\phi}$.

We note that for the ith release, the probability that a tag is recovered in the jth year after release, given that it is recovered, is

$$f(j) = \frac{\phi^{j-1}(1-\phi)\lambda}{(1-\phi^{t-i+1})\lambda} \quad (j = 1, 2, \ldots, t-i+1)$$

$$\approx \phi^{j-1}(1-\phi) \tag{5.42}$$

for large t. Haldane [1953] also discussed the case of large t using (5.42) as the basis for his approach. He assumed that we have $\sum_j D_j$ ($= N$ say) observations from this geometric distribution and hence obtained $\widetilde{\phi}$ and $\widetilde{V}[\widetilde{\phi}]$ above. He also suggested a goodness-of-fit test of the data by comparing the observed frequencies D_j with the expected frequencies

$$E_j = N\widetilde{\phi}^{j-1}(1-\widetilde{\phi}). \tag{5.43}$$

CHOOSING A FIRST APPROXIMATION. The main difficulty in solving (5.41) lies in the choice of a suitable first approximation for ϕ. One method for doing this consists of guessing the value of ϕ, say ϕ_1, then truncating the table of m_{ij} horizontally at release number s', say, so that $\phi_1^{t-s'+1} \approx 0$, and then estimating ϕ from this reduced table using Lack's estimate $\widetilde{\phi}$.

APPLICATIONS TO BIRD POPULATIONS. The above methods of Haldane and Lack have been widely used for bird populations where the birds are banded as nestlings before they leave the nest. However, as mortality both in the nest and in the first few months of independent life is heavy, bands recovered from these birds should be ignored, as far as the above methods are concerned, for a suitable initial period. For example, as British birds are mostly hatched from April to July, Lack recommends that tables of band returns be based on recoveries during each calendar year (cf. Farner [1949: 68; 1955]); that is, birds whose corpses are discovered between the first and second New Year's Day of their lives are said to have died at the age of one year, and so on; thus D_j is the number of bands recovered from birds that die at age j. But a number of authors suggest earlier dates (Hickey [1952: 15–16]), while for the case of wood-pigeons, Murton [1966] recommends extending the initial period until March of the following year on the grounds that inexperienced birds tend to be more susceptible to shooting.

TABLE 5.18

Application of Lack's method to recoveries from lapwings ringed as nestlings from 1909 to 1939 inclusive: from Haldane [1955: Table 1].

Recovery age (j)	Number recovered (D_j)	jD_j	E_j	$\dfrac{(D_j - E_j)^2}{E_j}$
1	194	194	206·38	0·74
2	145	290	136·21	0·57
3	90	270	89·90	0·00
4	54	216	59·33	0·48
5	48	240	39·16	2·00
6	25	150	25·85	0·03
7	24	168	17·06	2·82
8	9	72	11·25	ʼ 0·45
9	6	54	7·43	0·28
10	5	50	4·90	0·00
11	5 ⎫	55		
12	1 ⎪ 7	12	9·53	0·67
13	0 ⎪	0		
14	1 ⎭	14		
Total	607	1785	607·00	8·04

Example 5.11   Lapwing (*Vanellus vanellus*): Haldane [1955]

The recoveries from lapwings ringed as nestlings from 1909 to 1939 inclusive are given in Table 5.18. Using Lack's method we have

$$\tilde{\phi} = 1 - (607/1785) = 0{\cdot}6599$$

and

$$\tilde{\sigma}[\tilde{\phi}] = (1-\tilde{\phi})[\tilde{\phi}/\underset{j}{\textstyle\sum} D_j]^{\frac{1}{2}} = 0{\cdot}0112.$$

The expected frequencies E_j (cf. 5.43) were also calculated, and from Table 5.18 we have

$$\underset{j}{\textstyle\sum} (D_j - E_j)^2/E_j = 8{\cdot}04,$$

which, for 9 degrees of freedom, indicates an excellent fit.

Table 5.19 gives the recovery data for lapwings ringed from 1940 to 1951; here $s = t = 12$ and Haldane's method of analysis can be used. As a first step we have to solve $g(\phi) = c$, where

$$c = \sum_{i=1}^{s} r_i(s-i+1) - \sum_{j=1}^{s} jD_j = 640 - 237 = 403$$

and, from (5.41),

$$g(\phi) = \frac{(-120+9)}{1-\phi} + \frac{36}{1-\phi^2} + \frac{45}{1-\phi^3} + \cdots + \frac{84}{1-\phi^{12}}.$$

As a first approximation we can use $\tilde{\phi} = 0{\cdot}66$; thus Haldane calculates

TABLE 5.19
Application of Haldane's method to recoveries from lapwings ringed as nestlings from 1940 to 1951: from Haldane [1955: Table 2].

Year ringed	i	Age on recovery (j) 1	2	3	4	5	6	7	8	9	10	11	12	(1) r_i	(2) $12-i+1$	(1) × (2)
1940	1	2	1	0	2	1	0	1	0	0	0	0	0	7	12	84
1941	2	1	0	1	0	0	0	1	0	0	0	0		3	11	33
1942	3	2	2	1	0	2	0	0	0	0	0			7	10	70
1943	4	2	1	0	0	0	0	0	0	0				3	9	27
1944	5	4	1	3	2	0	0	0	1					11	8	88
1945	6	7	3	0	0	0	1	0						11	7	77
1946	7	6	0	0	0	0	0							6	6	36
1947	8	4	3	5	2	1								15	5	75
1948	9	6	4	2	3									15	4	60
1949	10	10	2	3										15	3	45
1950	11	15	3											18	2	36
1951	12	9												9	1	9
Totals D_j		68	20	15	9	4	1	2	1	0	0	0	0	120		640
	jD_j	68	40	45	36	20	6	14	8	0	0	0	0	237		

$g(0\cdot66) = 367\cdot7$, $g(0\cdot60) = 457\cdot9$ and $g(0\cdot63) = 398\cdot7$, which is a close enough approximation when one considers $\hat{\sigma}[\hat{\phi}]$. Thus

$$\hat{\phi} = 0\cdot63, \quad \hat{V}[\hat{\phi}] = 0\cdot001\ 567,$$

and hence

$$\hat{\sigma}[\hat{\phi}] = 0\cdot039\ 58.$$

Also the expectation of life at birth is estimated to be (cf. (1.3))

$$-1/\log\hat{\phi} = 2\cdot2 \text{ years.}$$

To compare $\hat{\phi}$ with $\tilde{\phi}$ we compute

$$z = \frac{|\hat{\phi} - \tilde{\phi}|}{\sqrt{\hat{V}[\hat{\phi}] + \tilde{V}[\tilde{\phi}]}} = \frac{0\cdot03}{\sqrt{0\cdot001\ 693}} = 0\cdot75$$

and enter this value in the tables of the unit normal distribution. As $z < 1\cdot96$ we do not reject the hypothesis that ϕ is different for the two periods 1909–1939 and 1940–1951. The two estimates may then be combined, using weights (cf. 1.3.2) $w_1 = \{\hat{V}[\hat{\phi}]\}^{-1}$ and $w_2 = \{\tilde{V}[\tilde{\phi}]\}^{-1}$, thus

$$\phi' = (w_1\hat{\phi} + w_2\tilde{\phi})/(w_1 + w_2) = 0\cdot658,$$

$$V[\phi'] = (w_1 + w_2)^{-1} = 0\cdot000\ 112,$$

and

$$\sigma[\phi'] = 0\cdot0106.$$

A similar example is set out in some detail in Murton [1966: 185–191].

TABLE 5.20

Estimates of the average annual adult survival rates of starlings according to cause of death: from Coulson [1960: Table 1].

	Recoveries (D_j)										$\tilde{\phi} \pm \tilde{\sigma}[\tilde{\phi}]$
$j =$	1	2	3	4	5	6	7	8	9	10	(%)
"Found dead"	482	233	101	57	26	14	9	1	1	1	48·9 ± 1·2
Cat	57	26	19	8	2	2	1	1	0	0	50·6 ± 3·3
Shot	157	50	27	11	4	2	1	0	0	0	40·2 ± 2·4
Falling down chimney	22	9	5	4	1	0	0	0	0	0	46·0 ± 5·7
Other predators	12	4	1	2	0	1	0	0	0	0	46·0 ± 8·1
Other causes	7	15	4	3	0	0	0	1	0	0	
Total	737	337	157	85	33	19	11	3	1	1	47·2 ± 1·0

Example 5.12   Starling (*Sturnus vulgaris*): Coulson [1960]

In Great Britain there have been more starlings ringed and recovered than any other bird. Coulson has used the methods of Lack and Haldane to analyse some of these extensive data. With the large numbers it was possible to break down the data into various sub-categories, such as cause of death, regions where ringed and recovered, sex, and even country where recovered. For example, Table 5.20 uses Lack's method for starlings ringed before 1950 to relate survival estimates to cause of death. The estimates of ϕ are given in the final column, together with their standard deviations; there is a significant difference between the survival rates for those "found dead" and those shot.

ESTIMATING REPORTING RATES. Four methods have been suggested for estimating the band reporting rate (Tomlinson [1968], Henny [1967]):

(i) If the birds are hunted then the number of bands reported can be compared with the estimated number (from kill surveys) of banded birds bagged. This method was used by Geis and Atwood [1961], Martinson [1966] and Martinson and McCann [1966] to determine band-reporting rates for recovered waterfowl in the late 1950's and 1960's.

(ii) If some of the birds carry a band inscribed "reward", then the proportion of reward bands returned can be compared with the proportion of ordinary bands. Bellrose [1955] used this method to determine reporting rates for waterfowl, while Tomlinson [1968] applied it to mourning doves (cf. Table 5.21). However, Tomlinson points out that the reward band method is unsatisfactory in two respects: (a) If it becomes known that there are two types of bands, one that yields a reward and another that does not, there may be fewer non-reward bands reported with a consequent reduction in the overall band-reporting rate. (b) There is not necessarily a 100 per cent reporting of all reward bands recovered, so that the actual reporting rates may be less than the rates estimated by this method.

TABLE 5.21

The estimation of band reporting rates in U.S.A., using reward banding for states whose rates of recovery are probably not influenced by local game-management activities: from Tomlinson [1968: Table 2].

Location	Year	Reward bands			Ordinary bands		
		Bandings	Recoveries No.	Recoveries %	Bandings	Recoveries No.	Recoveries %
California	1965	35	2	5·71	71	3	4·23
Idaho	1966	43	2	4·65	147	2	1·36
Maryland	1965	50	9	18·00	159	7	4·40
Maryland	1966	50	10	20·00	254	13	5·12
Missouri	1966	50	3	6·00	99	4	4·04
South Carolina	1965	49	8	16·33	97	1	1·03
South Dakota	1966	50	3	6·00	150	1	0·67
Texas	1966	25	2	8·00	75	4	5·33
Virginia	1965	50	4	8·00	97	4	4·12
Washington	1965	50	6	12·00	100	3	3·00
Average				10·47			3·33

Band reporting rate $= 100 (3\cdot33)/10\cdot47 = 31\cdot81$ per cent

(iii) A comparison of band serial numbers taken at check stations with band serial numbers reported will give an estimate of the reporting rate. Stair [1957] used this method for mourning doves in Arizona.

(iv) The reporting rate can be estimated by comparing two independent estimates of hunting mortality where one estimate is not affected by incomplete reporting. Details of one such method are given in Henny [1967].

BAND LOSSES. A serious problem in bird-banding studies is the loss of bands, loss of legibility of numbers and directions printed on the bands, etc. Hickey [1952: 18 ff.], Farner [1955: 422], Paynter [1966]). Apart, however, from the double-tagging methods of 6.3.3 for estimating tag loss, very little has been done in setting up suitable models for the estimation of such losses. An interesting model based on the decline in band-weight through wear is used by Ludwig [1967] to correct for band loss.

5.4.3   Age-dependent survival

The case when survival depends on age rather than time has received little attention, from a mathematical point of view, in the literature so far. In the following discussion we shall consider a number of special cases only.

Suppose that the animals are banded and released at birth, and let ϕ_i now be defined as the probability that a live animal of exact age $i - 1$ years survives for a further year. Then, using the same notation as that given in

5.4.1 (apart from ϕ_i), the joint distribution of the $\{m_{ij}\}$, the number of tags recovered in year j from the ith release, is proportional to

$$[(1-\phi_1)\lambda_1]^{m_{11}}[\phi_1(1-\phi_2)\lambda_2]^{m_{12}}\ldots[\phi_1\phi_2\ldots\phi_{t-1}(1-\phi_t)\lambda_t]^{m_{1t}}(1-\theta_1)^{R_1-r_1}$$

$$\times[(1-\phi_1)\lambda_2]^{m_{22}}[\phi_1(1-\phi_2)\lambda_3]^{m_{23}}\ldots[\phi_1\phi_2\ldots\phi_{t-2}(1-\phi_{t-1})\lambda_t]^{m_{2t}}(1-\theta_2)^{R_2-r_2}$$

$$\times[(1-\phi_1)\lambda_3]^{m_{33}}[\phi_1(1-\phi_2)\lambda_4]^{m_{34}}\ldots[\phi_1\phi_2\ldots\phi_{t-3}(1-\phi_{t-2})\lambda_t]^{m_{3t}}(1-\theta_3)^{R_3-r_3}$$

$$\cdots\cdots\cdots\cdots\cdots\cdots\cdots\cdots\cdots\cdots\cdots\cdots\cdots\cdots\cdots\cdots\cdots\cdots\cdots$$

$$\times[(1-\phi_1)\lambda_s]^{m_{ss}}[\phi_1(1-\phi_2)\lambda_{s+1}]^{m_{s,\,s+1}}\ldots$$

$$[\phi_1\phi_2\ldots\phi_{t-s}(1-\phi_{t-s+1})\lambda_t]^{m_{st}}(1-\theta_s)^{R_s-r_s}$$

$$=\{\prod_{j=1}^{t}(1-\phi_j)^{D_j}\lambda_j^{m_j}\}\{\prod_{j=1}^{t-1}\phi_j^{D_{j+1}+D_{j+2}+\ldots+D_t}\}\{\prod_{i=1}^{s}(1-\theta_i)^{R_i-r_i}\}$$

where $\theta_i = (1-\phi_1)\lambda_i + \phi_1(1-\phi_2)\lambda_{i+1} + \ldots + \phi_1\phi_2\ldots\phi_{t-i}(1-\phi_{t-i+1})\lambda_t$,

$$D_j = m_{1j} + m_{2,j+1} + \ldots \qquad (j = 1, 2, \ldots, t),$$

and

$$m_j = \sum_{i=1}^{\min(j,s)} m_{ij}.$$

Here θ_i is the probability that a tag is eventually recovered from the ith release, D_j is the number of tags recovered from animals which die in their jth year of life, and m_j is the number of tags recovered during the jth year. Unfortunately the maximum-likelihood equations for the above model do not appear to have explicit solutions and must be solved iteratively, using, for example, the Newton–Raphson method (cf. **1.3.8**(3)).[*]

CONSTANT RECOVERY RATE. A number of methods have been proposed for estimating the ϕ_i when the recovery rate is constant from year to year (i.e. $\lambda_i = \lambda$). For example, if $\phi_1\phi_2\ldots\phi_{t-s+1} \approx 0$ (i.e. virtually all tagged animals are dead by the end of the experiment), then $(D_j + D_{j+1} + \ldots + D_t)/\lambda$ is approximately the total number of tagged animals surviving for $j-1$ years, (D_j/λ) is approximately the number of deaths of tagged animals in their jth year of life, and

$$q_j = D_j/(D_j + D_{j+1} + \ldots + D_t), \qquad (j = 1, 2, \ldots, t-s),$$

is an estimate of $1 - \phi_j$, the probability of a tagged animal dying in its jth year of life. Although the above formula can be used for $j = t-s+1, \ldots, t$, the resulting estimates will usually be inaccurate as they are based on few (if any) recoveries; for example if $D_j = 0$ then $q_j = 0$.

A number of authors have used the estimates q_j (usually in the form of a composite life-table; cf. **10.1.2**(3)), even when $\phi_1\phi_2\ldots\phi_{t-s+1}$ is not approximately zero. In this case some tagged animals may still be alive after the tth year, so that D_{t+1}, $D_{t+2}\ldots$, etc. should be added to the denominator of q_j, i.e. $1 - \phi_j$ will be overestimated. This can also be seen

---

[*]An explicit solution has been given by Seber [1971] for the case when $\lambda_i = \lambda$; see also Fordham [1970].

by noting from the relation

$$E[m_{ij}] = R_i\phi_1\phi_2\ldots\phi_{j-i}(1-\phi_{j-i+1})\lambda$$

that for $j \leqslant t - s + 1$

$$\frac{E[D_j]}{E[D_j+\ldots+D_t]} = \frac{(R_1+R_2+\ldots+R_s)\phi_1\phi_2\ldots\phi_{j-1}(1-\phi_j)}{(R_1+R_2+\ldots+R_s)\phi_1\phi_2\ldots\phi_{j-1} - K_j}$$

$$< 1 - \phi_j,$$

where $K_j = R_1\phi_1\phi_2\ldots\phi_t + R_2\phi_1\phi_2\ldots\phi_{t-1} + \ldots + R_s\phi_1\phi_2\ldots\phi_{t-s+1}.$

Another method for estimating $1 - \phi_j$ has been proposed independently by Bellrose and Chase [1950], Paludan [1951] and Hickey [1952: 11, Table 2]. These authors use D'_j (usually expressed as a percentage) obtained by dividing D_j by the total number of animals which could contribute to D_j. For example, if $s = 3$, $t = 5$ we have

$$D'_1 = \frac{D_1}{R_1 + R_2 + R_3} , \quad D'_2 = \frac{D_2}{R_1 + R_2 + R_3}, \quad D'_3 = \frac{D_3}{R_1 + R_2 + R_3},$$

$$D'_4 = \frac{D_4}{R_1 + R_2} \quad \text{and} \quad D'_5 = \frac{D_5}{R_1} .$$

Then $1 - \phi_j$ is estimated by

$$q'_j = D'_j/(D'_j + D'_{j+1} + \ldots + D'_t).$$

To check on the validity of this method of estimation we note that

$$E[D'_j] = \phi_1\phi_2\ldots\phi_{j-1}(1-\phi_j)\lambda, \tag{5.44}$$

$$E[D'_j+\ldots+D'_t] = \phi_1\phi_2\ldots\phi_{j-1}(1-\phi_j)\lambda + \ldots + \phi_1\phi_2\ldots\phi_{t-1}(1-\phi_t)\lambda$$

$$= (\phi_1\phi_2\ldots\phi_{j-1} - \phi_1\phi_2\ldots\phi_t)\lambda$$

and

$$\frac{E[D'_j]}{E[D'_j+\ldots+D'_t]} = \frac{1 - \phi_j}{1 - \phi_j\phi_{j+1}\ldots\phi_t}.$$

Therefore the above ratio is approximately equal to $1 - \phi_j$ if and only if $\phi_j\phi_{j+1}\ldots\phi_t \approx 0$, so that as j increases, q'_j will tend to overestimate $1 - \phi_j$; this is probably the case in Hickey's example set out in Table 5.22 (see also Imber and Williams [1968: 259, Table 2]).

Some authors compute the values $D'_j (R_1+\ldots+R_s)$, i.e. when $s = 3$, $t = 5$ we have

$$D_1, D_2, D_3, D_4\left(1 + \frac{R_3}{R_1 + R_2}\right), \quad D_5\left(1 + \frac{R_2 + R_3}{R_1}\right),$$

so that one only needs to calculate the corrections $D_4R_3/(R_1 + R_2)$ and $D_5(R_2+R_3)/R_1$ (e.g. Westerskov [1963]). The above series can also be written as

TABLE 5.22

Mortality estimates for a hypothetical population: from Hickey [1952: 11].

Year banded	Number banded (R_i)	Number reported dead by age intervals			
		0–1	1–2	2–3	3–4
1940	1000	100	30	10	3
1941	1000	95	25	11	–
1942	1000	100	20	–	–
Total $(D_j) =$		295	75	21	3
$\Sigma\, R_i =$		3000	3000	2000	1000
$100\, D'_j =$		9·83	2·50	1·05	0·30
$100\,(D'_j + \ldots + D'_t) =$		13·68	3·85	1·35	0·30
Ratio $(q'_j) =$		0·72	0·65	0·78	–
$\log_{10}(100\, D'_j) =$		0·993	0·398	0·021	−0·523
$j - 1 =$		0	1	2	3

$$D_1,\ D_2,\ D_3,\ D_4 \frac{(R_1 + R_2 + R_3)}{(R_1 + R_2 + R_3) - R_1}\ ,\qquad D_5 \frac{R_1 + R_2 + R_3}{(R_1 + R_2 + R_3) - R_1 - R_2}\ ,$$

which is the formulation used by Farner [1955: 402, equation (3)].

CONSTANT SURVIVAL RATE. A maximum-likelihood method for dealing with this special case of constant survival and reporting rates is given in **5.4.2**(2). However, from (5.44) with $\phi_j = \phi$ we have

$$E[100\, D'_j] \ =\ \phi^{j-1}(1-\phi)(100\,\lambda)$$

and we consider the linear regression model

$$Y_j \ =\ \log\,[100\,\lambda\,(1-\phi)] + (j-1)\log\phi + e_j, \tag{5.45}$$

where $Y_j = \log\,(100\, D'_j)$. Strictly speaking the e_j will be correlated, as each D'_j is based on random variables from the same multinomial distributions, though the degree of dependency will be small if the D'_j are small (which is usually the case). Therefore it is not unreasonable to assume that the e_i are approximately independently distributed with mean zero and constant variance, and to estimate ϕ and $100\,\lambda$ by least squares. We note that logarithms to the base 10 can be used in (5.45), and using Hickey's hypothetical data in Table 5.22 we find that the plot is approximately linear.

We should also mention briefly one other method of estimation sometimes used in the literature. From (5.44) we have

$$\frac{E[D'_{j+1}]}{E[D'_j]} \ =\ \frac{\phi_j(1-\phi_{j+1})}{(1-\phi_j)}\ ,$$

so that when $\phi_j = \phi$ $(j = 1, 2, \ldots, t)$, D'_2/D'_1, D'_3/D'_2 , etc. are all estimates of ϕ. A slight variation of this method which does not utilize all the data is to work with appropriate pairs of D_j (modified so as to be based on the same tag releases). For example, from Table 5.22 one can use the ratios 75/295, 21/55 and 3/10; a further example is given in Jenkins *et al.* [1967: Table 6]. It is stressed that such ratios only apply when ϕ_j is constant.

For a more comprehensive review of the above two sections cf. Seber [1972].

CHAPTER 6

OPEN POPULATION: MARK RELEASES BEFORE SAMPLING PERIOD

6.1 SINGLE RELEASE: INSTANTANEOUS SAMPLES

The multi-release methods considered in the previous chapter, although providing maximum information about population changes, involve a considerable expenditure of effort. Also, such multiple releases may be impractical or uneconomic, particularly in the study of commercially exploited populations. It is natural, therefore, to consider what can be learnt from just a single tag—release followed by a series of "instantaneous" random samples *removed* from the population. For example, in fisheries the scientist could release a single batch of tagged fish, and the series of random samples would then be the hauls by the fishermen. However, since the information from a single tag—release is limited, we find that we must build into our statistical models extra assumptions about the population processes of death, migration, etc. if the effects of these processes are to be estimated. Obviously the ideal model will be the one with the minimum of assumptions and which is robust with regard to departures from those assumptions. We shall now consider a number of possible models.

6.1.1 Constant instantaneous mortality and constant recruitment

The first model we shall consider, due to Parker [1963], is for a population in which there is mortality and recruitment but no migration as, for example, in a bounded population area. This model is not unrealistic because of the existence of localised populations, even, for example, in fisheries which do not have any physical boundaries. Let

N_0 = size of the catchable population at time zero,
M_0 = number of marked or tagged individuals released into the population at time zero,
U_0 = $N_0 - M_0$ (size of unmarked population at time zero),
s = number of samples removed,
n_i = size of the ith sample ($i = 1, 2, \ldots, s$),
m_i = number of marked individuals in the ith sample,
u_i = $n_i - m_i$,
N_i = size of total population just before the ith sample,
M_i = size of marked population just before the ith sample,
U_i = $N_i - M_i$,

and t_i = time at which the ith sample is taken.

It should be noted that we are departing slightly from the notation of Chapter 5 where, for reasons of symmetry, it was convenient to define $n_1 = M_0$, i.e. the first sample provided the mark release.

We shall assume the following:

(1) Immigration and emigration are negligible.

(2) The instantaneous natural mortality rate μ is constant throughout the whole experiment and is the same for both marked and unmarked; thus from (1.1) the probability of an individual surviving for time t between samples is exp $(-\mu t)$.

(3) The number recruited per unit time, r, is constant throughout the whole experiment.

(4) Every individual, whether marked or unmarked, has the same probability of being caught in the ith sample (given that it is alive at the time of the ith sample).

(5) Sampling is instantaneous, or at least takes a negligible period of time as far as population changes are concerned.

(6) Marked individuals do not lose their marks, and all marks are reported on recovery.

For $t < t_1$, let r_t be the number recruited and still surviving at time t. Then, if the recruits are subject to the same mortality rate as the original members of the population, we have the deterministic equation (cf. (1.2))

$$\frac{dr_t}{dt} = -\mu r_t + r$$

which, for the boundary condition $r_0 = 0$, has solution

$$r_t = r(1 - e^{-\mu t})/\mu.$$

Using this relationship, we can now build up the following sequence of deterministic equations:

$$M_1 = M_0 e^{-\mu t_1},$$
$$M_2 = (M_1 - m_1) e^{-\mu(t_2 - t_1)}$$
$$= M_0 e^{-\mu t_2} - m_1 e^{-\mu(t_2 - t_1)},$$
$$N_1 = N_0 e^{-\mu t_1} + r(1 - e^{-\mu t_1})/\mu,$$
$$N_2 = (N_1 - n_1) e^{-\mu(t_2 - t_1)} + r(1 - e^{-\mu(t_2 - t_1)})/\mu$$
$$= N_0 e^{-\mu t_2} - n_1 e^{-\mu(t_2 - t_1)} + r(1 - e^{-\mu t_2})/\mu; \quad \text{etc.}$$

Then, from assumption (4), if the n_i are fixed parameters and M_0/N_0 is small, m_i will be approximately distributed as Poisson with conditional mean

$$a_i = E[m_i | m_1, m_2, \ldots, m_{i-1}]$$
$$= n_i M_i / N_i$$

TABLE 6.1

Recapture data for a population of bluegills, common sunfish and hybrids between the two in Flora Lake, Wisconsin, 1953–1955: from Parker [1963: Table 1].

t	1953 n	1953 m	1954 n	1954 m	1955 n	1955 m	t	1953 n	1953 m	1954 n	1954 m	1955 n	1955 m
1			260	69			30	250	20	280	99	580	46
2			439	109			31	443	47	240	59		
3	86	11	447	99			32					151	24
4	733	41	366	89	509	179	33	580	61	164	31		
5	618	51	634	141			34			249	36		
6	703	54	543	100	340	63	35			90	11		
7	673	93	840	98			37					290	15
8	948	123	540	125			38	628	43	141	16		
9	244	24					39	151	7	269	51	821	19
10			808	84			40	305	56	97	23		
11			241	26	221	38	41	36	7	543	66	470	17
12	225	26	836	89			43			592	57		
13	622	93			173	29	45	91	11	406	34		
14	540	78					46	1025	51	283	17	226	19
15	650	100					47	308	14	302	22		
16	907	103					48			363	41		
17	252	17			151	37	49	308	14			223	15
18	1019	113					50			475	76	131	2
19	549	48					52	51	5			221	3
20	567	78					54			582	85	280	8
21	438	45			105	24	55	78	5				
22	183	11			152	20	56	193	10			465	6
23	537	60	599	146			57			376	56		
24	98	7	320	62	215	35	58	290	39				
25	795	77	195	33			59					148	6
26			301	63	287	25	60			177	21		
27	387	15	166	26			83	54	4				
28	233	24					104	132	5				
29			112	33									

$$= \frac{n_i \left(M_0 - \sum_{j=1}^{i-1} m_j e^{\mu t_j} \right) e^{-\mu t_i}}{\left[N_0 - \sum_{j=1}^{i-1} n_j e^{\mu t_j} + r(e^{\mu t_i} - 1)/\mu \right] e^{-\mu t_i}}$$

and the joint probability function of $m_1, m_2, \ldots, m_s$ is

$$f(\{m_i\} \mid \{n_i\}) = \prod_{i=1}^{s} e^{-a_i} \frac{a_i^{m_i}}{m_i!} = L, \text{ say.}$$

When the periods of time between samples are the same, say $t_j = j$, then this model reduces to that given by Parker.

The maximum-likelihood estimates $\hat{N}_0$, $\hat{\mu}$ and $\hat{r}$ are solutions of the equations

$$\partial \log L / \partial N_0 = \partial \log L / \partial \mu = \log L / \partial r = 0,$$

which must be obtained by some iterative procedure such as the Newton–Raphson method (1.3.8(3)). This method also provides an estimate of the asymptotic variance–covariance matrix from the last iteration.

Example 6.1   Bluegills and common sunfish: Parker [1963]

The above model was applied to recapture data collected from a population of bluegills, sunfish, and hybrids between the two in Flora Lake, Wisconsin, during the summers of 1953 through 1955 (Table 6.1). The fish have a growing season which extends from mid-May to mid-October and the catchable population consisted of fish over 50 millimetres in total length. The marking of 3229, 2768 and 1111 individuals caught with fyke nets was concluded on 6 July, 25 June, and 1 July of 1953, 1954 and 1955, respectively, and during the three following summers, 16 943, 14 155 and 6159 individuals respectively were removed. A day was taken as the unit of time, and although instantaneous sampling was assumed in applying the above theory, each sample actually represented a day's catch.

Starting values for the three maximum-likelihood estimates were selected by choosing those which provided the minimum sum of squared deviations of the expected m_i from the observed m_i, utilising a grid of various combinations of N_0, μ and r which could reasonably be expected to bracket the true values. The computations were performed on an IBM 709 system and Parker's results are given in Table 6.2 in the form *estimate ± estimated standard deviation*.

TABLE 6.2

Maximum-likelihood estimates and their standard deviations, using Parker's method: from Parker [1963: Table 2].

Year	M_0	$\hat{N}_0$	$\hat{\mu}$	$\hat{r}$
1953	3 229	28 149 ± 1 114	0·0161 ± 0·0067	89·7 ± 43·4
1954	2 768	14 347 ± 432	0·0054 ± 0·0048	50·7 ± 18·1
1955	1 111	3 089 ± 267	0·0212 ± 0·0022	145·5 ± 18·3

From Table 6.2 we see that the estimated standard deviations for $\hat{\mu}$ and $\hat{r}$ are very large for 1953 and 1954, while for 1955 all estimated standard deviations are small. A value of $\hat{\mu} = 0 \cdot 02$ operating for 60 days means that only $100 \exp(-60\hat{\mu}) = 30$ per cent of the initial population survived the experiment in 1955, and Parker infers from this that μ must be small for much of the year for the population to maintain itself.

6.1.2 Variable mortality and constant instantaneous recruitment

Using the same notation as in **6.1.1**, we shall now consider a different model based on the following assumptions:

(1) Immigration and emigration are negligible.
(2) Marked and unmarked have the same instantaneous natural mortality rate μ_i during the interval $[t_{i-1}, t_i)$ for $i = 1, 2, \ldots, s$ $(t_0 = 0)$.
(3) The instantaneous recruitment rate λ is constant, i.e. if there was no natural mortality or exploitation the unmarked population would increase exponentially, so that the expected unmarked population size at time t would be $U_0 \exp(\lambda t)$ and the expected number of new recruits $U_0 (\exp(\lambda t) - 1)$. This assumption of recruitment proportional to population size could reasonably apply, for example, to a population where the recruitment is from births (cf. Feller [1957: 405]).
(4) Sampling is instantaneous.
(5) Every individual, whether marked or unmarked, has the same probability p_i $(= 1 - q_i)$ of being caught in the ith sample (given that it is alive at the time of the ith sample (time t_i)).
(6) Marked individuals do not lose their marks or tags, and all marks are reported on recovery.

Given the above assumptions, we have the deterministic equations

$$U_i = U_0 e^{(\lambda - \mu_1)t_1} q_1 e^{(\lambda - \mu_2)(t_2 - t_1)} \cdots q_{i-1} e^{(\lambda - \mu_i)(t_i - t_{i-1})},$$

$$M_i = M_0 e^{-\mu_1 t_1} q_1 e^{-\mu_2(t_2 - t_1)} \cdots q_{i-1} e^{-\mu_i(t_i - t_{i-1})}$$

and hence

$$\frac{U_i}{M_i} = \frac{U_0 e^{\lambda t_i}}{M_0}.$$

Estimating U_i/M_i by u_i/m_i and taking logarithms, we are led to consider the regression model

$$E[y_i] \approx \log(U_0/M_0) + \lambda t_i \quad (i = 1, 2, \ldots, s)$$

$$= \beta_0 + \lambda t_i, \quad \text{say},$$

where $y_i = \log(u_i/m_i)$. Therefore, assuming the y_i to be approximately distributed as independent normal variables with constant variance, we can obtain a confidence interval for β_0 and hence for U_0.

If sampling is not instantaneous but takes a time Δ_i, say, for the ith sample, then u_i/m_i is now an estimate of $\bar{U}_i/\bar{M}_i$, the ratio of the average unmarked population size to the average marked population size during the ith period of sampling. Suppose that μ_{Ei} is the instantaneous rate of exploitation during the ith sample; then

$$q_i = e^{-\mu_{Ei}\Delta_i},$$

$$\bar{U}_i = \int_0^{\Delta_i} U_i \exp\{(\lambda - \mu_i - \mu_{Ei})t\}\, dt/\Delta_i$$

$$\approx U_i \exp\{\tfrac{1}{2}(\lambda - \mu_i - \mu_{Ei})\Delta_i\}$$

and similarly

$$\bar{M}_i \approx M_i \exp\{-\tfrac{1}{2}(\mu_i + \mu_{Ei})\Delta_i\}.$$

Hence

$$\frac{\bar{U}_i}{\bar{M}_i} \approx \frac{U_i}{M_i} \exp(\tfrac{1}{2}\lambda\Delta_i)$$

and our model becomes

$$E[y_i] \approx \log(U_0/M_0) + \lambda(t_i + \tfrac{1}{2}\Delta_i).$$

We note that if $\Delta_i = 1$, then rearranging,

$$E[y_i] \approx [\log(U_0/M_0) - \tfrac{1}{2}\lambda] + \lambda(t_i + 1),$$

which, apart from a change of sign throughout, is Fischler's model [1965: 304].

Example 6.2   Blue crab (*Callinectes sapidus*): Fischler [1965]

The above regression method was applied to a population of blue crabs in the Neuse River, North Carolina, during 1958. Here recruitment referred to the process whereby a crab grew into the legal size-range, so that N_0 was the initial size of the *legally* exploitable population. A total of 361 tagged crabs were available at the beginning of the experiment (4 July) and sampling was continued through to 30 July. The unit of time was a day, and a sample consisted of a day's catch ($\Delta_i = 1$). It was found necessary, for practical reasons, to measure the recaptures m_i in numbers, but the total catches n_i in thousands of pounds, so that the estimate of U_0 is also in thousands of pounds. As the number of recaptures was very small we have effectively $n_i = u_i$ and $y_i = \log(n_i/m_i)$.

The data are set out in Table 6.3, and Fischler showed that a plot of y_i against $t_i + 1$ (Fig. 6.1) was approximately linear, thus supporting the above theoretical analysis. He obtained an estimate of 722 000 pounds for U_0 with a 95 per cent confidence interval of (516 000, 1 009 000) pounds. From the slope of the regression line λ was estimated to be 0·086.

For a full discussion of the assumptions underlying the regression model the reader is referred to the original article, but the following points should be noted: (a) Natural mortality appeared to be negligible during the period of the experiment. (b) Estimates of the daily recruitment rate (e^λ) from both the ratios of soft to hard crabs and the ratios of precommercial

TABLE 6.3
Data used in Fischler's method for estimating the size of the crab population in the Neuse River on 4 July 1958: from Fischler [1965: Table 15].

Date	m_i (number)	n_i (1000 lb)	y_i ($\log n_i/m_i$)	t_i
7 July	13	21·8	0·546 96	3
8	10	16·5	0·500 77	4
9	7	16·0	0·826 68	5
10	7	27·5	1·368 28	6
11	6	15·0	0·916 29	7
14	3	32·0	2·367 13	10
15	1	22·0	3·091 04	11
16	6	22·3	1·312 83	12
17	2	23·4	2·459 59	13
18	3	18·0	1·791 76	14
21	3	31·4	2·348 20	17
22	1	16·6	2·809 40	18
23	3	18·6	1·824 55	19
24	1	22·7	3·112 36	20
25	2	17·8	2·186 05	21
28	3	34·6	2·445 24	24
29	3	31·7	2·357 71	25
30	1	18·4	2·912 35	26

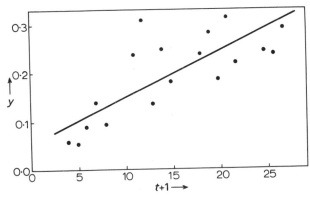

Fig. 6.1   Plot of y_i ($= \log (n_i/m_i)$) against $t_i + 1$ for a blue crab population: redrawn from Fischler [1965].

crabs (3·4 to 4·5 inches carapace width) to commercial size crabs (> 4·5 inches) did not seem to vary greatly, thus supporting the assumption of constant λ. However, it was felt that on the basis of these experiments

$\hat{\lambda} = 0\cdot086$ was too high, a more realistic value being something less than $0\cdot05$. This was put down to movement of the fishing fleet away from the release areas with the tagged crabs consequently becoming less catchable than the untagged. (c) From the length–frequency distribution of crabs in the catches there did not seem to be any size selectivity with the three major types of fishing gear used. (d) There was a lower percentage of tagged females recaptured than of tagged males. This was felt to be mainly due to the tendency of female crabs to move out of the river after mating into the more saline waters of Pamlico and Core Sounds. If this emigration was proportional to the population size then its effect would be to reduce λ. (e) The loss of carapace tags through moulting appeared to be negligible. (f) The tagging program was widely advertised and every effort was made to obtain a complete tag return. (g) The estimate of U_0 agreed favourably with two other estimates obtained from catch-effort data.

6.1.3   Variable mortality and recruitment

1   Parker's model

The model in **6.1.1** is effectively based on the simple equation

$$E\left[\frac{m_i}{n_i} \,|\, M_i, \, N_i\right] = \frac{M_i}{N_i}$$

and if recruitment is taking place, M_i/N_i will decrease as i increases. Parker [1955] suggests the simple expedient of plotting m_i/n_i against t_i and extending this curve to $t = 0$ to obtain an estimate of M_0/N_0, and hence of N_0. Provided there is a reasonably smooth trend in the m_i/n_i, and the above equation is valid, such a method will always give a graphical estimate of N_0 irrespective of the changes taking place in the population. In some circumstances, however, it may be possible to choose a suitable transform-ation of the m_i/n_i which will stabilise the variance, achieve some measure of normality and, hopefully, linearise the regression so that N_0 can be estimated by least squares. For example, in **6.1.2** it was found that a logarithmic transformation achieved linearity when the assumptions stated there were valid.

Example 6.3   Bluegills and common sunfish: Parker [1955]

Using a unit of time of one day, the above method was applied to marking data collected in 1953 from a population of bluegills, common sunfish, and hybrids between the two in Flora Lake, Wisconsin. The fish were caught in fyke nets between 2 and 6 July and marked by the removal of the left pelvic fin. Using the same net, subsequent samples were removed from the popu-lation between 9 July and 18 October, giving the data of Table 6.4.

Assuming the m_i to be approximately binomially distributed, the trans-formation $y_i = \arcsin \sqrt{(m_i/n_i)}$ is known to stabilise the variance and achieve some measure of normality (tables for the arcsin transformation are given in Hald [1952; $m/n = 0 \ (0\cdot001) \ 1$, radians] and Snedecor [1946: $m/n = 0 \ (0\cdot001) \ 1$,

TABLE 6.4

Data used in making a marked-fish regression analysis of a population of bluegills, common sunfish, and hybrids between the two: from Parker [1955: Table 1].

Date	t_i	Sample size n_i	Number marked m_i	Ratio $\left(\dfrac{m_i}{n_i}\right)$	arcsin $\sqrt{\text{ratio}}$ (degrees) y_i	Estimated nat. mort. of marked	No. marked remaining at day's end
July 6	0	—	—	—	—	9 [a]	3220
9	3	86	11	0·1279	20·96	15	3194
10	4	733	41	0·0559	13·69	5	3148
11	5	618	51	0·0825	16·64	5	3092
12	6	703	54	0·0768	16·11	5	3033
13	7	673	93	0·1382	21·81	5	2935
14	8	948	123	0·1298	21·13	4	2808
15	9	244	24	0·0984	18·24	4	2780
18	12	225	26	0·1156	19·91	13	2741
19	13	622	93	0·1495	22·79	4	2644
20	14	540	78	0·1444	22·30	4	2562
21	15	650	100	0·1538	23·11	4	2458
22	16	907	113	0·1246	20·70	4	2341
23	17	252	17	0·0675	15·12	4	2320
24	18	1 019	113	0·1109	19·46	4	2203
25	19	549	48	0·0874	17·16	3	2152
26	20	567	78	0·1376	21·81	3	2071
27	21	438	45	0·1027	18·72	3	2023
28	22	183	11	0·0601	14·18	3	2009
29	23	537	60	0·1117	19·55	3	1946
30	24	98	7	0·0714	15·45	3	1936
31	25	795	77	0·0969	18·15	3	1856
Aug 2	27	387	15	0·0388	11·39	6	1835
3	28	233	24	0·1030	18·72	3	1808
5	30	250	20	0·0800	16·43	6	1782
6	31	443	47	0·1061	19·00	3	1732
8	33	580	61	0·1052	18·91	5	1666
13	38	628	43	0·0685	15·12	13	1610
14	39	151	7	0·0464	12·39	2	1601
15	40	305	56	0·1836	25·40	2	1543
16	41	36	7	0·1944	26·13	2	1534
20	45	91	11	0·1209	20·36	9	1514
21	46	1 025	51	0·0498	12·92	2	1461
22	47	308	14	0·0454	12·25	2	1445
24	49	308	14	0·0454	12·25	4	1427
27	52	51	5	0·0980	18·24	7	1415
30	55	78	5	0·0641	14·65	6	1404
31	56	193	10	0·0518	13·18	2	1392
Sept 2	58	290	39	0·1345	21·47	4	1349
27	83	54	4	0·0741	15·79	51	1294
Oct 18	104	132	5	0·0379	11·24	41	1248
Totals	1203	16 930	1701		712·83	280	

[a] During marking

degrees]. A plot of y_i against t_i (Fig. 6.2) was found to be approximately linear, so that a linear regression model

$$y_i = \beta_0 + \beta t_i + e_i, \quad (i = 1, 2, \ldots, s),$$

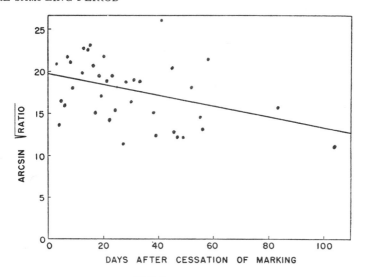

Fig. 6.2 Plot of y_i ($= \arcsin \sqrt{m_i/n_i}$) versus t_i, the time of the ith sample, for a population of bluegills and sunfish: from Parker [1955].

where the e_i are assumed to be independently and identically distributed as $\mathcal{N}[0, \sigma^2]$, could be fitted. The usual least-squares estimates of β and β_0 are

$$\hat{\beta} = \frac{\Sigma\, y_i\, (t_i - \bar{t})}{\Sigma\, (t_i - \bar{t})^2} \,,$$

$$\hat{\beta}_0 = \bar{y} - \hat{\beta}\bar{t} \,,$$

and, since $\hat{\beta}_0$ is an estimate of $\arcsin \sqrt{(M_0/N_0)}$, we have the estimate

$$\hat{N}_0 = M_0 / (\sin \hat{\beta}_0)^2 .$$

A $100(1-\alpha)$ per cent confidence interval for β_0 is given by

$$\hat{\beta}_0 \pm t_{s-1}[\alpha/2][\hat{\sigma}^2 \Sigma\, t_i^2 / n\, \Sigma\, (t_i - \bar{t})^2]^{1/2} \,,$$

where $(s-2)\hat{\sigma}^2 = \Sigma\, (y_i - \bar{y})^2 - \hat{\beta}^2 \Sigma\, (t_i - \bar{t})^2 .$

By taking sines, squaring and inverting, this interval can be converted into a confidence interval for N_0.

Without going through the details, we have from Parker $M_0 = 3220$, $s = 40$, and the fitted regression line is

$$Y = 19 \cdot 7614° - 0 \cdot 065\,269\,t .$$

Also

$$\hat{N}_0 = 3220 / (\sin 19 \cdot 7614°)^2 = 28\,159$$

and an approximate 95 per cent confidence interval for N_0 is (23 394, 34 681). If we ignore the effects of possible recruitment, we note that the conventional Petersen estimate given by Paloheimo in **3.7**, namely

$$M_0 \,\Sigma\, n_i / \Sigma\, m_i = 3220\,(16\,930)/(1701) = 32\,049 \,,$$

is about 14 per cent greater than $\hat{N}_0$.

An average annual survival rate of 0·573 was calculated from age data (cf. Chapter 10) and from this Parker was able to estimate the total recruitment for the whole experiment as follows. Assuming a constant instantaneous natural mortality rate μ throughout the year, a daily natural mortality rate of

$$1 - e^{-\mu} = 1 - (e^{-\mu t})^{1/t} = 1 - (0\cdot573)^{1/365} = 0\cdot001\,53$$

was used to estimate the number of marked fish dying between each sample. At the beginning of the experiment it was estimated that 9 of the 3229 tagged fish released died during the actual tagging period, so that M_0 was taken to be 3220. Putting $t = 104$ in the fitted regression line, we can estimate M_{40}/N_{40} which, using $M_{40} = 1248$, leads to $\hat{N}_{40} = 24\,468$ as the estimate of the total population size on October 18. Parker then estimated the total natural mortality throughout the experiment as 3457: we are not given the method of estimation, though it was no doubt similar to that used for estimating the mortality of marked fish (for example, an N_i sequence can be calculated from the M_i sequence using the fitted regression). Finally, assuming no migration, we have

$$recruitment = removal + natural\ mortality - (\hat{N}_0 - \hat{N}_{40})$$
$$= 16\,930 + 3457 - (28\,159 - 24\,468)$$
$$= 16\,696.$$

2   Chapman's model

If θ_i is the probability that a member of N_0 survives up to time t_i then, using the notation of **6.1.1**,

$$E[N_i] = N_0\,\theta_i + R_i$$

and

$$E[M_i] = M_0\,\theta_i,$$

where R_i is the net recruitment to time t_i (i.e. recruits that come into the population after time zero and are still alive at time t_i). Applying the Petersen method (**3.1.1**) to the ith sample, we have

$$E\left[\frac{n_i M_i}{m_i}\,\Big|\,M_i,\,N_i\right] \approx N_i,$$

which, combined with the above two equations, gives us

$$E[y_i] \approx N_0 + R_i/\theta_i,$$

where $y_i = n_i M_0/m_i$. If θ_i is not too small, a small adjustment for bias can be made, such as (Chapman [1965: 536]) $y_i = (n_i + 1)(M_0 + 1)/(m_i + 1)$.

Since R_i and θ_i will both be functions of t_i, we are led to consider the regression model

$$E[y_i] = N_0 + g(t_i), \quad i = 1, 2, \ldots, s,$$

where g is a function of t_i. Chapman recommends approximating g by a polynomial function and estimating N_0 by least squares.

The basic assumptions underlying the above approach are as follows:

(1) All members of N_0 have the same probability of survival θ_i to time t_i. This means that the tagged and untagged members of N_0 have the same mortality rates and are removed by the sampling process at the same rate.

(2) The assumptions underlying the Petersen method must hold for each sample, i.e. samples are random or, if the samples are systematic, there is uniform mixing of tagged and untagged (including new recruits).

(3) Immigration and emigration are negligible. Alternatively these can be included if θ_i is defined to be the probability that a member of N_0 survives and is in the population at t_i, and R_i is defined to be the net influx of new life (which may be negative) rather than just net recruitment. In the two examples below, migration is negligible (lake population).

(4) Individuals do not lose their tags, and all tags found are reported.

Example 6.4   Bluegills and sunfish: Chapman [1965].

Using the 1953 data in Table 6.1 (p. 258), the values of y_i were calculated and entered in Table 6.5. A plot of y_i versus t_i (Fig. 6.3) revealed a

TABLE 6.5

Values of $y_i = (n_i + 1)(M_0 + 1)/(m_i + 1)$ calculated from Parker's 1953 data: from Chapman [1965: Table 2].

t_i (days)	y_i (thousands)	t_i (days)	y_i (thousands)	t_i (days)	y_i (thousands)
3	23·4	19	36·3	38	46·2
4	56·4	20	23·2	39	61·4
5	38·4	21	30·8	40	17·3
6	41·3	22	49·5	41	14·9
7	23·2	23	28·5	45	24·8
8	28·7	24	40·0	46	63·7
9	32·6	25	33·0	47	66·5
12	27·0	27	78·3	49	66·5
13	21·4	28	30·2	52	28·0
14	22·1	30	38·6	55	42·5
15	20·8	31	29·9	56	57·0
16	28·2	33	30·3	58	23·5
17	45·4				
18	28·9				

slight trend and, because of the wide variation in the y_i, the linear model

$$E[y_i] = N_0 + \beta t_i \qquad (6.1)$$

seemed most appropriate. To test for linearity, Chapman suggested grouping the y_i into classes, so that the members of a class may be regarded as

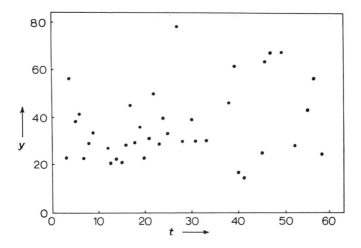

Fig. 6.3   Plot of y_i $(= (n_i + 1)(M_0 + 1)/(m_i + 1))$ versus t_i, the day of recapture, for Parker's 1953 data.

approximate replicates. A working unit of 7 days was used, and the class divisions are indicated by lines in the t_i column of Table 6.5. A straight line was then fitted to the class means, and from Table 6.6 the usual analysis of variance test for linearity was carried out (cf. Draper and Smith [1966: 26] or Keeping [1962: 342]). Chapman pointed out that since the

TABLE 6.6

Test for linearity of Parker's 1953 data: from Chapman [1965: Table 3].

Source	d.f.	Sum of squares	Mean square
Linear regression	1	843·88	843·88
Deviations from linear regression	5	1408·09	281·62
Error	31	6994·89	225·64

Test for linearity of regression: $F = 1·25$ with (5, 31) d.f.

replicates are not true replicates, there is a trend in each class, and the error sum of squares in Table 6.6 is inflated. An attempt to correct for this can be made by fitting a straight line within each class and then pooling the residual sums of squares to give a residual mean-square error of 183·06 with 24 degrees of freedom. If deviations from linear regression are tested against this, the F value is 1·54 which is still not significant.

From Fig. 6.3 the variance of y_i does not appear to depend on t_i, so that, assuming constant variance and making the usual normality assumptions, we have the least-squares estimates $\hat{\beta} = 0·2954$ and $\hat{N}_0 = 28·73$ *thousand* with a 95 per cent confidence interval for N_0 of 28 730 ± 9820. With stronger but unverifiable assumptions, Parker (Example 6.1) obtained the much narrower limits 28 149 ± 2183.

If we assume the linear regression model with $R(t)/\theta(t) = \beta t$ and define $r(t)$ to be the net recruitment on the tth day, then

$$
\begin{aligned}
r(t) &= R(t) - R(t-1) \cdot \frac{\theta(t)}{\theta(t-1)} \\
&= \theta(t)[\beta t - \beta(t-1)] \\
&= \beta\theta(t).
\end{aligned}
$$

Since $\theta(t)$ is the survival from both natural mortality and the sampling or catching process, it will be a decreasing function of t, and $r(t)$ will also be decreasing. If estimates of the natural mortality rate are available then it is possible to calculate $\theta(t)$, and hence $R(t)$, in a stepwise fashion for each value of t. Alternatively one could endeavour to represent $r(t)$ by some simple function such as

$$
r(t) = r_0 e^{-r_1 t}.
$$

We note in passing that if $\theta(t)$ is near unity for t near zero, then $\hat{r}(0) = \hat{\beta} = 0\cdot2954$ *thousand*, i.e. the initial recruitment is estimated at about 300 per day.

Finally Chapman points out that a test of $\beta = 0$ for (6.1) is not significant at the 5 per cent level (though it is at the 10 per cent level) and it is therefore possible to assert that the hypothesis $R(t) = 0$ is consistent with the data. However, he adds that, in spite of this, the regression model (6.1) seems to be the best simple model.

Example 6.5 Bluegills and sunfish: Chapman [1965]

Using the 1955 data of Table 6.1, Chapman constructed Table 6.7. In contrast to Example 6.4, we find that a plot of y_i versus t (Fig. 6.4) is now

TABLE 6.7

Values of $y_i = (n_i + 1)(M_0 + 1)/(m_i + 1)$ calculated from Parker's 1955 data: from Chapman [1965: Table 2].

t_i (days)	y_i (thousands)	t_i (days)	y_i (thousands)	t_i (days)	y_i (thousands)
4	3·2	24	6·7	46	12·6
6	5·9	26	12·3	49	15·6
11	6·3	30	13·7	50	48·9
13	6·4	32	6·8	52	61·7
17	4·4	37	20·2	54	34·7
21	4·7	39	45·7	56	74·0
22	8·1	41	29·1	59	23·7

non-linear and the variance of y_i increases with t_i. We can therefore either try to fit, say, a quadratic using weighted least squares or else look for a

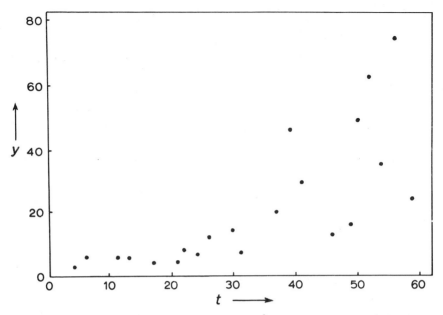

Fig. 6.4   Plot of y_i $(= (n_i + 1)(M_0 + 1)/(m_i + 1))$ versus t_i, the day of recapture, for Parker's 1955 data.

suitable transformation of the data which stabilises the variance and (hope-fully) linearises the regression.

Chapman once again grouped the data into classes (the choice of class divisions in Table 6.7 being somewhat more arbitrary than for Table 6.5) and

TABLE 6.8

Preliminary analysis of Parker's 1955 data: from Chapman [1965: Table 4].

Group	Mean day of sampling	Mean of y_i	Standard deviation
1	8·5	5·45	1·52
2	21·5	7·24	3·21
3	30·5	23·10	15·08
4	49·0	34·70	24·39
5	56·3	44·13	26·44

TABLE 6.9

Test for linearity of 1955 data (log. transformation): from Chapman [1965: Table 5].

Source	d.f.	Sum of squares	Mean square
Linear regression	1	2·0752	2·0752
Deviations from linear regression	3	2·0709	0·6903
Error	16	10·7728	0·6733

Test for linearity of regression: $F = 1·02$ with (3, 16) d.f.

calculated Table 6.8. On the evidence of this table an analysis was first attempted with the standard deviation of y_i assumed proportional to t_i. However, this did not provide a satisfactory fit using either a linear or a quadratic regression, and a logarithmic transformation was subsequently tried. It was found that for the transformed data the variances within classes did not differ significantly from class to class, and the analysis of variance test for linearity (Table 6.9) was not significant. Therefore the model

$$\log y_i = \beta_0 + \beta t_i + e_i.$$

was chosen, where the e_i are assumed to be independently and identically distributed as $\mathfrak{N}[0, \sigma^2]$. The least-squares estimates of β_0 and β are

$$\hat{\beta}_0 = 1 \cdot 2485, \quad \hat{\beta} = 0 \cdot 043\ 96$$

and a 95 per cent confidence interval for β_0 is $(0 \cdot 673, 1 \cdot 836)$. Now, from the above model we have the relationship

$$N_0 + g(t) = \exp(\beta_0 + \beta t),$$

so that putting $t = 0$ and $g(0) = 0$ (since $R(0) = 0$), we have $N_0 = \exp(\beta_0)$ and $\hat{N}_0 = \exp(\hat{\beta}_0) = 3 \cdot 485$ *thousand*. A 95 per cent confidence interval for N_0 is $(\exp(0 \cdot 673), \exp(1 \cdot 836))$ or $(1 \cdot 96, 6 \cdot 27)$ *thousand*. Since

$$R(t)/\theta(t) = g(t) = e^{\beta_0}(e^{\beta t} - 1),$$

the daily net recruitment rate is given by

$$\begin{aligned} r(t) &= R(t) - R(t-1) \cdot \theta(t)/\theta(t-1), \\ &= \theta(t)e^{\beta_0}(e^{\beta t} - e^{\beta(t-1)}) \\ &= e^{\beta_0}(1 - e^{-\beta})\theta(t)e^{\beta t} \end{aligned}$$

which is estimated by

$$\hat{r}(t) = (0 \cdot 1502)\theta(t)e^{0 \cdot 043\ 96\ t}.$$

Knowledge of the natural mortality rate would then enable one to calculate $\theta(t)$ and hence $\hat{r}(t)$. For example, if

$$\theta(t) = e^{-\mu t - h(t)},$$

where μ is the instantaneous natural mortality rate and $h(t)$ represents the mortality due to the catching process, then $\hat{r}(t)$ is increasing or decreasing according as $(0 \cdot 043\ 96 - \mu)t$ is greater or less than $h(t)$. Chapman estimates the daily instantaneous removal rate to be about $0 \cdot 02$ (i.e. $h(t) = 0 \cdot 02\ t$), so that for Parker's assumption of constant $r(t)$ to hold (cf. **6.1.1**), we must have $\mu \approx 0 \cdot 044 - 0 \cdot 02 = 0 \cdot 024$. This may be compared with Parker's estimate of $0 \cdot 0212$ (Table 6.2).

6.2 SINGLE RELEASE AND CONTINUOUS SAMPLING: MORTALITY ESTIMATES

6.2.1 Maximum-likelihood estimates

In **6.1** sampling was considered to be a discrete instantaneous process as, for example, in short intensive fisheries. We shall now consider a model discussed by Gulland [1955b], Chapman [1961] and Paulik [1963a] in which sampling is regarded as a continuous process and recaptures are comparatively so few that the recapture times of marked animals can be recorded. The assumptions we shall make are the following:

(1) Immigration and emigration are negligible.
(2) For tagged individuals the instantaneous natural mortality rate and the instantaneous exploitation rate are both constant and equal to μ and μ_E respectively (more commonly M and F in fishery research). If we can also assume that these rates are the same for tagged and untagged then the estimates obtained below can be applied to the whole population.
(3) Individuals do not lose their tags, and all tags found are reported.

Let $Z = \mu + \mu_E$ be the total instantaneous mortality rate and let M_0 be the size of the tagged population at time zero. Then the probability of a tagged individual remaining alive at time t is $\theta(t) = \exp(-Zt)$ and the probability of recapture in the interval $(t, t + dt)$ is $\mu_E \theta(t)dt$. Suppose that m tagged individuals are recaptured at times $T_1, T_2, \ldots, T_m$ respectively, and that these are the only recaptures up to time τ when exploitation ceases. Then the probability that a tagged individual is caught during the whole period is

$$\int_0^\tau \mu_E \,\theta(t)dt \;=\; \mu_E[1-\theta(\tau)]/Z \;=\; P_\tau, \quad \text{say}. \tag{6.2}$$

Here P_τ can be regarded as the proportion of the initial tagged population M_0 expected to be finally caught, and it is therefore called the exploitation rate for time τ. For example, if τ is one year, then P_1 is the annual exploitation rate.

Assuming that tagged individuals are independent of each other, m has a binomial distribution

$$f_1(m) \;=\; \binom{M_0}{m} P_\tau^m (1-P_\tau)^{M_0 - m}.$$

Also, given that a tagged individual is caught during the experiment, then the conditional density function for its recovery time is

$$f_2(t) \;=\; \begin{cases} \mu_E \theta(t)/P_\tau & 0 \leqslant t \leqslant \tau, \\ 0 & \text{otherwise}. \end{cases} \tag{6.3}$$

Hence the joint density function of $T_1, T_2, \ldots, T_m$ and m is

$$\begin{aligned} f(T_1, T_2, \ldots, T_m, m) &= f(T_1, T_2, \ldots, T_m \mid m)f_1(m) \\ &= \left(\prod_{i=1}^m f_2(T_i)\right) f_1(m) \\ &= \binom{M_0}{m} (1-P_\tau)^{M_0 - m} \prod_{i=1}^m \{\mu_E \theta(T_i)\}. \end{aligned} \tag{6.4}$$

For the case when τ is so large that $\theta(\tau)$ can be neglected (i.e. $P_\tau \approx P_\infty = \mu_E/Z$), Gulland [1955b] derives maximum-likelihood estimates of μ and μ_E

and shows that they are biased: his estimates are slightly modified (cf. Seber [1962: 346]) when a short period of time is allowed at the beginning for the dispersal of the tagged individuals. Chapman [1961: 156] notes that for this special case of τ large, no minimum variance unbiased estimates of μ and μ_E exist. However, he shows that for $m \geqslant 2$ (i.e. neglecting the possibility of no recaptures or just a single recapture) approximately unbiased estimates are given by

$$\mu_E^* = m(m-1)/(M_0 T.)$$

and

$$\mu^* = (M_0 - m)(m-1)/(M_0 T.),$$

where $T. = \sum\limits_{i=1}^{m} T_i$. Approximate variances are given by

$$V[\mu_E^*] = \frac{\mu_E(\mu_E + 2\mu)}{M_0} + \frac{2(\mu_E + \mu)}{M_0^2} + O\!\left(\frac{1}{M_0^3}\right)$$

and

$$V[\mu^*] = \frac{\mu(\mu_E^2 + \mu_E \mu + \mu^2)}{M_0} + O\!\left(\frac{1}{M_0^2}\right).$$

If we simply wished to estimate $Z = \mu + \mu_E$ then it transpires that

$$Z^* = (m-1)/T.$$

is a minimum variance unbiased estimate of Z with variance

$$V[Z^*] = Z^2/(m-2).$$

As this estimate and its variance do not depend on knowing M_0 we can use any well-defined class in the population at the beginning of the experiment as the "tagged" portion, provided there is no further recruitment into this class during the sampling period τ. For example, the class of animals of a particular age or size can be used, and the times T_i would then refer to the capture times of animals from this class.

When $\theta(\tau)$ cannot be neglected, Paulik [1963a] shows that it is more convenient to work with P_τ and S ($= e^{-Z}$), the probability of survival for one unit of time, than with μ and μ_E. By writing μ_E in terms of P_τ and S (cf. (6.2)), we find from (6.4) that the maximum-likelihood estimates of P_τ and S are given by

$$\hat{P}_\tau = m/M_0$$

and

$$-\frac{1}{\tau \log \hat{S}} - \frac{\hat{S}^\tau}{1 - \hat{S}^\tau} = \frac{\bar{T}}{\tau}$$

where $\bar{T} = T./m$ Writing $\hat{Z} = -\log \hat{S}$, this last equation becomes

$$\frac{1}{\tau \hat{Z}} - \frac{1}{\exp(\tau \hat{Z}) - 1} = \frac{\bar{T}}{\tau} \tag{6.5}$$

and when $0 \leqslant \bar{T} < \tau/2$, (6.5) can be solved using a table of $\bar{T}/\tau$ as a function of $\tau \hat{Z}$ given by Deemer and Votaw [1955: 50] and reproduced in A5; for

$\bar{T} \geqslant \tau/2$ the maximum-likelihood estimate of Z is 0. For m large Paulik shows that

$$V[\hat{S}] \approx \frac{1}{M_0 P_\tau} \left\{ \frac{1}{[S \log S]^2} - \frac{\tau^2 S^{\tau-2}}{[1-S^\tau]^2} \right\}^{-1},$$

$$\to (S \log S)^2/(M_0 P_\tau) \quad \text{as } \tau \to \infty$$

and

$$V[\hat{P}_\tau] = P_\tau(1-P_\tau)/M_0.$$

GROUPED OBSERVATIONS. Very often the recovery times are not known exactly. For example, in a trap fishery the traps may be lifted only when a certain number of fish have been accumulated, so that a tagged fish could have been caught at any time since the last lift. Also in many fisheries the catch is not thoroughly examined for tags until the vessels are unloaded in port or until the fish are being processed. Suppose, then, that the recovery period $[0, \tau)$ is divided into J small intervals $[0, d_1), [d_1, d_2), \ldots, [d_{J-1}, d_J)$, where $d_J = \tau$, and suppose that m_j tagged individuals are caught in the jth interval $(j = 1, 2, \ldots, J)$. Paulik suggests taking the middle point as representative of each interval and approximating $\bar{T}$ by

$$\sum_{j=1}^{J} m_j(d_j + d_{j-1})/2m$$

in (6.5). However, when the intervals are large, he gives a more efficient method as follows.

Suppose that the intervals are of equal length so that $d_j = j$ units of time $(j = 1, 2, \ldots, J : J = \tau)$ and, from (6.2),

$$P_J = \mu_E(1-e^{-ZJ})/Z \tag{6.6}$$

$$= -\mu_E(1-S^J)/\log S. \tag{6.7}$$

Then, neglecting the complications of sampling without replacement within each interval, the $\{m_j\}$ have a joint multinomial distribution

$$f(\{m_j\}) = \frac{M_0!}{\{\prod_j m_j!\}(M_0-m)!}(1-P_J)^{M_0-m} \prod_{j=1}^{J} \{p_j^{m_j}\}, \tag{6.8}$$

where p_j = (probability of surviving for $(j-1)$ intervals) × (probability of capture in the jth interval)

$$= e^{-Z(j-1)}P_1 \tag{6.9}$$

$$= S^{j-1}P_J \Big/ \sum_{j=0}^{J-1} S^j, \tag{6.10}$$

since, from (6.7),

$$P_1 = P_J(1-S)/(1-S^J).$$

Hence, substituting for p_j in (6.8) we find that $f(\{m_j\})$ is proportional to

$$S^x P_J^m (1-P_J)^{M_0-m} \left(\sum_{j=0}^{J-1} S^j \right)^{-m}$$

where $x = \sum_{j=1}^{J} (j-1)m_j$. The maximum-likelihood estimates of P_J and S are therefore given by

$$\widetilde{P}_J = m/M_0$$

and

$$\left(\sum_{j=0}^{J-1} j\widetilde{S}^j\right)\Big/\left(\sum_{j=0}^{J-1} \widetilde{S}^j\right) = x/m. \tag{6.11}$$

The solution of (6.11) is discussed in **A6** where a table is available for $3 \leqslant J \leqslant 9$. The asymptotic variances are given by

$$V[\widetilde{P}_J] = P_J(1-P_J)/M_0$$

and

$$V[\widetilde{S}] = \frac{S^2\left(\sum\limits_{j=0}^{J-1} S^j\right)^2}{M_0 P_J\left[\left(\sum\limits_{j=0}^{J-1} S^j\right)\left(\sum\limits_{j=0}^{J-1} j^2 S^j\right) - \left(\sum\limits_{j=0}^{J-1} j S^j\right)^2\right]}$$

$$= \frac{1}{M_0 P_J}\left[\frac{1}{S(1-S)^2} - \frac{J^2 S^{J-2}}{(1-S^J)^2}\right]^{-1}.$$

If the number of recovery periods J is sufficiently large to make $S^J \to 0$, then

$$V[\widetilde{S}] \to S(1-S)^2/(M_0 P_J).$$

When estimates of μ and μ_E are required, we can obtain maximum-likelihood estimates by solving for μ and μ_E in terms of S and P_J. For example, from (6.7) we have

$$\widetilde{\mu}_E = -\widetilde{P}_J \log \widetilde{S}/(1-\widetilde{S}^J)$$

and, using the delta method,

$$V[\widetilde{\mu}_E] \approx \left\{\frac{\log S}{1-S^J}\right\}^2 V[\widetilde{P}_J] + \left\{\frac{P_J}{S(1-S^J)}\right\}^2 \cdot \left\{1 + \frac{JS^J \log S}{1-S^J}\right\}^2 V[\widetilde{S}].$$

Paulik points out that the actual recovery data may be a mixture of "exact" times T_i and interval approximations as above. If only a few of the recovery times are known exactly, it does not seem worthwhile to treat them separately. However, if the exact times are known for a sizeable proportion of the recoveries he suggests that two estimates should be made: one for the exact time and one for the interval data. A suitable estimate of S would be a weighted mean of the two estimates where the weights are proportional to the estimated variances (cf. **1.3.2**) or more simply to the numbers of tagged individuals used in each estimate.

In concluding this section we note that $M_0 P_J$ can be estimated by m, so that $\hat{S}$, $\widetilde{S}$ and estimates of their variances do not depend on knowledge of M_0.

Example 6.6 Plaice: Paulik [1962, 1963a]

Using a time-interval of three months and $J = 4$ recovery periods, the following recovery data on tagged plaice were obtained:

j	1	2	3	4	M_0	=	1000
m_j	139	91	52	40	m	=	322

Then $x/m = [91 + 2(52) + 3(40)]/322 = 0\cdot9783$, and interpolating in Table **A6** with $K = J - 1 = 3$, (6.11) can be solved to give the *three-monthly* estimate $\tilde{S} = 0\cdot644$ with asymptotic variance $0\cdot001\ 207$; the *annual* survival rate is estimated to be $\tilde{S}^4 = 0\cdot172$. Also the *annual* exploitation rate is estimated by $\tilde{P}_4 = 322/1000 = 0\cdot322$ with variance estimate $0\cdot000\ 218\ 3$. Assuming approximate normality for $\tilde{S}$ and $\tilde{P}_4$, then approximate 95 per cent confidence intervals for S and P_4 are therefore $(0\cdot58, 0\cdot71)$ and $(0\cdot292, 0\cdot352)$ respectively. Finally we note that

$$\tilde{\mu}_E = -\tilde{P}_4 \log \tilde{S}/(1-\tilde{S}^4)$$
$$= (0\cdot322)(0\cdot440)/(1-0\cdot172)$$
$$= 0\cdot171$$

is the three-monthly instantaneous exploitation rate (or an annual rate of $0\cdot684$). The large-sample variance of $\tilde{\mu}_E$ is estimated to be $0\cdot000\ 882\ 5$ and an approximate 95 per cent confidence interval for μ_E is $(0\cdot112, 0\cdot230)$.

6.2.2   Regression estimates

Using grouped observations and J equal time-intervals once again, we have, from (6.8), $E[m_j] = M_0 p_j$. Therefore, taking logarithms and using (6.9), we are led to consider the linear regression model (Beverton [1954])

$$\log m_j = \log (M_0 P_1) + Z - Zj + e_j.$$

If we could assume the e_j to be independently and identically distributed as $\mathcal{N}[0, \sigma^2]$ (Winsor and Clark [1940] give some support for this), then a straightforward confidence interval for the slope $-Z$ is available: in addition, since M_0 is known, P_1 (and hence μ_E) can be estimated. However, the m_j are approximately multinomially distributed (cf. (6.8)) and therefore correlated, so that Paulik [1963a] suggests the following analysis:

Consider the regression model (Paulik's model with the sign changed to make y_j positive)

$$E[y_j] \approx \beta_0 + \beta j, \tag{6.12}$$

where $y_j = \log (M_0/m_j)$, $\beta_0 = -(Z + \log P_1)$ and $\beta = Z$. Then, using the delta method, it can be shown that for large $M_0 p_i$ (greater than 10, say)

$$V[y_j] = V[\log m_j] \approx (1-p_j)/(M_0 p_j)$$

and

$$\text{cov}[y_j, y_k] \approx -1/M_0 \quad (j \neq k),$$

so that the variance–covariance matrix of the y_j is approximately $\sigma^2 \mathbf{B}$, where

$$\mathbf{B} = \frac{1}{M_0} \begin{bmatrix} p_1^{-1} - 1 & -1 & \cdots & -1 \\ -1 & p_2^{-1} - 1 & \cdots & -1 \\ \cdots & \cdots & \cdots & \cdots \\ -1 & -1 & \cdots & p_J^{-1} - 1 \end{bmatrix}$$

and $\sigma^2 = 1$. Using the method of **1.3.5** (2) with

$$\mathbf{B}^{-1} = \frac{M_0}{1 - \Sigma p_j} \begin{bmatrix} p_1(1-\Sigma p_j+p_1), & p_1 p_2 & , & \ldots, & p_1 p_J \\ p_2 p_1 & , & p_2(1-\Sigma p_j+p_2), & \ldots, & p_2 p_J \\ \ldots & , & \ldots & , & \ldots, \ldots \\ p_J p_1 & , & p_J p_2 & , & \ldots, & p_J(1-\Sigma p_j+p_J) \end{bmatrix}$$

$$\mathbf{y} = \begin{bmatrix} y_1 \\ y_2 \\ \ldots \\ y_J \end{bmatrix}, \quad \mathbf{X} = \begin{bmatrix} 1 & 1 \\ 1 & 2 \\ & \ldots \\ 1 & J \end{bmatrix}, \quad \boldsymbol{\beta} = \begin{bmatrix} \beta_0 \\ \beta \end{bmatrix},$$

σ^2 unknown, and p_j estimated by m_j/M_0 in $\mathbf{B}^{-1}$, it transpires that the least-squares estimates of Z and β_0 are given by

$$\hat{Z} = \hat{\beta} = \frac{\Sigma\, jy_j m_j - (\Sigma\, y_j m_j)(\Sigma\, jm_j)/m}{\Sigma\, j^2 m_j - (\Sigma\, jm_j)^2/m} = \frac{C}{D}, \quad \text{say},$$

and

$$\hat{\beta}_0 = (\hat{\beta}\, \Sigma\, jm_j - \Sigma\, y_j m_j)/m.$$

Variance estimates are:

$$\hat{V}[\hat{Z}] = \hat{\sigma}^2/D,$$

$$\hat{V}[\hat{\beta}_0] = \hat{\sigma}^2 \left[\frac{1}{m} + \frac{(\Sigma\, jm_j)^2}{m^2 D} \right],$$

and

$$\widehat{\text{cov}}\,[\hat{Z}, \hat{\beta}_0] = \hat{\sigma}^2 \Sigma\, jm_j/(mD),$$

where $(J-2)\hat{\sigma}^2 = \Sigma\, y_j^2 m_j - (\Sigma\, y_j m_j)^2/m - C^2/D$.

All the above summations are for $j = 1, 2, \ldots, J$; $\Sigma\, m_j = m$.

From a lemma of Chapman [1956] it is readily shown that for large M_0, $\mathbf{y}$ is approximately distributed as the multivariate normal. Hence an approximate $100(1-\alpha)$ per cent confidence interval for Z is

$$\hat{Z} \pm t_{J-2}[\alpha/2]\,\hat{\sigma}\, D^{-1/2}. \tag{6.13}$$

It transpires that this interval is functionally independent of M_0 and therefore does not depend on a correct knowledge of M_0 (though $\hat{\beta}_0$ does). This is what we would expect, as $\log M_0$ could have been included under β_0 in (6.12).

If estimates of S and P_1 are required we have

$$\hat{S} = e^{-\hat{Z}},$$

$$\hat{P}_1 = e^{-(\hat{Z}+\hat{\beta}_0)}$$

and, using the delta method,

$$V[\hat{S}] \approx S^2 V[\hat{Z}]$$

and

$$V[\hat{P}_1] \approx P_1^2(V[\hat{Z}] + 2 \text{ cov } [\hat{Z}, \hat{\beta}_0] + V[\hat{\beta}_0]),$$

which can be estimated from the formulae given above.

Estimates of μ and μ_E are given by (cf. (6.7))

$$\hat{\mu}_E = \frac{\hat{Z}\hat{P}_1}{1-\hat{S}} = \frac{\hat{Z}e^{-\hat{\beta}_0}}{e^{\hat{Z}} - 1}$$

and

$$\hat{\mu} = \hat{Z} - \hat{\mu}_E.$$

The asymptotic variances can be calculated in the usual manner using the delta method.

In comparing the above regression method with the maximum-likelihood approach of **6.2.1**, Paulik mentions that one objection to the regression method might be that for a small number of recovery periods J, the large t-value may lead to overwide confidence intervals. However, he feels that the wider regression interval is probably more realistic in practice, especiall as the sampling may not be truly random (as required for (6.8)). When some of the m_j are small or equal to zero, regression estimates cannot be used, whereas maximum-likelihood estimates can.

Example 6.7   Plaice: Paulik [1963a]

Using the data of Example 6.6, Table 6.10 was calculated. A plot of y_j versus j is roughly linear, thus suggesting that the above regression

TABLE 6.10

Application of Paulik's regression method to tagged plaice.

j	m_j	jm_j	j^2m_j	y_j	y_jm_j	jy_jm_j
1	139	139	139	1·9733	274·29	274·29
2	91	182	364	2·3969	218·12	436·24
3	52	156	468	2·9565	153·74	461·22
4	40	160	640	3·2189	128·76	515·04
Total	322	637	1611		774·91	1686·79

method can be used. Hence

$$C = 1686\cdot79 - (774\cdot91)(637)/322 = 153\cdot82,$$
$$D = 1611 - (637)^2/322 = 350\cdot85,$$
$$\hat{Z} = 153\cdot82/350\cdot85 = 0\cdot4384,$$
$$\hat{\beta}_0 = [637(0\cdot4384) - 774\cdot91]/322 = -1\cdot5393,$$
$$\hat{\sigma}^2 = 0\cdot397\ 86, \quad \hat{V}[\hat{Z}] = 0\cdot001\ 134,$$
$$\hat{V}[\hat{\beta}_0] = 0\cdot005\ 676 \text{ and } \widehat{\text{cov}}[\hat{Z}, \hat{\beta}_0] = 0\cdot002\ 243.$$

The 95 per cent confidence interval for Z, the three-monthly total instantaneous mortality rate, is $(0\cdot2935, 0\cdot5833)$. The survival and exploitation rates per three-month period are $\hat{S} = 0\cdot645$ and $\hat{P}_1 = 0\cdot1384$, and we note that these estimates are very close to the maximum-likelihood estimates obtained in Example 6.6. Finally, from the confidence interval for Z we obtain the approximate 95 per cent confidence interval $(0\cdot56, 0\cdot75)$ for $S = e^{-Z}$.

DISCONTINUITIES IN SAMPLING EFFORT. In exploited populations the sampling may not be completely continuous, as for example in some fisheries which close down for the weekend. However, in such a case if the unit of time is taken as a *working* week and Δ represents the length of the closed period (say the weekend), then (6.9) becomes

$$p_j = P_1 \exp\{-[\mu(1+\Delta) + \mu_E](j-1)\} \qquad (6.14)$$

and $\hat{\beta}$, the estimate of the slope of (6.12), will now be an estimate of $\mu(1+\Delta) + \mu_E$ instead of $Z = \mu + \mu_E$. Thus the calculation of $\hat{\beta}_0$, $\hat{\beta}$ and $\hat{P}_1$ ($= \exp(-\hat{\beta}_0 - \hat{\beta})$) will be the same as before, but $\hat{S} = \exp(-\hat{\mu} - \hat{\mu}_E)$ ($\neq \exp(-\hat{\beta})$) will be different. Least-squares estimates of μ and μ_E are now obtained by solving

$$\hat{\beta} = Z(1+\Delta) - \mu_E\Delta$$

and

$$\hat{P}_1 = \frac{\mu_E}{Z}(1 - e^{-Z})$$

for Z and μ_E; $\hat{S}$ can then be calculated.

If, however, the lengths of the closure periods are irregular but the actual sampling or working periods are still of equal length (say one time-unit) then

$$p_j = P_1 \exp\{-\mu_E(j-1) - \mu\gamma_{j-1}\},$$

where γ_{j-1} is the total number of time-units from time zero to the beginning of the jth sampling period. This leads to the regression model

$$E[y_j] = -\log P_1 + \mu_E(j-1) + \mu\gamma_{j-1}, \qquad (6.15)$$

and least-squares estimates of $-\log P_1$, μ and μ_E can be obtained once again, using the method of **1.3.5**(2). Paulik points out that this model is only useful in practice if there is sufficient variability in the lengths of the closure, so that γ_{j-1} is not approximately proportional to $j-1$. If this is not the case the matrix to be inverted for the least-squares estimation (namely $\mathbf{X'B^{-1}X}$ in the notation of **1.3.5**(2)) will be ill-conditioned.

6.3   SINGLE RELEASE: UNDERLYING ASSUMPTIONS

6.3.1   Initial mortality of tagged

In many fishery experiments a large proportion, $1 - \nu$ say, of the tagged fish die immediately after release through the traumatic experience of being handled and tagged (cf. Paulik [1963b]). For the models in **6.1**, using M_0 instead of νM_0 leads to the overestimation of N_0, though in **6.1.2** the estimation of the instantaneous recruitment rate (λ) is unaffected. However, in the continuous sampling models of **6.2** the estimates of Z and S do not depend on knowing M_0, so that the above type of error will not affect these estimates or their variance estimates.

The parameter ν can sometimes be estimated by retaining some of the tagged fish in tanks and noting the proportion $\hat{\nu}$ which survive an initial period. Alternatively, all the tagged fish can be retained for a short period after tagging to allow for any tagging mortality (North Atlantic Fish Marking Symposium [1963: 8]).

6.3.2 Non-reporting of tags

Another source of error which also has the effect of reducing M_0 is the non-reporting of tags by the fishermen. Suppose that ρ is the proportion of tags recaptured which are subsequently reported to the research agency by the fishermen, and assume that this proportion remains constant during the entire sampling period. Let $c = \nu\rho$ (= r in Paulik's notation), the effective proportion of M_0 utilised. Then, as already mentioned above, the estimates of Z and S for the models in **6.2** do not depend on knowing M_0 and are therefore not affected by a *constant* reporting rate of less than unity. However, as far as the other parameters are concerned, Paulik [1963a] shows that the procedures of **6.2** now estimate cP_J (or cP_1 in the regression models) instead of P_J and, since

$$\mu_E = P_J Z/(1 - e^{-ZJ}),$$

$c\mu_E$ instead of μ_E. For example, if $\nu = 0\cdot7$ and $\rho = 0\cdot7$ then $c \approx 0\cdot5$ and the true μ_E would be about twice the estimated μ_E. Unfortunately c cannot be estimated separately, except for the model (6.15) which now becomes

$$E[y_j] = -(\log c + \log P_1) + \mu_E(j-1) + \mu\gamma_{j-1}.$$

Since P_1 is a function of μ and μ_E only, we can now estimate the three parameters μ, μ_E and c.

Paulik suggests that for seasonal fisheries, an obvious way to separate c and μ_E is to take advantage of the change in the total mortality rate that occurs when the fishery season begins. This may be done by making one or more additional releases of tagged fish before fishing begins and using the methods of **6.4**.

By letting n_2 denote the catch from a particular sample or sampling period and n_1 and N denote the size of the marked and total population, respectively, just before the sample is taken, then ρ can be estimated for each sample using the theory of **3.2.4**. The method described there consists of dividing the catch from each sample into two parts; one part is inspected by trained observers ($\rho = 1$) and the other is inspected by the fishermen ($\rho \leqslant 1$). ρ is then estimated by the ratio of the marked fraction of the catch as reported by the fishermen to the marked fraction reported by the observers; a test for $\rho = 1$ can also be carried out. If the same proportion of the catch is inspected by trained observers on each occasion, then, since ρ is related to a binomial parameter p ($p = (1 + \rho\gamma)^{-1}$ in the notation of **3.2.4**), a contingen table can be set up to test for constant p (and hence for constant ρ).

6.3.3 Tag losses

Tags eventually become detached through wear and tear and the rate of tag loss can be estimated using, for example, double tagging (cf. **3.2.3**). Suppose that the two types of tag are denoted by A and B and let

$\pi_k(t)$ = probability that a tag of type k comes off by time t after the initial tag release $(k = A, B)$,

$\pi_{AB}(t)$ = probability that both tags come off by time t,

$m_k(t)$ = number caught at time t bearing tag k only,

$m_{AB}(t)$ = number caught at time t bearing both tags,

$m_0(t)$ = number caught at time t which have lost both tags,

and

$m(t) = m_A(t) + m_B(t) + m_{AB}(t) + m_0(t)$ (the number belonging to the release M_0 that are caught at time t).

Then assuming that the tags are independent of each other (i.e. $\pi_{AB}(t) = \pi_A(t)\pi_B(t)$), we can estimate $\pi_A(t)$, $\pi_B(t)$ and $m(t)$ using the moment estimates (cf. **3.2.3**(2)):

$$\hat{\pi}_A(t) = m_B(t)/[m_B(t) + m_{AB}(t)],$$

$$\hat{\pi}_B(t) = m_A(t)/[m_A(t) + m_{AB}(t)],$$

and

$$\hat{m}(t) = [m_A(t) + m_{AB}(t)][m_B(t) + m_{AB}(t)]/m_{AB}(t)$$

$$= c_t[m_A(t) + m_B(t) + m_{AB}(t)], \text{ say.}$$

This means that the observed recaptures given in the above square bracket must be corrected by a factor of

$$c_t = \left[1 - \frac{m_A(t)m_B(t)}{[m_A(t) + m_{AB}(t)][m_B(t) + m_{AB}(t)]} \right]^{-1}$$

to give an estimate of the actual number of recaptures $m(t)$. The estimates $\hat{\pi}_A(t)$, $\hat{\pi}_B(t)$ and $\hat{m}(t)$ (or suitable transformations) can also be plotted against t; in the case of continuous sampling one would normally take t as the midpoint of the sampling interval.

If the loss of tags follows a Poisson process with parameter L_k, say, for type k tag, then (cf. (1.1))

$$1 - \pi_k(t) = e^{-L_k t} \tag{6.16}$$

and the plot of $\log (1 - \hat{\pi}_k(t))$ against t should be approximately linear. This type of model has been used, for example, by Beverton and Holt [1957: 204], Gulland [1963] and Backiel [1964].

If single tagging is used (say tag A) and (6.16) holds, then

$$observed\ recaptures = m(t)\,e^{-L_A t}.$$

Hence, for this case, if y_j is calculated from observed recaptures then the only change in the regression model (6.12) is that $\beta = Z + L_A$.

TAGS INDISTINGUISHABLE. In some situations the only information recorded is the number of tags for each tagged individual, so that just the numbers $m_{AB}(t)$ and $m_C(t)$ $(= m_A(t) + m_B(t))$ are available. For this case we can still estimate $m(t)$ if we can assume that the tags are independent and that $\pi_A(t) = \pi_B(t)$ $(= \pi(t)$ say). Then from **3.2.3** (2), $m(t)$ and $\pi(t)$ are estimated by

$$\widetilde{m}(t) = [m_C(t) + 2m_{AB}(t)]^2 / 4m_{AB}(t)$$

and

$$\widetilde{\pi}(t) = m_C(t)/[m_C(t) + 2m_{AB}(t)],$$

CONSTANT LOSS RATE. If one of the tags or marks is permanent, then Robson and Regier [1966] give the following method for estimating rate of tag retention for the other tag.

Suppose that tag A is permanent, so that $\pi_A(t) = \pi_{AB}(t) = 0$, $m_B(t) = m_0(t) = 0$ and $m(t) = m_A(t) + m_{AB}(t)$. Assuming that r, the probability of retaining a type B tag for unit time, is constant, then

$$1 - \pi_B(t) = r^t$$

and $m_{AB}(t)$ is a binomial variable with parameters $m(t)$ and r^t. Hence, if a series of instantaneous samples is taken at times t_j $(j = 1, 2, \ldots, J)$ and we define $m_j = m(t_j)$ and $b_j = m_{AB}(t_j)$, then the joint distribution of the $\{b_j\}$ given the $\{m_j\}$ is

$$f(\{b_j\} | \{m_j\}) = \prod_{j=1}^{J} \binom{m_j}{b_j} (r^{t_j})^{b_j} (1 - r^{t_j})^{m_j - b_j}. \qquad (6.17)$$

The maximum-likelihood estimate $\hat{r}$ of r is readily shown to be the solution of

$$\sum_{j=1}^{J} \left\{ \frac{(b_j - m_j r^{t_j}) t_j}{(1 - r^{t_j})} \right\} = 0 \qquad (6.18)$$

which must be solved iteratively. If $\hat{r}$ is close to unity, Robson and Regier suggest that an approximate solution r_0 is given by

$$\log(1 - r_0) = \frac{\sum m_j t_j \log\left(\dfrac{m_j - b_j}{m_j t_j}\right)}{\sum m_j t_j}, \qquad (6.19)$$

which may be used as a trial value in the iterative process. The successive approximations r_i are then given by

$$r_{i+1} = r_i \left\{ 1 + \frac{\sum_j (b_j - m_j r_i^{t_j}) t_j/(1 - r_i^{t_j})}{\sum_j m_j t_j^2 r_i^{t_j}/(1 - r_i^{t_j})} \right\}. \qquad (6.20)$$

From standard maximum-likelihood theory, $\hat{r}$ is asymptotically normal with mean r and variance

$$V[\hat{r}] = r^2 / \sum_j \left(\frac{m_j t_j^2 r^{t_j}}{1 - r^{t_j}} \right). \qquad (6.21)$$

When r is close to unity and t_j is small, as when t is measured in years, then

$$V[\hat{r}] \approx \frac{r(1-r)}{\Sigma\, m_j t_j}.$$

Robson and Regier infer from this that the estimator $\hat{r}$ behaves approximately as the proportion of tag retentions occurring in a single sample of size $\Sigma\, m_j t_j$ taken one year after release. That is, the sample of size m_j taken t_j years after release contributes as much information as a sample of size $m_j t_j$ taken one year after release.

Since the b_j are binomial variables, a goodness-of-fit statistic for the validity of the model (6.17) is given by

$$T = \sum_{j=1,}^{J} \left\{ \frac{(b_j - m_j \hat{r}^{t_j})^2}{m_j \hat{r}^{t_j}(1 - \hat{r}^{t_j})} \right\}, \tag{6.22}$$

where T is approximately distributed as χ^2_{J-1}.

For the special but not atypical case of only two samples taken $t_1 = 1$ and $t_2 = 2$ years after the tag—release, an explicit solution to equation (6.18) is available in the form

$$\hat{r} = \frac{b_1 - m_1}{2(m_1 + 2m_2)} + \sqrt{\left[\frac{b_1 - m_1}{2(m_1 + 2m_2)}\right]^2 + \frac{b_1 + 2b_2}{m_1 + 2m_2}}$$

In conclusion we note that if (6.16) holds, then $r = \exp(-L_B)$. Cucin and Regier [1965: 255] used this relationship to correct recaptures for tag losses in an application of the Petersen method.

Example 6.8 Lake whitefish (*Coregonus clupeaformis*): Robson and Regier [1966]

The above method of estimation was applied to data (Table 6.11) from a study of lake whitefish reported by Cucin and Regier (unpublished). Type B

TABLE 6.11

Recoveries of fish that were both tagged and fin-clipped at time of release: from Robson and Regier [1966: Table 1].

Year of release	Year of capture					
	1962		1963		1964	
	Tag retained	Tag lost	Tag retained	Tag lost	Tag retained	Tag lost
1961	79	4	20	0	3	1
1962	–	–	66	1	5	1

tag consisted of a plastic streamer type attached through the anterior base of the dorsal fin by means of monofilament nylon. Fish tagged in 1961 had their left pectoral fins removed by clipping, while those tagged in 1962 had their left pelvics removed. A biologist carefully examined over 10 000 whitefish taken in gill nets and pound nets from 1961 to 1964 and noted the number of clipped fish that had lost tags. A number of lines of evidence, all indirect, indicated that tagged fish were not appreciably more vulnerable to the gear than untagged fish. From previous study it was known that very few, if any, of the fish in this population were missing fins due to natural causes.

The fishing season was 7 months long each year, and fish were examined throughout the season. For purposes of this example the authors treated the data as though all sampling within a year was considered to have been instantaneous and on anniversaries of the mean tagging date. For greater accuracy and with sufficiently large samples, the recapture data could be summarised by months.

According to the binomial model used for the distribution of b_j (cf. (6.17)), the fractions $79/(79+4)$ and $66/(66+1)$ are both estimates of r, the annual tag retention rate; the fractions $20/(20+0)$ and $5/(5+1)$ are both estimates of r^2; and $3/(3+1)$ is an estimate of r^3. A test of the model (6.17) for the two releases can be carried out in two stages: first, testing the homogeneity of the two estimates of r and the two estimates of r^2 in separate 2-by-2 contingency tables, and, second, testing the goodness of fit of the pooled estimates of 1-, 2-, and 3-year retention rates to the predicted values of $\hat{r}$, $\hat{r}^2$ and $\hat{r}^3$ respectively, using T (6.22).

The homogeneity of the two estimates of r was tested in Table 6.12, giving a chi-square value of $X_1^2 = 1.27$ $(P > 0.20)$. Homogeneity of the two

TABLE 6.12

A test of homogeneity for the two estimates of r: from Robson and Regier [1966].

	1961–62	1962–63	Total
Tag retained	79	66	145
Tag lost	4	1	5
Total	83	67	150

TABLE 6.13

A test of homogeneity for the two estimates of r^2: from Robson and Regier [1966].

	1961–63	1962–64	Total
Tag retained	20	5	25
Tag lost	0	1	1
Total	20	6	26

estimates of r^2 was tested in Table 6.13. Here the numbers are too small for the use of the chi-square approximation, but the exact probability is readily computed in this case. Given that one tag loss occurred among the 26 fish, the probability that it occurred among the 6 fish of the 1962–64 sample rather than among the 20 fish of the 1961–63 sample is simply 6/26 or $0 \cdot 23$; hence once again $P > 0 \cdot 20$. For later reference we note from the chi-square table that $P = 0 \cdot 23$ corresponds to a χ_1^2 value of approximately $1 \cdot 47$.

Since homogeneity was not rejected at this stage of the test, $\hat{r}$ can be calculated using the pooled data in the form:

	$t_1 = 1$	$t_2 = 2$	$t_3 = 3$
m_j	150	26	4
b_j	145	25	3

Clearly $\hat{r}$ must be near to unity, so that a first approximation r_0 can be calculated from (6.19). Thus

$$\log (1-r_0) = \frac{150(1) \log \left(\dfrac{150-145}{150(1)} \right)}{150(1) + 26(2) + 4(3)}$$

$$+ \frac{26(2) \log \left(\dfrac{26-25}{26(2)} \right) + 4(3) \log \left(\dfrac{4-3}{4(3)} \right)}{150(1) + 26(2) + 4(3)}$$

$$= -1 \cdot 512,$$

or $r_0 = 0 \cdot 969$. Substituting this value into (6.18) we get

$$\frac{[145 - 150(0 \cdot 969)](1)}{1 - 0 \cdot 969} + \frac{[25 - 26(0 \cdot 969)^2](2)}{1 - (0 \cdot 969)^2} + \frac{[3 - 4(0 \cdot 969)^3](3)}{1 - (0 \cdot 969)^3}$$

$$= -11 \cdot 2903(1) + 9 \cdot 6170(2) - 7 \cdot 0930(3),$$

which adds to $-13 \cdot 335$ instead of 0. The value $0 \cdot 969$ is therefore slightly larger than $\hat{r}$, and applying the iteration formula (6.20) we divide $-13 \cdot 335$ by

$$\sum_j \frac{m_j t_j^2 r_0}{(1 - r_0^{t_j})} = \frac{150(1)^2(0 \cdot 969)}{1 - 0 \cdot 969} + \frac{26(2)^2(0 \cdot 969)^2}{1 - (0 \cdot 969)^2} + \frac{4(3)^2(0 \cdot 969)^3}{1 - (0 \cdot 969)^3}$$

$$= 6651 \cdot 86$$

to give

$$r_1 = 0 \cdot 969 \left[1 - \frac{13 \cdot 335}{6651 \cdot 86} \right] = 0 \cdot 967 .$$

Substituting this value into (6.18) we get $+2 \cdot 6374$ instead of 0, indicating that $0 \cdot 967$ is slightly smaller than $\hat{r}$. Going through the iteration once again using r_1 as the trial value leads to

$$r_2 = 0 \cdot 967 \left[1 + \frac{2 \cdot 6374}{6233 \cdot 54} \right] = 0 \cdot 9674 ,$$

so that to three decimal places, $\hat{r} = 0 \cdot 967$.

The estimated standard deviation of $\hat{r}$ calculated from (6.21) is

$$\hat{\sigma}[\hat{r}] = 0 \cdot 967 / \sqrt{6233 \cdot 54} = 0 \cdot 0122 ,$$

so that an approximate 95 per cent confidence interval for r is $\hat{r} \pm 1.96(0.0122)$ or $(0.943, 0.991)$.

From (6.22)

$$T = \frac{[145 - 150(0.967)]^2}{150(0.967)(1 - 0.967)} + \frac{[25 - 26(0.967)^2]^2}{26(0.967)^2[1 - (0.967)^2]} + \frac{[3 - 4(0.967)^3]^2}{4(0.967)^3[1 - (0.967)^3]}$$

$$= 1.399 ,$$

which, for two degrees of freedom, is non-significant $(P > 0.70)$. We can now pool all the tests of significance by summing the three independent chi-square values which we have computed, namely

$$1.27 + 1.47 + 1.40 = 4.14$$

with $1 + 1 + 2 = 4$ degrees of freedom. This overall test $(P \approx 0.40)$ is also non-significant, so that we have found no grounds for rejecting the underlying model.

6.3.4   Exponential distribution of recapture times

From (6.3) we find that the density function for the recovery time of a tagged individual, given that it is caught in the time-interval $[0, \tau]$, is

$$f_2(t) = \frac{Z e^{-Zt}}{1 - e^{-Z\tau}}, \quad 0 \leqslant t \leqslant \tau.$$

Therefore, given a random sample of m recapture times from this truncated exponential distribution, one can carry out a standard chi-squared goodness-of-fit test based on comparing observed and expected frequencies. For example, if the interval $[0, \tau]$ is split up into J intervals $[d_0, d_1), [d_1, d_2), \ldots, [d_{J-1}, d_J)$ (where $d_0 = 0$, $d_J = \tau$) and m_j tagged are caught in the jth interval, then the goodness-of-fit statistic is

$$\sum_{j=1}^{J} (m_j - E_j)^2/E_j ,$$

where $\quad E_j = m \left\{ \dfrac{\exp(-\hat{Z}d_{j-1}) - \exp(-\hat{Z}d_j)}{1 - \exp(-\hat{Z}\tau)} \right\}$

and $\hat{Z}$ is a suitable estimate of Z. When $\hat{Z}$ is the solution of (6.5) then, for large m, the above goodness-of-fit statistic is approximately distributed as χ^2_{J-1} (one degree of freedom is not subtracted for the estimation of Z as $\hat{Z}$ is calculated from the ungrouped data: Kendall and Stuart [1973: 447]). We note that if τ is large, so that the effect of truncation on $f_2(t)$ above is negligible, then a wide variety of tests for the negative exponential distribution are available (see p. 46).

When μ is zero, or negligible compared to μ_E, the recovery process of tagged individuals is Poisson with parameter μ_E. This means that the distribution of recapture times is negative exponential and, as Paulik [1963a] has pointed out, the theory of life-testing can be applied here. For example, we start with M_0 tagged individuals, and an individual is said to have "failed" in time $[0, \tau]$ if it is recaptured. Hence all the techniques associated with the exponential distribution and life-testing given in Epstein and Sobel [1953, 1954] and Epstein [1954, 1960a, b, c, d] (see also Buckland [1964] for a general bibliography) can be used here.

6.4   MULTI-RELEASE MODELS: CONTINUOUS SAMPLING

6.4.1   Notation and assumptions

In **6.3.2** it was mentioned that one way of separating μ_E and c is to take advantage of the change in the total mortality rate that occurs when the season begins. This can be accomplished by making one or more tag releases before the fishing begins. Let

I = number of pre-season releases,

R_i = number of tagged fish released τ_i time-units before the opening date ($i = 1, 2, \ldots, I$),

m_{ij} = number of tagged fish from the ith release of R_i individuals caught in the jth sampling period ($j = 1, 2, \ldots, J$),

$m_{i.} = \sum_i m_{ij}$,

$m_{..} = \sum_i \sum_j m_{ij}$,

and

$$x_. = \sum_i \sum_j (j-1)m_{ij} = \sum_j (j-1)m_{.j}$$

We shall assume that

(1) The instantaneous natural mortality rate and the instantaneous exploitation rate are constant and equal to μ and μ_E, respectively; we define $Z = \mu + \mu_E$.

(2) The pre-season instantaneous natural mortality rate is the same as the during-season rate μ.

(3) Immigration and emigration are negligible.

(4) All tagged fish have the same probability of being caught in a particular sampling period (given that they are alive at the beginning of the period), irrespective of which tag release they are from.

(5) Sampling periods are each of one time-unit duration as in **6.2**.

(6) Fish do not lose their tags.

(7) A constant proportion ν of each tag release survives the tagging process. It is assumed further that the tagging mortality is complete within a short time after tagging.

(8) The probability ρ that a recaptured tag is reported remains constant during the entire season for all types of recapture gear used.

In the following models it is convenient to use the parameters

$$c = \nu\rho, \quad S = e^{-Z}, \quad \phi = e^{-\mu}, \quad \text{and} \quad \beta = cP_J,$$

where P_J is given by (6.7); β is the probability that a tagged individual survives any initial tagging mortality, is recaptured, and is reported on capture.

6.4.2 Maximum-likelihood estimates

1 General theory

Working with the parameters S, ϕ and β instead of μ, μ_E and c, Paulik [1963a] showed that the joint distribution of the $\{m_{ij}\}$ is proportional to (cf. (6.8))

$$\prod_{i=1}^{I}\left\{\left[\frac{\beta\phi^{T_i}}{\sum_{j=0}^{J-1}S^j}\right]^{m_{i1}}\left[\frac{\beta\phi^{T_i}S}{\sum_{j=0}^{J-1}S^j}\right]^{m_{i2}}\cdots\left[\frac{\beta\phi^{T_i}S^{J-1}}{\sum_{j=0}^{J-1}S^j}\right]^{m_{iJ}}(1-\beta\phi^{T_i})^{R_i-m_{i.}}\right\}$$

and the maximum-likelihood estimates are the solutions of

$$\left(\sum_{j=0}^{J-1}j\hat{S}^j\right)\left(\sum_{j=0}^{J-1}\hat{S}^j\right) = x_./m_{..}, \tag{6.23}$$

$$\sum_{i=1}^{I}\left\{\frac{T_i(m_{i.}-R_i\hat{A}_i)}{1-\hat{A}_i}\right\} = 0, \tag{6.24}$$

and

$$\sum_{i=1}^{I}\left\{\frac{(m_{i.}-R_i\hat{A}_i)}{1-\hat{A}_i}\right\} = 0, \tag{6.25}$$

where $\hat{A}_i = \hat{\beta}\hat{\phi}^{T_i}$. Equation (6.23) can be solved for $\hat{S}$ using the method described in **A6**, while (6.24) and (6.25) must be solved iteratively for $\hat{\phi}$ and $\hat{\beta}$, using, for example, the methods of **1.3.8**.

The estimate $\hat{S}$, which is uncorrelated with $\hat{\phi}$ and $\hat{\beta}$, has asymptotic variance

$$V[\hat{S}] = \left\{\left[\sum_i R_i A_i\right]\left[\frac{1}{S(1-S)^2} - \frac{J^2 S^{J-2}}{(1-S^J)^2}\right]\right\}^{-1}, \tag{6.26}$$

while the asymptotic variance–covariance matrix for $\hat{\phi}$ and $\hat{\beta}$ is $\mathbf{V}^{-1}$, where (cf. (1.7))

$$\mathbf{V} = \begin{bmatrix} \sum_i\left\{\dfrac{T_i^2 R_i A_i}{\phi^2(1-A_i)}\right\}, & -\sum_i\left\{\dfrac{T_i R_i \phi^{T_i}}{\phi(1-A_i)}\right\} \\ \cdots & , & \sum_i\left\{\dfrac{R_i \phi^{T_i}}{\beta(1-A_i)}\right\} \end{bmatrix}.$$

(The (1, 1) element in $\mathbf{V}^{-1}$ is $V[\hat{\phi}]$).

The maximum-likelihood estimates of μ, μ_E and c are given by

$$\hat{\mu} = -\log\hat{\phi},$$
$$\hat{\mu}_E = -\log\hat{S}-\hat{\mu},$$

and, writing

$$\hat{P}_J = -\hat{\mu}_E(1-\hat{S}^J)/\log\hat{S},$$
$$\hat{c} = \hat{\beta}/\hat{P}_J.$$

Approximate variances of these estimates can be calculated using the delta method and V^{-1}. Thus

$$V[\hat{\mu}] \approx \frac{1}{\phi^2} V[\hat{\phi}]$$

and

$$V[\hat{\mu}_E] \approx \frac{1}{S^2} V[\hat{S}] + \frac{1}{\phi^2} V[\hat{\phi}].$$

DISCONTINUITIES IN SAMPLING EFFORT. The above theory can be readily modified, as on p. 279, to deal with the case when the sampling is not completely continuous: for example, some fisheries close down for the weekend. Suppose that the unit of time is now taken as the *working* week and let Δ be the length of the closed period (assumed to be constant from week to week). Then $\hat{S}$, $\hat{\phi}$, $\hat{\beta}$, $\hat{\mu} = -\log \hat{\phi}$, and their variances, are calculated as above, but now S and β are different functions of μ, μ_E and c. For example, S, the probability of surviving for one calendar week, is given by

$$S = \exp(-\mu_E - (1+\Delta)\mu).$$

Now P_1, the probability of capture in a *working* week, is still given by (cf. (6.6))

$$P_1 = \mu_E(1-e^{-Z})/Z$$

and the probability that a tagged individual is recaptured in the jth week of the season *given* that it is alive at the beginning of the season is

$$p_j = P_1 S^{j-1}.$$

Therefore P_J, the probability that a tagged individual is recaptured during the season given that it is alive at the beginning of the season, is given by

$$P_J = \sum_{j=1}^{J} p_j$$
$$= P_1(1-S^J)/(1-S).$$

Hence from $\hat{S}$, $\hat{\mu}$ and $\hat{\beta}$ we obtain

$$\hat{\mu}_E = -\log\hat{S} - (1+\Delta)\hat{\mu}, \tag{6.27}$$
$$\hat{Z} = \hat{\mu} + \hat{\mu}_E,$$

$$\hat{P}_J = \frac{\hat{\mu}_E}{\hat{Z}} \cdot (1-e^{-\hat{Z}}) \cdot \frac{(1-\hat{S}^J)}{(1-\hat{S})} \tag{6.28}$$

and finally

$$\hat{c} = \hat{\beta}/\hat{P}_J. \tag{6.29}$$

Also, from the delta method,

$$V[\hat{\mu}_E] \approx \frac{1}{S^2} V[S] + \left(\frac{1+\Delta}{\phi}\right)^2 V[\hat{\phi}]. \tag{6.30}$$

2   One release before the season

ONE RELEASE AT START OF SEASON. For the simple case of just two releases, when the first release is made some time (τ_1) before the season begins and the second immediately preceding the season ($T_2 = 0$), equations (6.24) and (6.25) have solutions (Paulik [1962])

$$\hat{\phi}^{T_1} = \left(\frac{m_{1.}}{R_1}\right) \Big/ \left(\frac{m_{2.}}{R_2}\right) \quad (= \hat{a} \text{ say})$$

and

$$\hat{\beta} = m_{2.}/R_2.$$

Defining $a = \phi^{T_1} = e^{-\mu T_1}$, the probability of survival up to the beginning of the season, it is readily seen that $m_{1.}$ and $m_{2.}$ have independent binomial distributions with parameters $(R_1, a\beta)$ and (R_2, β) respectively. Hence, using the delta method

$$V[\hat{a}] \approx a^2 \left[\frac{1-a\beta}{R_1 a\beta} + \frac{1-\beta}{R_2\beta}\right], \tag{6.31}$$

$$V[\hat{\beta}] = \beta(1-\beta)/R_2, \tag{6.32}$$

$$V[\hat{\mu}] \approx \frac{1}{(T_1 a)^2} V[\hat{a}], \tag{6.33}$$

and

$$V[\hat{\mu}_E] \approx \frac{1}{S^2} V[\hat{S}] + (1+\Delta)^2 V[\hat{\mu}]. \tag{6.34}$$

By setting $a = a_1$ and $\beta = \beta_2$ we see that the above model is a special case of Ricker's two-release method in **5.1.3**(4). Hence an approximately unbiased estimate of a is

$$\tilde{a} = \frac{m_{1.}(R_2+1)}{R_1(m_{2.}+1)},$$

and

$$v[\tilde{a}] = \tilde{a}^2 - \frac{m_{1.}(m_{1.}-1)(R_2+1)(R_2+2)}{R_1(R_1-1)(m_{2.}+1)(m_{2.}+2)}$$

is an approximately unbiased estimate of $V[\tilde{a}]$.

Example 6.9   Pink salmon: Paulik [1962]

Paulik applied the above two-release method to some data reported by Elling and Macy [1955]; the m_{ij} for the three sampling periods ($J = 3$) are given in Table 6.14. Here the basic unit of time is the number of days (5.5)

TABLE 6.14

Tag recoveries, m_{ij}, from two releases where the second release is made at the beginning of the season: from Paulik [1962: Table 1].

| | Recoveries in jth period | | | |
	1	2	3	$m_{i.}$
$R_1 = 1195$	304	63	12	379
$R_2 = 1083$	566	86	26	678
$m_{.j}$	870	149	38	1057

fished during 1 week, and as the first tag release was made 6 days before the fishing began, $\tau_1 = 6/5\cdot5 = 1\cdot0909$. From Table 6.14,

$$x_./m_{..} = [149 + 2(38)]/1057 = 0\cdot2129$$

and, from **A6** with $K = 2$, (6.23) has solution $\hat{S} = 0\cdot1825$. We also have

$$\hat{\beta} = 678/1083 = 0\cdot6260$$

$$\hat{\alpha} = \frac{379}{1195} \cdot \frac{1083}{678} = 0\cdot5066$$

and

$$\hat{\mu} = -(1/\tau_1) \log(\hat{\phi}^{\tau_1}) = -(1/\tau_1) \log \hat{\alpha} = 0\cdot6234.$$

The period of closure is $\Delta = 1\cdot5/5\cdot5 = 0\cdot2727$, so that from (6.27), (6.28) and (6.29),

$$\hat{\mu}_E = 1\cdot7010 - 0\cdot7934 = 0\cdot9076,$$

$$\hat{P}_J = \frac{0\cdot9076}{1\cdot5310} (1 - e^{-1\cdot5310}) \frac{(1 - 0\cdot1825^3)}{(1 - 0\cdot1825)}$$

$$= 0\cdot5648,$$

and

$$\hat{c} = 0\cdot6260/0\cdot5648 = 1\cdot11.$$

To find the variance estimates we note first of all that

$$\sum_i R_i \hat{A}_i = R_1 \hat{\alpha} \hat{\beta} + R_2 \hat{\beta} = m_{1.} + m_{2.} = m_{..}. \tag{6.35}$$

Hence, replacing parameters by estimates, we have from (6.26), (6.32) and (6.31),

$$\hat{V}[\hat{S}] = 0\cdot000\ 144\ 7,$$

$$\hat{V}[\hat{\beta}] = \hat{\beta}(1 - \hat{\beta})/R_2 = 0\cdot000\ 216\ 2,$$

and

$$\hat{V}[\hat{\alpha}] = \hat{\alpha}^2 \left[\frac{1}{m_{1.}} + \frac{1}{m_{2.}} - \frac{1}{R_1} - \frac{1}{R_2} \right]$$

$$= 0\cdot000\ 604\ 0$$

Therefore, from (6.33),

$$\hat{V}[\hat{\mu}] = \frac{1}{(\tau_1 \hat{\alpha})^2} \hat{V}[\hat{\alpha}] = 0\cdot001\ 978$$

and, from (6.34),

$$\hat{V}[\hat{\mu}_E] = \frac{1}{\hat{S}^2} \hat{V}[\hat{S}] + (1 + \Delta)^2 \hat{V}[\hat{\mu}]$$

$$= 0\cdot007\ 549.$$

The approximate 95 per cent confidence intervals, *estimate* $\pm 1\cdot96$ (*standard deviation*), for S, μ and μ_E are $(0\cdot159, 0\cdot206)$, $(0\cdot536, 0\cdot711)$ and $(0\cdot737, 1\cdot078)$, respectively. (In the original example Paulik neglected the effect of the closed period Δ in estimating μ_E and c.)

ONE RELEASE DURING SEASON. Paulik [1962] points out that the above method can still be used if the second release is made some time during the season instead of at the beginning of the season. In this case the recaptures from the second release are recorded according to the sampling period numbered from the time of this release (the coded period) rather than from the beginning of the season, and the data from the first release are truncated, so that the number of recording periods is the same for each release. For example, suppose that the second release in the previous example had been made at the beginning of the second week of fishing. Then the number of recovery periods is now 2 and the recoveries m_{ij} are arranged as in Table 6.15. The analysis is the same as above, the only difference being that $J = 2$ instead of 3.

TABLE 6.15

Tag recoveries, m_{ij}, from two releases where the second release is made during the season: from Paulik [1962].

	Recoveries in jth coded period		$m_{i.}$
	1	2	
$R_1 = 1195$	304 (first week)	63 (second week)	367
$R_2 = 1083$	566 (second week)	86 (third week)	652
$m_{.j}$	870	149	1019

SEVERAL RELEASES DURING SEASON. If several releases are made during the season, then the recoveries from these releases can be grouped according to the coded recovery period. Paulik [1962] points out that if the same number J of coded recovery periods are used for each release then the grouped data can be treated as a single release, and the problem reduces to the above case of just one release during the season.

6.4.3 Regression estimates

Returning to the problem of **6.4.1**, where all releases are made before the fishing season begins, we see that p_{ij}, the probability that a tagged fish from the ith release is recaptured and reported in the jth sampling interval, is given by

$$p_{ij} = cP_1 \exp\{-Z(j-1) - \mu\tau_i\}.$$

Now $E[m_{ij}] = R_i p_{ij}$, and putting $y_{ij} = \log(R_i/m_{ij})$ we are led to consider the regression model

$$E[y_{ij}] \approx -(\log c + \log P_1 + Z) + \mu\tau_i + Zj. \tag{6.36}$$

This model is an extension of (6.12) and can be analysed the same way.

Thus, by writing

$$\mathbf{Y}' = (y_{11}, y_{12}, \ldots, y_{1J}, \ldots, y_{I1}, y_{I2}, \ldots, y_{IJ})$$

$$\mathbf{X} = \begin{bmatrix} 1 & \tau_1 & 1 \\ 1 & \tau_1 & 2 \\ \cdots & \cdots & \cdots \\ 1 & \tau_1 & J \\ 1 & \tau_2 & 1 \\ 1 & \tau_2 & 2 \\ \cdots & \cdots & \cdots \\ 1 & \tau_2 & J \\ \cdots & \cdots & \cdots \\ 1 & \tau_I & 1 \\ 1 & \tau_I & 2 \\ \cdots & \cdots & \cdots \\ 1 & \tau_I & J \end{bmatrix}, \qquad \beta = \begin{bmatrix} -(\log c + \log P_1 + Z) \\ \mu \\ Z \end{bmatrix}$$

and given $\sigma^2 \mathbf{B}$, the variance–covariance matrix of $\mathbf{Y}$, the least-squares estimate of β can be found using the method of **1.3.5**(2). Paulik [1963a] shows that

$$\mathbf{B}^{-1} = \begin{bmatrix} \mathbf{B}_1^{-1} & \mathbf{0} & \cdots & \mathbf{0} \\ \mathbf{0} & \mathbf{B}_2^{-1} & \cdots & \mathbf{0} \\ \cdots & \cdots & \cdots & \cdots \\ \mathbf{0} & \cdots & \cdots & \mathbf{B}_I^{-1} \end{bmatrix}, \qquad (6.37)$$

where $B_i^{-1} =$

$$\frac{R_i}{1-\sum_j p_{ij}} \begin{bmatrix} p_{i1}(1-\sum_j p_{ij}+p_{i1}), & -1 & , & \cdots, & -1 \\ -1 & , & p_{i2}(1-\sum_j p_{ij}+p_{i2}), & \cdots, & -1 \\ \cdots & , & \cdots & , & \cdots, & \cdots \\ -1 & , & -1 & , & \cdots, & p_{iJ}(1-\sum_j p_{ij}+p_{iJ}) \end{bmatrix}$$

$$(6.38)$$

and p_{ij} is estimated by m_{ij}/R_i. A qualitative check on the linearity of the regression model is given by plotting y_{ij} against j for each i.

Explicit least-squares estimates of $\log(cP_1)$, μ and μ_E, together with the estimate of their variance–covariance matrix, are given by Paulik in the appendix to his paper. However, it should be noted that he uses the model

$$E[\log(m_{ij}/R_i)] = \log c + \log P_1 + Z - \mu\tau_i - Zj$$

or, in terms of his notation,

$$E[\log(n_{ij}/N_i)] = \log r + \log\left\{\frac{F}{F+X}(1-e^{-(F+X)}\right\} + F + X - X\rho_i - (F+X)j.$$

Example 6.10   Pink salmon: Paulik [1963a, b]

In 1950 Elling and Macy [1955] conducted a tagging programme in the northern part of south-eastern Alaska, when daily releases of tagged pink salmon were made for $I = 12$ consecutive days immediately preceding the opening of the commercial fishery. During the fishery, recaptures m_{ij} from each of the releases were recorded for three consecutive weeks $(J = 3)$, and the data are set out in Table 6.16. In this experiment the fishery was closed

TABLE 6.16

Weekly recoveries of tagged pink salmon in the commercial fishery for twelve pre-season releases: from Paulik [1963a: Table 1].

i	$5.5\tau_i$	R_i	m_{i1}	m_{i2}	m_{i3}	i	$5.5\tau_i$	R_i	m_{i1}	m_{i2}	m_{i3}
1	12	784	96	5	2	7	6	351	93	19	5
2	11	574	88	15	5	8	5	1 509	475	96	24
3	10	862	137	29	6	9	4	1 003	352	81	20
4	9	1 097	219	33	6	10	3	1 938	705	155	30
5	8	1 146	305	51	17	11	2	1 661	783	117	34
6	7	1 195	304	63	12	12	1	1 083	566	86	26
	$m_{..} = 5\ 060$					Total		13 203	4 123	750	187

for $1\frac{1}{2}$ days in each week, so that a working week of 5.5 days was taken as the unit of time. However, apart from scaling the values of τ_i, the effect of the closures was otherwise ignored in the application of the maximum-likelihood and regression methods (though this point is raised in Paulik [1963b: 235]).

Using the regression method, Paulik shows that the least-squares estimate of

$$\begin{bmatrix} \log(cP_1) \\ \mu \\ \mu_E \end{bmatrix} \quad \text{is} \quad \begin{bmatrix} -0.6868 \\ 0.6410 \\ 0.9660 \end{bmatrix}$$

with variance—covariance matrix estimated by

$$\begin{bmatrix} 0.000\ 233\ 0 & 0.000\ 481\ 5 & -0.000\ 094\ 4 \\ \cdots & 0.000\ 927\ 2 & -0.000\ 933\ 3 \\ \cdots & \cdots & 0.002\ 706\ 5 \end{bmatrix}.$$

Assuming asymptotic normality, approximate 95 per cent confidence intervals for μ and μ_E are (0.5814, 0.7007) and (0.8641, 1.0681) respectively.

Estimates of S, β and c are obtained from the relations

$$S = \exp(-\mu - \mu_E),$$

$$\beta = cP_J,$$
$$= (cP_1)(1-S^J)/(1-S),$$

and

$$c = \beta/P_J.$$

Thus $\hat{S} = 0{\cdot}2005$, $\hat{\beta} = 0{\cdot}6243$ and $\hat{c} = 1{\cdot}046$. Using the delta method, an approximate 95 per cent confidence interval for c is $(0{\cdot}957, 1{\cdot}137)$.

The above estimates may be compared with the corresponding maximum-likelihood estimates (cf. **6.4.2**)

$$\hat{S} = 0{\cdot}1970, \quad \hat{\beta} = 0{\cdot}6230 \quad \text{and} \quad \hat{c} = 1{\cdot}0515.$$

The closeness of the two sets of estimates provides quantitative support for the validity of the assumptions underlying the two models. Graphical evidence for the validity of the regression model is given in Paulik [1963b].

In conclusion we note that the hypothesis $c = 1$ is not rejected by the above data as the confidence interval for c contains unity.

DISCONTINUITIES IN SAMPLING EFFORT. If the unit of time is the working week and the length of the closed period is constant (Δ say) from week to week, then

$$p_{ij} = c e^{-\mu \tau_i} S^{j-1} P_1,$$

where $S = e^{-(\mu_E + (1+\Delta)\mu)}$.

Hence the regression model now becomes

$$E[y_{ij}] = -\log(cP_1) + \mu \tau_i + (-\log S)(j-1),$$

where $P_1 = \mu_E(1-e^{-Z})/Z$.

If the closed period is variable but the working period is constant (of duration one time-unit), then the appropriate regression model is

$$E[y_{ij}] = -\log(cP_1) + \mu(\tau_i + \gamma_{j-1}) + \mu_E(j-1),$$

where γ_{j-1} is the total number of time-units from the beginning of the season to the beginning of the jth recovery period. When the natural mortality rate is not constant but changes from μ' to μ when the season begins, then

$$E[y_{ij}] = -\log(cP_1) + \mu' \tau_i + \mu \gamma_{j-1} + \mu_E(j-1)$$

and we can use this model to test the hypothesis that $\mu' = \mu$.

RELEASES DURING THE SEASON. Paulik [1963a] also indicates briefly how the above regression methods can be simply extended to deal with releases actually made during the fishing season. For such releases we put $\tau_i = 0$ and $j = 1, 2, \ldots, J_i$ where, for a given i, j always refers to the recovery period *relative* to the ith release (i.e. the coded recovery period); this model is also mentioned in **5.3.2**.

CHAPTER 7

CATCH-EFFORT METHODS: CLOSED POPULATION

7.1 VARIABLE SAMPLING EFFORT

7.1.1 Introduction

Catch-effort methods are based on the general assumption that the size of a sample caught from a population is proportional to the effort put into taking the sample. More specifically, this means that one unit of sampling effort is assumed to catch a fixed proportion of the population, so that if samples are permanently removed, the decline in population size will produce a decline in catch per unit effort. Such techniques, first used in 1914 for bears in Norway (Hjort and Ottestad [1933]), are now widely used in the study of fish and small-mammal populations, where effort is usually measured in such units as line or trap per unit time. Let

N = initial population size,

n_i = size of ith sample removed from the population ($i = 1, 2, \ldots, s$),

$$x_i = \sum_{j=1}^{i-1} n_j \ (i = 2, 3, \ldots, s + 1; \ x_1 = 0),$$

f_i = units of effort expended on the ith sample,

and

$$F_i = \sum_{j=1}^{i-1} f_j \ (i = 2, 3, \ldots, s + 1; \ F_1 = 0).$$

We shall assume that:

(1) The population is closed.

(2) Sampling is a Poisson process with regard to effort. Mathematically this means that the probability of a given individual being caught when the population is subjected to δf units of sampling effort is $k\delta f + o(\delta f)$. Here k, usually called the (Poisson) *catchability coefficient* (or q in fishery research), is assumed to be constant throughout the whole experiment and the same for each individual. Also the units of effort are assumed to be independent, i.e. traps do not compete with each other.

(3) All individuals have the same probability p_i ($= 1 - q_i$) of being caught in the ith sample. It can be shown from assumption (2) that (cf. (1.1)) $q_i = \exp(-kf_i)$.

Using the above assumptions we shall now consider a number of possible models.

7.1.2 Maximum-likelihood estimation

The joint distribution of the $\{n_i\}$ is given by

$$f(\{n_i\}) = \prod_{i=1}^{s} \binom{N-x_i}{n_i} p_i^{n_i} q_i^{N-x_{i+1}} \tag{7.1}$$

or, rearranging, the multinomial distribution

$$\frac{N\,!}{\left(\prod_{i=1}^{s} n_i!\right)(N-x_{s+1})!} p_1^{n_1} (q_1 p_2)^{n_2} \ldots (q_1 q_2 \ldots q_{s-1} p_s)^{n_s} (q_1 q_2 \ldots q_s)^{N-x_{s+1}} \tag{7.2}$$

It is readily shown that the maximum-likelihood estimates $\hat{N}$ and $\hat{k}$ are solutions of

$$kF_{s+1} = -\log(1-(x_{s+1}/N)) \tag{7.3}$$

and

$$NF_{s+1} = \sum_{i=1}^{s} f_i(x_i + n_i p_i^{-1}). \tag{7.4}$$

These two equations can be solved iteratively, using, for example, the regression estimates of **7.1.3** as first approximations. Usually kf_i is small so that $p_i = 1 - \exp(-kf_i) \approx kf_i(1-\frac{1}{2}kf_i)$ and (7.4) reduces to

$$NF_{s+1} \approx \sum_{i=1}^{s} f_i (x_i + \tfrac{1}{2} n_i)(x_{s+1}/k).$$

Using standard maximum-likelihood theory (cf. (1.7)) and Stirling's approximation for large factorials, it can be shown that the asymptotic variance–covariance matrix of $\hat{N}$ and $\hat{k}$ is $\mathbf{V}^{-1}$, where

$$\mathbf{V} = \begin{bmatrix} E\left[\dfrac{x_{s+1}}{N(N-x_{s+1})} + \dfrac{1}{2N^2} - \dfrac{1}{2(N-x_{s+1})^2} \right], & F_{s+1} \\ F_{s+1}, & E\left[\sum_{i=1}^{s} n_i f_i^2 q_i/p_i^2 \right] \end{bmatrix}$$

The $(1, 1)$ element of $\mathbf{V}^{-1}$ is the asymptotic variance of $\hat{N}$. $\mathbf{V}$ can be estimated by replacing the expectations by random variables, and the unknown parameters by their estimates.

A suitable statistic for testing the adequacy of the model (7.2) is (cf. **1.3.6**(3))

$$\sum_{i=1}^{s} \left\{ \frac{(n_i - \hat{N}\hat{q}_1 \hat{q}_2 \ldots \hat{q}_{i-1}\hat{p}_i)^2}{\hat{N}\hat{q}_1 \hat{q}_2 \ldots \hat{q}_{i-1}\hat{p}_i} \right\}$$

which is asymptotically distributed as chi-squared with $s - 2$ degrees of freedom.

7.1.3 Regression estimates

1 Leslie's method

From (7.1) we have

$$\begin{aligned} E[n_i \mid x_i] &= p_i(N-x_i) \\ &\approx kf_i(N-x_i), \end{aligned}$$

when kf_i is small. The relationship $p_i \approx kf_i$ can be made exact if we redefine k to be the average probability that an individual is captured with one unit of effort and assume units of effort to be independent and additive (cf. **3.2.2**(3)): we shall call this new coefficient, K, the "binomial catchability coefficient". Therefore, defining $Y_i = n_i/f_i$, the catch per unit effort, we have

$$E[Y_i \mid x_i] = K(N - x_i), \quad (i = 1, 2, \ldots, s). \tag{7.5}$$

This linear regression model was first given by Leslie and Davis [1939] and De Lury [1947], and then developed more fully by De Lury [1951] and Chapman [1954: 10].

Now, conditional on $\{x_i\}$, the $\{Y_i\}$ are independently distributed and

$$V[Y_i \mid x_i] = (N - x_i)p_i q_i / f_i^2$$
$$= (N - x_i)K q_i / f_i$$
$$\approx (N - x_i)K / f_i .$$

Therefore, setting $\theta = N$ and $\gamma = K$, we can carry out a weighted least-squares analysis using the method of **1.3.5**(3) with, for example, $w_i = f_i/(N - x_i)$. However, the weights contain N and must be estimated iteratively. To start the cycle we can set $w_i = 1$, estimate N, and then substitute this estimate in w_i, etc. However, as Ricker [1958: 147] points out, factors other than sample size (e.g. day-to-day variations in the catchability) can also play a large part in producing the scatter of points about the regression line. It is therefore recommended that one carries out an unweighted least-squares analysis, i.e. the variance of Y_i is assumed to be constant ($= \sigma^2$, say). Then, setting $w_i = 1$ ($i = 1, 2, \ldots, s$) in **1.3.5**(3), the least-squares estimates of K and N are

$$\tilde{K} = - \sum_{i=1}^{s} Y_i (x_i - \bar{x}) / \sum_{i=1}^{s} (x_i - \bar{x})^2$$

and

$$\tilde{N} = \bar{x} + (\bar{Y}/\tilde{K}).$$

The estimate $\tilde{N}$ is approximately unbiased, and from (1.18)

$$V[\tilde{N}] \approx \frac{\sigma^2}{K^2} \left[\frac{1}{s} + \frac{(N - \bar{x})^2}{\Sigma (x_i - \bar{x})^2} \right].$$

A plot of Y_i versus x_i will provide a rough visual check on the adequacy of the regression model, including the assumption of constant variance. However, such graphical evidence should not be taken as final, for a straight-line fit is still possible in some situations, even when the assumptions do not hold. For example, a linear model is still possible even with natural mortality (cf. p. 307) or migration (Ketchen [1953]) taking place.

We note that $N = x$ when $Y = 0$, so that a simple graphical estimate of N is obtained by extending the fitted regression line to cut the x-axis. Also the intercept on the Y-axis will give an estimate of KN.

Assuming the Y_i to be approximately normally distributed, we can use the general method of **1.3.5**(3) to obtain an approximate $100(1 - \alpha)$ per cent confidence interval for N, namely $(\bar{x}+d_1,\ \bar{x}+d_2)$. Here d_1 and d_2 are the roots of -

$$d^2\left(\tilde{K}^2 - \frac{\tilde{\sigma}^2 t_{s-2}^2[\alpha/2]}{\Sigma(x_i-\bar{x})^2}\right) - 2d\bar{Y}\tilde{K} + \left(\bar{Y}^2 - \frac{\tilde{\sigma}^2 t_{s-2}^2[\alpha/2]}{s}\right) = 0, \quad (7.6)$$

where $(s-2)\tilde{\sigma}^2 = \Sigma[Y_i - \tilde{K}(\tilde{N}-x_i)]^2$

$$= \Sigma(Y_i-\bar{Y})^2 - [\Sigma Y_i(x_i-\bar{x})]^2/\Sigma(x_i-\bar{x})^2.$$

Also a $100(1-\alpha)$ per cent confidence interval for the slope K is

$$\tilde{K} \pm t_{s-2}[\alpha/2]\,(\tilde{\sigma}^2/\Sigma(x_i-\bar{x})^2)^{\frac{1}{2}}. \quad (7.7)$$

Apart from one or two isolated examples (e.g. Eberhardt *et al.* [1963], Van Etten *et al.* [1965], Lewis and Farrar [1968]), the Leslie regression method has been used mainly in fishery research. For example, a number of interesting graphs for fish populations are given in Omand [1951].

Example 7.1   Lobsters: De Lury [1947]

Table 7.1 gives a day-by-day record for 17 days of the lobster catch in a particular area. The unit of effort is taken as a thousand traps fished for one day, and the catch n_i is measured in units of 1000 lb. De Lury mentions

TABLE 7.1

Leslie's regression method applied to data from a lobster population: from De Lury [1947: Table 1].

Date	n_i (1000 lb)	f_i (1000 traps)	y_i $(= n_i/f_i)$	x_i (1000 lb)
May 23	6·995	8·470	0·8259	0
24	5·851	7·770	0·7530	6·995
25	3·221	3·430	0·9391	12·846
26	6·345	7·970	0·7961	16·067
27	3·035	4·740	0·6403	22·412
29	6·271	8·144	0·7700	25·447
30	5·567	7·965	0·6989	31·718
31	3·017	5·198	0·5804	37·285
June 1	4·559	7·115	0·6408	40·302
2	4·721	8·585	0·5499	44·861
5	3·613	6·935	0·5210	49·582
6	0·473	1·060	0·4462	53·195
7	0·928	2·070	0·4483	53·668
8	2·784	5·725	0·4863	54·596
9	2·375	5·235	0·4537	57·380
10	2·640	5·480	0·4818	59·755
12	3·569	8·300	0·4300	62·395

that the distribution of lobster size remained fairly constant during the 17 days, so that identifying "pounds" with "number" of individuals in the above theory is not a serious misrepresentation. This is equivalent to redefining K as the average probability that 1000 pounds of lobster are caught by 1000 traps in one day.

Assumption (1) that the population is closed is a reasonable one, as migration and natural mortality were negligible and the sampling period excluded moulting times when there would be recruitment through growth to legal size. A plot of Y_i versus x_i (Fig. 7.1) is approximately linear, thus suggesting that the assumptions underlying (7.5) are approximately satisfied.

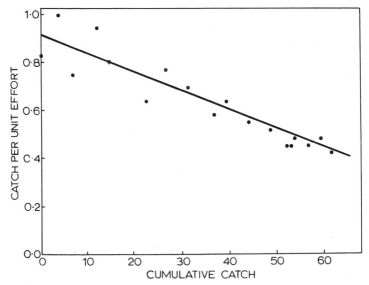

Fig. 7.1 Plot of catch per unit effort (n_i/f_i) versus the cumulative catch (x_i) for a lobster population: redrawn from De Lury [1947].

From Table 7.1 it was found that $\tilde{N} = 120 \cdot 5$ *thousand* pounds with 95 per cent confidence interval (77, 327), and $\tilde{K} = 0 \cdot 0074$ per 1000 traps with 95 per cent confidence interval $(0 \cdot 0058, 0 \cdot 0090)$.

Example 7.2   Blue crab (*Callinectes sapidus*): Fischler [1965]

The total catches per week, for 12 weeks, from a commercial-size male crab population are given in Table 7.2. Three different kinds of gear were used, namely trot-lines, crab trawls and shrimp trawls. From the ratios of total catch over total effort for each gear, the relative "fishing powers" of the three gears were found to be 1·00, 1·42 and 1·06 respectively; for example, one crab trawl per day was equivalent to 1·42 trot-lines operating for one day. These powers were then used to convert the weekly fishing efforts of the crab and shrimp trawls into standard units of trot-lines per day.

The assumption of a closed population seems reasonable because of the following: (i) Natural mortality was negligible. (ii) The catches n_i in

TABLE 7.2

Catch-effort data for a population of commercial-size male crabs: from Fischler [1965: Table 11].

i (week)	n_i (lb)	f_i (effort)	y_i $(= n_i/f_i)$	x_i (lb)
1	33 541	194	172·9	0
2	47 326	248	190·8	33 541
3	36 460	243	150·0	80 867
4	33 157	301	110·2	117 327
5	29 207	357	81·8	150 484
6	33 125	352	94·1	179 691
7	14 191	269	52·8	212 816
8	9 503	244	38·9	227 007
9	13 115	256	51·2	236 510
10	13 663	248	55·1	249 625
11	10 865	234	46·4	263 288
12	9 887	227	43·6	274 153

Table 7.2 have been corrected for recruitment from precommercial to commercial size, so that the population under study was the population of
commercial-size male crabs at the beginning of the experiment. (iii) Only
the male population is considered, as tagging studies indicated that
commercial-size females were emigrating out of the area while the fishery
was operating.

The approximate linearity of Y_i versus x_i in Fig. 7.2 gives qualitative
support for the adequacy of the regression model, and in particular for the

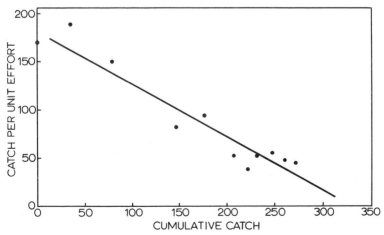

Fig. 7.2   Plot of catch per unit effort (n_i/f_i) versus the cumulative catch (x_i) for a
population of male crabs: redrawn from Fischler [1965].

assumption of constant catchability K. Obviously K could be affected by weather conditions, as adverse weather would reduce successful fishing time each day and consequently reduce the daily catch per unit of effort. Fischler felt that this was the case at the beginning and end of the 12 weeks and also for weeks 7 and 8, when there were north-east winds of from 5 to 11 knots. In order to study some of the weather effects the trot-line was singled out as the gear most likely to be affected. It was found that the multiple regression of trot-line catch per unit effort with respect to average wind speed and average wind direction each day during the season was not significant. This suggests that weather generally had little effect on catchability, except possibly at the three periods mentioned above.

Fischler points out that if any increase or decrease in the number of units of effort is followed by an immediate decrease or increase in catch per unit effort, then competition between fishing units may be indicated. As there was no evidence of this for all the catch-effort data examined, it was assumed that gear competition did not affect catchability.

From Table 7.2 it was found that $\tilde{N} = 330\,300$ with (cf. (7.6)) 95 per cent confidence interval (299 600, 373 600), and $\tilde{K} = 0 \cdot 000\,56$ with 95 per cent confidence interval ($0 \cdot 000\,45$, $0 \cdot 000\,67$).

2   Ricker's method

Treating the whole experiment as F_{s+1} samples, each with one unit of effort, then

$$E[N - x_i] = N(1-K)^{F_i}. \tag{7.8}$$

Therefore, combining this equation with (7.5), and taking logarithms, leads to Ricker's regression model [1958: 149].

$$E[y_i] \approx \log(KN) + [\log(1-K)]F_i, \tag{7.9}$$

where $y_i = \log Y_i = \log(n_i/f_i)$. For this model the least-squares estimates K' and N' are given by

$$\log(1-K') = \sum_{i=1}^{s} y_i(F_i - \bar{F}) / \sum_{i=1}^{s} (F_i - \bar{F})^2$$

and

$$\log N' = \bar{y} - \bar{F} \log(1-K') - \log K',$$

where $\bar{F} = \sum_{i=1}^{s} F_i/s$. If the y_i are assumed to be independently distributed with constant variance σ_y^2 then, using the delta method,

$$V[K'] \approx (1-K)^2 \; V[\log(1-K')]$$
$$= (1-K)^2 \; \sigma_y^2 / \Sigma(F_i - \bar{F})^2$$

and

$$V[N'] \approx N^2 \left\{ \frac{\sigma_y^2}{s} + \left(\frac{\bar{F}}{1-K} - \frac{1}{K} \right)^2 \cdot V[K'] \right\},$$

where σ_y^2 can be estimated by

$$\sigma_y'^2 = \frac{1}{s-2} \{\Sigma (y_i - \bar{y})^2 - [\Sigma y_i (F_i - \bar{F})]^2 / \Sigma (F_i - \bar{F})^2\}.$$

Assuming normality, a confidence interval for K can be obtained from the corresponding interval for $\log(1-K)$, the slope of the regression line.

Ricker also suggests the moment-type estimate (cf. (7.8) with $i = s+1$)

$$N^* = \frac{x_{s+1}}{1 - (1-K')^{F_{s+1}}}.$$

This estimate is used, for example, by Libosvárský [1962] for estimating the weight of fish stock in small streams.

3   *De Lury's method*

F rom (7.2),

$$E[n_i] = Nq_1 q_2 \ldots q_{i-1} p_i$$

$$= Np_i \exp(-kF_i). \tag{7.10}$$

If kf_i is small, so that $p_i \approx kf_i$, then, taking logarithms, we are led to consider De Lury's [1947] regression model

$$E[y_i] \approx \log(kN) - kF_i. \tag{7.11}$$

We note that when K is small, $\log(1-K) \approx -K$, $K \approx k$ and (7.9) reduces to (7.11).

The least-squares estimates k'' and N'' for this model are given by

$$k'' = -\sum_{i=1}^{s} y_i (F_i - \bar{F}) / \sum_{i=1}^{s} (F_i - \bar{F})^2$$

and

$$\log(k''N'') = \bar{y} + k''\bar{F}.$$

Assuming the y_i to be independently distributed with constant variance σ_y^2, then, using the delta method,

$$V[N''] \approx \sigma_y^2 N^2 \left[\frac{1}{s} + \left(\frac{k\bar{F}-1}{k}\right)^2 \cdot \frac{1}{\Sigma(F_i - \bar{F})^2}\right],$$

where σ_y^2 is estimated by $\sigma_y'^2$ in section 2, above.

We note that the $\{n_i\}$ have the multinomial distribution (7.2), so that the $\{y_i\}$ are not strictly independent and the method of **6.2.2** should be used. However, as $N \to \infty$, the joint distribution of the $\{y_i\}$ tends to the multivariate normal, and for small $\{p_i\}$ the covariances are small compared with the variances; the combination of these two implies approximate independence. Also, as in the Leslie method, the day-to-day fluctuations in catchability, etc. would generally rule out a weighted least-squares analysis, so that the above assumptions of independence and constant variance are perhaps not unreasonable.

The model (7.11) can be modified using an approximation suggested by

Paloheimo [1961] and Chapman [1961]. From (7.10),

$$E[Y_i] = Nf_i^{-1} p_i e^{-kF_i}$$
$$= kN e^{-kF_i} (1 - e^{-kf_i})/(kf_i),$$

and using the approximation

$$\log[(1 - e^{-w})/w] \approx -w/2$$

for small w, we have

$$E[Y_i] \approx kN \exp\{-k(F_i + \tfrac{1}{2}f_i)\}.$$

Therefore, taking logarithms, our regression model is now

$$E[y_i] \approx \log(kN) - k(F_i + \tfrac{1}{2}f_i).$$

The robustness of this model (and of (7.11)) with respect to changes in k has been studied by Braaten [1969], using simulation.

4 Comparison of methods

Ricker recommends the use of (7.11) when k and K are small — say less than $0·02$ — and (7.9) when K is larger. The reason for this is that when K is of moderate size, so that one unit of effort catches a reasonable proportion of the population (say, greater than 2 per cent), then the effects of sampling without replacement *during* each sample can no longer be ignored.

Although Leslie's method is superior to De Lury's in that it is less dependent on the underlying assumptions and provides a graphical estimate of N, both lines could be plotted as a check on the underlying assumptions. De Lury points out that any departures from the underlying assumptions are likely to be reflected in different ways in the two models, and a substantial agreement between the two would give qualitative support to the assumptions.

In conclusion we make the obvious remark that the above methods will only work if sufficient individuals are removed from the population, so that there is a significant decline in the catch per unit effort. For example, Omand [1951] found no such decline in several fish populations.

7.1.4 Estimating total catch

In this section we mention briefly a slightly different problem from that discussed above. Suppose that the total catch for a given period of time is unknown (e.g. as in a sports fishery), then, if the catch per unit effort is constant during the period and can be estimated by means of a suitable sampling procedure, and the total effort can be estimated, the product of these two estimates will provide an estimate of the total catch (e.g. Havey [1960]). A brief review and examples of this method as applied to a sports fishery are discussed in Rose and Hassler [1969]; a general review covering game-kill and creel census procedures has been prepared by Schultz [1959]. Two detailed models for carrying out a creel census of fishermen are given by Robson [1960, 1961].

7.1.5 Departures from underlying assumptions

We shall now consider the main types of departure from the underlying assumptions of **7.1.1**; much of the following discussion is based on Ricker [1958: Chapters 1 and 6], which should be consulted for further details.

1 Population not closed

Any recruitment, natural mortality, or migration in the population could introduce serious errors into the above methods unless opposing tendencies happened to balance out. Obviously these effects will be minimised if the whole experiment is concentrated in one area for as short a period of time as possible. Such considerations should be taken into account in determining s, the number of samples. A model allowing for a constant instantaneous natural mortality rate μ is discussed in section 7 below.

2 Variation in catchability coefficient

In many populations the catchability may vary with size (or species, sex, time, locality etc.; cf. **3.2.2**(*1*)), so that K is effectively the average coefficient over all sizes appearing in the catch. This average will be constant only if the distribution of size in the population remains constant, a condition that may not be met. One could overcome this difficulty by stratifying the sample according to size-classes and estimating each class separately.

De Lury [1951: 296] points out that the constancy of k or K does not refer to day-to-day variation, which may be treated as error, but to the absence of trends during the whole experiment. However, environmental conditions can produce trends in the catchability coefficient, and Paloheimo [1963: 73], for example, considers a model for a lobster population where the catchability is assumed to vary linearly with water temperature. Thus, if K_i is the catchability coefficient for the ith sample, he assumes on the basis of experimental studies (McLeese and Wilder [1958]) that $K_i = K(T_i - T)$, where T_i is the temperature at the time of the ith sample and T is the "threshold" temperature at which the lobsters become inactive. Leslie's regression model now becomes

$$E[Y_i \mid x_i] = K(T_i - T)(N - x_i)$$

and K, T and N can be estimated by non-linear least-squares (cf. Draper and Smith [1966]) or by regressing Y_i on x_i, T_i and $T_i x_i$. However, the T_i are subject to error and Paloheimo suggests an alternative method of estimation. For a certain range of temperatures lobster activity is linearly related to temperature, so that an alternative model is $K_i = K(a_i - a)$ where a_i is an index of activity (say walking rate).

3 Variation in trap efficiency

In fisheries many kinds of gear decrease in efficiency the longer they are left before being lifted and reset, so that the catch per unit time begins to fall off. For example, as more fish are caught on a set line, the less vacant hooks there are, and eventually a point of saturation may be reached. On the other hand, with net fishing, the presence of some fish already in the net may tend to scare others away, so that saturation may be reached long before the net is full.

In some situations the catches can be corrected for gear saturation. For example, Murphy [1960] gives a comprehensive deterministic model for doing this for the case of the baited long line. His model is developed in terms of instantaneous rates of bait loss, of hooking fish, and of losing hooked fish; the reader is referred to his article for details.

4   Units of effort not independent

This can happen when traps are so close together that "physical" competition exists between them which is independent of population size. For example, too many anglers at a pool may frighten the fish; or setting a new gill net near one already in operation may scare the fish away from the latter.

5   Non-random sampling

In fisheries, for example, not all the population may be subjected to the catching process; some parts of the fishing grounds may be inaccessible or too sparsely populated to warrant fishing there.

6   Incomplete record of effort

In commercially exploited populations information on effort may be missing or unavailable for some parts of the catch. To overcome this, Ricker suggests computing the catch per unit effort for as much data as possible and then dividing this value into the residual catch to give an estimate of effort for this residual. Adding this estimate to the known effort will then give the total effort f_i.

Sometimes effort records are complete and catch records incomplete, so that the above procedure can be used in reverse.

7   Different catching methods

When different fishing gears (e.g. trawl and drift-net) are used for the same population, the problem arises of converting all the different units of effort into a standard unit. This is a major problem, and for a discussion of some of the difficulties involved the reader is referred to Gulland [1955a], Beverton and Holt [1957: 172–7] and Parrish [1962]. In some cases standardising is impossible because the gears are so unlike, e.g. they may select different sizes of fish or may operate at different times of the year.

To examine the effect of two types of gear operating in the same area at the same time, let f_i', n_i', x_i' and Y_i' be the effort, catch, cumulated catch, and catch per unit effort, respectively, for a second type of gear. Then, using Leslie's model (7.5) we have

$$E[Y_i \mid x_i, x_i'] = K(N - x_i - x_i'),\qquad (7.12)$$

$$E[Y_i' \mid x_i, x_i'] = K'(N - x_i - x_i')$$

and

$$E[n_i'] = \frac{f_i' K'}{f_i K} E[n_i].$$

If $(f_i' K')/(f_i K)$ remains approximately constant throughout the whole experiment as indicated by the constancy of the sample ratios n_i'/n_i then, following De Lury [1951: 296],

$$E[Y_i \mid x_i] \approx KN - K\left(1 + \frac{K' f_i'}{K f_i}\right) x_i \tag{7.13}$$

$$= KN - K_0 x_i \text{ say.} \tag{7.14}$$

Let $\hat{K}_0$ be the least-squares estimate of the slope K_0 of (7.14), then estimating $(K' f_i')/(K f_i)$ by x_{s+1}'/x_{s+1} $(= \Sigma n_i'/\Sigma n_i)$, we have the estimate (Dickie [1955: 810])

$$\hat{K} = \hat{K}_0 / [1 + (x_{s+1}'/x_{s+1})].$$

N can then be estimated from the least-squares estimate of KN.

Similar equations can also be developed for De Lury's model as follows. Let p_i, p_i' be the probabilities of capture in the ith sample by the two gears, and assume that the two gears are operating independently. Then

$$E[N - x_i - x_i'] = N q_1 q_1' q_2 q_2' \ldots q_{i-1} q_{i-1}'$$

$$= N \exp(-k F_i - k' F_i'),$$

and when k is small, so that $k \approx K$ (cf. (7.10) and (7.11)), the above equation may be combined with (7.12) to give De Lury's model

$$E[\log Y_i] \approx \log(kN) - k F_i - k' F_i'. \tag{7.15}$$

If f_i'/f_i is constant (and therefore equal to F_i'/F_i), then (7.15) becomes

$$E[\log Y_i] \approx \log(kN) - k_0 F_i, \tag{7.16}$$

where $k_0 = k[1 + (k' F_i'/k F_i)]$.

The above theory can be applied to two important situations (Dickie [1955]). The first is when several gears are used but one is dominant and the rest are lumped together under f_i'; if x_i and x_i' are known for each sample, then the appropriate model for estimating N is (7.12). The second is when data are available for only part of the effort, or when good records are kept for part of the catch only; in this case (7.14) and (7.16) are appropriate.

When sampling is continuous, so that n_i is the number caught in the ith sample interval, then n_i' may also be interpreted as the number dying from natural mortality in this interval. If the instantaneous natural mortality rate is constant and equal to μ say, then (cf. (1.1))

$$E[n_i' \mid x_i, x_i'] = (1 - e^{-\mu t_i})(N - x_i - x_i')$$

$$\approx \mu t_i (N - x_i - x_i'),$$

where t_i is the length of the ith sampling interval. Therefore, if t_i/f_i remains constant $(= \Sigma t_i/\Sigma f_i = T_{s+1}/F_{s+1}$ say), K_0 in (7.14) is given by

$$K_0 = K\left(1 + \frac{\mu T_{s+1}}{KF_{s+1}}\right) = K + \frac{\mu T_{s+1}}{F_{s+1}},$$

and, given an estimate of μ, K can be estimated.

When there are two types of gear as well as natural mortality, then

$$K_0 = K\left(1 + \frac{K' f_i'}{K f_i} + \frac{\mu t_i}{K f_i}\right).$$

In this case we can also use the model (Dickie [1955])

$$E[Y_i \mid x_i, x_i'] = KN - K\left(1 + \frac{\mu t_i}{K f_i + K' f_i'}\right)(x_i + x_i')$$

$$= KN - K_1(x_i + x_i'), \qquad (7.17)$$

where $K_1 = K\left\{1 + \frac{\mu t_i}{K f_i} \cdot \frac{1}{[1 + (K' f_i')/(K f_i)]}\right\}.$

Therefore, if the ratios $f_i' : f_i : t_i$ are constant, then, given an estimate $\hat{\mu}$ of μ, K can be estimated by

$$\hat{K} = \hat{K}_1 - \frac{\mu T_{s+1}}{F_{s+1}} \cdot \frac{1}{[1 + (x_{s+1}'/x_{s+1})]},$$

where $\hat{K}_1$ is the least-squares estimate of the slope of (7.17). Some examples using the above model are given in Dickie [1955: 816–19].

7.1.6   Single mark release

1   Regression methods

Suppose that M individuals are captured, marked and released at the beginning of the experiment, so that $U = N - M$ is the initial unmarked population size. We shall let the suffixes m and u denote membership of the marked and unmarked populations respectively. Then, treating these populations separately, two Leslie regression lines can be considered:

$$E[m_i/f_i] = K_m(M - x_{mi}) \qquad (7.18)$$

and

$$E[u_i/f_i] = K_u(U - x_{ui}), \qquad (7.19)$$

where m_i and u_i $(= n_i - m_i)$ are the numbers of marked and unmarked in the ith sample respectively. With the usual normality assumptions we can now derive a t-statistic (e.g. Wetherill [1967: 236]) for testing the hypothesis that the two lines have the same slope, i.e. that marked and unmarked are equicatchable. We note that apart from actual variations in catchability, any migration or recruitment will affect the two models differently. For example, recruitment and immigration will have no effect on the marked population; under certain conditions immigration and emigration rates can be estimated from the two slopes (Ketchen [1953]).

Since M is known, the values $m_i/[f_i(M-x_{mi})]$ can be used to provide a check on the constancy of K_m. Sometimes, however, there is an initial mortality of d marked individuals through the effects of handling and marking, so that the number of marked alive before the first sample is actually $M-d$. Fitting the line (7.18) will then provide an estimate of $M-d$, and hence of d.

In conclusion we note that De Lury's method can also be generalised in the same way to give two regression lines.

2 Maximum-likelihood estimation

When $K_m = K_u$ ($= K$ say) and the assumptions underlying the Petersen method (cf. **3.1.1**) hold for each sample, then

$$E[n_i \mid x_i] = Kf_i(N-x_i)$$

and

$$E[m_i \mid n_i] = n_i(M-x_{mi})/(N-x_i)$$

$$= n_i M_i/N_i \text{ say.}$$

If we can assume that n_i has a Poisson distribution and that m_i, conditional on n_i, is also Poisson, then we have Chapman's model [1954: 12]

$$\prod_{i=1}^{s} \left\{ e^{-Kf_iN_i}\frac{(Kf_iN_i)^{n_i}}{n_i!} e^{-n_iM_i/N_i}\frac{(n_iM_i/N_i)^{m_i}}{m_i!} \right\}$$

For this model the maximum-likelihood estimates of N and K are solutions of

$$KF_{s+1} = \sum_{i=1}^{s} \left\{ \left(\frac{n_i-m_i}{N-x_i}\right) + \frac{n_iM_i}{(N-x_i)^2} \right\}$$

and

$$NF_{s+1} = \sum_{i=1}^{s} x_if_i + (x_{s+1}/K),$$

with asymptotic variance–covariance matrix (expressed in terms of the N_i and M_i) the inverse of

$$\begin{bmatrix} \sum_{i=1}^{s} \left\{ \frac{Kf_i}{N_i}\left(1+\frac{M_i}{N_i}\right) \right\}, & F_{s+1} \\ F_{s+1} & \sum_{i=1}^{s} f_iN_i/K \end{bmatrix}$$

7.2   CONSTANT SAMPLING EFFORT: REMOVAL METHOD

7.2.1   Maximum-likelihood estimation

In **7.1** we assumed that p_i, the probability of capture in the ith sample, varied from sample to sample. However, in some carefully controlled experiments in which the same effort is used for each sample under almost identical conditions it is reasonable to assume that p_i is constant and equal to p, say. We therefore consider a model for which the following assumptions are true:

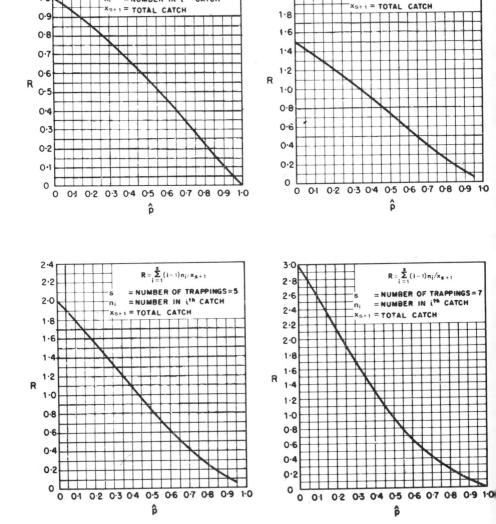

Fig. 7.3   Graphs for determining $\hat{p}$ for a given value of R and $s = 3, 4, 5$ and 7: from Zippin [1956].

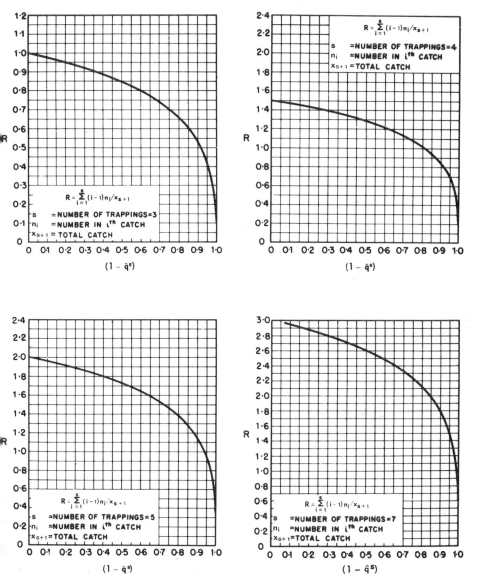

Fig. 7.4 Graphs for determining $(1 - \hat{q}^s)$ for a given value of R and $s = 3$, 4, 5 and 7: from Zippin [1956].

(1) The population is closed.

(2) The probability of capture in the ith sample is the same for each individual exposed to capture.

(3) The probability of capture p remains constant from sample to sample.

Given the above assumptions, Moran [1951] and Zippin [1956, 1958] obtained maximum-likelihood estimates of N and p; the following discussion is from Zippin [1956].

Setting $p_i = p$ in (7.2), the joint probability function of the $\{n_i\}$ becomes

$$\frac{N!}{(N - x_{s+1})! \; \prod\limits_{i=1}^{s} n_i!} \; p^{x_{s+1}} \, q^{sN - \sum\limits_{i=1}^{s+1} x_i} . \qquad (7.20)$$

The maximum-likelihood estimates $\hat{N}$ and $\hat{p}$ are given by the equations

$$\hat{N} \;=\; x_{s+1}/(1 - \hat{q}^s) \qquad (7.21)$$

and

$$\frac{\hat{q}}{\hat{p}} - \frac{s\hat{q}^s}{(1 - \hat{q}^s)} \;=\; \frac{\sum\limits_{i=1}^{s} (i-1)n_i}{x_{s+1}} \quad (= R \text{ say}), \qquad (7.22)$$

and Zippin gives graphs (reproduced as Fig. 7.3 and 7.4) for facilitating their solution. It can be shown (cf. Seber and Whale [1970]) that (7.22) has a unique solution $\hat{q}$ in the range $[0, 1]$ for $0 \leqslant R \leqslant (s-1)/2$; when $R > (s-1)/2$ the above method is not applicable and the experiment "fails". This condition for failure can be rewritten as

$$\sum\limits_{i=1}^{s} (s + 1 - 2i) n_i < 0 ,$$

and the probability of this happening will generally be small for large N and will decrease as s increases.

For N large, $\hat{N}$ and $\hat{p}$ are asymptotically unbiased with asymptotic variances

$$V[\hat{N}] \;=\; \frac{N(1 - q^s) q^s}{(1 - q^s)^2 - (ps)^2 q^{s-1}} \qquad (7.23)$$

and

$$V[\hat{p}] \;=\; \frac{(qp)^2 (1 - q^s)}{N[q(1 - q^s)^2 - (ps)^2 q^s]} .$$

Assuming asymptotic normality, confidence limits for N and p can be calculated in the usual manner.

In designing an experiment for a particular population the biologist will be interested in knowing what proportion of the total population he must catch in order to estimate N with a given precision. Also, given that a certain proportion is to be caught, the question arises as to whether it is better to set a large number of traps and take a few samples or set a few traps and take a large number of samples. In studying this latter question first, we shall assume that the home ranges of the animals overlap sufficiently for the number of animals exposed to capture during the first trapping to be the same, regardless of the number of traps set.

Suppose that t traps are set on each sampling occasion, and let p_0 ($= 1 - q_0$) represent the probability that an animal will be captured by a single trap during a single sampling period. Assuming that p_0 is the same for each trap and that the traps operate independently, the probability that an animal is not caught in a particular sample is $q = q_0^t$. Thus fixing $1 - q^s$ ($= 1 - q_0^{st}$), the total proportion of the population expected to be caught, is equivalent to fixing st, the total number of traps set throughout the whole experiment. Zippin shows that as t and s vary subject to fixed st, $V[\hat{N}]$ decreases only slightly as s increases beyond 3. This means that for practical purposes $V[\hat{N}]$ depends almost entirely on N, the trap efficiency p_0, and the total trapping effort st. Hence, for a given accuracy or precision the biologist would normally choose t and s to minimise the cost. However, if migration and natural death are possible factors, a short intensive programme would be the most appropriate.

Zippin gives a table (reproduced as Table 7.3) showing the total proportion of the population which must be captured in order that $\hat{N}$ may have

TABLE 7.3

Proportion of total population required to be trapped for a specified coefficient of variation of $\hat{N}$: from Zippin [1956: Table 1].

N	Coefficient of variation			
	30%	20%	10%	5%
	Proportion (to nearest 0·05) of population to be captured (in 100 or fewer trappings)			
200	0·55	0·60	0·75	0·90
300	0·50	0·60	0·75	0·85
500	0·45	0·55	0·70	0·80
1 000	0·40	0·45	0·60	0·75
10 000	0·20	0·25	0·35	0·50
100 000	0·10	0·15	0·20	0·30

a specified coefficient of variation. A cursory glance at the table reveals that a relatively large proportion of the population must be captured (particularly when $N < 200$) in order to obtain reasonably accurate estimates. This is a serious limitation on the above so-called "removal method", except in the situation where it is desirable to reduce a population sharply, as in the case of a crop-damaging rodent. Alternatively the problem of depletion can be avoided if the animals are live-trapped and "removed" by tagging before releasing them. In this case N can also be estimated from the recapture data using the methods of Chapter 5. However, on the debit side, the trapping effort used in catching untagged animals may fall off significantly as tagged animals take up more and more of the traps.

If the population is physically depleted, other problems may arise. For example, p may not remain constant as the more catchable animals are caught first (Libosvárský [1962: 517]). Also, as numbers decrease, territorial competition on the boundary of the population may lead to immigration into the depopulated sampling area; Chew and Butterworth [1964] give an example where home range appears to vary inversely with population density.

To test the validity of the model (7.20) we can use a standard goodness-of-fit statistic (cf. **1.3.6** (*3*))

$$T_1 = \sum_{i+1}^{s} (n_i - E_i)^2 / E_i ,$$

where $E_i = \hat{N}\hat{p}\hat{q}^{i-1}$ ($i = 1, 2, \ldots, s$). When (7.20) is valid, T_1 is asymptotically distributed as χ^2_{s-2}. Zippin also suggests the test statistic (cf. (7.1))

$$T_2 = \sum_{i=1}^{s} [n_i - (\hat{N} - x_i)\hat{p}]^2 / [(\hat{N} - x_i)\hat{p}\hat{q}]$$

and demonstrates that it is asymptotically equivalent to T_1.

Example 7.3   Zippin [1956, 1958]

In a three-night trapping programme the following catches were made: $n_1 = 165$, $n_2 = 101$, $n_3 = 54$. Then

$$R = \sum_{i=1}^{s} \frac{(i-1)n_i}{x_{s+1}} = \frac{n_2 + 2n_3}{n_1 + n_2 + n_3} = \frac{20}{320} = 0\cdot 65 ,$$

and entering the graphs, Fig. 7.3 and 7.4, with $s = 3$ and $R = 0\cdot 65$ gives $\hat{p} = 0\cdot 42$, $(1 - \hat{q}^3) = 0\cdot 80$ and

$$\hat{N} = x_{s+1}/(1 - \hat{q}^3) = 320/0\cdot 80 = 400 .$$

Estimating N and q by $\hat{N}$ and $\hat{q}$ in (7.23) leads to

$$\hat{V}[\hat{N}] = \frac{400(0\cdot 80)(0\cdot 20)}{(0\cdot 80)^2 - (1\cdot 26)^2 (0\cdot 20/0\cdot 58)} = 691\cdot 1$$

and

$$\hat{\sigma}[\hat{N}] = 26\cdot 3 .$$

Hence an approximate 95% confidence interval for N is $400 \pm 1\cdot 96 \ (26\cdot 3)$ or (348, 452).

To calculate T_1 we compare

$$(n_1, n_2, n_3) = (165, 101, 54)$$

and

$$(E_1, E_2, E_3) = (\hat{N}\hat{p}, \ N\hat{p}\hat{q}, \ \hat{N}\hat{p}\hat{q}^2)$$
$$= (168, 97\cdot 44, 56\cdot 52) .$$

Hence

$$T_1 = \Sigma(n_i - E_i)^2 / E_i = 0\cdot 29$$

which, at one degree of freedom, indicates an excellent fit.

Further examples of the above method are given in Davis [1957: mice], Menhinick [1963: insects], Kikkawa [1964: 262, small rodents] and Johnson [1965: freshwater fish]. The removal method has also been widely used by Polish ecologists under the name of "Standard Minimum" method (Grodzinski et al. [1966]).

7.2.2 Maximum-likelihood estimation: special cases

As the above method is widely used in studying small mammal populations and (with the advent of efficient electric fishing methods — cf. Moore [1954]) fish populations in streams, we shall now consider the three cases $s = 3, 2, 1$ in more detail.

1 Three-sample method

It was mentioned above that under certain conditions $V[\hat{N}]$ decreases only slightly as s increases beyond 3. Since standard trap-lines operating for three nights have been widely used, and three samples is the minimum number of samples for which the preceding goodness-of-fit statistics T_1 and T_2 can be applied, it is of practical interest to consider this special case more closely.

When $s = 3$, (7.21) and (7.22) now have explicit solutions (Junge and Libosvársky [1965])

$$\hat{N} = \frac{6X^2 - 3XY - Y^2 + Y(Y^2 + 6XY - 3X^2)^{\frac{1}{2}}}{18(X-Y)} \qquad (7.24)$$

and

$$\hat{p} = \frac{3X - Y - (Y^2 + 6XY - 3X^2)^{\frac{1}{2}}}{2X}$$

where $X = 2n_1 + n_2$ and $Y = n_1 + n_2 + n_3$; we note that the experiment "fails" if $n_3 > n_1$, for then $\hat{N}$ is negative. From sampling experiments using binomial tables and a table of random numbers, Zippin [1956] came to the conclusion that for $p = 0.4$ and $N > 200$ the large-sample normal theory gave reasonable confidence intervals for N. For $50 < N < 200$ he found that $V[\hat{N}]$ of (7.23) was too small, and the distribution of N was skewed, so that $\hat{N} \pm 1.96\,\hat{\sigma}[\hat{N}]$ was more like a 90 per cent than a 95 per cent confidence interval. However, the adequacy of this confidence interval needs further investigation for other values of p, particularly when p is small.

Another aspect worthy of investigation is the effect of varying catchability in the population on the validity of $\hat{N}$ and $V[\hat{N}]$. Junge and Libosvársky [1965] indicated by means of an example that $\hat{N}$ will not be greatly affected by any variation in catchability if the range of p over the population is less than 0.05 and $p > \frac{1}{2}$. Their conclusion is borne out by the following theoretical analysis given by Seber and Whale [1970].

Let x_j ($j = 1, 2, \ldots, N$) be the probability that the jth member of the population is caught in a sample, given that it is in the population, and suppose that this probability remains constant from sample to sample. Then, assuming that the population represents a random selection of such probabilities with regard to the species as a whole and the particular trapping method used, $x_1, x_2, \ldots, x_N$ may be regarded as a random sample from a probability density function $f(x)$. Hence the probability of obtaining a given outcome of the experiment is

$$P_u =$$

$$\left(\prod_{i=1}^{n_1} x_i \right) \left\{ \prod_{j=n_1+1}^{n_1+n_2} (1-x_j)x_j \right\} \left\{ \prod_{k=n_1+n_2+1}^{n_1+n_2+n_3} (1-x_k)^2 x_k \right\} \left\{ \prod_{l=n_1+n_2+n_3+1}^{N} (1-x_l)^3 \right\}$$

and the conditional probability function of n_1, n_2 and n_3 given the particular sample of probabilities $\{x_j\}$ is

$$f(n_1, n_2, n_3 \mid \{x_j\}) = \sum_u P_u,$$

where $\sum_u$ denotes summation over all possible groupings of the N animals such that n_1, n_2 and n_3 are the numbers in each of the three categories, i.e. u represents a permutation of N objects such that n_i fall in the ith category ($i = 1, 2, 3$). Since the $\{x_j\}$ are independent they can be integrated out and the unconditional probability function is given by:

$$
\begin{aligned}
f(n_1, n_2, n_3) &= E[f(n_1, n_2, n_3 \mid \{x_j\})] \\
&= E[\sum_u P_u] \\
&= \sum_u E[P_u] \\
&= \sum_u a_1^{n_1}(a_1-a_2)^{n_2}(a_1-2a_2+a_3)^{n_3}(1-3a_1+3a_2-a_3)^{N-Y} \\
&= \frac{N!}{\left(\prod_{i=1}^{3} n_i! \right)(N-Y)!} \, a_1^{n_1}(a_1-a_2)^{n_2}(a_1-2a_2+a_3)^{n_3} \\
&\qquad \times (1-3a_1+3a_2-a_3)^{N-Y},
\end{aligned}
$$

$$(7.25)$$

where $a_k = E[x^k]$. This distribution can be used to examine the asymptotic mean of $\hat{N}$ for different density functions $f(x)$. For example, if $f(x)$ is uniform on $[c, d]$ ($0 \leqslant c \leqslant d \leqslant 1$) and $w = c/d$, then $a_1 = \frac{1}{2}d(1+w)$, $a_2 = \frac{1}{3}d^2(1+w+w^2)$ and $a_3 = \frac{1}{4}d^3(1+w)(1+w^2)$. The asymptotic mean of $\hat{N}$, obtained by replacing X and Y in (7.24) by their expected values with respect to (7.25), is given by BN, where B is tabulated for different values of w and d in Table 7.4. When $w = 1$, then $B = 1$; this is to be expected since a maximum-likelihood estimate is asymptotically unbiased. From the table we find that for $w < 1$, $B < 1$, and B is sensitive to the value of w; when $w \geqslant 0 \cdot 5$ the asymptotic bias is negligible, and when $c > 0 \cdot 01$, $B > 0 \cdot 8$. If $f(x)$ has a unimodal (e.g. beta type) distribution, defined on $[c, d]$, then the corresponding values of B will be larger. Therefore we can conclude that, asymptotically, $\hat{N}$ is fairly insensitive to variations in catchability.

    If there is a considerable variation in catchability, the more catchable individuals will be caught first, so that the average probability of capture will decrease from one trapping to the next and $\hat{N}$ will underestimate N.

TABLE 7.4

Asymptotic value of $B = E[\hat{N}]/N$ when the catchability distribution $f(x)$ is uniform on $[c, d]$: from Seber and Whale [1970: Table 2].

$w \, (= c/d)$	d	c	B
0·5	0·25	0·125	0·968
	0·50	0·250	0·972
	0·75	0·375	0·976
	1·00	0·500	0·983
0·2	0·25	0·050	0·879
	0·50	0·100	0·888
	0·75	0·150	0·899
	1·00	0·200	0·913
0·01	0·25	0·0025	0·766
	0·50	0·005	0·776
	0·75	0·0075	0·788
	1·00	0·010	0·803
0·0	0·25	0	0·758
	0·50	0	0·768
	0·75	0	0·780
	1·00	0	0·795

TABLE 7.5

Expected percentage underestimation of N by $\hat{N}$ when the probability of capture decreases by a constant amount on three successive trappings: from Zippin [1958: Table 2].

p (First trapping)	Absolute decrease in p on successive trappings				
	0·01	0·02	0·03	0·05	0·10
0·1	47·4	65·7	75·3	85·2	–
0·2	15·7	27·6	37·3	51·4	71·6
0·3	6·1	12·0	17·4	26·9	45·7
0·4	2·8	5·6	8·4	13·9	26·6
0·5	1·4	2·8	4·2	7·2	14·8
0·6	0·7	1·4	2·2	3·7	7·9

For example,

$$\frac{E[n_2]}{E[n_1]} = \frac{a_1 - a_2}{a_1}$$

$$= (1 - a_1)\left(\frac{a_1 - a_2}{a_1 - a_1^2}\right)$$

$$\leqslant (1 - a_1),$$

317

since $a_2 = E[x^2] \geqslant (E[x])^2 = a_1^2$, with equality if and only if x is constant (i.e. $a_k = p^k$). Such a decrease in average catchability can also occur when climate or other factors decrease the activity of the animals during the course of the experiment (e.g. Gentry and Odum [1957]). Zippin [1958] gives a table (Table 7.5) showing the expected percentage underestimation of N when p decreases by a constant amount.

2   Two-sample method

ESTIMATION. In some situations it may be unwise or impracticable to take more than just two samples. For example, if climatic conditions have considerable effect on p then a short intensive survey may be needed if p is to remain constant. Again, if p is large, three samples may severely deplete the population unless tagging can be used to represent "removal". On the other hand, if one wishes to reduce the population as much as possible, then for p large it may be a waste of time or uneconomic to take a third sample. For example, Le Cren points out (Seber and Le Cren [1967]) that the removal method can also be applied to quite different situations such as the sorting of animals out of samples of mud, soil or grain. Here a third sorting may be a waste of effort when p, the probability of being found in a given searching, is large.

When $s = 2$ and $n_1 > n_2$, the maximum-likelihood equations (7.21) and (7.22) have simple solutions

$$\hat{N} = n_1^2/(n_1 - n_2)$$

and

$$\hat{q} = n_2/n_1,$$

which are also the moment estimates. Using the delta method, we have (Seber and Le Cren [1967])

$$E[\hat{N}] \approx N + \frac{q(1+q)}{p^3} = N + b[N], \text{ say,} \tag{7.26}$$

$$V[\hat{N}] \approx \frac{Nq^2(1+q)}{p^3}, \tag{7.27}$$

$$E[\hat{p}] \approx p - \frac{q}{Np} \tag{7.28}$$

and

$$V[\hat{p}] \approx \frac{q(1+q)}{Np}. \tag{7.29}$$

Replacing N and p by their estimates leads to the variance estimates

$$\hat{V}[\hat{N}] = \frac{n_1^2 n_2^2 (n_1 + n_2)}{(n_1 - n_2)^4} \tag{7.30}$$

and

$$\hat{V}[\hat{p}] = \frac{n_2(n_1 + n_2)}{n_1^3}. \tag{7.31}$$

We note that $\hat{p}$ may be adjusted for bias; thus

$$p^* = 1 - [n_2/(n_1 + 1)]$$

is almost unbiased as

$$E[p^*] = \mathop{E}_{n_1} E[p^* | n_1]$$

$$= \mathop{E}_{n_1} \left[1 - \frac{(N - n_1)p}{(n_1 + 1)} \right]$$

$$= \mathop{E}_{n_1} \left[1 - \frac{(N + 1)p}{(n_1 + 1)} + p \right]$$

$$= p + 1 - (1 - q^{N+1})$$

$$= p + q^{N+1}$$

$$\approx p .$$

Robson and Regier [1968] have considered a slight modification of $\hat{N}$, namely

$$N^* = (n_1^2 - n_2)/(n_1 - n_2) .$$

This estimate, based on the relationship

$$E[n_1^2 - n_2] = (E[n_1])^2 ,$$

is also biased, though the bias will probably be less than $b[N]$ of (7.26) for small samples.

Example 7.4   Trout: Seber and Le Cren [1967]

Two successive electric fishings in a section of a small stream yielded 79 and 28 young trout respectively. Thus,

$$\hat{N} = 79^2/(79 - 28) = 122 ,$$

$$\hat{q} = 28/79 = 0{\cdot}35 ,$$

$$\hat{\sigma}[\hat{N}] = 79(28)(\sqrt{107})/51^2 = 8{\cdot}8, \quad \hat{b}[\hat{N}] = 2 ,$$

and

$$\hat{N} \pm 1{\cdot}96\,\hat{\sigma}[\hat{N}] = 122 \pm 18 \quad \text{or} \quad 104 < N < 140 .$$

VALIDITY OF VARIANCE FORMULAE.   The bias terms are included in (7.26) and (7.28) to indicate the accuracy of the delta method. Thus by considering the proportional biases ($b[N]/N$, etc.) we find that (7.26) and (7.27) only hold for large values of Np^3, while (7.28) and (7.29) are valid for large Np^2. This means that for quite reasonable values of N and p, the formula for $V[\hat{N}]$ may be unsatisfactory. For example, if $n_1 = 27$ and $n_2 = 20$, then $\hat{N} = 104$, $\hat{p} = 0{\cdot}25$, $\hat{b} = 74$ and $\hat{\sigma}[\hat{N}] = 76$, the last two estimates indicating that (7.26) and (7.27) may not be valid.

When Np^3 is small, not only is the formula for $V[\hat{N}]$ unreliable but the estimates $\hat{N}$ and $\hat{\sigma}[\hat{N}]$ may be totally misleading. For example, Robson and Regier [1968: 147] in a simulation experiment with $N = 500$ and $p = 1/10$ obtained "catches" of $n_1 = 60$ and $n_2 = 35$, thus yielding $\hat{N} = 144$ and

$\hat{\sigma}[\hat{N}] = 35\cdot7$. Here N is drastically underestimated as the first catch happened to exceed substantially its expected value of $Np = 50$, and the second catch was smaller than expected. Also $\sigma[\hat{N}] = 877$ and $b[\hat{N}] = 1710$, so that the asymptotic formulae are open to question and the value of $\hat{\sigma}[\hat{N}]$ is completely misleading.

By considering a coefficient of variation of 25 per cent it is suggested, as a rough guide, that the $100(1-\alpha)$ per cent confidence interval

$$\hat{N} \pm z_{\alpha/2}\, \hat{\sigma}[\hat{N}] \tag{7.32}$$

will be satisfactory if

$$Np^3 \geqslant 16q^2(1+q). \tag{7.33}$$

Also computer studies indicate that on account of the skewness of the distribution of $\hat{N}$, the bias correction $\hat{b}[\hat{N}]$ is best ignored when calculating this confidence interval. However, when (7.33) is not satisfied the following method can be used:

Let

$$u = \frac{n_1 - n_2 + 2}{(n_1 + 1)(n_1 + 2)},$$

then from Seber and Whale [1970]

$$E[u] = \frac{1}{N+1} + \frac{Nq^{N+1}}{N+1},$$

$$\approx \frac{1}{N+1},$$

and

$$E[u^2] \approx (1+p)^2 b_2 - [2Np(1+p) + 2 + 9p + 5p^2]b_3 +$$
$$+ (N+2)p[Np+9p+9]b_4 - 6(N+2)^2 p^2 b_5,$$

where $\quad b_k = \dfrac{1}{N^k} \cdot \dfrac{(N-2)(N-3)\ldots(N-k-1)}{[(N-1)p-1][(N-1)p-2]\ldots[(N-1)p-k]}.$

Hence

$$V[u] \approx E[u^2] - (N+1)^{-2},$$

and estimating Np, p, N and $(N+1)^{-1}$ by $n\cdot$, p^*, $u^{-1}-1$ and u respectively leads to an estimate $\hat{V}[\hat{u}]$. Assuming u to be approximately normal, an approximate $100(1-\alpha)$ per cent confidence interval for $(N+1)^{-1}$ is given by

$$u \pm z_{\alpha/2}\, \hat{\sigma}[\hat{u}] \cdot \tag{7.34}$$

To compare this interval with (7.32) the following probabilities were computed for different values of N and p and for $z_{\alpha/2} = 1\cdot64,\ 1\cdot96$ (i.e. 90 per cent and 95 per cent confidence intervals respectively):

$$A_1 = \Pr\left[\frac{|u-(N+1)^{-1}|}{\hat{\sigma}[\hat{u}]} < z_{\alpha/2} \quad \text{and} \quad n_1 > n_2\right],$$

$$A_2 = \Pr\left[\frac{|\hat{N}-N|}{\hat{\sigma}[\hat{N}]} < z_{\alpha/2} \quad \text{and} \quad n_1 > n_2\right]$$

and

$$C = \Pr[n_1 > n_2],$$

where C is the probability that the experiment is "successful". Since the confidence intervals are only calculated when the experiment is successful, the true "worth" of an interval should be measured by the conditional probabilities A_i/C; these are given in Table 7.6. From this table we see that

TABLE 7.6

A comparison of the exact probabilities (A_i) for two large-sample confidence intervals based on the two-sample removal method for different population sizes (N) and probabilities of capture (p): C is the probability that the first sample is greater than the second ($n_1 > n_2$).

N	p	Np^3	$16q^2(1+q)$	C	90% interval			95% interval		
					A_1	A_2	A_1/C	A_1	A_2	A_1/C
20	0·50	2·5	6	0·88	0·82	0·67	0·93	0·84	0·68	0·95
	0·75	8·5	1·3	0·98	0·98	0·83	1·00	0·98	0·85	1·00
40	0·25	0·6	15·8	0·68	0·57	0·44	0·84	0·58	0·45	0·85
	0·50	5	6	0·96	0·88	0·79	0·91	0·93	0·80	0·96
	0·75	17	1·3	1·00	1·00	0·84	1·00	1·00	0·89	1·00
80	0·05	0·0	28·2	0·44	0·43	0·06	1·00	0·43	0·06	1·00
	0·10	0·1	24·6	0·53	0·49	0·21	0·93	0·49	0·21	0·93
	0·25	1·3	15·8	0·78	0·60	0·56	0·77	0·64	0·59	0·83
	0·50	10	6	1·00	0·88	0·84	0·88	0·89	0·88	0·90
	0·75	34	1·3	1·00	0·96	0·87	0·96	0·99	0·91	0·99
100	0·05	0·0	28·2	0·46	0·46	0·09	0·99	0·46	0·10	0·99
	0·10	0·1	24·6	0·55	0·45	0·23	0·82	0·45	0·25	0·82
	0·25	1·6	15·8	0·81	0·64	0·60	0·79	0·68	0·63	0·84
	0·50	12·5	6	1·00	0·85	0·86	0·85	0·90	0·90	0·90
	0·75	42	1·3	1·00	0·92	0·89	0·92	0·97	0·91	0·97
200	0·05	0·0	28·2	0·50	0·40	0·13	0·79	0·40	0·16	0·80
	0·10	0·2	24·6	0·60	0·40	0·31	0·66	0·44	0·33	0·74
	0·25	3	15·8	0·90	0·74	0·72	0·82	0·78	0·75	0·86
	0·50	25	6	1·00	0·79	0·89	0·79	0·85	0·92	0·85
	0·75	85	1·3	1·00	0·71	0·90	0·71	0·80	0·93	0·80
400	0·05	0·1	28·2	0·53	0·33	0·20	0·62	0·38	0·21	0·71
	0·10	0·4	24·6	0·66	0·50	0·40	0·76	0·54	0·42	0·82
	0·25	6	15·8	0·97	0·82	0·82	0·85	0·86	0·85	0·89
	0·50	50	6	1·00	0·74	0·90	0·74	0·81	0·93	0·81
	0·75	169	1·3	1·00	0·36	0·90	0·36	0·43	0·94	0·43

the values of A_1 fluctuate (particularly as N increases), while those for A_2 steadily increase up to 0.90 and 0.94. This shows that the maximum-likelihood method is satisfactory for $N > 200$ and p not too small. In particular the simple rule (7.33) seems reasonable, though the coefficient 16 may be too small. We also note from Table 7.6 that (7.34), although requiring considerable calculation without a computer, can be used for small values of p, and for $N < 200$. However, such cases should be avoided in practice, as when p is small, C is also small, and the resulting confidence interval may be so wide as to be almost useless.

VARIABLE PROBABILITY OF CAPTURE.   The effect of variation in p from animal to animal on the maximum-likelihood estimate $\hat{N}$ can be investigated using the method of p. 316. From (7.25) the joint distribution of n_1 and n_2 is

$$\frac{N!}{n_1!\, n_2!\, (N-n_1-n_2)!}\, a_1^{n_1}(a_1 - a_2)^{n_2}(1 - 2a_1 + a_2)^{N-n_1-n_2}. \qquad (7.35)$$

Using the delta method, we find that, asymptotically,

$$E_B\,[\hat{N}] \;=\; \frac{(E[n_1])^2}{E[n_1 - n_2]} \;=\; \frac{Na_1^2}{a_2} \;=\; NB, \text{ say}, \qquad (7.36)$$

$$E_B\,[\hat{p}] \;=\; a_2/a_1 \qquad (7.37)$$

and

$$V_B\,[\hat{N}] \;=\; Na_1^3(4a_2^2 - 5a_1 a_2 - a_1 a_2^2 + 2a_1^2)/a_2^4\,,$$

where the suffix B denotes expectation with respect to (7.35). We note that when x_j is constant $(= p$, say$)$, $a_k = p^k$, and $V_B\,[\hat{N}]$ reduces to $V[\hat{N}]$ as expected.

From (7.27) our estimate of the variance of $\hat{N}$ is

$$\hat{V}[\hat{N}] \;=\; \hat{N}\hat{q}^2(1+\hat{q})/\hat{p}^3$$

and replacing each estimate by its asymptotic expected value (cf. (7.36) and (7.37)), we find that, asymptotically,

$$E[\hat{V}[\hat{N}]] \;=\; Na_1^3(4a_2^2 - 5a_1 a_2 - a_1 a_2^2 B^{-1} + 2a_1^2)/a_2^4.$$

Since

$$1 - B \;=\; \frac{a_2 - a_1^2}{a_2} \;=\; \frac{V[x]}{a_2} \;>\; 0,$$

B is less than unity and the net effect of variable catchability is that $\hat{N}$ and $\hat{V}[\hat{N}]$ will (asymptotically) underestimate N and $V_B\,[\hat{N}]$ respectively, while $\hat{p}$ will overestimate a_1 $(= E[x])$.

The effect of the shape of $f(x)$ on B has already been discussed in **3.2.2**(2). For example, when $f(x)$ is uniform on $[c,\, d]$ $(0 \leqslant c < d \leqslant 1)$ and $w = c/d$, then $B = 3(1+w)^2/[4(1+w+w^2)]$: for $w = 1$, $\frac{1}{2}$, $\frac{1}{5}$ and 0, the respective values of B are 1, $\frac{27}{28}$ $\frac{27}{31}$ and $\frac{3}{4}$. Unfortunately, $\hat{V}[\hat{N}]$ is also sensitive to the value of w; for the "worst" case $c = 0$, $d = 1$ we have

asymptotically
$$G = E[\hat{V}[\hat{N}]]/V_B[\hat{N}] = \tfrac{2}{3}.$$

Sometimes there are definite subgroups in the population due to age, sex, habitat preference, etc., so that $f(x)$ may be regarded as a probability function, namely

$$f(x_j) = \theta_j \quad (j = 1, 2, \dots, J; 0 \leqslant x_j \leqslant 1; \textstyle\sum_j \theta_j = 1),$$

where θ_j is the proportion of the population with probability of capture x_j. Then

$$1 - B = 1 - (\textstyle\sum x_j\theta_j)^2/(\textstyle\sum x_j^2\theta_j)$$

$$= (\textstyle\sum_{j<k}\textstyle\sum \theta_j\theta_k(x_j-x_k)^2)/(\textstyle\sum x_j^2\theta_j),$$

which will be small if the x_j are not too different. For example, when $J = 3$, $\theta_j = \tfrac{1}{3}$ ($j = 1, 2, 3$) and $x_1 = \tfrac{1}{4}$, $x_2 = \tfrac{1}{2}$, $x_3 = \tfrac{3}{4}$, then $B = \tfrac{6}{7}$ and $G = 0.90$.

REMOVAL BY MARKING. We shall now briefly consider the use of marking as a means of "removal". Suppose that n_1 animals caught in the first sample are marked and released back into the population and that m_2 are recaptured in the second sample of n_2 animals. Let p be the constant probability of capture of an unmarked animal and let p_0 be the probability that a marked animal is caught in the second sample. Then using (7.20) with $s = 2$, and $u_2 (= n_2 - m_2)$ instead of n_2, the joint probability function of n_1, u_2 and m_2 is

$$f(n_1, u_2, m_2) = f(n_1, u_2)f(m_2 \mid n_1, u_2)$$

$$= \frac{N!}{n_1!\, u_2!\, (N-n_1-u_2)!}\, p^{n_1+u_2}q^{2N-2n_1-u_2}\binom{n_1}{m_2}p_0^{m_2}q_0^{n_1-m_2}$$

which can be rearranged as the multinomial distribution

$$\frac{N!}{u_1!u_2!m_2!(N-n_1-u_2)!}(pq_0)^{u_1}(qp)^{u_2}(pp_0)^{m_2}(q^2)^{N-n_1-u_2}, \qquad (7.38)$$

where $u_1 = n_1 - m_2$. Estimates of N and p are the same as before, with u_2 replacing n_2; p_0 is estimated by m_2/n_1. However, the above model can be used for testing the hypothesis H that $p = p_0$. Setting $p = p_0$ in (7.38), the maximum-likelihood estimates of N and p are

$$\tilde{N} = (n_1+n_2)^2/4m_2$$

and

$$\tilde{p} = 2m_2/(n_1+n_2),$$

and from **1.3.6** (3) the multinomial goodness-of-fit statistic for testing H is

$$T = \frac{(u_1-\tilde{N}\tilde{p}\tilde{q})^2}{\tilde{N}\tilde{p}\tilde{q}} + \frac{(u_2-\tilde{N}\tilde{q}\tilde{p})^2}{\tilde{N}\tilde{p}\tilde{q}} + 0 + 0$$

$$= \frac{(u_1-u_2)^2}{(u_1+u_2)},$$

which is approximately χ_1^2 when H is true.

When $p = p_0$ we can compare the efficiencies of the removal estimate $\hat{N} = n_1^2/(n_1 - n_2)$ and the Petersen estimate $\hat{N}_p = n_1 n_2/m_2$. Thus, from **4.1.6**(1) (with $s = 2$) and (7.27), we have

$$\frac{V[\hat{N}_p]}{V[\hat{N}]} \approx \frac{Nq^2/p^2}{Nq^2(1+q)/p^3} = \frac{p}{1+q} < 1,$$

and when p is small the Petersen estimate is much more efficient than the removal estimate (see also **12.1.2**(5)). However, the cost of marking the first sample increases with p as more individuals are caught in the first sample, so that for large p the removal estimate may give more value for money. If the removal is not permanent, as is sometimes the case for large p, then the cost of marking is offset to some extent by the cost of storing the first sample until the second sample is taken.

Example 7.5   Hypothetical data: Seber and Le Cren [1967]

Electric fishing is carried out in an enclosed section of a small brook on two occasions under identical conditions. On the first fishing 49 trout are caught, marked and released alive. On the second fishing 50 trout are caught of which 24 are found to be marked. Thus $u_1 = 25$, $u_2 = 26$ and $m_2 = 24$ and the two-sample removal estimate of N is

$$\hat{N} = n_1^2/(n_1 - u_2) = 49^2/23 = 104$$

with

$$\hat{\sigma}[\hat{N}] = \frac{n_1 u_2 \sqrt{(n_1 + u_2)}}{(n_1 - u_2)^2} = \frac{49(26)\sqrt{75}}{23^2} = 20.$$

From the recapture data we can also calculate the modified Petersen estimate (cf. **3.1.1**)

$$N^* = \frac{(n_1 + 1)(n_2 + 1)}{m_2 + 1} - 1 = \frac{50(51)}{25} - 1 = 101$$

with standard deviation estimated by

$$\sqrt{v^*} = \left\{ \frac{(n_1 + 1)(n_2 + 1)u_1 u_2}{(m_2 + 1)^2 (m + 2)} \right\}^{\frac{1}{2}} = 10.$$

We note that N^* has a smaller standard deviation than $\hat{N}$, as expected.

To test $p = p_0$ we calculate

$$T = (25 - 26)^2/(25 + 26) = 0 \cdot 02$$

which, for one degree of freedom, indicates an excellent fit.

3   One-sample method

Suppose we take a single catch of size n from a population of size N. If an estimate $\hat{p}$ of p ($= 1/\theta$), the probability of capture, is available from other sources, then, assuming n to be binomially distributed, an estimate

of N is
$$\hat{N} = n/\hat{p} = n\hat{\theta}, \text{ say.}$$
Assuming $\hat{\theta}$ to be statistically independent of n, we have
$$E[\hat{N}] = NpE[\hat{\theta}] = N + Np\, E[\hat{\theta} - \theta]$$
and
$$V[\hat{N}] = (N^2p^2 + Npq)\, E[(\hat{\theta} - \theta)^2] + Nq(2E[\hat{\theta} - \theta] + \theta) - N^2p^2(E[\hat{\theta} - \theta])^2.$$
Usually $\hat{\theta}$ is obtained from another population with approximately the same value of p or from different areas of the same population. However, as pointed out by Seber and Le Cren [1967: 636], unless fairly precise estimates of θ are available, $\hat{N}$ will have a large coefficient of variation. The reader is referred to that article for a number of different methods of estimating θ; in each case $E[(\hat{\theta} - \theta)^2]$ is given (asymptotically) and $E[\hat{\theta} - \theta]$ is assumed to be negligible.

7.2.3 Regression estimates

When the probability of capture, p, is constant we have from (7.1) that
$$E[n_i \mid x_i] = p(N - x_i), \qquad (7.39)$$
a model considered by Hayne [1949b]. If a plot of n_i versus x_i, the accumulated catch, is approximately linear, then we have some qualitative evidence as to the stability of p. Once again the methods of **1.3.5** (3) provide point and interval estimates of p and N; the intercept of the fitted line on the x-axis will provide a graphical estimate of N.

Since
$$E[n_i] = Npq^{i-1},$$
then taking logarithms we have an alternative regression model
$$E[y_i] \approx \log(pN) + (i-1)\log q, \qquad (7.40)$$
where $y_i = \log n_i$. This model is of the same form as (7.9), so that the approximate formulae given in that section apply here. We also note that logarithms to the base 10 can be used in (7.40).

Example 7.6 Whitefish (*Coregonus clupeaformis*): Ricker [1958: 150]

A small lake on an island in Lake Nipigon, Ontario, was fished by gill nets in an identical manner for 7 successive weeks; the same sizes of nets, positions and lengths of sets were repeated each week. For whitefish of fork length 13–14 inches (33–35 cm) the weekly catches and their logarithms are given in Table 7.7.

A plot of n_i against x_i (Fig. 7.5) suggests that the model (7.39) is not unreasonable. Therefore least-squares estimates of p and N are
$$\tilde{p} = -\Sigma\, n_i(x_i - \bar{x})/\Sigma\, (x_i - \bar{x})^2 = 0.1895$$
and
$$\tilde{N} = \bar{x} + (\bar{n}/\tilde{p}) = 142.$$

325

TABLE 7.7
Weekly catches (n_i) from a whitefish population: data from Ricker [1958: 150].

Week (i)	1	2	3	4	5	6	7
Catch (n_i)	25	26	15	13	12	13	5
Cumulative catch (x_i)	0	25	51	66	79	91	104
$\log_{10} n_i$	1·40	1·42	1·18	1·11	1·08	1·11	0·70
$i - 1$	0	1	2	3	4	5	6

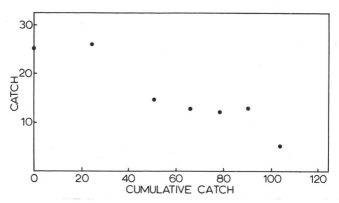

Fig. 7.5   Plot of catch (n_i) versus cumulative catch (x_i) for a population of whitefish: data from Ricker [1958].

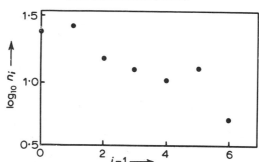

Fig. 7.6   Plot of $\log_{10} n_i$ versus $i - 1$ for a population of whitefish: data from Ricker [1958].

Using (7.6) and (7.7) with n_i and p instead of Y_i and K, approximate 95 per cent confidence intervals for N and p are (114, 207) and (0·108, 0·271), respectively.

The plot of $y_i = \log_{10} n_i$ against ($i - 1$) is approximately linear (Fig. 7.6), so that (7.40) can also be used. Least-squares estimates of $\log_{10} (1 - p)$ and $\log_{10} (pN)$ can be obtained in the usual manner, leading to the estimates

$$p' = 0·2070 \quad \text{and} \quad N' = 134.$$

Assuming normality, and taking the antilogarithm of the confidence interval for the slope of (7.40), gives a confidence interval for $1 - p$; thus the 95

per cent confidence interval for p is found to be (0·101, 0·300). Also from
p. 302,

$$V[N'] \approx N^2 \sigma_y^2 \left[\frac{1}{s} + \left(\frac{\frac{1}{2}(s+1)p - 1}{p} \right)^2 \cdot \frac{12}{s(s-1)(s+1)} \right],$$

where σ_y^2 is estimated from the usual residual sum of squares about the
regression line. Assuming N' to be approximately normally distributed, an
approximate 95 per cent confidence interval for N is given by $N' \pm 1 \cdot 96 \sqrt{v'}$,
where v' is the estimate of $V[N']$. This interval (which is of doubtful
validity) is found to be (122, 146).

Finally, applying Zippin's maximum-likelihood method (cf. Example
7.3 on p. 314),

$$R = \frac{n_2 + 2n_3 + \ldots + 6n_7}{n_1 + n_2 + \ldots + n_7} = \frac{238}{109} = 2 \cdot 18,$$

and entering this value in Fig. 7.3 and 7.4 with $s = 7$, we find that $\hat{p} = 0 \cdot 19$
and $(1 - \hat{q}^7) = 0 \cdot 77$. Hence

$$\hat{N} = 109/0 \cdot 77 = 142$$

and an approximate 95 per cent confidence interval for N is (109, 175).

Further examples of the above regression methods, together with graphs,
are given in Hayne [1949b], Barbehenn [1958], Menhinick [1963], Webb
[1965], Golley et al. [1965], Southwood [1966: 182], Andrzejewski and
Jezierski [1966], Buchalczyk and Pucek [1968], and Pucek [1969].

UTILISING TAG DATA. If a tag release is made at the beginning of the
experiment then the respective regression lines from the tagged and untagged
populations can be compared as in 7.1.6. For example, when the two graphs
are approximately linear we can test the hypothesis that the two regression
lines are parallel, i.e. that p is the same for tagged and untagged.

327

CHAPTER 8

CATCH-EFFORT METHODS: OPEN POPULATION

8.1 CONTINUOUS SAMPLING

8.1.1 Mortality only

1 Catch equations

Consider a population which is closed, apart from depletion through continuous exploitation and natural mortality. Let

$$N(t) = \text{total population size at time } t,$$
$$n(t) = \text{number caught in the interval } [0, t],$$

and

$$\overset{o}{f}(t) = \text{effort } per\ unit \text{ time operating at time } t.$$

We shall assume that the natural mortality and mortality due to exploitation are both Poisson processes with parameters μ and $k\overset{o}{f}(t)$ respectively. We note in passing that in **7.1.1** we defined the sampling (exploitation) process as one in which the probability of an individual being caught when subject to δf units of effort is $k\delta f + o(\delta f)$, where k is the Poisson catchability coefficient (cf. **7.1.1**). When the effort depends on the time spent, we have $\delta f = \overset{o}{f}(t)\delta t$.

From the above definitions and assumptions Chapman [1961] shows that (cf. (1.1))

$$E[N(t)] = N(0) \exp(-\mu t - k\overset{o}{F}(t)), \tag{8.1}$$

where $\overset{o}{F}(t) = \int_0^t \overset{o}{f}(x)\, dx$ is the accumulated effort up to time t, and

$$E[n(t)] = \int_0^t k\overset{o}{f}(x)N(x)\, dx. \tag{8.2}$$

Suppose that the whole sampling period can be divided up into s periods $[0, t_1), [t_1, t_2), \ldots, [t_{s-1}, t_s)$ during which the effort per unit time is constant and equal to $\overset{o}{f_1}, \overset{o}{f_2}, \ldots, \overset{o}{f_s}$ respectively; thus $\overset{o}{f}(t) = \overset{o}{f_i}$ for $t_{i-1} \leqslant t < t_i$ ($i = 1, 2, \ldots, s$; $t_0 = 0$). Let

N_i = size of population at the *beginning* of the ith period,
n_i = size of catch in the ith period,
$\Delta_i = t_i - t_{i-1}$,
$f_i = \Delta_i \overset{o}{f_i}$ (total effort in ith period),

μ_{Ei} = $k\overset{o}{f_i}$ (instantaneous exploitation rate in the ith period),
Z_i = $\mu + \mu_{Ei}$ (total instantaneous mortality rate),

and

F_i = $\sum\limits_{j=1}^{i} f_j$ (accumulated effort up to *and including* the ith period,
　　　　i.e. accumulated effort up to time t_i).

(We note that $N_1 = N(0)$, and the definition of F_i differs from that in **7.1.1** where the ith period is not included. A number of authors (e.g. Chapman [1961]) use the symbol f_i instead of $\overset{o}{f_i}$ to represent the effort per unit time, and sometimes the distinction is not clearly made). Then, from (8.1) and (8.2) it is readily shown that

$$E[N_i] = N_1 \exp(-\mu t_{i-1} - kF_{i-1}),\qquad(8.3)$$

$$E[N_{i+1}|N_i] = N_i \exp(-Z_i\Delta_i),\qquad(8.4)$$

and

$$E[n_i|N_i] = k\overset{o}{f_i} \int_0^{\Delta_i} N_i \exp(-Z_ix)\,dx$$

$$= k\overset{o}{f_i}N_i[1 - \exp(-Z_i\Delta_i)]/Z_i.\qquad(8.5)$$

We note that (8.5) can also be written in the form

$$E[n_i|N_i] = kf_i\bar{N}_i,\qquad(8.6)$$

where 　$\bar{N}_i = \int_0^{\Delta_i} N_i \exp(-Z_ix)\,dx / \int_0^{\Delta_i} dx$

$$= N_i[1 - \exp(-Z_i\Delta_i)]/(Z_i\Delta_i)$$

is the average population size during the ith period. The above equations, often referred to as the "catch" equations, have been widely used in fishery work and go back to Baranov [1918] and Ricker [1940, 1944].

2　Estimation

MORTALITY AND CATCHABILITY. We shall now consider a number of methods for estimating μ and k given the $\{n_i\}$ and $\{\overset{o}{f_i}\}$.

From (8.5), with $i + 1$ instead of i, and (8.4), we have

$$E[n_{i+1}|N_i] = \frac{\Delta_{i+1}k\overset{o}{f}_{i+1}}{\Delta_{i+1}Z_{i+1}} \cdot N_i \exp(-Z_i\Delta_i)\{1 - \exp(-Z_{i+1}\Delta_{i+1})\}.$$

Dividing by (8.5) and taking logarithms, we are led to consider the model

$$E[(y_i/\Delta_i)] \approx \frac{1}{\Delta_i} \log\left[\frac{Z_{i+1}\Delta_{i+1}\{1 - \exp(-Z_i\Delta_i)\}}{Z_i\Delta_i\{1 - \exp(-Z_{i+1}\Delta_{i+1})\}}\right] + \mu + k\overset{o}{f_i},\quad(8.7)$$

where $y_i = \log(n_i/f_i) - \log(n_{i+1}/f_{i+1})$. Widrig [1954] and, independently, Beverton and Holt [1957: 192] first developed this model and suggested regressing y_i/Δ_i on $\overset{o}{f_i}$ by least squares; Beverton and Holt also gave an

iterative procedure to take into account the non-linear term. However, if
$W = Z\Delta$ is small for each period then

$$\log\left[1 - e^{-W}\right)/W] \approx -\tfrac{1}{2}W, \tag{8.8}$$

and, to a first order of approximation, the non-linear term of (8.7) is given by

$$\tfrac{1}{2}(Z_{i+1}\Delta_{i+1} - Z_i\Delta_i)/\Delta_i = \tfrac{1}{2}\left[(\Delta_{i+1} - \Delta_i)\mu + (f_{i+1} - f_i)k\right]/\Delta_i.$$

Paloheimo [1961] shows that (8.8) is a satisfactory approximation for the
likely range $0{\cdot}10 \leqslant W \leqslant 1{\cdot}00$. Using the above approximation, (8.7) becomes

$$E[y_i] \approx \mu(\Delta_i + \Delta_{i+1})/2 + k(f_i + f_{i+1})/2, \quad i = 1, 2, \ldots, s-1, \tag{8.9}$$

which, for the common situation $\Delta_i = 1$ (e.g. one year), reduces to Paloheimo's
[1961] linear model:

$$E[y_i] \approx \mu + k(f_i + f_{i+1})/2. \tag{8.10}$$

We note that (8.9) can be analysed as a multiple linear regression, provided
the $\overset{o}{f_i}$ vary. When $\overset{o}{f_i} = \overset{o}{f}$, say, then (8.9) reduces to

$$E[y_i] \approx (\mu + k\overset{o}{f})(\Delta_i + \Delta_{i+1})/2$$

and μ and k cannot be estimated separately; only $Z = \mu + k\overset{o}{f}$ is estimable.
In this case, when $\Delta_i = 1$, the instantaneous natural mortality and exploitation
rates are both constant and the method of **6.2.2** can also be used here for
estimating Z (with m_i and M_0 replaced by n_i and N_1; (6.13) does not depend
on knowing M_0).

In analysing (8.9) and (8.10) we have to take into account the fact that
the y_i will be correlated even if the n_i are assumed to be independent.
Several approaches are possible; if n_i is not small compared with N_1 we can
assume the n_i to be multinomially distributed and use the delta method to
derive the variance—covariance matrix (cf. **6.2.2**); if n_i is small compared
with N_i we can ignore the dependence between the n_i and, following Chapman
[1961: 160], assume the $\log n_i$ to be independently normally distributed with
common variance σ^2. Then, since

$$V[y_i] = V[\log n_i - \log n_{i+1}] = 2\sigma^2, \text{ etc.,}$$

we find that the variance—covariance matrix of $\mathbf{y}' = (y_1, y_2, \ldots, y_{s-1})$ is
$\sigma^2\mathbf{B}$, where

$$\mathbf{B} = \begin{bmatrix} 2 & -1 & 0 & 0 & \cdots & 0 & 0 \\ -1 & 2 & -1 & 0 & \cdots & 0 & 0 \\ 0 & -1 & 2 & -1 & \cdots & 0 & 0 \\ \cdots & \cdots & \cdots & \cdots & \cdots & \cdots & \cdots \\ 0 & 0 & 0 & 0 & \cdots & -1 & 2 \end{bmatrix}$$

and

$$B^{-1} = \frac{1}{s} \begin{bmatrix} s-1, & s-2, & s-3 & \ldots & 3, & 2, & 1 \\ s-2, & 2(s-2), & 2(s-3) & \ldots & 6, & 4, & 2 \\ s-3, & 2(s-3), & 3(s-3) & \ldots & 9, & 6, & 3 \\ \ldots & \ldots & \ldots & \ldots & \ldots & \ldots & \ldots \\ 3, & 6, & 9 & \ldots & (s-3)3, & (s-3)2, & s-3 \\ 2, & 4, & 6 & \ldots & (s-3)2, & (s-2)2, & s-2 \\ 1, & 2, & 3 & \ldots & (s-3), & s-2, & s-1 \end{bmatrix}.$$

Thus, in vector notation (8.10) is $y = X\beta + e$, where

$$X = \begin{bmatrix} 1, & (f_1+f_2)/2 \\ 1, & (f_2+f_3)/2 \\ \ldots, & \ldots \\ 1, & (f_{s-1}+f_s)/2 \end{bmatrix}, \qquad \beta = \begin{bmatrix} \mu \\ k \end{bmatrix},$$

and e has a multivariate normal distribution $\mathcal{N}[0, \sigma^2 B]$. This model can now be analysed using the general method of 1.3.5 (2).

INITIAL POPULATION SIZE. We now turn our attention to the problem of estimating N_1, the initial population size. From (8.5) and (8.3) we have

$$E[n_i] = \Delta_i \, k\overset{o}{f_i}N_1 \exp(-\mu t_{i-1} - kF_{i-1})[1 - \exp(-Z_i\Delta_i)]/Z_i\Delta_i. \qquad (8.11)$$

Taking logarithms and using the approximation (8.8), we arrive at the multiple linear regression

$$E[\log(n_i/f_i)] \approx \log(N_1 k) - \mu(t_{i-1} + \tfrac{1}{2}\Delta_i) - k(F_{i-1} + \tfrac{1}{2}f_i)$$

$$= \log(N_1 k) - \mu(t_{i-1} + t_i)/2 - k(F_{i-1} + F_i)/2, \qquad (8.12)$$

for which least-squares estimates of k, μ, $\log(N_1 k)$ and hence N_1 can be obtained in the usual way. Estimates of the N_i can then be calculated iteratively from (8.4). The model (8.12) was first derived by Chapman [1961: 161], though there is a small misprint; in terms of our notation the left-hand side of his equation is $E[\log(n_i/\overset{o}{f_i})]$.

3 *Seasonal fishery*

Chapman [1961: 155] and Muir and White [1963] have discussed the problem of a seasonal fishery where exploitation takes place only for a fraction τ_i of each period at the beginning of the period. Equations (8.4) and (8.5) become

$$E[N_{i+1} | N_i] = N_i \exp(-Z_i\tau_i\Delta_i) \exp\{-\mu(1-\tau_i)\Delta_i\}$$

$$= N_i \exp\{-(\mu + k\overset{o}{f_i}\tau_i)\Delta_i\}$$

and
$$E[n_i \mid N_i] \;=\; k \, \overset{o}{f_i} N_i \{1 - \exp\left(-Z_i \tau_i \Delta_i\right)\}/Z_i.$$

Using the same method as above and redefining the total effort in the ith period, namely $f_i = \Delta_i \tau_i \overset{o}{f_i}$, (8.9) becomes

$$E[y_i] \;\approx\; \mu \left[\Delta_i + \tfrac{1}{2}\left(\Delta_{i+1}\tau_{i+1} - \Delta_i \tau_i\right)\right] + k(f_i + f_{i+1})/2.$$

When $\Delta_i = 1$ and $\tau_i = \tau$ this equation reduces to (8.10) which, as pointed out by Muir and White [1963], can therefore be used for a seasonal fishery, provided the efforts f_i are correctly calculated.

4   Age-composition data

In using (8.10) the different y_i values can be based on different segments of the population. For example, let N_1 be the size of the age-class of age a or older (designated $a+$) and let N_2 be the size of the same age-class a year later (i.e. of age $(a+1)+$). If n_1 and n_2 are the respective catches from this class in the first two years, then from (8.10) we have

$$E[\log(n_1 f_2 / n_2 f_1)] \;\approx\; k + \mu(f_1 + f_2)/2.$$

For the second and third years we now redefine N_2 to be the size of the age-class $a+$ in the second year, and N_3 the size of this class one year later. If n_2' and n_3' are the respective catches from this class in the second and third years, then

$$E[\log(n_3' f_2 / n_2' f_3)] \;\approx\; k + \mu(f_2 + f_3)/2.$$

Thus by recording the number caught in the age-classes $a+$ and $(a+1)+$ for each year, a regression model of the form (8.10) can be used.

If there is a large number of individuals growing into the age-class a each year, the statistical dependence between the ratios $n_1 f_2 / n_2 f_1$, $n_2' f_3 / n_3' f_2$, etc. will not be as strong as between $n_1 f_2 / n_2 f_1$, $n_2 f_3 / n_3 f_2$, etc. For this situation an unweighted least-squares estimation of k and μ may be more appropriate than Chapman's method using the variance–covariance matrix $\sigma^2 \mathbf{B}$. However, although equations (8.4) and (8.5) apply in this situation, (8.3) (and hence (8.12)) no longer apply as N_i refers to a different population for each pair of consecutive years.

The essential feature of the above approach is that the same cohort is considered for each pair of consecutive catches. Also the same efforts are used throughout, though strictly speaking f_i is the effort applied to the whole population and not just the cohort. This means that if trapping is used to catch the individuals, the effort with respect to the cohort will begin to fall off as the traps fill up with individuals not belonging to the cohort. It would seem desirable, therefore, to choose the age a so as to include as much of the total catch data as possible in the analysis, or else modify the effort data f_i in some reasonable manner. For example, one possibility is to estimate the "effective" number of traps as the average of the total number of traps and the number of traps at the end of a sampling period not filled by individuals from outside the cohort.

If some of the age-classes are not fully recruited (i.e. catchable) then a should be chosen as the youngest fully recruited age-group.

When sample sizes are large, as in commercial fishing, it requires considerable time and expense to age each individual in the sample. However, if age and some easily measured quantity such as length are correlated, then a simpler method is to measure the length of each individual and then either choose a random subsample from the whole sample for age determinations, or else stratify the sample according to length-classes and choose a subsample from each class (Tanaka [1953]). Let

X_i = number in the sample belonging to the ith length-class ($i = 1, 2, \ldots, I$),

X_{ai} = number of age a in the ith length-class,

X_a = $\sum_i X_{ai}$ (number of age a in the sample),

x_i = number in the subsample belonging to the ith length-class,

and x_{ai} = number from the group of x_i individuals of age a.

We shall introduce a random variable y_a which is defined to be equal to unity if an individual chosen at random is of age a and zero otherwise. Let y_{aij} be the value of y_a for the jth member of the ith class in the subsample; then, irrespective of which of the above two subsampling methods is used, an unbiased estimate of X_a is given by

$$\hat{X}_a = \sum_{i=1}^{I} \frac{X_i}{x_i} \sum_{j=1}^{x_i} y_{aij}.$$

5   Variable catchability coefficient

If the catchability k is not constant, we can work with the parameters μ and μ_{Ei} instead. By the same method as was used in deriving (8.5) it is readily shown that

$$E[n_i \mid N_i] = \mu_{Ei} N_i [1 - \exp(-Z_i \Delta_i)]/Z_i. \qquad (8.13)$$

From the ratios n_i/n_{i+1} Murphy [1965] gives an iterative method for estimating all the μ_{Ei}, given that one has estimates of μ and just one of the μ_{Ei}. If we define $\mu_{Ei} = k_i \overset{o}{f}_i$ then k_i can also be estimated.

AGE DATA. If the catchability varies with age (e.g. Muir [1963]), Muir [1964]) shows how to modify the method in section 4 above to allow for this. Let k_{a+} and $k_{(a+1)+}$ be the (constant) catchabilities for the age-classes $a+$ and $(a+1)+$ respectively; then, using the general catch equations we find that (8.10) now becomes

$$E[y_i] \approx \mu - \log\left(\frac{k_{(a+1)+}}{k_{a+}}\right) + \tfrac{1}{2} k_{a+} \left(f_i + \frac{k_{(a+1)+}}{k_{a+}} f_{i+1}\right). \qquad (8.14)$$

If an independent estimate of $k_{(a+1)+}/k_{a+}$ is available, then the above linear regression can be analysed as in section 4.

Paloheimo and Kohler [1968: 564, 568] suggest setting up a linear regression for each pair of age-classes a and $a+1$ and give a combined nonlinear model which allows the natural mortality to be age-dependent as well as the catchability.

Example 8.1   Maskinonge (*Esox masquinongy*): Muir [1964]

Working with the age-group $a =$ IV and an estimate $k_{V+}/k_{IV+} = 1\cdot2$, Muir used (8.14) to analyse the data in Table 8.1. Here

$$y_i = \log\left(\frac{n_{i,IV+}/f_i}{n_{i+1,V+}/f_{i+1}}\right) \quad \text{and} \quad x_i = \tfrac{1}{2}(f_i + 1\cdot2\, f_{i+1});$$

for example, $y_1 = \log(1\cdot42/0\cdot82) = 0\cdot5481$ and $x_1 = \tfrac{1}{2}(90 + 1\cdot2(102)) = 106$.

TABLE 8.1

Catch and effort data for age-groups IV+ and V+ for a population of maskinonge: from Muir [1964: Table 4].

Year i	Age-class	Catch n_i	Effort f_i	$\dfrac{n_i}{f_i}$	y_i	x_i
1952	IV+	128	90	1·42	0·5481	106
	V+	89		0·99		
1953	IV+	103	102	1·01	0·5710	114
	V+	84		0·82		
1954	IV+	104	105	0·99	−0·0726	144
	V+	60		0·57		
1955	IV+	215	152	1·41	0·4187	168
	V+	164		1·07		
1956	IV+	217	153	1·42	1·0438	352
	V+	143		0·93		
1957	IV+	369	458	0·81	1·3987	475
	V+	231		0·50		
1958	IV+	138	410	0·34	0·8198	385
	V+	81		0·20		
1959	IV+	84	299	0·28	0·2390	292
	V+	44		0·15		
1960	IV+	129	237	0·54		
	V+	53		0·22		

Writing (8.14) in the form

$$E[y_i] = \beta_0 + \beta x_i,$$

where $\beta_0 = \mu - \log 1\cdot2$ and $\beta = k_{a+}$, we find that the least-squares estimates are

$$\hat{\mu} \ = \ \hat{\beta}_0 + \log 1\cdot 2 \ = \ 0\cdot 0008 + 0\cdot 1823 \ = \ 0\cdot 1831$$
and
$$\hat{k}_{a+} \ = \ 0\cdot 0024.$$

Assuming the y_i to be independently normally distributed with constant variance, the usual 95 per cent confidence interval for the slope of the regression line yields $0\cdot 0024 \pm 0\cdot 0022$. This wide interval reflects the wide dispersion about the fitted line (Fig. 8.1); from the graph the dispersion appears to be less when the fishing effort is greater (the trend in the graph is accentuated by the compression of the x-axis).

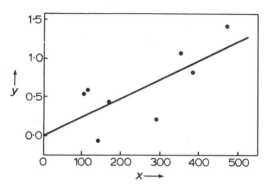

Fig. 8.1   Application of Muir's regression model (8.14) to a population of maskinonge: redrawn from Muir [1964].

8.1.2   Recruitment only

Consider a population in which recruitment and removal through exploitation are the only processes affecting the population size. Then from Fischler [1965] we have the following deterministic modification of Leslie's regression model $(7.1.3\,(1))$ to allow for recruitment:

Using the notation of **8.1.1**, let r_i be the number recruited in the ith interval. Then

$$N_{i+1} \ = \ N_i - n_i + r_i \tag{8.15}$$

and from (8.6) we have the deterministic relationship

$$n_i \ = \ k \, f_i \bar{N}_i \tag{8.16}$$

where $\bar{N}_i$ is the average population size during the ith interval. Assuming recruitment to be proportional to population size, i.e. $r_i = a_i \bar{N}_i$, and approximating $\bar{N}_i = \frac{1}{2}(N_i + N_{i+1})$, we have from (8.15) the recurrence relationship

$$N_{i+1} \ = \ \left(\frac{2 + a_i}{2 - a_i} \right) N_i - \frac{2n_i}{2 - a_i} \, . \tag{8.17}$$

Substituting for N_{i+1} in (8.16) leads to

$$\frac{n_i}{f_i} \ = \ k \left[\frac{2N_i}{2 - a_i} - \frac{n_i}{2 - a_i} \right] \tag{8.18}$$

335

and applying (8.17) recursively gives us:

$$\frac{n_1(2-a_1)}{f_1\,2} = k\left(N_1 - \frac{n_1}{2}\right),$$

$$\frac{n_2}{f_2}\frac{(2-a_2)}{2}\frac{(2-a_1)}{(2+a_1)} = k\left(N_1 - \frac{2n_1}{2+a_1} - \frac{(2-a_1)n_2}{(2+a_1)2}\right),$$

and, for $i > 2$,

$$\frac{n_i}{f_i}\cdot\frac{(2-a_i)(2-a_{i-1})\ldots(2-a_1)}{2(2+a_{i-1})\ldots(2+a_1)} = k\left(N_1 - \frac{2n_1}{2+a_1} - \frac{2(2-a_1)}{(2+a_2)(2+a_1)}\cdot n_2\right.$$

$$-\ldots- \frac{2(2-a_{i-2})(2-a_{i-3})\ldots(2-a_1)}{(2+a_{i-1})(2+a_{i-2})(2+a_{i-3})\ldots(2+a_1)}\cdot n_{i-1}$$

$$\left.- \frac{(2-a_{i-1})(2-a_{i-2})\ldots(2-a_1)}{(2+a_{i-1})(2+a_{i-2})\ldots(2+a_1)}\cdot\frac{n_i}{2}\right).$$

If independent estimates of the a_i are available, such as from age data, then substituting these estimates in the above equations leads to a regression model of the form

$$y_i = k(N_1 - x_i) + e_i, \tag{8.19}$$

which can be analysed as in **1.3.5**(*3*). Once least-squares estimates of N_1 and k are calculated, estimates of the N_i can be obtained iteratively from (8.17).

Example 8.2   Blue crab (*Callinectes sapidus*): Fischler [1965]

Using a time-interval of one week, estimates $\hat{a}_i$ of the weekly recruitment rate were calculated from such information as the ratio of pre-commercial to commercial size crabs in catches and the moulting rate. From these estimates the values of y_i and x_i above were calculated and set out in Table 8.2. A plot of y_i against x_i is fairly linear (Fig. 8.2), thus supporting the use of (8.19). Therefore, using the method of **1.3.5**(*3*) with $w_i = 1$, we find that the least-squares estimates of k and N_1 are $\hat{k} = 0\cdot000\,427$ and $\hat{N}_1 = 454\,413$ with 95 per cent confidence intervals $(0\cdot000\,375, 0\cdot000\,479)$ and $(432\,749, 480\,583)$ respectively.

8.1.3   **Variable catchability: recruitment estimates**

If the catchability k is not constant or effort data are not available, then we must work with the instantaneous natural mortality and exploitation rates μ and μ_{Ei} respectively. Allen [1966, 1968] gives the following deterministic method for estimating the level of recruitment each year: by recruitment we shall mean the process whereby an individual becomes catchable.

Suppose that exploitation is carried out continuously for s consecutive periods of time, which we shall call years for convenience, and suppose that

TABLE 8.2
Weekly crab catch in pounds weight per standard unit of effort, and estimates of weekly recruitment rate: from Fischler [1965: Table 13].

i (week)	n_i/f_i (pounds)	$\hat{a}_i$ (recruitment rate)	y_i	x_i
1	186·0	0·147	172·33	2 167·00
2	244·5	0·147	195·51	72 032·12
3	226·7	0·161	155·27	124 512·05
4	203·9	0·245	113·42	170 864·93
5	196·4	0·245	85·40	212 897·56
6	292·5	0·245	99·43	256 123·54
7	210·0	0·245	55·80	287 210·05
8	194·5	0·175	42·02	302 997·27
9	293·3	0·161	53·57	318 657·09
10	370·0	0·126	58·60	336 895·08
11	350·6	0·112	49·32	353 735·97
12	373·7	0·112	46·99	368 371·92
13	266·1	0·112	29·91	380 038·30
14	274·2	0·105	27·66	389 870·93
15	283·7	0·084	26·04	399 069·71
16	258·8	0·084	21·84	405 681·19
17	238·4	0·175	17·62	410 620·66
18	246·6	0·175	15·29	415 131·36
19	202·3	0·091	11·01	448 464·97
20	170·2	0·077	8·52	420 692·93

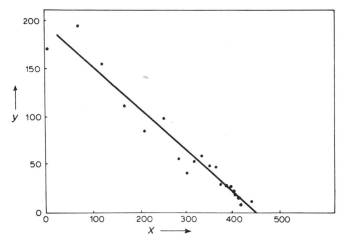

Fig.8.2 Application of Fischler's regression model (8.19) to a population of blue crabs: redrawn from Fischler [1965].

all animals caught can be assigned to one of $J + 1$ age-groups. We shall assign an individual to the age-class a if its age lies in the interval $[a, a + 1)$. Although recruitment is sometimes a continuous process, we shall assume that all recruiting takes place at the beginning of each year.

For $i = 1, 2, \dots, s$ and $a = 0, 1, \dots, J$ let:

N_i = size of the total population, including both catchable and non-catchable (i.e. non-recruited) members, at the beginning of year i,

$N_{i,a}$ = size of the age-class a at the beginning of year i,

n_i = total catch in year i,

$n_{i,a}$ = number caught in year i which belonged to age-class a at the beginning of the year,

$P_{i,a}$ = $n_{i,a}/n_i$,

$\rho_{i,a}$ = proportion of the group of size $N_{i,a}$ which is catchable at the beginning of year i,

$R_{i,a}$ = proportion of the catchable group of $\rho_{i,a} N_{i,a}$ individuals newly recruited at the beginning of year i,

$\bar{r}_{i,a}$ = average number of catchable individuals alive during year i from the group $\rho_{i,a} N_{i,a}$ individuals,

p_i = probability that a catchable member of the N_i individuals is caught in the ith year,

Z_i = $\mu_i + \mu_{Ei}$ = total instantaneous mortality rate for the catchable part of the partially recruited age-classes in year i,

Z'_i = $\mu'_i + \mu'_{Ei}$ = total instantaneous mortality rate for the fully re-cruited age-classes in year i,

and μ''_i = instantaneous natural mortality rate for the unrecruited part of the population in year i.

Assuming that the age-classes $a = A, A + 1, \dots, J$ are fully recruited (i.e. $\rho_{i,a} = 1$), we also define

$$N_{i(A)} = \sum_{a=A}^{J} N_{i,a} ;$$

$n_{i(A)}$, $P_{i(A)}$, $\rho_{i(A)}$ and $\bar{r}_{i(A)}$ are all similarly defined. The aim of the following analysis is to provide estimates of the $R_{i,a}$.

For $a < A$ we have the deterministic equation

$$N_{i+1,a+1} = \rho_{i,a} N_{i,a} e^{-Z_i} + (1 - \rho_{i,a}) N_{i,a} e^{-\mu''_i} ,$$

while for the fully recruited age-classes,

$$N_{i+1(A+1)} = N_{i(A)} e^{-Z'_i} .$$

Also

$$\bar{r}_{i,a} = \left(\int_0^1 \rho_{i,a} N_{i,a} e^{-Z_i t} dt \right) \Big/ \left(\int_0^1 dt \right)$$

$$= \rho_{i,a} N_{i,a} (1 - e^{-Z_i})/Z_i$$

and, by the same argument,

$$\bar{N}_{i(A)} \;=\; N_{i(A)}\,(1-e^{-Z'_i})/Z'_i \;,$$

where $\bar{N}_{i(A)}$ is the average number alive, during year i, from the group represented by $N_{i(A)}$. Therefore

$$E[n_{i,\,a}] \;=\; p_i\,\bar{r}_{i,\,a}, \qquad E[n_{i(A)}] \;=\; p_i\,\bar{N}_{i(A)}, \tag{8.20}$$

and equating random variables to the expectations, we have the deterministic equations

$$\frac{P_{i,\,a}}{P_{i(A)}} \;=\; \frac{n_{i,\,a}}{n_{i(A)}} \;=\; \frac{\bar{r}_{i,\,a}}{\bar{N}_{i,\,a}}.$$

For the next year we have a similar expression

$$\frac{P_{i+1,\,a+1}}{P_{i+1(A+1)'}} \;=\; \frac{\bar{r}_{i+1,\,a+1}}{\bar{N}_{i+1\,(A+1)}}.$$

Using the proportions $P_{i,\,a}$ rather than the actual numbers $n_{i,\,a}$ means that these proportions can be estimated from a subsample of the n_i without having to collect age data for the whole sample. From the above equations we have (introducing $B_{i,\,a}$ as a convenient intermediary statistic)

$$
\begin{aligned}
B_{i,\,a} \;&=\; \frac{P_{i(A)}\,P_{i+1,\,a+1}}{P_{i,\,a}\,P_{i+1(A+1)}} \\[2mm]
&=\; \frac{\rho_{i+1,\,a+1}\big[\rho_{i,\,a}\,e^{-Z_i} + (1-\rho_{i,\,a})\,e^{-\mu''_i}\big]\left[\dfrac{1-e^{-Z_{i+1}}}{Z_{i+1}}\right]\left[\dfrac{1-e^{-Z'_i}}{Z'_i}\right]}{\rho_{i,\,a}\,e^{-Z'_i}\left[\dfrac{1-e^{-Z_i}}{Z_i}\right]\left[\dfrac{1-e^{-Z'_{i+1}}}{Z'_{i+1}}\right]}.
\end{aligned}
\tag{8.21}
$$

Allen points out that although Z_i may vary considerably from year to year, it is unlikely that the relation between the mortality rates for the partially recruited age-classes and the fully recruited age-classes will change appreciably from one year to the next. In particular, since $[1 - \exp(-Z'_i)]/Z'_i$ is the average annual probability of survival for the fully recruited age-classes, we would expect

$$\left(\frac{1-e^{-Z'_i}}{Z'_i}\right)\bigg/\left(\frac{1-e^{-Z_i}}{Z_i}\right) \;\approx\; \left(\frac{1-e^{-Z'_{i+1}}}{Z'_{i+1}}\right)\bigg/\left(\frac{1-e^{-Z_{i+1}}}{Z_{i+1}}\right)$$

and these terms can be cancelled out in (8.21). Also it is not unreasonable to assume that $\mu''_i = \mu_i$, so that

$$
\begin{aligned}
B_{i,\,a} \;&=\; \frac{\rho_{i+1,\,a+1}\big[\rho_{i,\,a}\,e^{-Z_i} + (1-\rho_{i,\,a})\,e^{-\mu_i}\big]}{\rho_{i,\,a}\,e^{-Z'_i}} \\[2mm]
&=\; \frac{\rho_{i+1,\,a+1}}{\rho_{i,\,a}}\,T_i\,[\rho_{i,\,a} + (1-\rho_{i,\,a})\,e^{\mu E i}],
\end{aligned}
\tag{8.22}
$$

where $T_i = e^{Z_i' - Z_i}$.

From the definition of $R_{i+1, a+1}$, we have

$$1 - R_{i+1, a+1} = \frac{\rho_{i,a} N_{i,a} e^{-Z_i}}{\rho_{i+1, a+1} N_{i+1, a+1}}$$

$$= \frac{\rho_{i, a}}{\rho_{i+1, a+1}} \cdot \frac{1}{\rho_{i, a} + (1 - \rho_{i,a}) e^{\mu E_i}} \qquad (8.23)$$

and substituting from (8.22) leads to

$$1 - R_{i+1, a+1} = T_i / B_{i,a}.$$

For most practical purposes it is probably satisfactory to take $T_i = 1 \cdot 0$, though an approximate estimate is available if certain assumptions can be made. For example, assuming that the total instantaneous mortality rate of the youngest fully recruited age-class A is the same as the total instantaneous mortality rate of the recruited part of the partially recruited age-classes, we have, using (8.20),

$$\frac{P_{i+1, A+1}}{P_{i, A}} = \frac{n_{i+1, A+1}}{n_{i+1}} \cdot \frac{n_i}{n_{i, A}}$$

$$= \frac{n_i}{n_{i+1}} \cdot \frac{p_{i+1} \bar{N}_{i+1, A+1}}{p_i \bar{N}_{i, A}}$$

$$= \frac{n_i p_{i+1}}{n_{i+1} p_i} \cdot \frac{(1 - e^{-Z_i'}) Z_i'}{(1 - e^{-Z_{i+1}'}) Z_{i+1}'} \cdot \frac{N_{i+1, A+1}}{N_{i, A}}$$

$$= c \, e^{-Z_i} \text{, say,}$$

since we have assumed that

$$N_{i+1, A+1} = N_{i, A} \, e^{-Z_i}.$$

Using a similar argument, we find that

$$\frac{P_{i+1(A+1)}}{P_{i(A)}} = c \, e^{-Z_i'}$$

so that T_i can be estimated by

$$\hat{T}_i = \frac{P_{i+1, A+1} \, P_{i(A)}}{P_{i, A} \, P_{i+1(A+1)}}.$$

This leads to the estimate $(B_{i, a} - \hat{T}_i)/B_{i, a}$ for $R_{i+1, a+1}$.

Allen [1966] also gives a number of methods for estimating k and N_i, and the reader is referred to his article for further details.

8.2   INSTANTANEOUS SAMPLES

8.2.1   Mortality only: regression method

We shall now consider a model in which samples may be regarded as "instantaneous" or point samples removed at times $t_1, t_2, \ldots, t_s$. Let

N = size of the population just before the first sample,

p_i = probability that an individual in the population at time t_i is caught in the ith sample,

n_i = size of ith sample removed from the population, and

ϕ_i = probability of survival of an individual from time t_i to time t_{i+1} given that it is alive just after the ith sample ($i = 1, 2, \ldots, s - 1$).

Assuming that sampling and natural mortality are the only processes affecting the population size, then the joint probability function of the n_i is given by (cf. (7.2))

$$\frac{N!}{(\prod_{i=1}^{s} n_i!)(N-n)!} \; p_1^{n_1}(q_1\phi_1 p_2)^{n_2} \cdots (q_1\phi_1 q_2\phi_2 \cdots q_{s-1}\phi_{s-1} p_s)^{n_s} \; \theta^{N-n}$$

where $\quad n = \sum_{i=1}^{s} n_i$ and

$$\theta = 1 - p_1 - q_1\phi_1 p_2 - \cdots - q_1\phi_1 q_2\phi_2 \cdots q_{s-1}\phi_{s-1} p_s.$$

However, we have more parameters than random variables, so that additional assumptions must be made if all the parameters are to be estimated. Possible assumptions are:

(i) $q_i = \exp(-k f_i)$ where f_i is the effort expended on the ith sample.

(ii) Natural mortality is a Poisson process with parameter μ, so that $\phi_i = \exp[-\mu(t_{i+1} - t_i)]$.

With these assumptions it transpires that the maximum-likelihood equations for N, μ and k do not have explicit solutions and must be solved iteratively. However,

$$E[n_i] = N \, q_1\phi_1 q_2\phi_2 \cdots q_{i-1}\phi_{i-1} p_i$$

$$= N \, q_1 q_2 \cdots q_{i-1} p_i \phi_1\phi_2 \cdots \phi_{i-1}$$

$$= N \exp(-kF_{i-1})\{1 - \exp(-kf_i)\} \exp\{-\mu(t_i - t_1)\},$$

where $F_{i-1} = \sum_{j=1}^{i-1} f_j$, so that using the approximation (8.8) and taking logarithms, we obtain the linear regression model

$$E[\log(n_i/f_i)] \approx \log(Nk) - k(F_{i-1} + F_i)/2 - \mu(t_i - t_1),$$

which is of the same form as (8.12).

If assumption (i) is replaced by (i)': $p_i = p$ ($i = 1, 2, \ldots, s$), then

$$E[n_i] = N q^{i-1} p \exp\{-\mu(t_i - t_1)\}$$

and a possible regression model is

$$E[\log n_i] \approx \log(Np) + (i-1)\log q - \mu(t_i - t_1).$$

This can be analysed by least squares in the usual manner, provided t_i is not proportional to i.

8.2.2 Mortality only: pairwise removal method

Another method for estimating survival probabilities is to take a series of pairs of removals. For example, Seber and Le Cren [1967] discuss the following special case of just two pairs which may be useful for studying small populations.

1 First two catches returned

Consider two two-sample experiments in which the time taken by each experiment is sufficiently short for natural mortality to be negligible during the experiment. Let N_1, N_2 be the respective sizes of the population before each experiment, let n_i ($i = 1, 2, 3, 4$) be the respective catches, let p_1 be the constant probability of capture for the first two samples, and let p_2 be the constant probability of capture for the second two samples. After the first experiment the total catch $n_1 + n_2 - d$ is returned to the population, d being the number which are deliberately removed or die accidentally through trapping. Then the joint probability function of the n_i is (cf. (7.20) with $s = 2$)

$$\frac{N_1!}{n_1! \, n_2! \, (N_1 - n_1 - n_2)!} \, p_1^{n_1+n_2} \, q_1^{2N_1 - 2n_1 - n_2}$$

$$\times \frac{N_2!}{n_3! \, n_4! \, (N_2 - n_3 - n_4)!} \, p_2^{n_3+n_4} \, q_2^{2N_2 - 2n_3 - n_4}$$

and the maximum-likelihood estimates of the unknown parameters are

$\hat{q}_1 = n_2/n_1$, $\hat{q}_2 = n_4/n_3$, $\hat{N}_1 = n_1^2/(n_1 - n_2)$ and $\hat{N}_2 = n_3^2/(n_3 - n_4)$.

The probability of survival ϕ between the two experiments is estimated by

$$\hat{\phi} = \hat{N}_2/(\hat{N}_1 - d).$$

Using the delta method it is readily shown that

$$V[\hat{\phi}] \approx \frac{N_1 \, q_1^2 (1 + q_1) \phi^2}{p_1^3 (N_1 - d)^2} + \frac{N_2 \, q_2^2 (1 + q_2)}{p_2^3 (N_1 - d)^2}.$$

Example 8.3 Trout: Seber and Le Cren [1967]

Suppose the following catches are observed: 91, 54, 53 and 12 ($d = 0$); then

$$\hat{q}_1 = 54/91, \quad \hat{q}_2 = 12/53, \quad \hat{N}_1 = 91^2/37, \quad \hat{N}_2 = 53^2/41$$

and

$$\hat{\phi} = 53^2(37)/91^2(41) = 0{\cdot}31.$$

Replacing unknown parameters by their estimates, $\hat{V}[\hat{\phi}] = 0{\cdot}0037$ and, assuming $\hat{\phi}$ to be approximately normal, an approximate 95 per cent confidence interval for ϕ is $0{\cdot}31 \pm 0{\cdot}12$.

If $p_1 = p_2 = p$, say, the maximum-likelihood estimates are now given by (Seber and Le Cren [1967: 639])

$$\tilde{q} = (n_2 + n_4)/(n_1 + n_3),$$
$$\tilde{N}_1 = (n_1 + n_2)/(1 - \tilde{q}^2),$$
$$\tilde{N}_2 = (n_3 + n_4)/(1 - \tilde{q}^2),$$

and

$$\tilde{\phi} = \tilde{N}_2/(\tilde{N}_1 - d).$$

When d is negligible we have from the delta method

$$V[\tilde{\phi}] = V[(n_3 + n_4)/(n_1 + n_2)]$$
$$\approx \frac{q^2 N_2 (N_1 + N_2)}{N_1^3 (1 - q^2)}.$$

If we can assume that $p_1 = p_2$ we would expect intuitively that the estimate of ϕ for this model would have a smaller variance than that obtained for the model with $p_1 \neq p_2$, since the information from the catches is shared among fewer parameters, i.e. we would expect $V[\tilde{\phi}] < V[\hat{\phi}]$.

One can test the hypothesis $p_1 = p_2$ by assuming that $\hat{p}_i$ is asymptotically normal. Since $\hat{p}_1$ and $\hat{p}_2$ are statistically independent, $\hat{p}_2 - \hat{p}_1$, is then asymptotically normally distributed with mean $p_2 - p_1$ and variance estimated by (cf. (7.31))

$$[n_2 (n_1 + n_2)/n_1^3] + [n_4 (n_3 + n_4)/n_3^3].$$

Example 8.4 Using the data of the previous example,

$$\hat{p}_2 - \hat{p}_1 = 0.3670, \quad \hat{V}[\hat{p}_2 - \hat{p}_1] = 0.0156$$

and

$$z = \frac{0.3670}{\sqrt{(0.0156)}} = 2.93.$$

Since $|z| > 2.58$ we reject the hypothesis that $p_1 = p_2$ at the 1 per cent level of significance.

2   First two catches not returned

It was assumed above that after the first experiment the total catch was returned to the population. If, however, this catch is not returned, the joint probability function of the n_i now becomes (with $n = \Sigma n_i$)

$$\frac{N_1!}{(\prod_{i=1}^{4} n_i!)(N_1 - n)!} p_1^{n_1} (q_1 p_1)^{n_2} (q_1^2 \phi p_2)^{n_3} (q_1^2 \phi q_2 p_2)^{n_4} [q_1^2 (1 - \phi(1 - q_2^2))]^{N_1 - n},$$

$$(8.24)$$

and the maximum-likelihood estimates (also the moment estimates) are

$$\hat{q}_1 = n_2/n_1, \quad \hat{q}_2 = n_4/n_3, \quad \hat{N}_1 = n_1^2/(n_2 - n_1)$$

and

$$\hat{\phi} = \frac{n_3^2(n_1 - n_2)}{n_2^2(n_3 - n_4)}.$$

Using the delta method we find that

$$V[\hat{\phi}] \approx \phi^2 \left[\frac{4}{N_1 q_1^2 \phi p_2} + \frac{5}{N_1 p_1^2} + \frac{4}{N_1 p_1 q_1} - \frac{3}{N_1 q_1^2 \phi p_2^2} \right].$$

If $p_1 = p_2 = p$ the maximum-likelihood estimates are now

$$\tilde{q} = (n_2 + n_4)/(n_1 + n_3),$$

$$\tilde{N}_1 = (n_1 + n_2)/(1 - \tilde{q}^2),$$

and

$$\tilde{\phi} = (n_3 + n_4)/[\hat{q}^2(n_1 + n_2)];$$

also

$$V[\tilde{\phi}] \approx \frac{\phi^2}{N_1 pq} \left[\frac{1 - \phi q^2}{\phi q(1 + q)} + \frac{4(1 + q)}{(1 + \phi q^2)} \right].$$

Once again we would expect $V[\tilde{\phi}] < V[\hat{\phi}]$.

Standard goodness-of-fit statistics for the multinomial distribution (8.24), and for (8.24) with $p_1 = p_2 = p$, can be calculated using the method of **1.3.6**(*3*). If both statistics are calculated, then the latter minus the former (with one degree of freedom) will give a test of the hypothesis $p_1 = p_2$, *given* that (8.24) is valid.

8.3   SINGLE TAG RELEASE

8.3.1   Continuous sampling

Consider an open population in which continuous exploitation, natural mortality, and possibly recruitment, are the only processes affecting the population size. If a single tag release is made at the beginning of the sampling, then by studying the tagged population only, the effects of recruitment can be ignored. When Z, the total instantaneous mortality rate, is constant throughout the whole experiment, then the methods of **6.2** are applicable here. If Z is not constant, then effort data can be used along with tag returns to estimate k and μ in $Z_i = \mu + k\overset{o}{f_i}$ as follows.

Adopting the same notation and assumptions as in **8.1.1** with the additions

M_i = size of tagged population at the beginning of the ith period,

and

m_i = number of tagged individuals caught in the ith period,

we have from (8.12) Chapman's [1961: 157] regression model

$$E[\log(m_i/f_i)] \approx \log(M_1 k) - \mu(t_{i-1} + t_i)/2 - k(F_{i-1} + F_i)/2.$$

If the proportion of the release recaptured is small, then the dependence between the m_i can be neglected. Therefore, assuming the $\log(m_i/f_i)$ to be independently distributed with approximately constant variance, least-squares estimates of k and μ can be obtained by a straightforward multiple linear regression. Chapman points out that we do not have to use the information on M_1, the size of the tag release, which is an advantage if there is (i) initial tag loss, (ii) initial mortality caused by the tagging procedure, or (iii) a time-lapse before tagged individuals are fully "vulnerable" to the catching process.

Chapman also gives an alternative method when the actual recapture times of tagged individuals are recorded as in **6.2.1.** For each period of length Δ_i, the distribution of the recapture times T_{ij} ($j = 1, 2, \dots, m_i$) measured from the start of the period is the truncated exponential (cf. p. 272)

$$f(T) = \frac{Z_i \exp(-Z_i T)}{1 - \exp(-Z_i \Delta_i)}, \quad 0 \leqslant T \leqslant \Delta_i.$$

If $\bar{T}_i = \sum_{j=1}^{m_i} T_{ij}/m_i$, then, from Deemer and Votaw [1955], the maximum-likelihood estimate $\hat{Z}_i$ of Z_i is the solution of

$$\frac{1}{Z_i \Delta_i} - \frac{1}{\{\exp(Z_i \Delta_i)\} - 1} = \frac{\bar{T}_i}{\Delta_i}$$

when $0 \leqslant \bar{T}_i < \Delta_i/2$ and 0 otherwise. The authors also provide a table (reproduced as Table **A5**) to facilitate the solution of this equation. For large m_i we have approximately

$$E[\hat{Z}_i] = Z_i = \mu + k\overset{o}{f_i} \tag{8.25}$$

and

$$V[\hat{Z}_i] = \frac{1}{m_i}\left[\frac{1}{Z_i^2} - \frac{\Delta_i^2}{\exp(Z_i\Delta_i)\{1 - \exp(-Z_i\Delta_i)\}^2}\right]^{-1}. \tag{8.26}$$

Therefore, assuming the $\hat{Z}_i$ to be independent, $\hat{Z}_i$ can be plotted against $\overset{o}{f_i}$, and μ and k can be estimated by weighted least squares (cf. **1.3.5**(*1*)), the appropriate weights being estimated from (8.26).

We note that for both the above methods, separate estimation of μ and k is possible only if the effort per unit time $\overset{o}{f_i}$ varies from period to period; in general, the greater the variation, the smaller the variances of the estimates. It was also mentioned above that M_1 can be estimated, so that any well-defined class in the population at the beginning of the sampling can be used as a "tagged" population. In particular, if ageing can be done accurately, a fully recruited age-class can be used. This ensures that M_1 is a reasonable fraction of the population, which is a difficult goal to achieve with tagging experiments in large populations. It also means that the group M_1 is well mixed with the rest of the population, and the assumption that individuals are not affected by tagging as far as catchability is concerned is avoided. However, methods of ageing are time-consuming, and some other correlated variable such as weight or length is often used for age classification (cf. p. 333). This will mean that in applying the second method above, the variances given by (8.26) will be underestimates due to errors in the ageing process and the loss of information through the inevitable grouping of the catch-times T_{ij} which goes with large m_i. Also there is the problem of obtaining f_i, the effort associated with just the age-class only; one such method is suggested in **8.1.1**(4).

NON-REPORTING OF TAGS. If the total effort f_i can be classified into two categories, one which has a known proportional tag return of unity or nearly unity, and the other with an unknown reported ratio ρ_i, then a technique due to Paulik [1961] can be used to test whether ρ_i is significantly less than unity. In fisheries the former category would usually be the fraction of the effort inspected by special observers, or it may be the units of gear operated by fishermen who, by virtue of special training, long experience, or a keen interest in the experiment, could be relied upon to turn in all the tags that appear in their catches.

Adopting the notation of **8.1.1**, let $f_i = f_{ia} + f_{ib}$ ($i = 1, 2, \ldots, s$), where the suffixes a and b denote the above two categories "inspected" and "uninspected" respectively; m_i the number of tags recaptured and r_i the number of tags reported in the ith interval are split up in the same way ($r_{ia} = m_{ia}$). From equations (8.3) and (8.5), with M_i and m_i instead of N_i and n_i, the probabilities that a particular tag is reported from the f_{ia} and f_{ib} units of gear are $kf_{ia}A_i/(Z_i\Delta_i)$ and $k\rho_i f_{ib}A_i/(Z_i\Delta_i)$ respectively, where

$$A_i = \exp(-\mu t_{i-1} - k F_{i-1})[1 - \exp(-Z_i \Delta_i)].$$

Therefore, given that a tag has been reported in the ith interval, the probability that it comes from the f_{ia} units of gear is

$$p_i = f_{ia}/(f_{ia} + \rho_i f_{ib}), \tag{8.27}$$

and the conditional probability function of r_{ia} given r_i is

$$f(r_{ia} \mid r_i) = \binom{r_i}{r_{ia}} p_i^{r_{ia}} q_i^{r_{ib}}.$$

This model is of the same form as (3.28), and the discussion in **3.2.4** concerning the estimation of ρ_i and testing $\rho_i = 1$ is applicable here with p_0 now defined to be f_{ia}/f_i.

If $f_{ib}/f_{ia} = \gamma$ ($i = 1, 2, \ldots, s$) then $p_i = (1 + \rho_i \gamma)^{-1}$ and we can test the hypothesis $H : \rho_i = \rho$ using the statistic

$$\sum_{i=1}^{s} \frac{(r_{ia} - r_i \hat{p})^2}{r_i \hat{p} \hat{q}},$$

where $\hat{p} = (\Sigma r_{ia})/\Sigma r_i$ ($= r_a/r$ say). When H is true, the above statistic is approximately χ^2_{s-1} and the recoveries for the whole experiment can be pooled. In this case

$$f(r_a \mid r) = \binom{r}{r_a} p^{r_a} q^{r_b},$$

where $p = (1 + \rho\gamma)^{-1}$, and we can test the hypothesis that $\rho = 1$ using the same approach as in **3.2.4** (with $p_0 = (1 + \gamma)^{-1}$).

As p_i in (8.27) is unchanged if k is replaced by k_i, the above theory is still valid when the catchability varies from interval to interval.

Example 8.5   Trout: adapted from Paulik [1961]

A biologist wishes to estimate the percentage of tags detected and reported by sportsmen and would like to be 90 per cent sure of discovering a reported ratio of 80 per cent or lower. 2000 marked trout are released in a large lake shortly before the opening of the fishing season, and an unknown proportion of them are expected to survive until opening day. It is assumed that (i) the population is closed, apart from natural mortality and removals by the fishermen, (ii) the catchability of the marked trout is constant within a week but may vary from week to week, and (iii) ρ is constant throughout the season. If at least 1000 tags are expected to be returned, what percentage of the fishermen should be sampled each day? From **A2** with $\alpha = 0\cdot05$, $1 - \beta = 0\cdot90$, $\rho = 0\cdot8$ and $r = 1000$, we have $p_0 = 0\cdot20$ ($= f_{ia}/f_i$), so that 20 per cent of the fishermen should be interviewed.

If it can also be assumed that marked and unmarked are equally vulnerable to capturing and have the same natural mortality rate, so that the proportion of tagged in the population remains constant throughout the season, then the method of **3.2.4** is directly applicable here. The biologist must then decide whether to sample the catch or the effort, and this can be done by computing appropriate "cost" curves for the two methods as in Paulik [1961: 826].

8.3.2   Instantaneous samples

Suppose now that "instantaneous" or point samples are removed from the population at times $t_1, t_2, \ldots, t_s$. For a population in which there is just mortality and recruitment (i.e. no migration) Chapman [1965] gives the following method, based on a single release of M_0 marked individuals at time zero, for estimating the various unknown parameters. Let

M_i = size of marked population just before the ith sample,
N_i = size of total population just before the ith sample,
$U_i = N_i - M_i$,
n_i = size of the ith sample,
m_i = number found marked in the ith sample,
$u_i = n_i - m_i$,
K_m = binomial catchability of marked individuals (cf. **7.1.3**(I)),
K_u = binomial catchability of unmarked individuals,

f_i = effort applied to the ith sample,

r_i = number of recruits added to the population in the interval $[t_{i-1}, t_i)$ which are alive at time t_i ($t_0 = 0$),

and

$\Delta_i = t_i - t_{i-1}$

U_0 = size of unmarked population at time zero.

We shall assume that:

(1) $K_m = K_u$ (= K say), i.e. all animals are equally catchable, whether marked or unmarked, old or new recruits. Chapman points out that this assumption would not hold if the animals acquire some skill in avoiding being caught, for then new recruits would be more easily caught than "old" members of the population. We also note that for this assumption to hold, either the marked animals or the sampling effort must be distributed randomly.

(2) All individuals, whether marked or unmarked, have the same instantaneous natural mortality rate μ throughout the whole experiment.

(3) Units of effort are additive and are distributed randomly over the population, so that

$$E[u_i | U_i] = K f_i U_i$$

and

$$E[m_i | M_i] = K f_i M_i.$$

(4) Conditional on fixed M_i, the random variable m_i has a Poisson distribution ($i = 1, 2, \ldots, s$).

(5) There is no loss of marks, and all marks are reported on recovery.

From the above assumptions we have the following deterministic relationships:

$$U_i = (U_{i-1} - u_{i-1}) e^{-\mu \Delta_i} + r_i \qquad (8.28)$$

and

$$M_i = (M_{i-1} - m_{i-1}) e^{-\mu \Delta_i}, \qquad (8.29)$$

for $i = 1, 2, \ldots, s$ ($u_0 = m_0 = 0$). The basic parameters to be estimated are k, U_0, μ and $r_1, r_2, \ldots, r_s$ via the intermediary unknowns $\{M_i\}$ and $\{U_i\}$. We note that although μ and U_1 (= $U_0 \exp(-\mu \Delta_1) + r_1$) are estimable, U_0 and r_1 cannot be estimated separately without further assumptions or additional information. For example, Chapman suggests that it may be possible to estimate r_1 by extrapolation from $r_2, r_3, \ldots, r_s$ using a low-degree polynomial.

Since the m_i involve only the two parameters k and μ, Chapman derives maximum-likelihood equations from the joint probability function of the m_i:

$$\prod_{i=1}^{s} \left\{ e^{-K f_i M_i} \frac{(K f_i M_i)^{m_i}}{m_i!} \right\},$$

where the M_i satisfy (8.29). It is readily shown that $\hat{K}$ and $\hat{\mu}$, the maximum-likelihood estimates, are solutions of

$$K = (\sum_{i=1}^{s} m_i)/(\sum_{i=1}^{s} f_i M_i) \qquad (8.30)$$

and

$$\left(\sum_{i=1}^{s} m_i\right)\left(\sum_{i=1}^{s} f_i \frac{dM_i}{d\mu}\right)\bigg/\left(\sum_{i=1}^{s} f_i M_i\right) = \sum_{i=1}^{s} \frac{m_i}{M_i}\frac{dM_i}{d\mu}, \qquad (8.31)$$

where $\dfrac{dM_1}{d\mu} = -\Delta_1 M_1$

and $\dfrac{-dM_i}{d\mu} = e^{-\mu\Delta_i}\left[\Delta_i(M_{i-1}-m_{i-1})\dfrac{dM_{i-1}}{d\mu}\right], \qquad (i = 2, 3, \ldots, s).$

Equation (8.31) may be solved iteratively as it involves just the single unknown μ and the known parameter M_0. The asymptotic variance–covariance matrix of $\hat{K}$ and $\hat{\mu}$ is the inverse of (cf. (1.7))

$$\begin{bmatrix} \sum_{i=1}^{s} \dfrac{f_i M_i}{K}, & \sum_{i=1}^{s} f_i \dfrac{dM_i}{d\mu} \\ \sum_{i=1}^{s} f_i \dfrac{dM_i}{d\mu}, & \sum_{i=1}^{s} K \dfrac{f_i}{M_i} \cdot \left(\dfrac{dM_i}{d\mu}\right)^2 \end{bmatrix} \qquad (8.32)$$

Since the observations u_i alone involve the unknown r_i they must provide all the information necessary to estimate the r_i. This suggests equating each u_i to its expected value, so that, from assumption (3),

$$\hat{U}_i = u_i/(\hat{K} f_i), \quad (i = 1, 2, \ldots, s).$$

and from (8.28)

$$\hat{r}_i = \hat{U}_i - (\hat{U}_{i-1} - u_{i-1}) \exp(-\hat{\mu}\Delta_i) \qquad (8.33)$$

$$= \frac{u_i}{\hat{K}f_i} - u_{i-1}\left(\frac{1}{\hat{K}f_{i-1}} - 1\right)\exp(-\hat{\mu}\Delta_i), \quad (i = 2, 3, \ldots, s).$$

Assuming u_i and m_i to be statistically independent, we have, using the delta method (there is a small misprint in Chapman's equation: one f_i should be f_{i-1}),

$$V[\hat{r}_i] \approx \frac{1}{K^2 f_i^2} V[u_i] + \frac{1}{K^4}\left(\frac{E[u_i]}{f_i} - \frac{E[u_{i-1}]}{f_{i-1}} e^{-\mu\Delta_i}\right)^2 V[\hat{K}] +$$

$$+ \left(\frac{1}{K f_{i-1}} - 1\right)^2 e^{-2\mu\Delta_i} V[u_{i-1}] + (E[u_{i-1}])^2 \Delta_i^2 \left(\frac{1}{K f_{i-1}} - 1\right)^2 e^{-2\mu\Delta_i} V[\hat{\mu}] -$$

$$- \left[\frac{E[u_i]E[u_{i-1}]\Delta_i}{K^2 f_i}\left(\frac{1}{K f_{i-1}} - 1\right)e^{-\mu\Delta_i} - \frac{(E[u_{i-1}])^2 \Delta_i}{K^2 f_{i-1}}\left(\frac{1}{K f_{i-1}} - 1\right)e^{-2\mu\Delta_i}\right] \text{cov}[\hat{K}, \hat{\mu}].$$

If the u_i are assumed to have a Poisson distribution then

$$V[u_i] = E[u_i] = K f_i U_i.$$

The N_i can be estimated either iteratively from

$$\hat{N}_i = \hat{U}_i + \hat{M}_i$$
$$= (\hat{N}_{i-1} - n_i) \exp(-\hat{\mu}\Delta_i) + \hat{r}_i,$$

or more simply by the moment estimate

$$N_i^* = n_i/(\hat{K} f_i).$$

If u_i is very much greater than m_i, so that n_i ($\approx u_i$) is approximately independent of the m_i (and hence of $\hat{K}$), then, using the delta method,

$$V[N_i^*] \approx N_i^2 \left(\frac{V[n_i]}{K^2 f_i^2 N_i^2} + \frac{V[\hat{K}]}{K^2} \right).$$

We also note that the above theory can be applied to unbounded populations if μ represents natural mortality plus permanent emigration and r_i represents recruitment plus immigration of unmarked individuals.

Finally Chapman discusses the problem of testing for what Ricker [1958: 122] calls type C error, in which the marked individuals do not have the same catchability as the unmarked immediately after marking but eventually settle down to have the same catchability at a later stage. Ricker mentions that this error may be due to (i) abnormal behaviour of the marked during the season of their marking, or (ii) initial non-random distribution of the marked combined with a (possibly only temporary) non-random distribution of sampling effort. If the catchability (K_1, say) of marked individuals is different for the first sample, then the above model is modified to the extent of just the single equation

$$E[m_1 | M_1] = K_1 f_1 M_1.$$

The maximum-likelihood estimates $\tilde{K}$ and $\tilde{\mu}$ for this modified model, together with asymptotic variances and covariances, are still obtained from (8.30), (8.31) and (8.32) but with all summations now running from 2 to s instead of 1 to s. Estimates $\tilde{M}_i$, $\tilde{U}_i$ and $\tilde{r}_i$ are the same functions of $\tilde{\mu}$ and $\tilde{K}$ as before, except for $i = 1$. For this case $\tilde{K}$ is replaced by $\tilde{K}_1 = m_1/(f_1 \tilde{M}_1)$, where $\tilde{M}_1 = M_0 \exp(-\tilde{\mu}\Delta_1)$, so that $\tilde{U}_1 = u_1/(\tilde{K}_1 f_1)$ in the equation (8.33) for $\tilde{r}_2 : V[\tilde{r}_2]$ has to be modified accordingly.

Chapman gives an approximate test of the hypothesis $K_1 = K$ based on the statistic

$$x = m_1 - \tilde{E}[m_1] = m_1 - \tilde{K}_1 f_1 M_0 \exp(-\tilde{\mu}\Delta_1).$$

When the hypothesis is true, then for large samples x will be approximately normally distributed with mean zero and approximate variance

$$V[x] = V[m_1] + (f_1 M_0)^2 e^{-2\mu\Delta_1} (V[\tilde{K}] - 2K\Delta_1 \text{cov}[\tilde{K}, \tilde{\mu}] + K^2 \Delta_1^2 V[\tilde{\mu}]).$$

Assuming that m_1 has a Poisson distribution then K, μ and $V[m_1]$ can be estimated by $\widetilde{K}$, $\widetilde{\mu}$ and $\widetilde{E}[m_1]$ respectively in $V[x]$.

Example 8.6   Halibut: Chapman [1965]

The data in Table 8.3 are taken from a report of the International Pacific Halibut Commission [1960: Appendix 2]. $M_0 = 1051$ marked fish were released in 1951 and the catches n_i were taken in the successive seasons 1952 to 1957. The catches may be treated as point samples since the duration of the fishing was less than 10 per cent of the year for the first four

TABLE 8.3

Catch and effort data for a halibut fishery (Upper Hecate Strait, 1951–1957): from Chapman [1965: Table 1].

t_i (years)	n_i (million lb)	m_i	f_i (thousand skates)	$\widetilde{r}_i$ (million lb)	$\widetilde{N}_i$ (million lb)	$\widetilde{M}_i$
0	(30·6)	–	(320·8)	–	(58·7)	–
1	30·8	214	251·8	(55·5)	75·3	739·9
2	33·0	139	228·6	57·5	88·8	370·2
3	36·7	65	244·2	53·2	92·5	162·8
4	28·7	22	219·9	41·0	80·3	68·9
5	35·4	14	263·2	46·5	82·8	33·0
6	30·6	7	283·6	33·0	66·4	13·4

$M_0 = 1051$, $\Delta_i = 1$ (year)

years and about 15 per cent of the year for the last two years, and $\Delta_i = 1$ year. The effort data f_i are given in thousands of "skates", while n_i, $\widetilde{r}_i$ and $\widetilde{N}_i$ in Table 8.3 are recorded in millions of pounds. Since n_i and m_i are measured in different units, namely millions of pounds and numbers respectively, the u_i cannot be calculated. However, Chapman points out that since the m_i by weight amount to no more than a few thousand pounds out of a catch of at least 30 million pounds, n_i is a good approximation to u_i which he used in the following calculations.

To test whether there was less than full availability of tagged fish in 1952 (the first season after tagging), K and μ were estimated on the basis of the 1953 to 1957 recoveries. Thus

$$\widetilde{K} = 1\cdot 6248 \times 10^{-3}, \quad e^{-\widetilde{\mu}} = 0\cdot 704 \quad \text{or } \widetilde{\mu} = 0\cdot 35,$$

and the asymptotic variance–covariance matrix is

$$10^{-8} \times \begin{bmatrix} 1\cdot 654\ 34 & 72\cdot 385\ 33 \\ 72\cdot 385\ 33 & 8948\cdot 254\ 7 \end{bmatrix}.$$

Also
$$\widetilde{E}[m_1] = 1\cdot 6248 \times 10^{-3} (251\cdot 8)(1051)(0\cdot 704) = 302\cdot 7,$$

and $V[x]$ is estimated by

$$\tilde{V}[x] = 302 \cdot 7 + 501 \cdot 7 = 804 \cdot 4.$$

Therefore, to test $K_1 = K$ against the alternative $K_1 < K$ we calculate

$$z = (214 - 302 \cdot 7)/\sqrt{804 \cdot 4} = -3 \cdot 12$$

and, since $z < -2 \cdot 58$, we reject the hypothesis at the 1 per cent level of significance. By comparison, $\tilde{K}_1 = 0 \cdot 001\ 15$, which is only about 70 per cent of $\tilde{K}$.

The estimates $\tilde{M}_i$ and $\tilde{r}_i$ are given in Table 8.3 and the N_i are estimated by

$$\tilde{N}_i = n_i/(\tilde{K} f_i) \quad (i = 1, 2, \dots, 6).$$

In this experiment catch and effort data were available for time zero (before the mark release), so that N_0 and r_1 could be estimated by

$$\tilde{N}_0 = n_0/(\tilde{K} f_0) = 58 \cdot 7$$

and

$$\tilde{r}_1 = \tilde{U}_1 - (\tilde{U}_0 - u_0)\, e^{-\tilde{\mu}}$$

$$\approx \tilde{N}_1 - (\tilde{N}_0 - n_0)\, e^{-\tilde{\mu}}$$

$$= 75 \cdot 3 - (58 \cdot 7 - 30 \cdot 6)(0 \cdot 704)$$

$$= 55 \cdot 5.$$

Finally we note that if the data for the first year are included, thus ignoring the differences in catchability, then the estimates of K and μ are

$$\hat{K} = 1 \cdot 257 \times 10^{-3} \quad \text{and} \quad \hat{\mu} = 0 \cdot 323.$$

CHAPTER 9

CHANGE IN RATIO METHODS

9.1 CLOSED POPULATION

9.1.1 Introduction

Changes in observed sex ratios, age ratios and marked-to-unmarked ratios (the Petersen method) have been widely used to estimate population abundance, productivity and survival probabilities (Hanson [1963], Kelker and Hanson [1964]). For example, the idea that population numbers could be estimated from a knowledge of the sex ratios before and after a differential kill of the sexes was first noted by Kelker [1940, 1944]. A number of variations of this method were then used by Allen [1942], Rasmussen and Doman [1943], Riordan [1948], Petrides [1949], Lauckhart [1950] and Dasmann [1952]; a summary of this early work is given in Hanson [1963]. However, these early methods were based on intuitive notions, so that no estimates of variance were given. It seems that the first stochastic models were developed by Chapman [1954, 1955] for the closed population, and this work was followed up by Lander [1962] and Chapman and Murphy [1965] who considered the problem of estimating mortality rates.

Recently Paulik and Robson [1969] gave an excellent survey of the whole field, pointing out the wide applicability of the change in ratio (CIR) method (variously known as the "dichotomy", "change in composition", "survey—removal" method) and demonstrating how a number of different methods are all variations on the same theme. They suggest that one factor which has inhibited the proper statistical treatment of CIR methods, particularly in the derivation of variances and confidence intervals, is the terminology associated with applications in fish and wildlife management. For example, there is the widespread practice of using one component as a base and expressing the other component or components in terms of this base, which is sometimes arbitrarily standardised; e.g. number of males per 100 females or number of subadults per adult. As also pointed out by Smirnov [1967: 216], this is in direct contrast to the statistical practice of formulating such problems in terms of fractions, probabilities or proportions, for which sampling errors of estimation are more readily calculated.

NOTATION. Consider a closed population consisting of two types of animals. These are designated as x-type and y-type animals; for example, x-type may be males and y-type females, or x-type may be sport fish and y-type coarse fish. Suppose there iš a differential change in the ratio of

x-type to y-type animals between time t_1 and time t_2. Let

X_i = number of x-type animals in the population at time t_i,
Y_i = number of y-type animals in the population at time t_i,
N_i = $X_i + Y_i$ (total population size at time t_i),
P_i = X_i/N_i ($= 1 - Q_i$),
R_x = $X_1 - X_2$,
R_y = $Y_1 - Y_2$,
R = $R_x + R_y$ ($= N_1 - N_2$)

and f = R_x/R (f is only defined when R_x and R_y have the same sign).

The quantities R_x and R_y, which are assumed to be known, are called the "removals", though, as pointed out by Rupp [1966], the theory below still holds when the removals are negative, that is additions. (Paulik and Robson [1969] define $R_x = X_2 - X_1$, etc., though they use the above notation when discussing the planning of CIR experiments.)

ESTIMATING POPULATION SIZE. Now

$$P_2 = \frac{X_1 - R_x}{N_1 - R} = \frac{P_1 N_1 - R_x}{N_1 - R},$$

or, solving for N_1,

$$N_1 = \frac{R_x - R P_2}{P_1 - P_2}. \tag{9.1}$$

Therefore, if $\hat{P}_1$ and $\hat{P}_2$ are estimates of P_1 and P_2, N_1, X_1 and N_2 can be estimated by

$$\hat{N}_1 = \frac{R_x - R\hat{P}_2}{\hat{P}_1 - \hat{P}_2}, \tag{9.2}$$

$$\hat{X}_1 = \hat{P}_1 \hat{N}_1,$$

and

$$\hat{N}_2 = \hat{N}_1 - R = \frac{R_x - R\hat{P}_1}{\hat{P}_1 - \hat{P}_2}.$$

Example 9.1   Mule deer: Rasmussen and Doman [1943]

In their words: "... on a crowded range near Logan, Utah, a severe loss occurred during the winter of 1938–39. Age counts made before the loss showed 83 fawns per 100 adults and after the loss 53 fawns per 100 adults. In a complete coverage of the area 248 dead fawns and 60 dead adults were counted". Let X and Y represent the fawn and adult populations respectively. Then,

$\hat{P}_1$ = $83/183$ = $0\cdot4536$, $\hat{P}_2$ = $53/153$ = $0\cdot3464$, R_x = 248, R_y = 60,

so that

$$\hat{N}_1 = \frac{248 - 308(0\cdot3464)}{0\cdot4536 - 0\cdot3464} = 1318$$

and
$$\hat{X}_1 = \hat{P}_1 \hat{N}_1 = 598 \text{ fawns.}$$

Example 9.2   Trout: Rupp [1966]

Suppose that 500 sub-catchable trout are stocked in a small pond the same day that anglers remove 160 catchable trout. The proportions of catchable-size trout before and after this day were estimated to be $\hat{P}_1 = 0.40$ and $\hat{P}_2 = 0.10$ respectively. If X = size of catchable population, then $R_x = 160$, $R_y = -500$, so that

$$\hat{N}_1 = \frac{160 - (160 - 500)(0.10)}{0.40 - 0.10} = 647$$

and
$$\hat{X}_1 = 0.40(647) = 259.$$

VARIANCES.   Using the delta method it is readily shown that for independent estimates of P_1 and P_2,

$$V[\hat{N}_1] \approx (P_1 - P_2)^{-2} \{N_1^2 V[\hat{P}_1] + N_2^2 V[\hat{P}_2]\} \tag{9.3}$$

and
$$V[\hat{X}_1] \approx (P_1 - P_2)^{-2} \{N_1^2 P_2^2 V[\hat{P}_1] + N_2^2 P_1^2 V[\hat{P}_2]\}. \tag{9.4}$$

We shall now consider various methods of obtaining the estimates $\hat{P}_i$.

9.1.2   Binomial sampling

1   Chapman's model

Suppose that a random sample of n_i animals is taken with replacement at time t_i, and x_i animals are found to be of x-type. Then $\hat{P}_i = x_i/n_i$ is an unbiased estimate of P_i with variance

$$V[\hat{P}_i] = P_i Q_i / n_i$$

and (9.3) reduces to

$$V[\hat{N}_1] \approx (P_1 - P_2)^{-2} \left\{ \frac{X_1 Y_1}{n_1} + \frac{X_2 Y_2}{n_2} \right\}. \tag{9.5}$$

Equation (9.2) can also be written as

$$\hat{N}_1 = \frac{n_1(R_x y_2 - R_y x_2)}{x_1 y_2 - x_2 y_1},$$

where $y_i = n_i - x_i$, which avoids the round-off errors which may arise in calculating the $\hat{P}_i$. However, the formula (9.2) is simple to use and is quite satisfactory, provided that sufficient decimal places are carried over; for simplicity we shall quote estimates to 4 decimal places throughout this chapter, though in practice 5 or 6 decimal places are often more appropriate.

For the case of binomial sampling, $\hat{N}_1$ and $\hat{X}_1$ were first obtained by Chapman [1954] who showed that they are maximum-likelihood estimates for the binomial model

$$f(x_1, x_2 \mid \{X_i, Y_i, n_i\}) = \prod_{i=1}^{2} \binom{n_i}{x_i}\left(\frac{X_i}{N_i}\right)^{x_i}\left(\frac{Y_i}{N_i}\right)^{y_i}. \tag{9.6}$$

If $\hat{Y}_i = \hat{N}_i - \hat{X}_i$, the estimates can also be derived from the intuitive relations

$$\frac{x_1}{\hat{X}_1} = \frac{y_1}{\hat{Y}_1}\left(=\frac{n_1}{\hat{N}_1}\right) \tag{9.7}$$

and

$$\frac{x_2}{\hat{X}_1 - R_x} = \frac{y_2}{\hat{Y}_{1,} - R_y}\left(=\frac{n_2}{\hat{N}_1 - R}\right). \tag{9.8}$$

The binomial model is applicable when, for example, animals are just observed, so that a particular animal may be seen more than once. It can also be used as a reasonable approximation to the more realistic situation of sampling without replacement, provided that less than about 10 per cent of the population is sampled. When sampling is without replacement the probability function (9.6) must be replaced by

$$\prod_{i=1}^{2} \binom{X_i}{x_i}\binom{N_i - X_i}{n_i - x_i} \Big/ \binom{N_i}{n_i}.$$

However, it transpires that the maximum-likelihood estimates of N_1 and X_1 for this model are still $\hat{N}_1$ and $\hat{X}_1$. The asymptotic variances are now given by (9.3) and (9.4), with

$$V[\hat{P}_i] = \frac{P_i Q_i}{n_i} \cdot \frac{(N_i - n_i)}{(N_i - 1)},$$

which can be estimated by

$$\hat{V}[\hat{P}_i] = \frac{\hat{P}_i \hat{Q}_i}{n_i - 1}\left(1 - \frac{n_i}{\hat{N}_i}\right).$$

So far it has been assumed that the sample sizes n_1 and n_2 are chosen in advance, and this will be the case if the experiment is planned using the methods of section 5 below. However, if the size of the sample is determined by the amount of effort available for sampling, then n_1 and n_2 are strictly random variables. In this case Chapman and Murphy [1965] assume that x_1, x_2, y_1 and y_2 are all independent Poisson variables, and we find that under this assumption the conditional distribution of x_i given $n_i = x_i + y_i$ is binomial (cf. **1.3.7**(1)). But when n_1 and n_2 are random variables it can be shown, using arguments similar to those given in **3.1.3**, that the variance estimates of $\hat{N}_1$ and $\hat{X}_{1,}$ are virtually unchanged.

Example 9.3 Pheasants: Paulik and Robson [1969]

For a hypothetical pheasant (*Phasianus colchicus*) population in a pre-season survey, 600 of 1400 mature birds seen are cocks. In a post-season survey of 2000 mature birds only 200 are cocks. From a complete check it was found that during the hunting season 8000 cocks and 500 hens were killed. Therefore if X and Y, respectively, represent the cocks and hens, we have

$$\hat{P}_1 = 600/1400 = 0\cdot4286, \quad \hat{P}_2 = 200/2000 = 0\cdot1000, \quad R_x = 8000, \quad R_y = 500.$$

Hence
$$\hat{N}_1 = \frac{800 - 8500(0\cdot1)}{0\cdot4286 - 0\cdot1000} = 21\,761,$$

$$\hat{X}_1 = (0\cdot4286)\,21\,761 = 9326,$$

and
$$\hat{N}_2 = 21\,761 - 8500 = 13\,261.$$

Assuming binomial sampling:

$$\hat{V}[\hat{P}_1] = (0\cdot4286)(0\cdot5714)/1400 = 0\cdot000\,174\,9,$$

and
$$\hat{V}[\hat{P}_2] = (0\cdot1000)(0\cdot9000)/2000 = 0\cdot000\,045\,0.$$

so that, substituting in (9.3), $V[\hat{N}_1]$ is estimated by

$$\hat{V}[\hat{N}_1] = (0\cdot4286 - 0\cdot1000)^{-2}[(21\,761)^2(0\cdot000\,174\,9) + (13\,261)^2(0\cdot000\,045\,0)]$$
$$= 840\,571$$

and hence
$$\hat{\sigma}[\hat{N}_1] = \sqrt{\hat{V}[\hat{N}_1]} = 917.$$

2   Underlying assumptions

The basic assumptions underlying (9.6) are:

(i) The population is closed.
(ii) All animals have the same probability of being caught in the ith sample $(i = 1,\ 2)$.
(iii) The removals R_x and R_y are known exactly.

ASSUMPTION (i). With regard to the first assumption, (9.6) will still hold when there is mortality both between time t_1 and the time of removal, and between the time of removal and time t_2, provided that in each period the survival rates are the same for both types of animal (i.e. P_1 and P_2 remain constant). However, N_1 now refers to the population size just prior to the removal, as in the following example.

Example 9.4   Silver salmon: Paulik and Robson [1969]

Suppose a hatchery marks and releases 250 000 silver salmon smolts that are about 18 months old. Silver salmon mature at age -2 and age -3, and the fish maturing at age -2 are predominantly males. These precocious males or "jacks" are usually too small to be taken by the fishery, so that the number returning to the hatchery represents the total number of males "removed" from the ocean population of marked age -2 silver salmon. Suppose that 5000 marked age -2 jacks and no age -2 females return to the hatchery, and the following year both the catch and the escapement to the hatchery are sampled to determine the fraction of marked age -3 silvers that are males; four hundred of 1000 marked age -3 silvers sampled are found to be males. Assume that the sex ratio in the smolts is known to be $1:1$ and assume further that there is no difference between the ocean survival of male and female marked silver salmon during both their first 6 months (i.e. P_1 constant) and their last 12 months (i.e. P_2 constant) in the ocean. Then $P_1 = \frac{1}{2}$, $\hat{P}_2 = 400/1000 = 0.40$, $R_x = 5000$, $R_y = 0$, and the estimated number of marked age -2 silver salmon in the ocean just before the jacks left the population is

$$\hat{N}_1 = \frac{5000(1-0.40)}{0.50 - 0.40} = 30\ 000.$$

Assuming binomial sampling and setting $V[\hat{P}_1] = 0$ (as P_1 is known), we have from (9.3)

$$\hat{V}[\hat{N}_1] = (0.10)^{-2}\{25\ 000^2 (0.4)(0.6)/1000\} = 15 \times 10^6,$$

and $\hat{\sigma}[\hat{N}_1] = 3873.$

An estimate of the probability of survival during early ocean life is

$$\hat{N}_1/(250\ 000) = 0.12.$$

(The above example is based on Murphy [1952].)

ASSUMPTION (ii). This assumption implies that (a) the probability of capture is the same for all animals of a given type, and (b) the two types of animal have the same probability of capture. (a) may be realistic if care is exercised in the sampling procedure, though (b) may not be true in some situations. For example, cock pheasants are more easily seen than hen pheasants; if the two types of animal refer to different species they may be sampled at different rates, using possibly different trapping methods. Suppose then that λ is the ratio of the probability of sampling y-type animals to the probability of sampling x-type animals, and let $d = 1 - \lambda$. Then x_i now has a binomial distribution with parameters n_i and $P_{i\lambda} = X_i/(X_i + \lambda Y_i) = X_i/(N_i - dY_i)$. Therefore, replacing random variables by their expected values, we find that, for large n_i, $\hat{N}_1$, $\hat{X}_1$ and $\hat{N}_2$ are asymptotically unbiased estimates of (cf. Chapman [1955: 281])

$$N_1' = \frac{R_x - RP_{2\lambda}}{P_{1\lambda} - P_{2\lambda}} = (X_1 + \lambda Y_1)\left(\frac{N_1 R_x - X_1 R - dY_1 R_x + dR_x R_y}{N_1 R_x - X_1 R - dY_1 R_x + dX_1 R_y}\right),$$

$$X_1' = P_{1\lambda} N_1' = X_1\left(\frac{N_1 R_x - X_1 R - dY_1 R_x + dR_x R_y}{N_1 R_x - X_1 R - dY_1 R_x + dX_1 R_y}\right),$$

and

$$N_2' = \frac{R_x - RP_{1\lambda}}{P_{1\lambda} - P_{2\lambda}} = (X_2 + \lambda Y_2)\left(\frac{N_1 R_x - X_1 R - dY_1 R_x}{N_1 R_x - X_1 R - dY_1 R_x + dX_1 R_y}\right).$$

In general, $X_1' \neq X_1$, except in one important case, i.e. when $R_y = 0$. Unfortunately, for this case $N_i' = X_i + \lambda Y_i$, so that N_i' may be widely different from N_i when $\lambda \neq 1$. However, when $R_y = 0$, $\hat{X}_1$ is not only asymptotically unbiased but the usual estimate of variance, namely (cf. (9.4))

$$\hat{V}[\hat{X}_1] = (\hat{P}_1 - \hat{P}_2)^{-2} \{\hat{N}_1^2 \hat{P}_2^2 \hat{V}[\hat{P}_1] + \hat{N}_2^2 \hat{P}_1^2 \hat{V}[\hat{P}_2]\},$$

is also an asymptotically unbiased estimate of

$$(P_1\lambda - P_2\lambda)^{-2} \{(X_1 + \lambda Y_1)^2 P_2^2 \lambda V[\hat{P}_1] + (X_2 + \lambda Y_2)^2 P_1^2 \lambda V[\hat{P}_2]\}$$

which, by the delta method, is the true asymptotic variance of $\hat{X}_1$.

Example 9.5   Brook trout and cisco: Paulik and Robson [1969]

Paulik and Robson consider the following hypothetical data for a lake containing brook trout (*Salvelinus fontinalis*) and cisco (*Coregonus* sp.). Gill nets set before the fishing season caught 150 brook trout and 50 cisco. During the first three weeks of the season 700 brook trout were taken out of the lake. Gill nets set at the end of the three weeks captured 30 cisco and 30 brook trout. As the brook trout are about three times as susceptible to the gill nets as the cisco, N_1 cannot be estimated. However, $\hat{P}_1 = 150/200 = 0 \cdot 75$, $\hat{P}_2 = 30/60 = 0 \cdot 50$, $R_x = 700$, $R_y = 0$ and the number of brook trout before the season is estimated to be

$$\hat{X}_1 = \frac{0 \cdot 75(700)(1 - 0 \cdot 50)}{0 \cdot 75 - 0 \cdot 50} = 1050.$$

Assuming binomial sampling, we have $\hat{\sigma}[\hat{X}_1] = 93$.

Example 9.6   Mule deer (*Odocoileus hemonius*): Riordan [1948]

The adult sex ratios were estimated before and after the hunting season using aerial counts. Letting X and Y denote the buck and doe populations respectively, the following data were obtained for a particular area: $n_1 = 657$, $x_1 = 220$; $n_2 = 1011$, $x_2 = 129$; $R_x = 5500$ and $R_y = 0$. Then

$$\hat{N}_1 = 23\,150, \quad \hat{X}_1 = 7752$$

$$\hat{\sigma}[\hat{N}_1] = 2240 \quad \text{and} \quad \hat{\sigma}[\hat{X}_1] = 398.$$

Using aerial sighting, the assumption of "equicatchability" for the two sexes becomes the assumption that bucks and does are equally distinguishable. Possible departures from this assumption can occur through the difficulty of distinguishing between yearling bucks with spike antlers and does, and between does and fawns when the deer are bunched together in a large group. However, as shown above, since $R_y = 0$, $\hat{X}_1$ and $\hat{\sigma}[\hat{X}_1]$ are both robust with regard to such departures.

For a number of areas, ground surveys were also carried out and compared with the aerial survey data. Although in some areas there were significant differences in the estimated proportions of bucks, there seemed to be little difference in overall accuracy between the two methods. However, the aerial census has the following advantages over the ground census:

(1) All areas are equally accessible by air, thus allowing for more uniform sampling; this is not true to the same extent for ground observers.

(2) The sampling is more likely to be random by air since ground observers tend to drive the animals ahead, thus increasing the opportunity for the more wary ones to get away without being counted.

(3) The aerial method provides a larger number of animals for classification in a given time or at a given cost.

(4) In classifying animals on the ground, using binoculars at long distances, there is the possible error introduced by the fact that antlers make the bucks distinguishable at a greater distance than does and fawns can be distinguished from one another. Since only animals positively sex-identified are tallied, this would lead to an underestimate of the does and fawns in comparison with the number of bucks. However, with aerial counts one can simply cruise over the area at altitudes between 30 and 200 feet.

(5) If sex-ratio studies are made in the rutting season, the bucks, and especially the older ones, appear to be less cautious and are much easier to approach on foot or by automobile than at other times of the year. This would lead to an overestimate of the proportion of bucks.

ASSUMPTION (iii). This assumption may be false as R_x and R_y could be underevaluated because of unknown natural mortality, unreported kills, and possible "crippling losses". The main method of determining the total kill in a hunting season is by questionnaire. However, this can introduce all kinds of bias which must be allowed for if one is to have any confidence in the final values of R_x and R_y. To begin with, there is the non-response bias due to some hunters not returning their questionnaires (cf. Sen [1971a, b]); this may be minimised using repeated mailings (MacDonald and Dillman [1968]). Secondly, there may be certain biases present in the completed questionnaires. For example, Atwood [1956] detected three sorts of bias: (1) Prestige bias, which arose from pride and caused hunters to claim a higher number of daily-bag limits than they had actually obtained. The effect of this bias was greatest where the daily bag limit was low. (2) Type I-Memory bias, which was due to failure of memory acting together with overstatement to cause hunters to report their bag as a number ending in zero or five, and somewhat above the true figure. (3) Type II-Memory bias, which was due to the fact that hunters recalled small numbers more accurately than large ones, and further, when the total seasonal bag was large, it was reported as larger still. The above types of response bias can be partly allowed for by, for example, disregarding the replies from hunters who reported their kill as multiples of five or the daily-bag limit. MacDonald and Dillman [1968] suggest evaluating the questionnaires in the light of those returned by hunters whose performance is known. Reward banding has also been used as a means of checking reporting rates (Tomlinson [1968]).

The hunting kill can also be determined from bag checks in the field or by using checking stations. Bag checks are useful in determining the kill from a particular population or area and are commonly used for small game with wardens or conservation officers making the check in the course of their patrols. Possible sources of error are: inadequacy of the selection of hunters to be questioned; inadequacy of the time of questioning with regard to the whole hunting season; and interviewer bias. Checking stations are essentially roadblocks where it may or may not be compulsory for the hunter to stop. This method is particularly useful if a large area is accessible by only a few roads. However, check-station operation is expensive in manpower if the season is long; also, since hunters tend to move *en masse*, time-consuming lineups may form at the station in the evenings. Where it is not compulsory for a hunter to stop at a checking station, the tendency will be for the successful hunters to stop, and for those who are unsuccessful, or have inferior animals or over-bag animals, to pass on.

In addition to the animals which are killed and taken home, there are those which are killed, but lost or abandoned. Some crippled animals are subsequently found or shot by other hunters and so appear in the bag, while other crippled animals subsequently recover. The "crippling loss", defined to be those animals which are shot, do not recover, and do not appear in the bag, can be estimated from hunter reports or by searching sample plots or transects for carcasses (e.g. Whitlock and Eberhardt [1956], Robinette *et al.* [1954, 1956]). The Petersen method using tagged carcasses randomly placed has also been used in conjunction with other sampling methods. Small game found dead after the hunting season can be X-rayed for the presence of lead shot; X-ray methods have been used for estimating illegal kill (e.g. Chesness and Nelson [1964]).

The above discussion is based on Geis and Taber [1963] and the reader is referred to their chapter and MacDonald and Dillman [1968] for further references and details. To evaluate the effect of bias on our estimates $\hat{N}_1$ and $\hat{X}_1$, suppose that the actual removals are $R_x + \Delta_x$ and $R_y + \Delta_y$. Then (from equation (9.1)),

$$
\begin{aligned}
N_1 &= \frac{R_x + \Delta_x - P_2(R + \Delta_x + \Delta_y)}{P_1 - P_2} \\[2mm]
&= \frac{R_x - P_2 R}{P_1 - P_2} + \frac{\Delta_x - P_2(\Delta_x + \Delta_y)}{P_1 - P_2} \\[2mm]
&= \frac{R_x - P_2 R}{P_1 - P_2} \left\{ 1 + \frac{\Delta_x - P_2(\Delta_x + \Delta_y)}{R_x - P_2 R} \right\} \\[2mm]
&= \frac{R_x - P_2 R}{P_1 - P_2} \{1 + b_1\}, \text{ say.}
\end{aligned}
$$

Therefore, asymptotically (i.e. large n_i),

$$
E[\hat{N}_1] = \frac{R_x - P_2 R}{P_1 - P_2} = \frac{N_1}{(1 + b_1)} \;,
$$

and

$$E[\hat{X}_1] = P_1 E[\hat{N}_1]$$

$$= \frac{X_1}{(1+b_1)}.$$

Also, using a similar argument, it can be shown that asymptotically

$$E[\hat{N}_2] = N_2/(1+b_2),$$

where

$$b_2 = \frac{\Delta_x - P_1(\Delta_x + \Delta_y)}{R_x - P_1 R}.$$

Chapman [1955] has evaluated b_1 for two special cases:

Suppose that the percentage of unreported kills is the same for both classes so that

$$\frac{\Delta_x}{R_x} = \frac{\Delta_y}{R_y} = \frac{(\Delta_x + \Delta_y)}{R} = k, \text{ say.}$$

Then $b_1 = b_2 = k$, and if k is small and positive, N_1, X_1 and N_2 are all slightly underestimated.

Alternatively, if $R_y = 0$, $\Delta_x/R_x = k$ and the percentage mortality or unknown removal is the same for both classes (i.e. $\Delta_x/X_1 = \Delta_y/Y_1$), then it can be shown that

$$(1+b_1)^{-1} = 1 - (\Delta_x/X_1) = 1 - (k R_x/X_1),$$

(which is of smaller order than $(1+k)^{-1} \approx 1 - k$) and $b_2 = 0$. Therefore if k is small, any unreported removals will have little effect on $\hat{N}_1$, $\hat{X}_1$ and their variance estimates (which require both $\hat{N}_1$ and $\hat{N}_2$; cf. (9.3) and (9.4)).

Example 9.7   Pheasants: Paulik and Robson [1969]

In Example 9.3 (p.356), $\hat{N}_1 = 21\ 761$. However, if the game biologist thinks that the estimated number of hens shot is about 15 per cent too low because of some unreported illegal kills and that the estimated number of cocks is 15 per cent low because of unobserved deaths of wounded birds, then $b_1 = k = 0 \cdot 15$ and, correcting $\hat{N}_1$ for bias, we have

$$\hat{N}_1(1+b_1) = 21\ 761\ (1 \cdot 15) = 25\ 069.$$

If the illegal kill of hens is considered to be much greater, so that $k_x = 0 \cdot 15$ and $k_y = 1 \cdot 00$, say, then

$$\Delta_x = k_x R_x = (0 \cdot 15)\ 8000 = 1200,$$

$$\Delta_y = k_y R_y = (1 \cdot 00)\ 500 = 500,$$

and recalling that $\hat{P}_2 = 0 \cdot 10$, b_1 is estimated by

$$\hat{b}_1 = \frac{1200 - (0 \cdot 10)(1200 + 500)}{8000 - (0 \cdot 10)(8000 + 500)} = 0 \cdot 144.$$

Hence

$$\hat{N}_1(1+\hat{b}_1) = 21\ 761(1 \cdot 144) = 24\ 895.$$

3   Optimum allocation

If N_1 is to be estimated with minimum variance, subject to $n_1 + n_2$ fixed, then, using (9.5), Chapman [1955] shows that the optimal sample allocation is

$$\frac{n_2}{n_1} = \left(\frac{X_2 Y_2}{X_1 Y_1}\right)^{\frac{1}{2}} = \left(1 - \frac{R_x}{X_1}\right)^{\frac{1}{2}} \left(1 - \frac{R_y}{Y_1}\right)^{\frac{1}{2}},$$

and n_2 should be chosen smaller than n_1. For the special case $R_y = 0$,

$$\frac{n_2}{n_1} = \left(1 - \frac{R_x}{X_1}\right)^{\frac{1}{2}}$$

and Chapman's table of values

R_x/X_1	0·5	0·4	0·3	0·2	0·1	0·05	0·02	0·01
n_2/n_1	0·71	0·77	0·84	0·89	0·95	0·975	0·990	0·995

suggests that n_2 should then be only slightly less than n_1.

However, if X_1 is to be estimated, $V[\hat{X}_1]$ is minimised, subject to $n_1 + n_2$ fixed, if

$$\frac{n_2}{n_1} = \frac{N_2}{N_1} \left(\frac{X_1 Y_2}{X_2 Y_1}\right)^{-\frac{1}{2}}.$$

For the case $R_y = 0$, $Y_1 = Y_2$ and

$$\frac{n_2}{n_1} = \left(1 - \frac{R_x}{N_1}\right) \left(1 - \frac{R_x}{X_1}\right)^{-\frac{1}{2}}.$$

Table 9.1 (reproduced from Chapman) gives the ratio n_2/n_1 for different values of R_x/N_1 and $P_1 = X_1/N_1$, and we see that when $P_1 < 0·5$, the second sample should be slightly larger than the first. Therefore if $R_y = 0$ and both $\hat{N}_1$ and $\hat{X}_1$ are required, it would seem that choosing $n_1 = n_2$ represents a near optimum allocation of effort.

4   Confidence intervals

Paulik and Robson [1969] present three methods of constructing confidence intervals for N_1 which we now discuss.

The simplest confidence interval for N_1 is obtained by assuming $\hat{N}_1$ to be asymptotically normal, so that an approximate $100(1 - \alpha)$ per cent confidence interval is given by

$$\hat{N}_1 \pm z_{\alpha/2} \, \hat{\sigma}[\hat{N}_1]. \tag{9.9}$$

However, when the samples are too small for the delta method to apply in the derivation of $V[\hat{N}_1]$, it is still possible to obtain confidence intervals by employing modified versions of the delta method. One such modification is to find a monotonic function of $\hat{N}_1$ that approaches normality faster than $\hat{N}_1$;

TABLE 9.1

Optimum sample ratio n_2/n_1 for the estimate of X_1 when there is no removal of y-type animals: from Chapman [1955: Table 2].

P_1	R/N_1						
	0·25	0·20	0·15	0·10	0·05	0·02	0·01
0·1	–	–	–	–	1·344	1·096	1·043
0·2	–	–	1·700	1·273	1·097	1·033	1·015
0·3	1·838	1·386	1·202	1·102	1·041	1·014	1·007
0·4	1·225	1·132	1·075	1·039	1·016	1·005	1·003
0·5	1·061	1·032	1·016	1·007	1·001	1·000	1·000
0·6	0·982	0·980	0·982	0·986	0·993	0·997	0·998
0·7	0·935	0·947	0·959	0·972	0·985	0·994	0·997
0·8	0·905	0·924	0·943	0·963	0·981	0·993	0·996
0·9	0·882	0·907	0·931	0·954	0·977	0·991	0·996

the monotonicity property is then employed to convert the limits for the same function of N_1 into limits for N_1. (This procedure has already been used a number of times, e.g. **3.1.4** and equation (4.8).) In the present situation the distribution of $1/\hat{N}_1$ approaches normality faster than the distribution of $\hat{N}_1$; as Paulik and Robson point out, it is generally advisable to avoid using estimates with high variability in the denominator of a fraction (in this case in $\hat{P}_1 - \hat{P}_2$). Therefore, using the delta method,

$$V[1/\hat{N}_1] \approx \frac{(R_x - P_1 R)^2}{(R_x - P_2 R)^4} V[\hat{P}_2] + \frac{1}{(R_x - P_2 R)^2} V[\hat{P}_1]$$

and the approximate $100(1-\alpha)$ per cent confidence interval for $1/N_1$ is

$$(1/\hat{N}_1) \pm z_{\alpha/2} \, \hat{\sigma}[1/\hat{N}_1]. \tag{9.10}$$

This can be inverted and reversed to give an interval for N_1.

A minor objection to the above interval is that when it is inverted it may be quite skewed about the point estimate $\hat{N}_1$. One way of circumventing this difficulty is to use asymmetrical limits for $(1/\hat{N}_1)$, so that when inverted, $\hat{N}_1$ will be in the centre of the interval obtained. To do this we must choose α_1 so that

$$\hat{N}_1 - \{(1/\hat{N}_1) + z_{\alpha-\alpha_1} \hat{\sigma}[1/\hat{N}_1]\}^{-1} = \{(1/\hat{N}_1) - z_{\alpha_1} \hat{\sigma}[1/\hat{N}_1]\}^{-1} - \hat{N}_1.$$

(Our definition of z_{α_1}, given in **1.2.1**, differs from that of Paulik and Robson.) Rearranging the above equation, we find that α_1 is the solution of

$$h(\alpha_1) = 2\hat{N}_1 \, \hat{\sigma}(1/\hat{N}_1) \quad (= d, \text{ say}) \tag{9.11}$$

where $h(\alpha_1) = \dfrac{z_{\alpha-\alpha_1} - z_{\alpha_1}}{z_{\alpha_1} z_{\alpha-\alpha_1}}.$

Paulik and Robson point out that since the denominator of $h(\alpha_1)$ changes slowly in comparison with the numerator, (9.11) can usually be solved by linear interpolation between $h(a_{11})$ and $h(a_{12})$ using two initial guesses a_{11} and a_{12}, such that $h(a_{11}) < d$ and $h(a_{12}) > d$. Table 1 of the Kelley Statistical Tables (Kelley [1948: 37]) tabulates tail probabilities such as a_1 by 0·0001 increments and this provides a useful aid in the solution of (9.11). We note that the symmetrical interval will always be wider than the unsymmetrical interval based on (9.10).

When R_x and R_y have the same sign, a more exact confidence interval can be derived. Let $f = R_x/R$; then, from (9.1),

$$N_1 = \frac{R(f - P_2)}{P_1 - P_2}.$$

Since

$$w = R(f - \hat{P}_2) - N_1(\hat{P}_1 - \hat{P}_2)$$

is a linear function of binomial random variables, it is approximately normally distributed with mean zero and variance

$$V[w] = N_1^2 V[\hat{P}_1] + (N_1 - R)^2 V[\hat{P}_2].$$

Therefore, from the probability statement

$$\Pr[w^2 \leqslant z_{\alpha/2}^2 V[w]] \approx 1 - \alpha, \tag{9.12}$$

the upper and lower limits for N_1 can be found as the roots of a quadratic equation, namely

$$N_L = (a - b)/c, \quad N_U = (a + b)/c,$$

where $\quad a = \hat{N}_1 - R\left(\dfrac{z_{\alpha/2}\,\hat{\sigma}[\hat{P}_2]}{\hat{P}_1 - \hat{P}_2}\right)^2,$

$$b = \left[\left(\frac{z_{\alpha/2}\,\hat{\sigma}[\hat{P}_2]}{\hat{P}_1 - \hat{P}_2}\right)^2 \left((\hat{N}_1 - R)^2 + \frac{\hat{V}[\hat{P}_1]}{\hat{V}[\hat{P}_2]}\left\{\hat{N}_1^2 - \left(\frac{z_{\alpha/2}\,R\,\hat{\sigma}[\hat{P}_2]}{\hat{P}_1 - \hat{P}_2}\right)^2\right\}\right)\right]^{\frac{1}{2}},$$

and

$$c = 1 - \left(\frac{z_{\alpha/2}}{\hat{P}_2 - \hat{P}_1}\right)^2 (\hat{V}[\hat{P}_1] + \hat{V}[\hat{P}_2]).$$

Example 9.8

Paulik and Robson demonstrate the calculation of the above three confidence intervals for $\alpha = 0·05$ using the data of Example 9.3. We recall that $R_x = 8000$, $R_y = 500$, $f = 0·9412$, $\hat{P}_1 = 0·4286$, $\hat{P}_2 = 0·1$, $\hat{V}[\hat{P}_1] = 0·000\ 174\ 9$, $\hat{V}[\hat{P}_2] = 0·000\ 045\ 0$, $\hat{N}_1 = 21\ 761$ and $\hat{\sigma}[\hat{N}_1] = 917$.

The interval (9.9) is $21\ 761 \pm 1·96(917)$, i.e. $(19\ 964 < N_1 < 23\ 558)$.

The interval (9.10) is $(0·4216 \times 10^{-4} < 1/N_1 < 0·4975 \times 10^{-4})$ which inverted leads to $(20\ 101 < N_1 < 23\ 720)$. To calculate the symmetrical 95 per cent confidence interval for N_1, we calculate $2\hat{N}_1\,\hat{\sigma}[1/\hat{N}_1] = 0·084\ 25$, and interpolating between $h(0·030) = 0·0448$ and $h(0·35) = 0·0911$ obtain

$a_1 = 0 \cdot 0343$ $(h(0 \cdot 0343) = 0 \cdot 0845)$. Table I of Kelley [1948: 131, 134] gives $z_{0 \cdot 0343} = 1 \cdot 8210$ and $z_{0 \cdot 0157} = 2 \cdot 1520$, so that the 95 per cent confidence interval

$$(1/\hat{N}_1) - z_{0 \cdot 0343} \, \hat{\sigma}[1/\hat{N}_1] < 1/N_1 < (1/\hat{N}_1) + z_{0 \cdot 0157} \, \hat{\sigma}[1/\hat{N}_1]$$

is found to be $(0 \cdot 4242 \times 10^{-4} < 1/N_1 < 0 \cdot 5013 \times 10^{-4})$ which, inverted, yields $(19\ 948 < N_1 < 23\ 574)$.

Using the third interval with $z_{\alpha/2} = 1 \cdot 96$, we find that $a = 21\ 747$, $b = 1797$ and $c = 0 \cdot 9922$; $N_L = 20\ 108$ and $N_U = 23\ 730$. Thus the quadratic method leads to the confidence interval $(20\ 108 < N_1 < 23\ 730)$.

COMPARISON OF INTERVALS. Of the three methods of interval estimation, the interval derived from (9.12) is more accurate than the intervals obtained from (9.9) and (9.10). The quadratic method is more accurate in the sense that if all the underlying assumptions are satisfied, particularly the assumption of binomial sampling, the probability that the interval (N_L, N_U) covers the true population size will be closer to the nominal probability of $1 - \alpha$ than either of the other two intervals. For most situations the interval found from using the inverse is only slightly less accurate than that derived from the quadratic. When sample sizes are large, the three methods will yield practically identical confidence intervals, while for small samples the accuracy of (9.9) is significantly poorer than either of the other two intervals.

Paulik and Robson give a helpful numerical comparison of the three intervals by using the estimates given in Example 9.8, but allowing n_1 and n_2 to vary. Suppose that L1 is the length of the interval (9.9), L2 is the length of the interval obtained by inverting (9.10) and L3 is the quadratic interval. Then from Paulik and Robson we have Table 9.2.

TABLE 9.2

Comparison of the lengths of three confidence intervals: from Paulik and Robson [1969: 22].

$n_1 = n_2$	L1	L2	L3
100	13 696	15 202	15 516
200	9 685	10 189	10 287
500	6 125	6 249	6 272
1000	4 331	4 374	4 382

5 Planning CIR experiments

We shall now discuss the problem of determining the number of animals that must be sampled in order to obtain an estimate of N_1 with a given accuracy when R_x and R_y are both positive and the x-category is chosen so that $f = R_x/R$ is greater than P_1 (i.e. $f > P_1 > P_2$). The following discussion is based on Paulik and Robson [1969: 22–27].

Let $(1 - \alpha)$ be the probability that $\hat{N}_1$ will not differ from N_1 by more than $100\ \epsilon$ per cent; that is

$$\Pr\left[-\ \epsilon < \frac{\hat{N}_1 - N_1}{N_1} < \epsilon\right] \geqslant 1 - \alpha, \tag{9.13}$$

where α and ϵ are to be chosen by the experimenter (cf. **3.1.5**(*1*) for suggested levels: there $\epsilon = A$). Then substituting for $\hat{N}_1$, (9.13) becomes

$$\Pr\left[-\ \epsilon < \frac{u(f - \hat{P}_2)}{(\hat{P}_1 - \hat{P}_2)} - 1 < \epsilon\right] \geqslant 1 - \alpha, \tag{9.14}$$

where $u = R/N_1$, the rate of exploitation. To evaluate the left-hand side of (9.14) we define a random variable

$$w = uf + \epsilon(P_1 - P_2) + \hat{P}_2(1 + \epsilon - u) - \hat{P}_1(1 + \epsilon),$$

which has mean zero and variance

$$V[w] = (1 + \epsilon - u)^2 V[\hat{P}_2] + (1 + \epsilon)^2 V[\hat{P}_1].$$

Then

$$\Pr\left[\frac{\hat{N}_1 - N_1}{N_1} < \epsilon\right] = \Pr\left[w < \epsilon(P_1 - P_2)\right]$$

and a similar random variable can be defined to deal with the negative ϵ in (9.14). Therefore, if the sampling is binomial, w is asymptotically normal, and for the case $n_1 = n_2 = n$ Paulik and Robson show that

$$\Pr\left[\left|\frac{\hat{N}_1 - N_1}{N}\right| < \epsilon\right] \approx \phi\left(\frac{\epsilon\sqrt{n}(P_1 - P_2)}{\sqrt{\{(1 + \epsilon - u)^2 P_2 Q_2 + (1 + \epsilon)^2 P_1 Q_1\}}}\right)$$

$$- \phi\left(\frac{-\ \epsilon\sqrt{n}(P_1 - P_2)}{\sqrt{\{(1 - \epsilon - u)^2 P_2 Q_2 + (1 - \epsilon)^2 P_1 Q_1\}}}\right)$$

$$= 1 - \alpha, \tag{9.15}$$

where ϕ is the cumulative unit normal distribution. This approximation is reasonably good even for small sample sizes. The choice of $n_1 = n_2$ is prompted by the discussion on p. 363.

Equation (9.15) cannot be solved explicitly, and Paulik and Robson programmed an iterative routine to obtain approximate solutions which were then used to construct the charts in Fig. 9.1–9.3. In using these charts it is convenient to think in terms of $\Delta P = P_1 - P_2$; thus, $f - P_2 > \Delta P$ and $u = \Delta P/(f - P_2)$. Also, writing $P_2 = P_1 - \Delta P$, then

$$u = 1/[1 + (f - P_1)/\Delta P] \tag{9.16}$$

and, since $f \leqslant 1$,

$$u \geqslant 1/[1 + (1 - P_1)/\Delta P] = u\ (\text{min}), \text{ say}. \tag{9.17}$$

367

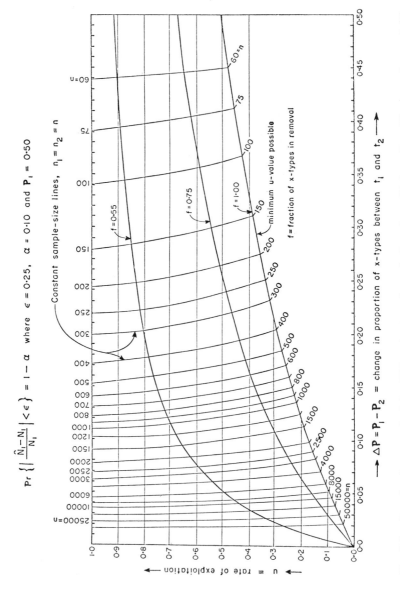

Fig. 9.1   Sample sizes required for CIR estimates of population size with less than 25 per cent error (with probability 0·90) when the initial proportion P_1 of x-types is 0·50: from Paulik and Robson [1969].

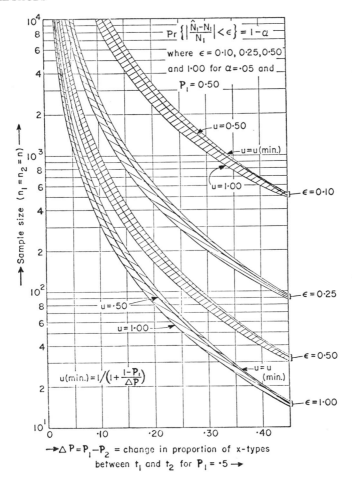

Fig. 9.2 Sample sizes required for CIR estimates of population size with less than 100 ϵ per cent error (with probability 0.95) for ϵ = 0.10, 0.25, 0.50 and 1.00, when the initial proportion P_1 of x-types is 0.50: from Paulik and Robson[1969].

For $P_1 = 0.50$, $\alpha = 0.10$, $\epsilon = 0.25$ and given ΔP, Fig. 9.1 gives combinations of n and u satisfying (9.15). If the investigator also knows the value of u, a unique n can be determined by interpolating between two sample size lines. A more conservative approach which avoids interpolation is to read n from the first line to the left of the $(u, \Delta P)$-point; a sample size so determined will be slightly larger than the minimum required. Alternatively, if f is known, the $(u, \Delta P)$-point can be found by interpolation using (9.16) and the u-lines for f values of 1.0, 0.75 and 0.55. For example, if the entire removal is of x-types ($f = 1$) and the expected ΔP is 0.25, u is approximately 0.33 and the sample size needed to estimate the population size within 25 per cent with 0.90 probability is 250 animals in each sample.

In using Fig. 9.1 it should be noted that n is much more sensitive to

369

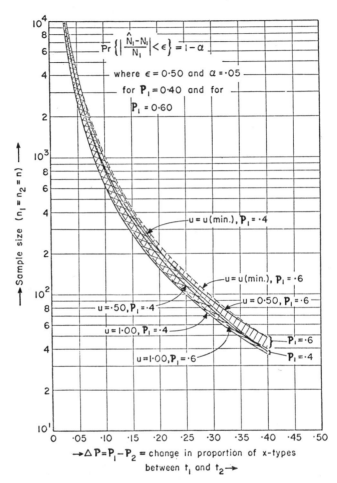

Fig. 9.3 Sample sizes required for CIR estimates of population size with less than 50 per cent error (with probability 0·95) for initial proportions of x-types $P_1 = 0.40$ and $P_1 = 0.60$: from Paulik and Robson[1969].

ΔP than to u. For example, when ΔP is in the vicinity of 0·015 to 0·020, sample sizes of around 50 000 are needed. This rapid increase in sample size as ΔP decreases below 0·05 indicates that the CIR method is so unreliable for ΔP's in this range that it is almost useless as a means of estimating N_1. For ΔP's between 0·05 and 0·10 use of this method is questionable, and the assumptions upon which the model is based become critical.

The four curves in Fig. 9.2 enable a planner to estimate sample sizes for $\epsilon = 0.10,\ 0.25,\ 0.50,\ 1.00,\ P_1 = 0.50$ and $\alpha = 0.05$. In each band $u = 1.00$ on the lower boundary and $u = u$ (min) (cf. (9.17)) on the upper boundary. As ΔP approaches its maximum value of $P_1 = 0.5$, the minimum value of u, u (min), approaches 0·50.

For the special case $\epsilon = 0·50$, Fig. 9.3 gives the same bands as Fig. 9.2 except for $P_1 = 0·4$ and $P_1 = 0·6$. We see that for ΔP's between $0·05$ and $0·30$, the range of most practical importance, these bands differ only slightly. This means that for a fixed ΔP, the sample size is insensitive to the value of P_1 for P_1 in the vicinity of $0·50$.

On p. 363 formulae are given for the optimum value of n_2/n_1. It was shown there that although a value of unity, which is used in Fig. 9.1–9.3, is not optimal it will be satisfactory in most cases. However. if u is large, the nominal accuracy of the estimate can be increased by partitioning the total sample size of $2n$ obtained from one of the three figures according to the optimum ratio.

Example 9.9

Consider the problem of designing an experiment to estimate the size of the pheasant population described in Example 9.3 (p. 356). How many pheasants should be sexed in pre- and post-season samples to ensure that, with probability $0·95$, $\hat{N}_1$ will not be in error by more than 25 per cent? Suppose we know from previous years that the kill of cocks can be expected to change the proportion of cocks by about $0·30$. Then assuming $P_1 = \frac{1}{2}$, u will be slightly greater than $u(\min) = 1/[1 + (0·5/0·3)] = 0·38$ as most of the removals will be males ($f \approx 1$). Therefore, entering Fig. 9.2 with $\epsilon = 0·25$, $\Delta P = 0·30$ and $u = 0·40$, we obtain $n \approx 250$, so that a total of 500 pheasants must be sexed in the two samples.

Alternatively, if $P_1 = 0·50$, $\Delta P = 0·2$, $f = 0·70$, then from (9.16) $u = 0·5$. Therefore for $\epsilon = 0·10$, $n \approx 3000$.

6 Removals estimated

In many situations the removals R_x, R_y are not known exactly and must be estimated. Therefore, if $\hat{R}_x$ and $\hat{R}_y$ are independent unbiased estimates of R_x and R_y, and $\hat{R} = \hat{R}_x + \hat{R}_y$, then N_1 can be estimated by

$$\hat{N}_1 = \frac{\hat{R}_x - \hat{P}_2 \hat{R}}{\hat{P}_1 - \hat{P}_2}.$$

Using the delta method,

$$V[\hat{N}_1] \approx (P_1 - P_2)^{-2} \{ N_1^2 V[\hat{P}_1] + N_2^2 V[\hat{P}_2] + (1 - P_2)^2 V[\hat{R}_x] + P_2^2 V[\hat{R}_y] \}.$$

Alternatively we may have independent estimates of R and $f = R_x/R$, so that

$$\hat{N}_1 = \frac{\hat{R}(\hat{f} - \hat{P}_2)}{\hat{P}_1 - \hat{P}_2}$$

and

$$V[\hat{N}_1] \approx (P_1 - P_2)^{-2} \{ N_1^2 V[\hat{P}_1] + N_2^2 V[\hat{P}_2] + R^2 V[\hat{f}] + (f - P_2)^2 V[\hat{R}] \}. \quad (9.18)$$

A possible situation is when R is known but f has to be estimated by means of a sample from the R removals. In this case $V[\hat{R}] = 0$ and $\hat{f}$ will usually be a binomial proportion.

7   The Petersen estimate

It was first pointed out by Rupp [1966] that the Petersen method can be regarded as a special case of the CIR method. To see this, let X and Y denote the marked and unmarked populations respectively ($X_1 = P_1 = 0$) and suppose that M animals are caught for marking and then released. Then $R_y = M$, $R_x = -M$, and

$$\hat{N}_1 = \frac{R_x - R\hat{P}_2}{\hat{P}_1 - \hat{P}_2}$$

$$= \frac{-M - (M-M)\hat{P}_2}{0 - \hat{P}_2}$$

$$= \frac{M}{\hat{P}_2},$$

the Petersen estimate (cf. **3.1.1**).

8   Several removals

Chapman [1955] has considered the less common situation in which there are several selective removals, each followed by a random sample. For $i = 1, 2, \ldots, s$ let R_{xi}, R_{yi} be the total (cumulative) removals from the X and Y classes respectively prior to the ith sample, and let

$$R_i = R_{xi} + R_{yi}, \quad (R_1 = 0),$$
$$X_i = X_1 - R_{xi},$$
$$N_i = N_1 - R_i,$$
and
$$P_i = X_i/N_i.$$

If the sampling is binomial then the maximum-likelihood estimates $\hat{X}_1$ and $\hat{N}_1$ are solutions of

$$\sum_{i=1}^{s} (x_i/X_i) = \sum_{i=1}^{s} (n_i/N_i)$$
and
$$\sum_{i=1}^{s} (y_i/Y_i) = \sum_{i=1}^{s} (n_i/N_i)$$

which can be obtained iteratively (cf. **1.3.8**). The asymptotic variances for these solutions are

$$V[\hat{X}_1] = \sum_{i=1}^{s} \left\{ \frac{n_i P_i^2}{X_i Y_i d} \right\}$$
and
$$V[\hat{N}_1] = \sum_{i=1}^{s} \left\{ \frac{n_i}{X_i Y_i d} \right\},$$

where $d = \left(\sum_{i=1}^{s} \frac{n_i}{X_i Y_i}\right)\left(\sum_{i=1}^{s} \frac{n_i P_i^2}{X_i Y_i}\right) - \left(\sum_{i=1}^{s} \frac{n_i}{N_i Y_i}\right)^2$

$$= \sum_i \sum_{<j} \left\{\frac{n_i n_j (P_i - P_j)^2}{X_i Y_i X_j Y_j}\right\}.$$

A pair of first approximations for the iterative process is given, for instance, by the first and last samples, i.e.

$$\tilde{N}_1 = \frac{R_s - R_{xs}\hat{P}_s}{\hat{P}_1 - \hat{P}_s}, \quad \tilde{X}_1 = \hat{P}_1 \tilde{N}_1.$$

If, however, R_{xi}, R_{yi} are small relative to X_1, Y_1, Chapman shows that approximations to the maximum-likelihood estimates are given by solving

$$(1 - \hat{P}_1) \sum x_i R_{xi} + \hat{P}_1 \sum y_i R_{yi} - \hat{P}_1(1 - \hat{P}_1) \sum n_i R_i = 0 \qquad (9.19)$$

and

$$\frac{\sum x_i}{\hat{P}_1} + \frac{\sum x_i R_{xi}}{\hat{N}_1 \hat{P}_1^2} - \frac{\sum y_i}{1 - \hat{P}_1} - \frac{\sum y_i R_{yi}}{\hat{N}_1 (1 - \hat{P}_1)^2} = 0, \qquad (9.20)$$

where the summation in each case is $i = 1, 2, \ldots, s$. $\hat{P}_1$ ($= \hat{X}_1/\hat{N}_1$) may be determined from the quadratic (9.19), and this substituted in (9.20) gives $\hat{N}_1$, and finally $\hat{X}_1$.

An interesting application of the above method is when the sampling process is combined with the removal process. For instance, the sampler might only return some of the animals examined (e.g. just y-types). Alternatively he can effectively "remove" animals by tagging them before they are released back into the population. Chapman suggests the possibility of setting up a sequential procedure for sampling and removing x-types from the sample for s steps, where s is determined by the actual observations.

9.1.3 Other sampling procedures

1 Subsampling

By regarding the x-type animals as "marked", the procedures given in **3.4** can be used for estimating P_i and $V[\hat{P}_i]$. For example, suppose that the total population area is divided up into K subareas of which k are selected at random. Let the suffix j ($j = 1, 2, \ldots, k$) denote membership of the jth subarea sampled; thus x_{ij} is defined to be the number of x-types in the ith sample from the jth subarea, and P_{ij} is the proportion of x-types in the jth subarea at time t_i. Then two estimates of P_i are available, namely the pooled estimate

$$\hat{P}_i = \left(\sum_{j=1}^{k} x_{ij}\right)\bigg/\left(\sum_{j=1}^{k} n_{ij}\right) = x_i/n_i$$

and the average

$$\bar{P}_i = \frac{1}{k}\sum_{j=1}^{k}\left(\frac{x_{ij}}{n_{ij}}\right) = \frac{1}{k}\sum_{j=1}^{k}\hat{P}_{ij}, \quad \text{say.}$$

If $P_{ij} = P_i$ $(j = 1, 2, \ldots, k)$, as indicated by a chi-squared test of homogeneity using

$$T = \sum_{j=1}^{k} \left\{ \frac{(x_{ij} - n_{ij}\hat{P}_i)^2}{n_{ij}\hat{P}_i(1 - \hat{P}_i)} \right\}$$

with $k - 1$ degrees of freedom, then one would use $\hat{P}_i$ with variance estimate $\hat{P}_i(1 - \hat{P}_i)/n_i$, and we are back to the case considered in **9.1.2**. However, if the hypothesis of homogeneity is strongly rejected, then $\bar{P}_{i.}$ should be used along with an empirical estimate of variance. For example, if k/K is small then an approximately unbiased estimate of $V[\bar{P}_{i.}]$ is (cf. **3.4.2**)

$$\hat{V}[\bar{P}_{i.}] = \frac{1}{k(k-1)} \sum_{j=1}^{k} (\hat{P}_{ij} - \bar{P}_{i.})^2.$$

We note that when $n_{ij} = n$, say, for all i, j, the two estimates of P_i are identical.

The above subsampling procedure can be generalised by allowing k_1 sampling locations in the pre-season survey and k_2, possibly different, locations in the post-season survey. In this case the above discussion still holds, except that for given i, $j = 1, 2, \ldots, k_i$. Paulik and Robson [1969: 15] point out that if the variability of the $\hat{P}_{ij}$ about $\bar{P}_{i.}$ is the same for each survey then the two sets of data can be combined to give an overall estimate of the variance of $\hat{P}_{ij}$, namely

$$v = \sum_{i=1}^{2} \sum_{j=1}^{k_i} (\hat{P}_{ij} - \bar{P}_{i.})^2 / (k_1 + k_2 - 2).$$

Then $V[\bar{P}_{i.}]$ is estimated by v/k_i.

2   Using a single sample

Suppose that the removal is due to a hunting season and that the two classes are males and females. With some populations it may not be possible to take a pre-season sample, and P_1 must be estimated by some sort of extrapolation from a post-season sample of the previous year. One such method is as follows:

Let P = last year's post-season proportion of males,

   b = proportion of births since the last season which are males,

and L = average number of young per female.

Assuming that the mortality in the period from the end of last season to the beginning of this season is negligible (or proportionally the same for males and females), then it is readily shown that the proportion of males before this season is

$$P_1 = \frac{P + QbL}{1 + QL},$$

where $Q = 1 - P$. Therefore, if the above parameters are all estimated independently by unbiased estimates $\hat{P}$, $\hat{b}$, $\hat{L}$ respectively, and $\hat{P}_1$ is the resulting estimate of P_1, then, using the delta method,

$$V[\hat{P}_1] \approx \frac{(1 + L - bL)^2}{(1 + QL)^4} V[\hat{P}] + \frac{Q^2 L^2}{(1 + QL)^2} V[\hat{b}] + \frac{Q^2 (Q + b - 1)^2}{(1 + QL)^4} V[\hat{L}].$$

However, in general, such estimates of P_1 will not be very accurate.

3 Utilising age ratios

Occasionally the P_i can be estimated from age ratios using the methods of Kimball [1948: 303] and Severinghaus and Maguire [1955] (cf. Hanson [1963: 42]) as follows. Let X, Y, X_s, Y_s be the number of male and female adults, and male and female subadults in the population (cf. 9.2.1 for a definition of subadult); then if $r = X/Y$ we have the identity

$$r = \frac{X_s}{Y_s} \cdot \frac{(Y_s/Y)}{(X_s/X)}.$$

Therefore if X_s/Y_s, the subadult sex ratio, is known (say equal to 1) and the ratios of subadults to adults for both males and females are estimated from a trapped sample, say, then r, and hence

$$P = \frac{X}{X + Y} = \frac{r}{r + 1},$$

can be estimated. This method could be particularly useful if sample estimates of adult sex ratios are strongly biased through a differential catchability between sexes or through the difficulty of obtaining a proper random sample with regard to sexes (particularly if the sexes tend to segregate). The main assumption underlying the method is that for each sex, the subadults and adults have the same probability of being caught, so that although sexes may have different catchabilities, unbiased estimates of Y_s/Y and X_s/X are obtainable.

Suppose that a sample from the population yielded numbers x, y, x_s, y_s in each of the four categories respectively. Then P is estimated by

$$\hat{P} = \hat{r}/(\hat{r} + 1),$$

where $\quad \hat{r} = \left(\frac{X_s}{Y_s} \right) \cdot \frac{(y_s/y)}{(x_s/x)}.$

Assuming the random variables x, y, x_s and y_s to be independent Poisson variables, and using the delta method, an estimate of $V[\hat{P}]$ is

$$\hat{V}[\hat{P}] = \frac{r^2}{(\hat{r} + 1)^4} \left\{ \frac{1}{x} + \frac{1}{y} + \frac{1}{x_s} + \frac{1}{y_s} \right\}.$$

4 Known sampling effort

Let

$$E[x_i] = a_i X_i \quad \text{and} \quad E[y_i] = b_i Y_i, \tag{9.21}$$

then in 9.1.2, with binomial sampling, we effectively assumed that

$a_i = b_i = n_i/N_i$. However, if the sample sizes are determined by sampling effort, so that the n_i are now random variables, we can make the alternative assumption (Chapman and Murphy [1965]) that $a_i = b_i = K f_i$, where K is the binomial catchability (cf. **7.1.3**(1)) and the sampling efforts f_1 and f_2 are known. Then replacing $E[n_i]$ by n_i we have the equations

$$\frac{n_1/f_1}{N_1} = \frac{n_2/f_2}{N_1 - R} = K \left(= \frac{(n_1/f_1) - (n_2/f_2)}{R} \right),$$

leading to the estimates

$$\hat{N}_1 = \frac{Rc_1}{c_1 - c_2} \quad \text{and} \quad \hat{N}_2 = \frac{Rc_2}{c_1 - c_2}, \tag{9.22}$$

where $c_i = n_i/f_i$. The estimate $\hat{N}_i$ was first given by Petrides [1949] and is particularly useful when it is not possible to distinguish between the two classes, e.g. deer tracks in the snow (Davis [1963]). We also note that

$$\hat{N}_1 = R \Big/ \left(1 - \frac{c_2}{c_1} \right),$$

so that animal signs can be used to measure the number of animals, provided the index signs-per-animal is the same for both samples.

Using the delta method, we have

$$V[\hat{N}_1] \approx (E[c_1] - E[c_2])^{-2} \{ N_1^2 V[c_2] + N_2^2 V[c_1] \}.$$

If n_1 and n_2 are assumed to be independent Poisson random variables then $E[c_i]$ and $V[c_i]$ can be estimated by c_i and c_i/f_i, respectively.

When the two classes are distinguishable, X_1 is estimated by

$$\hat{X}_1 = \frac{x_1}{n_1} \hat{N}_1.$$

Example 9.10 Pheasants (adapted from Petrides [1949])

A pre-season roadside count over 50 miles yields 150 pheasants, while a post-season count over 30 miles yields 45 pheasants. The season's kill is known to be 5600 birds. Then $f_1 = 50$, $f_2 = 30$, $c_1 = 150/50 = 3$ and $c_2 = 45/30 = 1\cdot 5$, so that

$$\hat{N}_1 = \frac{5600\,(3)}{3 - 1\cdot 5} = 11\,200$$

and

$$\hat{N}_2 = 11\,200 - 5600 = 5600.$$

Assuming Poisson sampling,

$$\hat{V}[\hat{N}_1] = (3 - 1\cdot 5)^{-2} \left\{ 11\,200^2 \left(\frac{1\cdot 5}{30} \right) + 5600^2 \left(\frac{3}{50} \right) \right\} = \cdot 3\,623\,800$$

and

$$\hat{\sigma}[\hat{N}_1] = 1903.$$

Since c_i is the sample mean of so many birds per mile, $V[c_i]$ can be estimated from replicated data using (1.9) if the number of birds seen each mile is recorded. This should be done where possible.

5 Constant class probability of capture

Chapman and Murphy consider the experimental situation in which $a_1 = a_2 = a$ and $b_1 = b_{2:} = b$ in (9.21), i.e. the classes X and Y are not equally catchable, but the chances of catching an x-type animal remain the same for the two surveys, and similarly for y-type animals. Then our estimating equations become (cf. (9.7) and (9.8) on p. 356)

$$\frac{x_1}{\hat{X}_1} = \frac{x_2}{\hat{X}_1 - R_x} \quad \text{and} \quad \frac{y_1}{\hat{Y}_1} = \frac{y_2}{\hat{Y}_1 - R_y},$$

which have solutions

$$\hat{X}_1 = \frac{R_x x_1}{(x_1 - x_2)} \quad \text{and} \quad \hat{Y}_1 = \frac{R_y y_1}{(y_1 - y_2)}.$$

Using the delta method,

$$V[\hat{X}_1] \approx (E[x_1 - x_2])^{-2} \{X_1^2 V[x_2] + X_2^2 V[x_1]\}$$

and

$$V[\hat{Y}_1] \approx (E[y_1 - y_2])^{-2} \{Y_1^2 V[y_2] + Y_2^2 V[y_1]\}.$$

Here $E[x_1 - x_2]$ is estimated by $x_1 - x_2$, and if x_1, x_2 are independent Poisson variables then $V[x_i]$ is estimated by x_i.

When the classes are equally catchable ($a = b$) then the data can be pooled to give

$$\frac{n_1}{\hat{N}_1} = \frac{n_2}{\hat{N}_1 - R}$$

or

$$\hat{N}_1 = \frac{R n_1}{n_1 - n_2}.$$

This estimate is the same as $\hat{N}_1$ of (9.22) when $f_1 = f_2$.

9.1.4 Utilising tag information

Chapman [1955: 284] showed that unless tagging is very much more expensive than the cost of classification, the Petersen recapture method will yield more information, that is, provide estimates with smaller variances, for the same amount of effort (cf. 12.1.2(6)). The recapture method in this situation would consist of tagging the n_1 animals from the pre-season survey and then observing the number of tagged animals, m ($= m_x + m_y$ say), in the removal R. The tagging estimate of N_1 would then be (cf. 3.1.1)

$$N_{1t}^* = \frac{(n_1 + 1)(R + 1)}{m + 1} - 1$$

with $V[N_{1t}^*]$ estimated by

$$v[N_{1t}^*] = \frac{(n_1+1)(R+1)(n_1-m)(R-m)}{(m+1)^2(m+2)}.$$

If R has to be estimated independently by $\hat{R}$, say, then we simply add $N_1^2 V[\hat{R}]/R^2$ to $V[N_{1t}^*]$.

When the animals in the first sample and in the m recaptures are also classified according to whether they are x-type or y-type, then several estimates of X_1 are available (cf. **3.2.5**(1)). For example, if the probabilities of capture in the first sample are the same for both types then the appropriate estimate of X_1 is

$$\tilde{X}_{1t} = \frac{x_1}{n_1} \cdot N_{1t}^* = \hat{P}_1 N_{1t}^*$$

with variance (cf. (1.11))

$$V[\tilde{X}_{1t}] \approx P_1^2 V[N_{1t}^*] + N_1^2 V[\hat{P}_1] + V[N_{1t}^*] V[\hat{P}_1].$$

But if the probabilities of capture in the first sample are different then X_1 should be estimated by

$$X_{1t}^* = \frac{(x_1+1)(R_x+1)}{(m_x+1)} - 1$$

with variance estimate

$$v[X_{1t}^*] = \frac{(x_1+1)(R_x+1)(x_1-m_x)(R_x-m_x)}{(m_x+1)^2(m_x+2)}.$$

If both R and $f = R_x/R$ are estimated independently by $\hat{R}$ and $\hat{f}$, then

$$X_{1t}^* = \frac{(x_1+1)(\hat{f}\hat{R}+1)}{(m_x+1)} - 1$$

and the variance is increased by approximately

$$X_1^2 \left\{ \frac{V[\hat{f}]}{f^2} + \frac{V[\hat{R}]}{R^2} \right\}.$$

When tagging is used along with classification, the first problem of interest is to determine whether there is a reasonable agreement between the two estimates $\hat{N}_1$ and N_{1t}^*. Since the two sources of information are essentially independent (for a given first sample size n_1), a large-sample test for agreement is to calculate

$$\frac{|\hat{N}_1 - N_{1t}^*|}{\{\hat{V}[\hat{N}_1] + \hat{V}[N_{1t}^*]\}^{\frac{1}{2}}} \tag{9.23}$$

and compare its value with the appropriate significance levels of the unit normal distribution. If the estimates are not compatible, it may be due to failure of the assumptions in one or both models. For example, as Chapman points out, the assumptions may be correct in regard to expectations, but the sampling may not be random, so that the variances are larger than those given above and the denominator of (9.23) is underestimated. Chapman also notes that the most important sources of error in $\hat{N}_1$, such as the under-estimation of R_x and R_y, will tend to bias $\hat{N}_1$ downwards, while factors such as trap-shyness and tag-mortality will tend to bias N_{1t}^* upwards. This means that if the estimates are compatible, there is a suggestion, but not neces-sarily proof, that the assumptions of the two models are fulfilled, and in this case we could obtain a new estimate, $\hat{N}_{1C}$ say, based on the combined tagging and removal information. The most efficient way of combining the two estimates is to take the weighted average (cf. **1.3.2**)

$$\hat{N}_{1C} = (w_1\hat{N}_1 + w_2 N_{1t}^*)/(w_1 + w_2),$$

where $w_1 = \{V[\hat{N}_1]\}^{-1}$ and $w_2 = \{V[N_{1t}^*]\}^{-1}$ have to be estimated. We then have

$$V[\hat{N}_{1C}] = (w_1 + w_2)^{-1}$$
$$= \frac{V[\hat{N}_1]V[N_{1t}^*]}{V[\hat{N}_1] + V[N_{1t}^*]}$$

and $\hat{N}_{1C}$ has smaller variance than either $\hat{N}_1$ or N_{1t}^*. Unfortunately the same method cannot be used to obtain $\hat{X}_{1C}$ since $\hat{X}_1$ and X_{1t}^* are not independent (both contain x_1).

If binomial sampling is used, an alternative approach suggested by Chapman is to use the combined binomial model (cf. (9.6) and (3.3))

$$\left\{\prod_{i=1}^{2}\binom{n_i}{x_i}\left(\frac{X_i}{N_i}\right)^{x_i}\left(1 - \frac{X_i}{N_i}\right)^{y_i}\right\}\binom{n_1}{m}\left(\frac{R}{N_1}\right)^{m}\left(1 - \frac{R}{N_1}\right)^{n_1 - m}$$

for which the maximum-likelihood estimates N'_{1C} and X'_{1C} are solutions of

$$\frac{x_1}{X_1} + \frac{x_2}{X_1 - R_x} = \frac{y_1}{Y_1} + \frac{y_2}{Y_1 - R_y} \qquad (9.24)$$

and

$$\frac{x_1}{X_1} + \frac{x_2}{X_1 - R_x} = \frac{2n_1}{N_1} + \frac{n_2 - n_1 + m}{N_1 - R}. \qquad (9.25)$$

These equations do not have simple solutions and must be solved iteratively (the solutions given by Chapman are incorrect). One method would be the following: start with a trial value of X_1 (say $\hat{X}_1$, X_{1t}^* or the average of the two), solve equations (9.24) and (9.25) as quadratics in Y_1 and N_1 respec-tively, choosing the larger root in each case, evaluate the new trial value $N_1 - Y_1$, and repeat the procedure. By inverting the information matrix, Chapman shows that asymptotically

$$V[N'_{1C}] = \left\{\frac{(P_1 - P_2)^2}{\dfrac{X_1Y_1}{n_1} + \dfrac{X_2Y_2}{n_2}} + \frac{n_1R}{N_1^2N_2}\right\}^{-1};$$

also

$$V[X'_{1C}] = \frac{\dfrac{X_1 Y_1}{n_1} P_2^2 + \dfrac{X_2 Y_2}{n_2} P_1^2 + \dfrac{n_1 R}{N_1^2 N_2} \left[\dfrac{X_1 Y_1}{n_1} \cdot \dfrac{X_2 Y_2}{n_2} \right]}{(P_1 - P_2)^2 + \dfrac{n_1 R}{N_1^2 N_2} \left[\dfrac{X_1 Y_1}{n_1} + \dfrac{X_2 Y_2}{n_2} \right]}$$

9.1.5 Exploitation rate

The rate of exploitation u is defined by Ricker [1958] as the fraction of the initial population removed by man during a specified time-interval. In terms of our notation, Paulik and Robson [1969: 16] show that

$$u = \frac{R}{N_1}$$

$$= \frac{R(P_1 - P_2)}{R_x - P_2 R}$$

$$= \frac{P_1 - P_2}{f - P_2},$$

where $f = R_x / R$. Similarly the rates of exploitation for x-type and y-type animals are respectively

$$u_x = \frac{R_x}{X_1} = \frac{fR}{P_1 N_1} = \frac{fu}{P_1}$$

and

$$u_y = \frac{(1-f)u}{(1-P_1)}.$$

To estimate u we do not need to know R_x; an estimate of f is sufficient. Thus

$$\hat{u} = \frac{\hat{P}_1 - \hat{P}_2}{\hat{f} - \hat{P}_2}, \quad \hat{u}_x = \frac{\hat{f}\hat{u}}{\hat{P}_1}$$

and, using the delta method,

$$V[\hat{u}] \approx (f - P_2)^{-4} \{ (f - P_2)^2 V[\hat{P}_1] + (f - P_1)^2 V[\hat{P}_2] + (P_1 - P_2)^2 V[\hat{f}] \}. \quad (9.26)$$

We note that the above estimates of u and u_x are essentially the same as those derived by Robinette [1949], Petrides [1954], Selleck and Hart [1957] and discussed by Hanson [1963]. These earlier papers, however, do not give any variance formulae.

Example 9.11   Pronghorn antelope (*Antilocapra americana*): Paulik and Robson [1969]

To estimate the rate of hunting mortality in a hypothetical pronghorn antelope population, a pre-season field survey yielded 550 mature females and 167 female fawns. In a sample of the kill, 63 mature females and 16 female fawns were found. The post-season survey yielded 370 mature females and 130 female fawns. Thus

$\hat{P}_1 = 550/717 = 0\cdot7671, \quad \hat{P}_2 = 370/500 = 0\cdot7400, \quad \hat{f} = 63/79 = 0\cdot7975,$

so that

$$\hat{u} = \frac{0\cdot7671 - 0\cdot7400}{0\cdot7975 - 0\cdot7400} = 0\cdot4713.$$

This result would commonly be expressed as either a hunting mortality of 47 per cent or an exploitation rate of $0\cdot47$.

Assuming binomial sampling in all three cases, we have

$$\hat{V}[\hat{P}_1] = (0\cdot7671)(0\cdot2329)/717 = 0\cdot000\ 249\ 2,$$
$$\hat{V}[\hat{f}] = (0\cdot7975)(0\cdot2025)/79 = 0\cdot002\ 044, \text{ etc.},$$

so that, substituting in (9.26),

$$\hat{V}[\hat{u}] = 0\cdot2455 \quad \text{and} \quad \hat{\sigma}[\hat{u}] = 0\cdot4955.$$

The estimate of $\sigma[\hat{u}]$ is very large, thus indicating, as pointed out by the authors, that $\hat{u}$ is almost useless as an estimate of u. The fault lies in the small observed value $\hat{P}_1 - \hat{P}_2 = 0\cdot0271$; for this difference it would be necessary to obtain impracticably large samples (cf. p. 370) to give an estimate of u with a reasonable coefficient of variation.

9.2 OPEN POPULATION

9.2.1 Estimating productivity

Hanson [1963] considers three age-classes in a population: juveniles, subadults, and adults. Juveniles are animals which are less than fully grown; subadults are essentially fully grown, but the majority of their cohort have not completed their first breeding season; adults are fully grown, and the majority of their cohort have completed one or more breeding seasons. Using these age-classes, Hanson defines productivity as the ratio of subadult to adult animals. Since the CIR method holds for additions to the population (that is, R_x and R_y negative), $\hat{u}$ defined in the previous section can be used as an estimate of productivity; the only difference being that $\hat{u}$ is now negative.

Example 9.12   Quail (*Lophortyx gambelii*): Hanson [1963].

A careful survey of Gambel's quail on a study area in early August (time t_1) found 325 adult males and 250 adult females, as well as a number of juvenile birds. A second survey in early October (time t_2), after all the juveniles had grown sufficiently to be considered subadults and to be part of the total population of mature birds, found 1230 mature males and 1145 mature females. It was assumed that the survival rates of adult males and females from August to October were the same, and that there was an equal sex ratio ($f = \frac{1}{2}$) in the juveniles recruited to the population of mature birds as subadults between August and October. Thus $\hat{P}_1 = 325/575 = 0\cdot565\ 22$, $\hat{P}_2 = 1230/2375 = 0\cdot517\ 89$ and

$$\hat{u} = \frac{0\cdot565\ 22 - 0\cdot517\ 89}{0\cdot5000 - 0\cdot517\ 89} = -2\cdot65.$$

In the calculation of $\hat{V}[\hat{u}]$ (cf. 9.26) we note that $V[\hat{f}] = 0$, as f is assumed known. Then, assuming binomial sampling,

$$\hat{V}[\hat{u}] = 5.70 \quad \text{and} \quad \hat{\sigma}[\hat{u}] = 2\cdot4.$$

The high value of $\hat{\sigma}[\hat{u}]$ reflects the general inaccuracy of productivity estimates when $f - P_2$ is small. In fact, as f is often near $0\cdot5$, the CIR method for estimating u is of limited usefulness; it will only apply when the sex ratio is widely different from unity.

9.2.2 Natural mortality

1 Survival ratios

Consider a population in which there are removals due to exploitation and possibly natural mortality. Let s_x and s_y be the fractions of x-type and y-type animals respectively surviving from time t_1 to time t_2. Then, using the notation of **9.1.1**, Paulik and Robson [1969] give the following equations:

$$X_2 = X_1 s_x, \quad Y_2 = Y_1 s_y,$$

and hence

$$\theta = \frac{s_x}{s_y} = \left(\frac{Y_1}{X_1}\right) \Big/ \left(\frac{Y_2}{X_2}\right) = \left(\frac{1 - P_1}{P_1}\right)\left(\frac{P_2}{1 - P_2}\right).$$

Therefore, if $\hat{P}_i$ is an estimate of P_i, θ is estimated by

$$\hat{\theta} = \left(\frac{1 - \hat{P}_1}{\hat{P}_1}\right)\left(\frac{\hat{P}_2}{1 - \hat{P}_2}\right), \tag{9.27}$$

and, using the delta method,

$$V[\hat{\theta}] \approx [P_1(1 - P_2)]^{-4}\{(1 - P_2)^2 P_2^2 V[\hat{P}_1] + (1 - P_1)^2 P_1^2 V[\hat{P}_2]\}. \tag{9.28}$$

If x_i and y_i are the numbers of x-type and y-type animals in the sample taken at time t_i, then (9.27) reduces to

$$\hat{\theta} = (y_1 x_2)/(y_2 x_1). \tag{9.29}$$

Paulik and Robson give several variations of the above method when additional information is available. For example, if it is known that the ratio of the two kinds of animals at time t_1 is $1:1$, then

$$\hat{\theta} = \hat{P}_2/(1 - \hat{P}_2), \tag{9.30}$$

and, setting $P_1 = \frac{1}{2}$, $V[\hat{P}_1] = 0$, in (9.28),

$$V[\hat{\theta}] \approx (1 - P_2)^{-4} V[\hat{P}_2]. \tag{9.31}$$

In some situations it is possible to ensure that $s_y = 1$, so that $\hat{\theta}$ is an estimate of s_x. One way of doing this is to introduce Y_1 animals into the population just before time t_2, so that no removals of y-types can occur before the survey for estimating P_2, i.e. $Y_2 = Y_1$. In practice this scheme is nearly always used under circumstances in which the values of X_1 and Y_1 are known (e.g. two tag releases) but either P_2 or the ratio (Y_2/X_2) must be estimated by sampling (Ricker [1958: 128–9], Geis and Taber [1963: 286], see also **5.1.3**(4) and **6.4.2**(2)). When X_1 and Y_1 are known,

$$\hat{\theta} = \frac{Y_1}{X_1} \frac{\hat{P}_2}{(1-\hat{P}_2)} = \left(\frac{x_2}{X_1}\right) \bigg/ \left(\frac{y_2}{Y_1}\right) \tag{9.32}$$

and setting $V[\hat{P}_1] = 0$ in (9.28),

$$V[\hat{\theta}] \approx (1-P_2)^{-4}\left(\frac{Y_1}{X_1}\right)^2 V[\hat{P}_2]. \tag{9.33}$$

Minor variations of this method are used by Eberhardt *et al.* [1963: 15] and Parker [1965: 1541]; further applications of the method are given in Meslow and Keith [1968: 822] and Barkalow *et al.* [1970].

Example 9.13   Pheasants: Paulik and Robson [1969]

On a hypothetical study area, field surveys of mature pheasants in the fall before hunting began found 830 cocks and 1000 hens. Similar surveys after the hunting season found 230 cocks and 1000 hens. Then, from (9.29) an estimate of the survival rate of cocks relative to hens is

$$\hat{\theta} = \frac{1000\ (230)}{830\ (1000)} = 0\cdot277.$$

Assuming binomial sampling with $\hat{V}[\hat{P}_i] = \hat{P}_i(1-\hat{P}_i)/n_i$, we find, using equation (9.28), that

$$\hat{V}[\hat{\theta}] = 0\cdot000\ 580\ 0 \quad \text{and} \quad \hat{\sigma}[\hat{\theta}] = 0\cdot024.$$

If there had been no pre-hunting season survey and it was known from past experience that the sex ratio in the pre-season mature pheasant population was $1:1$, then from (9.30) and (9.31)

$$\hat{\theta} = x_2/y_2 = 230/1000 = 0\cdot230,$$
$$\hat{V}[\hat{\theta}] = 0\cdot000\ 283\ 3 \quad \text{and} \quad \hat{\sigma}[\hat{\theta}] = 0\cdot017.$$

(The above example was adapted from Hanson [1963].)

Example 9.14   Striped bass (*Roccus saxatilis*): Paulik and Robson [1969]

The application of equation (9.32) is demonstrated using some mark–recapture data on striped bass. In the spring of 1958 (time t_1) $X_1 = 3891$ bass were tagged with disk-dangler tags and released in the Sacramento–San Joaquin delta area of San Francisco Bay. Approximately one year later (time t_2) a second group of $Y_1 = 2965$ bass were tagged using the same equipment and procedures. During the next 8 years from the spring of 1959 to the spring of 1967, $x_2 = 430$ recoveries from the first release and $y_2 = 1026$ recoveries from the second release were made. Assuming that both groups of tagged fish have the same probability of recovery and the same survival rate over the 8 years, so that P_2 remains constant, then y_2/x_2 will estimate Y_2/X_2, the ratio of the tagged population sizes at time t_2. Therefore, since $Y_2 = Y_1$ (i.e. $s_y = 1$), an estimate of s_x, the proportion of striped bass surviving for a year beginning in the spring of 1958 and ending in the spring of 1959 is

$$\hat{s}_x = \left(\frac{430}{3891}\right)\Big/\left(\frac{1026}{2965}\right) = 0\cdot3194.$$

Assuming binomial sampling with

$$\hat{P}_2 = 430/(430+1026) = 0\cdot2953,$$
$$\hat{V}[\hat{P}_2] = (0\cdot2953)(0\cdot7047)/1456 = 0\cdot000\ 142\ 9.$$

Hence, from (9.33),

$$\hat{V}[\hat{s}_x] = 0\cdot000\ 336\ 6 \quad\text{and}\quad \hat{\sigma}[\hat{s}_x] = 0\cdot0184.$$

EXPLOITATION AND NATURAL MORTALITY. In the above discussion no distinction was made between mortality due to exploitation and mortality due to natural causes. The next problem, then, is to separate the total mortality into these two components. To do this, Chapman and Murphy [1965] suggest that it is useful to consider two cases: "instantaneous" removal, when the removal takes place in a short period of time, and "continuous" uniform removal, when the removal takes place uniformly over an extended period. We shall now consider these two cases separately.

2   *Instantaneous removals*

Suppose that the removals R_x, R_y take place instantly at time zero. We shall now have a slight change in notation and assume that the two samples are taken time t_1 before the removal and time t_2 after the removal. Let:

ϕ_{1x} = fraction of x-types surviving from time $(-t_1)$ to time zero, and
ϕ_{2x} = fraction of x-types surviving from time zero to time t_2.

ϕ_{1y} and ϕ_{2y} are similarly defined. Then

$$X_2 = (X_1\phi_{1x} - R_x)\phi_{2x} \tag{9.34}$$

and

$$Y_2 = (Y_1\phi_{1y} - R_y)\phi_{2y}. \tag{9.35}$$

However, we now have too many parameters to estimate, and as a first step in simplifying the model we shall assume $\phi_{ix} = \phi_{iy} = \phi_i$ ($i = 1, 2$). Then from (9.34) and (9.35) we have

$$P_2 = \frac{X_2}{X_2 + Y_2} = \frac{P_1 N_1 \phi_1 - R_x}{N_1 \phi_1 - R} .$$

and rearranging,

$$N_1 \phi_1 = \frac{R_x - R P_2}{P_1 - P_2} .$$

This means that $\hat{N}_1$ defined in (9.2) is now an estimate of $N_1\phi_1$, the population size just prior to the removal, though the exploitation rate $u = R/(\phi_1 N_1)$ is still given by

$$u = \frac{f - RP_2}{P_1 - P_2},$$

a fact noted by Chapman and Murphy [1965]. However, these authors suggest a different model, using knowledge of sampling effort, which leads to the estimation of N_1, ϕ_1 and ϕ_2 as follows.

We shall assume that the instantaneous natural mortality rate is constant throughout the period under investigation and equal to μ; hence

$$\phi_i = e^{-\mu t_i}.$$

In addition we make the same assumptions as those given in **9.1.3** (4) with regard to sampling effort, namely,.

$$E[x_i] = Kf_i X_i \quad \text{and} \quad E[y_i] = Kf_i Y_i \quad (i = 1, 2), \tag{9.36}$$

where K is the binomial catchability and f_i, the sampling effort, is assumed known. We now have four unknown parameters X_1, Y_1, μ and K which can be estimated from the four observations x_1, y_1, x_2 and y_2. As a first step we substitute (9.36) in (9.34) and (9.35), and defining $\alpha = \phi_1/(Kf_1)$, $\beta = 1/(\phi_2 Kf_2)$, we have

$$\alpha E[x_1] - R_x = \beta E[x_2]$$

and

$$\alpha E[y_1] - R_y = \beta E[y_2].$$

Replacing the expectations by observed values, we have two equations in α and β which have solutions

$$\hat{\alpha} = \frac{R_x y_2 - R_y x_2}{x_1 y_2 - x_2 y_1}$$

and

$$\hat{\beta} = \frac{R_x y_1 - R_y x_1}{x_1 y_2 - x_2 y_1}.$$

Hence

$$\exp\{(-\hat{\mu}(t_1 + t_2)\} = \hat{\phi}_1 \hat{\phi}_2$$
$$= (f_1 \hat{\alpha})/(f_2 \hat{\beta}),$$

leading to the estimate

$$\hat{\mu} = -\frac{1}{(t_1 + t_2)} \log\left\{\frac{f_1}{f_2} \cdot \frac{R_x y_2 - R_y x_2}{R_x y_1 - R_y x_1}\right\}.$$

Also

$$1/\hat{K} = \{(f_1 \hat{\alpha})^{t_2} (f_2 \hat{\beta})^{t_1}\}^{1/(t_1 + t_2)},$$

so that from (9.36),

$$\hat{X}_i = x_i/(\hat{K}f_i), \quad \hat{Y}_i = y_i/(\hat{K}f_i)$$

and

$$\hat{N}_i = \hat{X}_i + \hat{Y}_i.$$

Assuming x_1, x_2, y_1 and y_2 to be independent Poisson variables, we have, using the delta method and estimating both $E[x_i]$ and $V[x_i]$ by x_i, the following variance estimates:

$$\hat{V}[\hat{\mu}] = \frac{1}{(t_1+t_2)^2}\left\{\frac{R_x^2 y_2 + R_y^2 x_2}{(R_x y_2 - R_y x_2)^2} + \frac{R_x^2 y_1 + R_y^2 x_1}{(R_x y_1 - R_y x_1)^2}\right\}$$

and

$$\frac{\hat{V}[\hat{N}_1]}{\hat{N}_1^2} = x_1\left\{\frac{(x_2+y_2)y_1}{(x_1+y_1)(x_1 y_2 - x_2 y_1)} + \frac{R_y t_1}{(t_1+t_2)(R_x y_1 - R_y x_1)}\right\}^2 +$$

$$+ y_1\left\{\frac{(x_2+y_2)x_1}{(x_1+y_1)(x_1 y_2 - x_2 y_1)} + \frac{R_x t_1}{(t_1+t_2)(R_x y_1 - R_y x_1)}\right\}^2 +$$

$$+ x_2\left\{\frac{y_1}{(x_1 y_2 - x_2 y_1)} - \frac{R_y t_2}{(t_1+t_2)(R_x y_2 - R_y x_2)}\right\}^2 +$$

$$+ y_2\left\{\frac{x_1}{(x_1 y_2 - x_2 y_1)} - \frac{R_x t_2}{(t_1+t_2)(R_x y_2 - R_y x_2)}\right\}^2 .$$

In conclusion we note that in the above theory n_i ($= x_i + y_i$) is no longer a fixed parameter but a random variable; one cannot fix both effort and sample size in advance.

3   Continuous removals

We shall now assume that the samples are taken at the beginning and end of a continuous uniform removal carried out over a period of time T. Let μ_x, μ_y be the instantaneous natural mortality rates and let μ_{Ex}, μ_{Ey} be the instantaneous removal (exploitation) rates for x-type and y-type animals respectively. We can now write the following equations:

$$E[x_i | X_i] = a_i X_i, \quad E[y_{ii} | Y_i] = b_i y_i, \tag{9.37}$$

$$E[X_2] = X_1 \exp\{-(\mu_x + \mu_{Ex})T\}, \tag{9.38}$$

$$E[Y_2] = Y_1 \exp\{-(\mu_y + \mu_{Ey})T\}, \tag{9.39}$$

and from (8.13) we have further

$$E[R_x] = \frac{X_1 \mu_{Ex}}{(\mu_x + \mu_{Ex})}(1 - \exp\{-(\mu_x + \mu_{Ex})T\}) \tag{9.40}$$

and

$$E[R_y] = \frac{Y_1 \mu_{Ey}}{(\mu_y + \mu_{Ey})}(1 - \exp\{-(\mu_y + \mu_{Ey})T\}). \tag{9.41}$$

A particular example of the above is the constant level fishery (a C.C.U. fishery in the terminology of Paulik [1963a]), where μ_{Ex} and μ_{Ey} are the instantaneous fishing rates. Typically the sampling might be done annually; each sample can be considered either as the second sample as far as the previous year is concerned or as the first sample for the next year.

Once again we have too many parameters to estimate, and as a first simplification Chapman and Murphy [1965] make the following assumptions:

(i) $\mu_x = \mu_y = \mu$, and
(ii) $a_i = b_i$ $(i = 1, 2)$.

Although we still cannot estimate all the parameters individually it is now possible to obtain approximate estimates of μ_{Ex} and μ_{Ey}. Replacing expected values by observed values and defining the unit of time so that $T = 1$, we have from equations (9.37) to (9.41):

$$\left(\frac{R_x}{x_1}\right)\bigg/\left(\frac{R_y}{y_1}\right) = \frac{[1 - \exp\{-(\mu+\mu_{Ex})\}]\,\mu_{Ex}/(\mu+\mu_{Ex})}{[1 - \exp\{-(\mu+\mu_{Ey})\}]\,\mu_{Ey}/(\mu+\mu_{Ey})} \qquad (9.42)$$

$$(= r_1, \quad \text{say}),$$

and

$$\left(\frac{x_2}{x_1}\right)\bigg/\left(\frac{y_2}{y_1}\right) = \exp\{-(\mu_{Ex}-\mu_{Ey})\} \quad (= r_2, \quad \text{say}). \qquad (9.43)$$

We note that (9.43) is equivalent to (9.29). If μ was known (or zero as in Lander [1962]) it would be possible to solve the above equations for μ_{Ex} and μ_{Ey}. Fortunately the dependence of r_1 on μ is slight, as seen from Table 9.3; for μ varying from $0{\cdot}0$ to $0{\cdot}4$ the variation in r_1 is about 1 per cent.

TABLE 9.3

r_1 evaluated for various μ_{Ex}, μ_{Ey} and μ: from
Chapman and Murphy [1965: Table 1].

μ_{Ex}	μ_{Ey}	μ	r_1	μ_{Ex}	μ_{Ey}	μ	r_1
0·35	0·05	0·0	6·05	0·50	0·20	0·0	2·17
0·35	0·05	0·2	6·09	0·50	0·20	0·2	2·18
0·35	0·05	0·4	6·11	0·50	0·20	0·4	2·19
0·50	0·10	0·0	4·13	0·60	0·20	0·0	2·49
0·50	0·10	0·2	4·16	0·60	0·20	0·2	2·51
0·50	0·10	0·4	4·19	0·60	0·20	0·4	2·52

In fact, approximating each exponential in (9.42) by the first three terms of its power-series expansion and using the further approximation $(1-w)^{-1} \approx 1 + w$, we find that μ cancels out of (9.42). This approximation for (9.42), together with (9.43), are readily solved to give the estimates

$$\hat{\mu}_{Ex} = \frac{r_1 \log r_2}{1 + \frac{1}{2}(\log r_2) - r_1} \qquad (9.44)$$

and

$$\hat{\mu}_{Ey} = \hat{\mu}_{Ex} + \log r_2. \qquad (9.45)$$

To find asymptotic variances of these estimates we assume once again that x_1, x_2, y_1 and y_2 are independent Poisson variables. In addition the catches or removals R_x and R_y are now random variables, and we assume them to be independent Poisson variables also. Therefore, using the delta method, Chapman and Murphy show that

$$\hat{V}[\hat{\mu}_{Ex}] = \left\{\frac{(2\ \log\ r_2)(2+\log\ r_2)}{(2+\log\ r_2-2r_1)^2}\right\}^2 \hat{V}[r_1] +$$

$$+ \left\{\frac{(8\ \log\ r_2)(2+\log\ r_2)r_1(1-r_1)}{r_2(2+\log\ r_2-2r_1)^4}\right\} \widehat{\text{cov}}\,[r_1,\ r_2] +$$

$$+ \left\{\frac{4r_1(1-r_1)}{r_2(2+\log\ r_2-2r_1)^2}\right\}^2 \hat{V}[r_2]$$

and

$$\hat{V}[\hat{\mu}_{Ey}] = \left\{\frac{(2\ \log\ r_2)(2+\log\ r_2)}{(2+\log\ r_2-2r_1)^2}\right\} \hat{V}[r_1] +$$

$$+ \frac{4}{r_2}\left\{\frac{(\log\ r_2)(2+\log\ r_2)}{(2+\log\ r_2-2r_1)}\right\}\left\{1+\frac{4r_1(1-r_1)}{(2+\log\ r_2-2r_1)^2}\right\} \widehat{\text{cov}}\,[r_1,\ r_2] +$$

$$+ \frac{1}{r_2^2}\left\{1+\frac{4r_1(1-r_1)}{(2+\log\ r_2-2r_1)^2}\right\}^2 \hat{V}[r_2],$$

where $\quad \hat{V}[r_1] = r_1^2\left\{\dfrac{1}{R_x}+\dfrac{1}{x_1}+\dfrac{1}{R_y}+\dfrac{1}{y_1}\right\},$

$$\hat{V}[r_2] = r_2^2\left\{\frac{1}{x_1}+\frac{1}{x_2}+\frac{1}{y_1}+\frac{1}{y_2}\right\}$$

and

$$\widehat{\text{cov}}\,[r_1,\ r_2] = r_1 r_2\left\{\frac{1}{x_1}+\frac{1}{y_1}\right\}.$$

As a check on the approximate solutions (9.44) and (9.45), and to provide more accurate estimates, Table 9.4 (reproduced from Chapman and Murphy [1965]) provides a tabulation of the function

$$r_1 = \frac{\mu_{Ex}(\mu+\mu_{Ex}+\log\ r_2)[1-\exp(-\mu-\mu_{Ex})]}{(\mu_{Ex}+\log\ r_2)(\mu+\mu_{Ex})[1-\exp(-\mu-\mu_{Ex}-\log\ r_2)]}$$

for $\mu_{Ex} = 0 \cdot 05\ (0 \cdot 05)\ 1$, $\mu = 0 \cdot 0,\ 0 \cdot 2,\ 0 \cdot 4$, and for admissible values of r_2. To use this table we choose the x-class as the more heavily exploited class, i.e. $\mu_{Ex} > \mu_{Ey}$ and $r_2 < 1$, then we choose μ, find r_1 in the appropriate r_2 column and read off μ_{Ex}, the required estimate; the estimate of μ_{Ey} is then given by (9.45). Although the most appropriate value of μ should be chosen, it is clear that the choice of μ makes little difference (this was first noted by Lander [1962]). In fact, if the assumption (i) of equal natural mortality was not true, we would expect the estimate of μ_{Ex} to be little affected, provided

TABLE 9.4

Tabulation of r_1 as a function of μ, μ_{Ex} and r_2: from Chapman and Murphy [1965: Table 2].

$\mu = 0.00$

$r_2 =$ 0·40	0·45	0·50	0·55	0·60	0·65	0·70	0·75	0·80	0·85	0·90	0·95	1·00	μ_{Ex}
												1·0000	0·05
											2·0018	1·0000	0·10
										3·1905	1·4820	1·0000	0·15
									4·9275	2·0074	1·3119	1·0000	0·20
								8·3474	2·6408	1·6426	1·2275	1·0000	0·25
							21·1709	3·5035	2·0178	1·4654	1·1771	1·0000	0·30
							4·8880	2·4787	1·7274	1·3608	1·1436	1·0000	0·35
						7·7755	3·1032	2·0338	1·5596	1·2918	1·1198	1·0000	0·40
					19·0385	4·0669	2·4186	1·7854	1·4504	1·2430	1·1021	1·0000	0·45
					5·8836	2·9467	2·0569	1·6270	1·3737	1·2067	1·0883	1·0000	0·50
				11·0121	3·7643	2·4066	1·8335	1·5173	1·3169	1·1786	1·0773	1·0000	0·55
			208·8195	5·2886	2·8983	2·0890	1·6820	1·4370	1·2733	1·1563	1·0684	1·0000	0·60
			9·4038	3·6787	2·4280	1·8801	1·5725	1·3756	1·2388	1·1381	1·0610	1·0000	0·65
		73·7130	5·1836	2·9208	2·1329	1·7324	1·4899	1·3273	1·2108	1·1231	1·0548	1·0000	0·70
		9·5470	3·7381	2·4804	1·9307	1·6225	1·4254	1·2884	1·1876	1·1105	1·0494	1·0000	0·75
	369·2827	5·4338	3·0085	2·1929	1·7837	1·5378	1·3736	1·2563	1·1682	1·0997	1·0449	1·0000	0·80
	11·4086	3·9442	2·5690	1·9906	1·6721	1·4704	1·3312	1·2294	1·1517	1·0904	1·0409	1·0000	0·85
	6·1488	3·1758	2·2756	1·8408	1·5846	1·4157	1·2960	1·2067	1·1375	1·0824	1·0374	1·0000	0·90
18·5009	4·3625	2·7073	2·0660	1·7254	1·5142	1·3704	1·2662	1·1872	1·1252	1·0754	1·0343	1·0000	0·95
7·8719	3·4639	2·3922	1·9090	1·6339	1·4564	1·3323	1·2407	1·1703	1·1145	1·0691	1·0316	1·0000	1·00

TABLE 9.4 (continued)

Tabulation of r_1 as a function of μ, μ_{Ex} and r_2: from Chapman and Murphy [1965: Table 2].

$\mu = 0.20$

$r_2 =$	0.40	0.45	0.50	0.55	0.60	0.65	0.70	0.75	0.80	0.85	0.90	0.95	1.00	μ_{Ex}
													1.0000	0.05
												2.0035	1.0000	0.10
											3.1961	1.4832	1.0000	0.15
										4.9408	2.0109	1.3130	1.0000	0.20
									8.3785	2.6479	1.6455	1.2285	1.0000	0.25
								21.2722	3.5165	2.0232	1.4680	1.1781	1.0000	0.30
								4.9114	2.4879	1.7321	1.3632	1.1446	1.0000	0.35
							7.8216	3.1180	2.0413	1.5638	1.2941	1.1208	1.0000	0.40
						19.1748	4.0910	2.4301	1.7920	1.4543	1.2452	1.1030	1.0000	0.45
						5.9256	2.9641	2.0667	1.6330	1.3774	1.2088	1.0892	1.0000	0.50
					11.1054	3.7911	2.4208	1.8422	1.5229	1.3205	1.1806	1.0782	1.0000	0.55
				210.8905	5.3333	2.9189	2.1013	1.6899	1.4422	1.2767	1.1583	1.0693	1.0000	0.60
				9.4968	3.7098	2.4452	1.8911	1.5800	1.3807	1.2421	1.1401	1.0619	1.0000	0.65
			74.5589	5.2347	2.9453	2.1480	1.7425	1.4969	1.3322	1.2140	1.1250	1.0556	1.0000	0.70
			9.6563	3.7749	2.5012	1.9444	1.6320	1.4320	1.2930	1.1908	1.1124	1.0503	1.0000	0.75
		374.1552	5.4958	3.0380	2.2112	1.7962	1.5467	1.3800	1.2608	1.1713	1.1016	1.0457	1.0000	0.80
		11.5587	3.9891	2.5942	2.0072	1.6838	1.4789	1.3374	1.2339	1.1547	1.0923	1.0417	1.0000	0.85
		6.2295	3.2118	2.2978	1.8560	1.5956	1.4238	1.3020	1.2110	1.1405	1.0842	1.0383	1.0000	0.90
	18.7797	4.4196	2.7380	2.0861	1.7397	1.5247	1.3782	1.2720	1.1914	1.1281	1.0771	1.0352	1.0000	0.95
	7.9901	3.5090	2.4192	1.9275	1.6474	1.4665	1.3399	1.2464	1.1744	1.1173	1.0709	1.0324	1.0000	1.00

TABLE 9.4 (concluded)

Tabulation of r_1 as a function of μ, μ_{Ex} and r_2: from Chapman and Murphy [1965: Table 2].

$\mu = 0.40$

$r_2 =$ 0·40	0·45	0·50	0·55	0·60	0·65	0·70	0·75	0·80	0·85	0·90	0·95	1·00	μ_{Ex}
												1·0000	0·05
											2·0052	1·0000	0·10
										3·2017	1·4845	1·0000	0·15
									4·9541	2·0144	1·3141	1·0000	0·20
								8·4094	2·6550	1·6483	1·2295	1·0000	0·25
							21·3734	3·5295	2·0286	1·4705	1·1791	1·0000	0·30
							4·9346	2·4970	1·7367	1·3655	1·1456	1·0000	0·35
						7·8675	3·1327	2·0488	1·5679	1·2963	1·1217	1·0000	0·40
					19·3109	4·1149	2·4416	1·7985	1·4581	1·2473	1·1039	1·0000	0·45
					5·9676	2·9815	2·0764	1·6389	1·3810	1·2108	1·0901	1·0000	0·50
				11·1986	3·8179	2·4349	1·8508	1·5284	1·3239	1·1826	1·0791	1·0000	0·55
			212·9613	5·3779	2·9394	2·1135	1·6978	1·4474	1·2801	1·1602	1·0702	1·0000	0·60
			9·5898	3·7407	2·4623	1·9020	1·5873	1·3856	1·2453	1·1420	1·0627	1·0000	0·65
		75·4052	5·2858	2·9698	2·1630	1·7525	1·5038	1·3369	1·2171	1·1269	1·0565	1·0000	0·70
		9·7656	3·8116	2·5219	1·9579	1·6414	1·4386	1·2976	1·1938	1·1142	1·0512	1·0000	0·75
	379·0338	5·5578	3·0675	2·2294	1·8087	1·5555	1·3864	1·2653	1·1743	1·1034	1·0466	1·0000	0·80
	11·7089	4·0339	2·6192	2·0237	1·6954	1·4873	1·3435	1·2382	1·1577	1·0941	1·0426	1·0000	0·85
	6·3101	3·2478	2·3198	1·8712	1·6066	1·4319	1·3079	1·2153	1·1434	1·0860	1·0391	1·0000	0·90
19·0589	4·4765	2·7685	2·1060	1·7538	1·5351	1·3860	1·2778	1·1956	1·1310	1·0789	1·0360	1·0000	0·95
8·1084	3·5541	2·4460	1·9458	1·6607	1·4764	1·3474	1·2520	1·1785	1·1201	1·0727	1·0332	1·0000	1·00

that μ_x and μ_y were not too different and were small relative to μ_{Ex}, which for example, is a common fishery situation.

If we replace assumption (ii) by the stronger assumption

(ii') $a_i = b_i = K f_i$ ($i = 1, 2$), f_1, f_2 known,

as in the instantaneous removal model, we can estimate μ from the relations (cf. (9.38) and (9.39))

$$\exp\{-(\mu + \mu_{Ex})\} = \frac{X_2}{X_1} = \frac{x_2 f_1}{x_1 f_2},$$

that is,

$$\hat{\mu} = \log\left(\frac{x_1 f_2}{x_2 f_1}\right) - \hat{\mu}_{Ex}, \tag{9.46}$$

and finally estimate X_1 and Y_1 from (9.40) and (9.41) (with $T = 1$). The procedure is iterative; μ is guessed, $\hat{\mu}_{Ex}$ determined from (9.44) or Table 9.4, and then $\hat{\mu}$ (from 9.46) may be used to re-enter the table. One cycle will usually be sufficient as the estimates of μ_{Ex} and μ_{Ey} are essentially independent of $\hat{\mu}$.

CHAPTER 10

MORTALITY AND SURVIVAL ESTIMATES FROM AGE DATA

10.1 LIFE TABLES

10.1.1 Introduction

A life table is basically a table used for determining the mortality rate and the life expectancy of an individual at a given age. Detailed techniques for constructing and analysing such tables (cf. Keyfitz [1968], Goodman [1969]) have been developed for human populations, where they are largely used for determining insurance premiums. The success of such methods is due to the fact that human populations are large and stable, and information on various categories such as age, occupation, sex, etc. is readily obtainable by census, sample survey, records, etc. In recent years life tables have also been used for investigating the dynamics of animal populations and for determining policies in population management (Quick [1963]). However, animal populations tend to fluctuate with changes in environment, and the data for the life table are usually based on samples which may not be as representative as one would hope. Therefore, although animal life tables may help to give an overall picture of the population, they are often of limited accuracy and, where possible, should be backed up by other methods of estimation (e.g. the methods of **5.4**).

Life table methods are discussed, for example, by Deevey [1947], Allee *et al.* [1949: 294–301], Hickey [1952: birds], Eberhardt [1969a: wildlife management], Southwood [1966: insects], Varley and Gradwell [1970: insects] and Caughley [1966: mammals]. General methods for ageing animals are described in Taber [1969], Tesch [1968: fish], Southwood [1966: insects].

NOTATION. An animal life table usually contains the following columns:

(i) age, x: measured in some convenient unit of time such as day, month, year, etc.

(ii) l_x: the number surviving to age x from a "cohort" of l_0 animals, i.e. from l_0 animals born at the same time. The usual convention of scaling l_x so that $l_0 = 1000$ or 1, although useful in standardising survivorship curves (see below), is not recommended when samples are small, as it can give a false impression of the accuracy of the table. If scaling is used then the raw data should also be quoted.

(iii) $d_x = l_x - l_{x+1}$: the number of deaths in the age-class $[x, x+1)$. If w is the last age in the table, then

$$l_x = (l_x - l_{x+1}) + (l_{x+1} - l_{x+2}) + \ldots + (l_w - 0)$$
$$= d_x + d_{x+1} + \ldots + d_w.$$

(iv) $q_x = d_x/l_x$: the observed mortality rate at age x. Here

$$q_x = d_x/(d_x + d_{x+1} + \ldots + d_w). \tag{10.1}$$

(v) e_x: the observed mean expectation of life remaining for animals of age x.

Two further columns can be included to facilitate the calculation of e_x:

(vi) L_x: the average number of animals alive during the interval $[x, x+1)$. Thus

$$L_x = \int_x^{x+1} l_t \, dt$$

which is usually approximated by $\frac{1}{2}(l_x + l_{x+1})$, the average of the numbers alive at the beginning and end of this interval (this approximation is used in Table 10.1).

(vii) T_x, where

$$T_x = \int_x^w l_t \, dt$$
$$= L_x + L_{x+1} + \ldots + L_w$$
$$\approx \frac{1}{2}(l_x + l_{x+1}) + \frac{1}{2}(l_{x+1} + l_{x+2}) + \ldots + \frac{1}{2}(l_w + 0)$$
$$= \frac{1}{2}l_x + \sum_{y=x+1}^w l_y. \tag{10.2}$$

This column is usually obtained by summing the L_x column cumulatively from the bottom up. As T_x is the total number of time-intervals lived by the group of l_x animals, we have from (10.2)

$$e_x = T_x/l_x \approx \frac{1}{2} + \sum_{y=x+1}^w (l_y/l_x). \tag{10.3}$$

An example of the calculations is given in Table 10.1.

TABLE 10.1

Artificial life table illustrating the calculation of q_x and e_x.

x	l_x	d_x	$100\,q_x$	L_x	T_x	e_x
0–1	1000	750	75	625	804	0·80
1–2	250	210	84	145	179	0·72
2–3	40	30	75	25	34	0·85
3–4	10	7	70	6·5	9	0·90
4–5	3	2	67	2	2·5	0·83
5–6	1	1	100	0·5	0·5	0·50
6–7	0	–	–	–	–	–

A number of other parameters are sometimes estimated from life tables. For example, if Z_x is the observed total instantaneous mortality rate during the interval $[x, x+1)$, then, from (1.1),

$$l_{x+1} = l_x \, e^{-Z\,x}. \qquad (10.4)$$

Hence the proportion surviving, $S_x \, (= 1 - q_x)$, is given by

$$S_x = l_{x+1}/l_x = e^{-Z\,x},$$

so that

$$Z_x = -\log S_x.$$

If q_x is small, then $-\log S_x \approx q_x$ and hence $q_x \approx Z_x$.

A plot of l_x versus x is called a "survival" or "survivorship curve", and the plot of d_x versus x is called a "mortality curve", or sometimes "kill curve". Deevey [1947] gives three basic types of survival curves (Fig. 10.1):

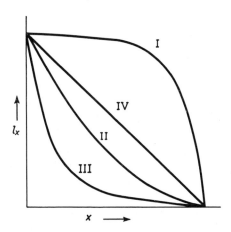

Fig. 10.1   Types of survival curves: after Slobodkin [1962].

Type I: mortality acts most heavily on the old individuals.

Type II: a constant fraction of the cohort dies at each age. This means that the survival rate is constant, so that $l_x = S \, l_{x-1} = S^x \, l_0$ and the plot of $\log l_x$ against x is a straight line. Such a survival curve is typical of the adult stages of many birds and fishes, though it may not hold strictly for all ages of a species' life. Indeed, as pointed out by Hickey [1952], many of the species that Deevey [1947] grouped under this category may exhibit a type I curve when their complete span of life is finally investigated.

Type III: mortality acts most heavily on the early stages of life. This curve reflects the common situation where a single female produces an enormous number of young, but there is a very high initial mortality through predation, etc. Although this type of curve is no doubt characteristic of many fishes and invertebrates it is very difficult to obtain accurate information on survival in the early stages of life.

The various types of curves are also discussed by Slobodkin [1962: Chapter 4] who adds a fourth curve, type IV, say, representing a population in which a constant number die per unit time, regardless of population size. This type of curve appears to be common in animals receiving some human protection, as for example in a laboratory or a zoo (Comfort [1957: 361]).

It should be noted that the above notation (which follows actuarial practice) differs from that commonly used in the construction of animal life tables. Usually x denotes the age-class $[x-1, x)$, and l_x is defined to be the number surviving at the beginning of this interval, i.e. of age $x - 1$: a survivorship curve is then the plot of l_x against $x - 1$. One reason for the popularity of the latter notation is that in applying life-table methods to tagging experiments it is convenient to think of d_x as the number of tagged animals found dead in their xth year of life (called D_x in **5.4.3**). Since both notations are used, they can in practice be combined by the simple expedient of labelling the age-intervals 0–1, 1–2, etc. instead of 0, 1, 2, etc. or 1, 2, 3, etc. For example, the first entry in the d_x column would refer to the number dying in the interval $[0, 1)$, irrespective of whether we call it d_0 or d_1.

LIFE AND FERTILITY TABLE. This table consists of the columns x and l_x (usually with $l_0 = 1$), as for the life table, but l_x now refers to the female population only. A new column is then added on the basis of observations: this is the m_x or *age-specific fertility* column that records the number of *living* females born per female in each age-interval; sometimes the sex at birth is not readily distinguishable, and m_x must be estimated from total births per female of age x and, say, a 50 : 50 sex ratio. If one is interested in numbers of births (and not just live births), e.g. numbers of eggs, then the fertility column is often called the *fecundity* column.

Columns l_x and m_x are then multiplied together to give the total number of females born in each age category; this is the $l_x m_x$ column of Table 10.2.

TABLE 10.2

Life and fertility table for the beetle, *Phyllopertha horticola*: **modified from Laughlin [1965].**

x (weeks)	l_x	m_x	$l_x m_x$	$x l_x m_x$
0	1·00	–		–
49	0·46	–	Immature	–
50	0·45	–	stages	–
51	0·42	1·0	0·42	21·42
52	0·31	6·9	2·13	110·76
53	0·05	7·5	0·38	20·14
54	0·01	0·9	0·01	0·54
Total			2·94	152·86

From this column one can calculate, for example, the *net reproductive rate*

$$R_0 = \sum_x l_x m_x / l_0 = 2 \cdot 94,$$

or estimate the *generation length* (i.e. mean lapse of time between a female's date of birth and the mean date of birth of her offspring) by

$$\frac{\sum_x x\, l_x m_x}{\sum_x l_x m_x} = \frac{152 \cdot 86}{2 \cdot 94} = 52 \cdot 0. \tag{10.5}$$

Further examples of such tables are given in Turner *et al.* [1970]. Other parameters which may be of interest are discussed in Southwood [1966: 287–91].

MEAN MORTALITY RATE. If the observed mortality rate is fairly constant, then we can combine the estimates q_x in several ways to give a mean mortality rate. For example, we can use the average

$$q_{\text{ave}} = \frac{1}{w} \sum_{x=0}^{w-1} q_x = \frac{1}{w} \sum_{x=0}^{w-1} \frac{d_x}{l_x} \tag{10.6}$$

(the last age w is not included as $d_w = l_w$ and $q_w = 1$), or the more popular pooled version

$$q_{\text{pool}} = \sum_{x=0}^{w-1} d_x \bigg/ \sum_{x=0}^{w-1} l_x. \tag{10.7}$$

Farner [1955: 405] discusses both these mortality rates and points out that unweighted averages such as q_{ave} give equal weight to age-classes of different sizes. To avoid this disproportionate emphasis one can take a weighted average of the q_x with weights proportional to the numbers l_x in the various classes; this leads to q_{pool}.

If Z_x is constant ($= Z$ say), then from (10.4) the mean expectation of life at birth is

$$(\int_0^\infty l_t\, dt)/l_0 = \int_0^\infty e^{-Zt}\, dt = 1/Z = -1/\log(1-q), \text{ say,}$$

which is commonly estimated by

$$e_0' = -1/\log(1 - q_{\text{pool}}).$$

A number of approximations have been widely used, particularly in ornithological work, e.g. when q is small $-\log(1-q) \approx q$ and $-1/\log(1-q) \approx 1/(q + \frac{1}{2}q^2) \approx (1 - \frac{1}{2}q)/q$; the latter approximation leads to Burkitt's [1926] estimate $(2-q)/2q$.

LIFE EQUATIONS. For some populations it is helpful to classify individuals according to stage of development rather than according to age. In this case, if a cohort of individuals (and possibly their offspring) are followed through the various stages, and the numbers in each stage recorded, one can set up for the cohort a so-called *life equation* (or commonly called "budget" in entomological studies; Southwood [1966: 277 ff.]); two examples of life equations are given in Tables 10.3 and 10.4. Quick [1963] describes

TABLE 10.3
Hypothetical life equation for a stationary turkey population: adapted from Mosby [1967: Table 6.12].

Stage	Factor causing change in population number	Population
(1)	Spring breeding population; 50 : 50 sex-ratio assumed.	1000
(2)	Number of successful nests (number females) × (% nesting success)/100 500 × 0·351 = 175·5	
(3)	Number poults hatched (successful nests) × (av. clutch) × (hatchability %)/100 175·5 × 12·3 × 0·93 = 2008	
(4)	Number of poults in fall (poults hatched) × (% poult survival)/100 2008 × 0·755 = 1516	
(5)	Fall adult population (adults in spring) × (% adult survival)/100 1000 × 0·993 = 993	
(6)	Total pre-hunting season population (5) + (4) =	2509
(7)	Hunting season harvest (pre-hunting season pop.) × (% harvest)/100 2509 × 0·333 = 836	
(8)	Post-hunting season population (6) − (7) =	1673
(9)	Overwinter loss (8) − (10) = 673	
(10)	Spring breeding population (assuming stationary population) (1) =	1000

the life equation as a "book-keeping system that involves intensive observation of living animals" and points out that the method requires a trained observer to spend a great deal of time in the field keeping a careful record of the development of the cohort.

As it is not always possible to count all the individuals in a given cohort, sampling methods may be needed for keeping track on the numbers entering each stage. In this case a difficulty can arise as the age-ranges for the different stages may overlap, so that different members of the cohort may be in different stages at the same time (e.g. Beaver [1966], Berryman

TABLE 10.4

Model life equation for spruce budworm, 1952 and 1953 (numbers per 10 sq.ft of branch surface): abridged from Morris and Miller [1954].

Stage	l_x	Mortality factor	d_x	100 q_x
Eggs	174	Parasites	3	2
		Predators	15	9
		Other	1	1
		Total egg loss	19	12
Instar I	155	Dispersion loss	74·4	48
Hibernacula	80·6	Winter loss	13·7	17
Instar II	66·9	Dispersion loss	42·2	63
Instars III and IV	24·7	Parasites	8·92	36
		Disease	0·54	2
		Birds	3·39	14
		Other	10·57	43
		Total larvae loss	23·42	95
Pupae	1·28	Parasites	0·10	8
		Predators	0·13	10
		Other	0·23	18
		Total pupae loss	0·46	36
Moths	0·82		0	0
Generation:		$\Sigma\, d_x = 173\cdot18$	$173\cdot18/174 = 99\cdot53$	

[1968], Hughes and Gilbert [1968], Ashford et al. [1970]). This means that a single sample taken at a time of overlap will not, in itself, provide useful information about any of the overlapping stages. Only by examining a series of samples, and perhaps using such methods as those described in Southwood [1966: 279], Kiritani and Nakasuji [1967], Read and Ashford [1968], McLaren and Pottinger [1969], Kiritani et al. [1970] and Pottinger and Le Roux [1971], can the numbers in each stage be estimated.

A life equation can also be described graphically in terms of a "survival curve"; examples of such curves are given in Quick [1963].

10.1.2   Types of life tables

1   Age-specific (horizontal) life tables

The age-specific life table (also known as the dynamic life table) is based on the fate of a given cohort of l_0 individuals. Data for such a table can be collected in two ways: (i) by recording the ages at death of animals in the cohort, thus giving a d_x series, or (ii) by recording the number of animals from the cohort still alive at various times, thus giving an l_x series.

The age-specific method is most readily applied to laboratory populations of relatively short-lived species such as voles (Leslie and Ranson [1940], Leslie *et al.* [1955]), or to zoo populations (Comfort [1957]). However, for natural as opposed to laboratory populations the initial cohort can be identified by tagging, and either of the above methods for collecting the data can be used; obviously the population must be enclosed, otherwise emigration is confounded with mortality (Brooks [1967]). If method (i) is used, then a problem can arise with regard to predation, as some of the tagged individuals may disappear without a trace or may be difficult to find; radioactive tagging can be useful here (Schnell [1968]). If live-trapping is used and the probability of capture is constant throughout the period of investigation, then the number caught on each trapping occasion will be a constant proportion of the live population, and the sequence of captures will form an l_x series, except for l_0 (e.g. Marsden and Baskett [1958: though emigration is confounded with mortality]).

2   Time-specific (vertical) life tables

If N_x, the number of animals of age x in a population, remains constant with respect to time for each x, then the population is said to be "stationary" and the age-structure of the population at any instant of time t will reflect the fate of a cohort of N_0 animals born at time t. Thus the cohort of N_0 individuals will reduce to N_1 individuals at time $t + 1$, to N_2 at time $t + 2$, etc., and the sequence $\{N_x\}$ forms an l_x series. This series can then be used to construct a life table as in the age-specific case above. Such a life table, determined from the age-structure of a stationary population at a given point in time, is called a "time-specific" life table. In practice the age structure must be determined by a sample, so that time-specific tables involve sampling errors. Therefore although a time-specific table is usually easier to obtain (particularly for long-lived animals), it is less accurate than an age-specific table and can only be used when the population is stationary. Some of the problems associated with obtaining a suitable sample are considered by Caughley [1966], and the following discussion is based on his article.

The concept of a stationary population has been developed from demographic research on man and is useful for species which, like man, have no seasonally restricted period of births. However, very few species breed at the same rate throughout the year, and the stationary age distribution must be redefined if it is to include seasonal breeders. For species with one restricted breeding season each year a stationary population can be defined as one in which the numbers in each age-group do not vary at intervals of one year. The stationary age distribution can then be defined for such populations as the distribution of ages at a given time of the year, and there will be an infinite number of different age distributions, depending on the time chosen. But the distribution of ages will only form an l_x series with integral values of x if all the births for the year occur instantaneously and the sample is taken at that instant. This situation does not occur in practice, though it is approximated when the population has a restricted season of births, the age structure is sampled over this period, and the ratio of live births to the numbers in the other age-groups can be determined, for example from the number of females either pregnant or suckling young, or from the size of egg-clutch. Birth rates can also be determined from dead animals using egg follicle counts (e.g. ducks, pheasants), corpora lutea counts (e.g. deer, elk, moose), placental scar counts (e.g. muskrat, beaver), and embryo counts (all mammals).

If the age distribution is sampled at time t $(0 < t < 1)$ after the breeding season then we obtain a series, l_t, l_{t+1}, ... In this case the series q_{x+t} can sometimes be fitted by a regression curve, and the values of q_x for integral values of x obtained by interpolation. For example, Hickey [1960] constructed a life table for the domestic sheep by recording the ages at death of 83 113 females on selected farms in the North Island of New Zealand. Using the age series 1½, 2½, 3½ years, etc. in preference to integral ages, he found that the q_x series conformed closely to the regression

$$\log q_x = 0 \cdot 156x + 0 \cdot 24 ;$$

q_0 was calculated from a knowledge of the number of lambs dying before 1 year of age out of 85 309 born alive.

SAMPLING METHOD. Caughley [1966] points out that because a sample consists of dead animals its age frequencies do not necessarily form a d_x series. Such a series is obtained only when the sample represents the frequencies of ages at death in a stationary population. Many published samples treated as if they formed a d_x series are not actually appropriate to this form of analysis. For instance, if the animals are obtained by shooting which is unselective with respect to age, the sample gives the age structure of the living population at that time and leads to an l_x series; that the animals were killed to get the data is irrelevant. Similarly, groups of animals killed by avalanches, fires or floods — catastrophic events that preserve a sample of the age-frequencies of animals during life — can sometimes be used for an l_x series.

A sample may include both l_x and d_x components. For example, it could consist of a number of dead animals, some of which have been unselectively shot, whereas the deaths of others are attributable to "natural" mortality. Alternatively it could be formed by a herd of animals killed by an avalanche in an area where carcasses of animals that died naturally were also present. In both these cases the l_x and d_x data are confounded and these heterogeneous samples of ages at death cannot be treated as either an l_x or a d_x series.

Summarising, there are three main methods for obtaining data for a time-specific life table:

(i) Recording ages at death by ageing a random sample of carcasses (d_x series).

(ii) Recording ages at death of a random sample killed by some catastrophic event; the frequencies form an l_x series (provided there are no carcasses from natural causes).

(iii) Recording the ages of a random sample obtained alive or dead by live-trapping or unselective hunting (l_x series). Common sources of such information with regard to hunting are (Quick [1963]): (a) Game inspection stations; particularly adaptable to procuring data on big game. These stations also provide opportunity to classify the source of data with respect to area and to the time-period in which the kills were made. (b) Game-bird wing and tail samples obtained by the co-operation of the hunter. This method is usable with upland game birds in the U.S.A., such as ruffed grouse, when the game department provides a stamped addressed and blood-proof envelope in which the hunter can send samples. (c) Hunter-bag checks. This procedure requires that the warden or game biologist personally examines the hunter's bag; it has been widely used to get data on water-fowl and to some extent on pheasants. (d) Fur buyer collections; examination of pelts and carcasses taken by trappers or in the possession of fur buyers.

BIAS. Unfortunately both methods (i) and (iii) are subject to bias with regard to the frequency of the first-year class. For example, dead immature animals, especially those dying soon after birth, tend to decay faster than the adults, so that they are under-represented in the count of carcases. Also the ratio of juveniles to adults in a shot sample is usually biased because the two age-classes have different susceptibilities to hunting. However, if a d_x series is used, we have from (10.1)

$$q_x = d_x \bigg/ \left(\sum_{y=x}^{w} d_y \right),$$

and the value of q at any age x, say, is independent of the frequencies of the younger age-classes. Alternatively, if an l_x series is used,

$$q_x = 1 - (l_{x+1}/l_x),$$

so that even if the l_x series is calculated from age-frequencies in which the initial frequency is inaccurate, the ratio l_{x+1}/l_x for the older ages will not be affected. Thus for methods (i) and (iii) the biases mentioned above will only affect the first one or two values of q_x.

When using an l_x series another possible source of error lies in the ageing technique. For example, l_x strictly refers to the number in the sample of *exact* age x years, whereas it may only be possible to age an animal to within one or two months. In this case l_x refers to the number in the age range $[x - \delta_1, x + \delta_2]$, where $\delta_i \geqslant 0$ ($i = 1, 2$). However, this error of age classification will have little effect on the ratios l_{x+1}/l_x, provided the mortality rate does not change too abruptly from one age-class to the next, so that

$$l_{x+1+t}/l_{x+t} \approx l_{x+1}/l_x$$

for all t in the interval $[-\delta_1, \delta_2]$. Even when the animals are aged exactly, the same problem can still arise because a sample is rarely instantaneous, often taking several months to collect.

TEST FOR STATIONARITY. Caughley [1966] mentions five methods used for determining whether the age structure of a sample is consistent with its having been drawn from a stationary age distribution:

(a) Comparison of the "mean mortality rate", calculated from the age distribution of the sample, with the proportion represented by the first age-class (Kurtén [1953: 51]).
(b) Comparison of the annual female fecundity of a female sample with the sample number multiplied by the life expectancy at birth, the latter statistic being estimated from the age structure (Quick [1963: 210]).
(c) Calculation of instantaneous birth rates and death rates, respectively, from a sample of the population's age distribution and a sample of ages at death (Hughes [1965]).
(d) Comparison of the age distribution with a prejudged notion of what a stationary age distribution should be like (Breakey [1963]).
(e) Examination of the l_x and d_x series, calculated from the sampled age distribution, for evidence of a common trend (Quick [1963: 204]).

Caughley points out that the above methods are invalid as they are based on circular arguments, e.g. methods (a) to (c) are tautological because they assume that the sampled age distribution is either an l_x or a d_x series. In fact, given no information other than a single age distribution, it is impossible to test whether the distribution is from a stationary population. Stationarity can only be tested by examining a chronological sequence of age distributions.

STABLE AGE DISTRIBUTION. When a population increases at a constant rate and when birth and mortality rates are constant, the age distribution eventually assumes a stable form (Slobodkin [1962: 49]). This stable age distribution should not be confused with the concept of a stationary age distribution which is a special case of the former. Caughley points out that a stable age distribution is not stationary, except when the rate of increase in population size is zero.

Example 10.1 Himalayan thar (*Hemitragus jemlahicus*): Caughley [1966]

The Himalayan thar, a hollow-horned ungulate introduced into New Zealand in 1904, now occupies 2000 square miles of mountainous country in the South Island. Thar were liberated at Mount Cook and have since spread mostly north and south along the Southern Alps. They are still spreading at a rate of about 1·1 miles a year, so that the populations

farthest from the point of liberation have been established only recently
and have not yet had time to increase greatly in numbers. Closer to the site
of liberation the density is higher, and around the point of liberation itself
there is evidence that the population has decreased.

The thar can be accurately aged from the growth rings on its horns
which are laid down in each winter of life other than the first. An l_x series
was obtained from a sample of 623 females, older than 1 year, shot in the
Godley and Macaulay valleys between November 1963 and February 1964.
Preliminary work on behaviour indicated that there was very little dispersal
of females into or out of this region, both because the females have distinct
home ranges and because there are few ice-free passes linking the valley
heads.

The first question Caughley considers is whether or not the population
is stationary. Obviously it is impossible to determine the stationary nature
of a population by examining the age structure of a single sample only, even
when birth rates are known. However, in some circumstances, a series of
age structures will give the required information. Caughley utilises this
fact in investigating the stability of the population sampled as follows.

The sample was taken about halfway between the point of liberation and
the edge of the range, that is, in the region between increasing and decreasing populations where one would expect to find a stationary population.
The animals came into the Godley valley from the southwest and presumably
colonised this side of the valley before crossing the two miles of river bed
to the northeast side. Having colonised the northeast side, the thar would
then cross the Sibald Range to enter the Macaulay valley, which is a further
six miles northeast. The sample can therefore be divided into three subsamples corresponding to the different periods of time that the animals have
been present in the three areas. A 10×3 contingency test for differences
between the three age distributions of females 1 year of age or older gave
no indication that the three subpopulations differed in age structure
($\chi^2_{18} = 22 \cdot 34$; $P = 0 \cdot 2$). This information can be interpreted in two ways:
either the three subpopulations were constant in size and were hence likely
to have stationary age distributions, or the subpopulations were increasing
at the same rate, so that they could have identical stable age distributions.
The second alternative implies that the subpopulations would have different
densities because they had been increasing for differing periods of time.
However, an analysis of the three densities gave no indication that they
were different and Caughley rejected the second alternative. Independent
subjective evidence based on observation also suggests that the populations
were not increasing or decreasing.

The second question Caughley considers is whether the age structure
of the sample is an unbiased estimate of the age structure of the population.
The most obvious source of bias is behavioural or range differences between
males and females. For instance, should males tend to occupy terrain which
is more difficult to hunt over than that used by females, they would be

under-represented in a sample obtained by hunting. During the summer, thar range in three main kinds of groups: one consisting of females, juveniles and kids, a second consisting of young males, and the third of mature males. The task of sampling these three groupings in the same proportion as they occur throughout the area is complicated by their preferences for terrain that differs in slope, altitude, and exposure. Consequently the attempt to take an unbiased sample of both males and females was abandoned, and the hunting was directed towards sampling only the nanny—kid herds in an attempt to take a representative sample of females.

To determine whether some age-classes were more susceptible than others to shooting, several chi-squared tests were carried out. Females other than kids were divided into two groups: those from herds in which some members were aware of the presence of the shooter before he fired, and those from herds which were undisturbed before shooting commenced. If any age-group is particularly wary, its members should occur more often in the "disturbed" category. However, a chi-squared test ($\chi_9^2 = 7 \cdot 28$; $P = 0 \cdot 6$) revealed no significant differences between the age structures of the two categories. The sample was also divided into those females shot at ranges less than 200 yards and those shot outside this range. If animals in a given age-class are more easily stalked than others, they will tend to be shot at closer ranges. Alternatively, animals which present small targets may be under-represented in the sample of those shot at ranges over 200 yards. This is certainly true of kids which are difficult to see, let alone to shoot, at ranges in excess of 200 yards. The kids, therefore, were not included in the analysis because their under-representation was an acknowledged fact. However, for the older females, a chi-squared test ($\chi_9^2 = 9 \cdot 68$; $P = 0 \cdot 4$) did not indicate any significant difference between the age structures of the two groups. This did not imply that no biases existed — the yearling class could well be under-represented beyond 200 yards — but that no bias could be detected from the sample of the size used.

The shooting yielded 623 females 1 year old or older, and the numbers (f_x) at each age are shown in Table 10.5. Here l_x refers to the (scaled) number of animals in the age interval x years $-\frac{1}{2}$ month to x years $+ 2\frac{1}{2}$ months rather than the number with exact age x. However, as pointed out on p. 402, this will not have much effect on the q_x series.

Up to the age of 12 years (beyond this age the frequencies dropped below 5) it was found that the age frequencies were closely fitted by a quadratic regression (Fig. 10.2)

$$E[\log f_x] = \beta_0 + \beta_1 x + \beta_2 x^2, \quad (x = 1, 2, \ldots, 12),$$

the cubic term being non-significant. The fitted regression is

$$\log Y_x = 1 \cdot 9673 + 0 \cdot 0246 x - 0 \cdot 01036 x^2,$$

and Caughley used it to obtain "smoothed" age frequencies Y_x. He felt that by using this curve he would greatly reduce the "noise" resulting from

TABLE 10.5

Life table and fecundity table for the thar *Hemitragus jemlahicus* (females only): from Caughley [1966: Table 2].

Age (years)	Frequency in sample f_x	Adjusted frequency Y_x	Female live births per female (m_x)	l_x	d_x	100 q_x
0–1	–	205	0·000	1000	533	53·3
1–2	94	95·83	0·005	467	6	0·3
2–3	97	94·43	0·135	461	28	6·1
3–4	107	88·69	0·440	433	46	10·6
4–5	68	79·41	0·420	387	56	14·5
5–6	70	67·81	0·465	331	62	18·7
6–7	47	55·20	0·425	269	60	22·3
7–8	37	42·85	0·460	209	54	25·8
8–9	35	31·71	0·485	155	46	29·7
9–10	24	22·37	0·500	109	36	33·0
10–11	16	15·04	0·500	73	26	35·6
11–12	11	9·64	} 0·470	47	18	38·2
12–13	6	5·90		29		
13–14	3		} 0·350			
14–15	4					
15–16	3					
16–17	0					
17–18	1					

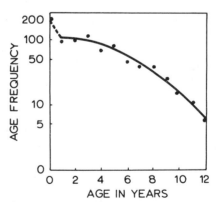

Fig. 10.2   Age frequency (on a logarithmic scale) plotted against age for a sample of female thar: from Caughley [1966].

sampling variation, the differential effect on mortality of different seasons, and the minor heterogeneities which, although not detectable, were almost certain to be present.

The frequencies of births were estimated from the observed mean numbers of female kids per female at each age (m_x in Table 10.5). They were calculated as the number of females at each age in the shot sample either carrying a foetus or lactating, divided by the number of females of that age. These values were then halved because the sex ratio of late foetuses and kids did not differ significantly from $1:1$ (93 males : 97 females). The method is open to a number of objections: it assumes that all kids were born alive, that all females neither pregnant nor lactating were barren for that season, and that twinning did not occur. The first assumption, if false, would give rise to a positive bias, and the second and third to a negative bias. However, the ratio of females older than 2 years that were neither pregnant nor lactating did not differ significantly between the periods November to December and January to February ($\chi_1^2 = 0{\cdot}79$, $P = 0{\cdot}4$), suggesting that still-births and mortality immediately after birth were not common enough to bias the calculations seriously. Errors were unlikely to be introduced by temporarily barren females suckling yearlings, because no female shot in November that was either barren (as judged by the state of the uterus), or pregnant, was lactating. Errors resulting from the production of twins would be very small: Caughley found no evidence of twinning in the area.

The potential number of female kids produced by the females in the sample was estimated by

$$Y_0 = \sum_{x=1}^{12} Y_x m_x = 205 \,,$$

and the series Y_x gave an l_x series. This series was standardised to give $l_0 = 1000$ by multiplying by $4{\cdot}878$; d_x and q_x were then calculated as in Table 10.1.

For a further study of the same population see Caughley [1970].

3 Composite life tables

These life tables are based on mortality data obtained by recording the ages of animals at death, irrespective of their year of birth. As the animals are born at different times they do not come from the same cohort, so that the numbers d_x dying in each age-class are not strictly amenable to an age-specific analysis. Also the d_x series does not refer to the numbers of deaths in each age-class at a given instant of time, so that a time-specific analysis is not appropriate. However, if we treat all the animals as though they were born at the same time, an age-specific life table can be constructed for this hypothetical population. Such a method was first used in 1902 by Karl Pearson who estimated life expectancies for ancient Egyptians using the age of deaths recorded on 141 mummy cases. This approach has also been widely used in analysing returns from bird-banding operations, and further details have been given in **5.4.3**.

10.2   AGE-SPECIFIC DATA: STOCHASTIC MODELS

10.2.1   General model

In **10.1** the approach to life-table analysis is basically deterministic in that one talks about observed rates rather than estimates of probabilities. However, this deterministic approach, although providing an intuitive basis for estimation (as we shall see below), does not provide variances; these are essential if mortality rates are to be compared either within or between populations. Chiang [1960a, b] has set up a stochastic model for the age-specific situation, and the following theory is based on his articles.

We shall assume that all individuals of age x from a given cohort have the same probability P_x ($= 1 - Q_x$) of surviving to age $x + 1$. Then, given l_x, l_{x+1} is a binomial variable with probability function

$$f(l_{x+1} \mid l_x) = \binom{l_x}{l_{x+1}} P_x^{l_{x+1}} Q_x^{l_x - l_{x+1}}, \tag{10.8}$$

so that the joint probability function of $l_1, l_2, \ldots, l_w$, given l_0 and w, is

$$f(l_1, l_2, \ldots, l_w \mid l_0, w) = \prod_{x=0}^{w-1} f(l_{x+1} \mid l_x).$$

Since $d_x = l_x - l_{x+1}$ and $l_x = d_x + d_{x+1} + \ldots + d_w$ (for $x = 0, 1, \ldots, w - 1$; $d_w = l_w$), we can readily obtain the joint probability function of the d_x, given l_0, from (10.8), namely

$$f(d_0, d_1, \ldots, d_w \mid l_0, w)$$

$$= \frac{l_0!}{\prod_{x=0}^{w} d_x!} Q_0^{d_0} (P_0 Q_1)^{d_1} \ldots (P_0 P_1 \ldots P_{w-2} Q_{w-1})^{d_{w-1}} (P_0 P_1 \ldots P_{w-1})^{d_w}. \tag{10.9}$$

The maximum-likelihood estimates $\hat{P}_x$ of the P_x for the multinomial distribution (10.9) are also the moment estimates, i.e. from (10.8)

$$\hat{P}_x = l_{x+1}/l_x.$$

This estimate is unbiased as

$$E[\hat{P}_x] = \underset{l_x}{E} E[\hat{P}_x \mid l_x]$$

$$= \underset{l_x}{E} [P_x]$$

$$= P_x,$$

and, using (1.12) and the delta method,

$$V[\hat{P}_x] = \underset{l_x}{E} \{V[\hat{P}_x \mid l_x]\} + \underset{l_x}{V} \{E[\hat{P}_x \mid l_x]\}$$

$$= \underset{l_x}{E} [P_x Q_x / l_x] + 0$$

$$= P_x Q_x E[1/l_x]$$

$$\approx \frac{P_x Q_x}{E[l_x]} \left\{ 1 + \frac{V[l_x]}{(E[l_x])^2} \right\}$$

$$= \frac{P_x Q_x}{E[l_x]} \left\{ 1 + \frac{1}{E[l_x]} - \frac{1}{l_0} \right\}. \tag{10.10}$$

The last step follows from the fact that l_x is (unconditionally) binomially distributed with parameters l_0 and $P_0 P_1 \ldots P_{x-1}$.

Turning our attention to Q_x, we note that

$$\hat{Q}_x = 1 - \hat{P}_x = d_x/l_x$$

is the same as q_x of **10.1.1**. Since

$$E[\hat{Q}_x] = 1 - E[\hat{P}_x] = Q_x,$$

and

$$V[\hat{Q}_x] = V[1 - \hat{P}_x] = V[\hat{P}_x],$$

we see that $\hat{Q}_x$ is unbiased and $V[\hat{Q}_x]$ can be estimated by

$$\hat{V}[\hat{Q}_x] = \frac{(1 - \hat{Q}_x)\hat{Q}_x}{l_x} \left\{ 1 + \frac{1}{l_x} - \frac{1}{l_0} \right\}, \quad (x = 0, 1, \ldots, w-1). \tag{10.11}$$

It is also readily shown that

$$\text{cov}[\hat{Q}_x, \hat{Q}_y] = 0 \quad (x \neq y) \tag{10.12}$$

so that the estimates $\hat{Q}_x$ are uncorrelated.

In conclusion we note that the above theory can still be applied when the sequence l_x is given for an arbitrary sequence of x values $x_0, x_1, \ldots, x_w$ (not necessarily integral) by simply replacing P_x by P_i, say, the probability of an individual of age x_i surviving to age x_{i+1}. For example, x may be integral, but the sequence of integers $0, 1, 2, \ldots$ may be incomplete as l_x may not be available for some values of x. Also, x could refer to the *coded* age, so that the above theory can be applied to any group of l_0 individuals of the same age.

Example 10.2 Patagonian cavy (*Dolichotis patagonica*): Comfort [1957]

The above theory was applied to a zoo population of $l_0 = 56$ animals (sexes combined), and the data and mortality estimates are set out in Table 10.6: here x is measured in units of 100 days. We note that the standard deviations of the estimates are comparatively large as l_0 is small.

INSTANTANEOUS MORTALITY RATE. If Z_x is the total instantaneous mortality rate for individuals of age x, then

$$P_x = \exp[- \int_0^1 Z_{x+t} \, dt]$$

$$\approx \exp[-Z_{x+\frac{1}{2}}].$$

Hence $Z_{x+\frac{1}{2}}$ can be estimated by

$$\hat{Z}_{x+\frac{1}{2}} = -\log \hat{P}_x = \log(l_x/l_{x+1})$$

TABLE 10.6
Age-specific life table for a zoo population of 56 Patagonian cavies: data from Comfort [1957: Table 5].

Age (days)	d_x	l_x	$\hat{Q}_x$	$\hat{\sigma}[\hat{Q}_x]$
0–99	13	56	0·2321	0·056
100–	4	43	0·0930	0·044
200–	3	39	0·0769	0·043
300–	1	36	0·0278	0·028
400–	4	35	0·1143	0·054
500–	3	31	0·0968	0·054
600–	1	28	0·0357	0·035
700–	1	27	0·0370	0·037
800–	4	26	0·1538	0·072
900–	3	22	0·1364	0·074
1000–	1	19	0·0526	0·052
1100–	2	18	0·1111	0·075
1200–	5	16	0·3125	0·12
1300–	1	11	0·0909	0·09
1400–	1	10	0·1000	0·10
1500–	4	9	0·4444	0·17
1600–	1	5	0·2000	0·19
1700–	0	4	0·0000	–
1800–	0	4	0·0000	–
1900–	3	4	0·7500	0·24
2000–	1	1	1·0000	–

and, using the delta method,

$$V[\hat{Z}_{x+\frac{1}{2}}] \approx V[\hat{P}_x]/P_x^2.$$

Kimball [1960] has compared this estimate with the classical actuarial estimate

$$Z'_{x+\frac{1}{2}} = d_x/[\tfrac{1}{2}(l_x+l_{x+1})].$$

He shows that both estimates are biased and that for small samples there is not a great deal to choose between them. However, if l_0 is large and $Z_{x+t} = Z_x$ for $0 \leqslant t < 1$ then $-\log \hat{P}_x$ is the maximum-likelihood estimate of Z_x and should be used instead of $Z'_{x+\frac{1}{2}}$.

For an arbitrary sequence of x values $x_0, x_1, \ldots, x_w$, and defining $l_i = l_{x_i}$ and $t_i = x_{i+1} - x_i$, we have

$$P_i = \exp[-\int_0^{t_i} Z_{i+t}\, dt]$$
$$\approx \exp[t_i Z_{i+\frac{1}{2}t_i}].$$

The corresponding estimates of $Z_{i+\frac{1}{2}t_i}$ are now

$$\hat{Z}_{i+\frac{1}{2}ti} = -(1/t_i)\log \hat{P}_i = (1/t_i)\log(l_i/l_{i+1})$$

and

$$Z'_{i+\frac{1}{2}t_i} = \frac{d_i}{\frac{1}{2}t_i(l_i + l_{i+1})},$$

where $d_i = l_i - l_{i+1}$.

When the exact age at death of each individual is known, as for example in some laboratory or zoo populations, and the sequence $x_0, x_1, \ldots, x_w$ is not chosen in advance but rather determined by the number dying in each interval $[x_i, x_{i+1})$ (i.e. the d_i are predetermined and the x_i are now random variables), Kimball gives two methods for estimating the instantaneous mortality rates: one method leads to a maximum-likelihood estimate of $\mu_{i+\frac{1}{2}t_i}$ while the other (due to Seal [1954]) leads to an unbiased estimate of μ_i when $\mu_{i+t} = \mu_i$ for $0 \leq t < t_i$.

EXPECTATION OF LIFE. If we choose an individual at random from a cohort of size l_0, then, given that the individual lives to age x, the *further* length of life lived by the individual will be a random variable with expected value E_x, say. From (10.3) the usual estimate of E_x is

$$e_x = \frac{1}{2} + \sum_{y=x+1}^{w} (l_y/l_x)$$

$$= \frac{1}{2} + \sum_{y=x+1}^{w} (d_y + d_{y+1} + \ldots + d_w)/l_x$$

$$= (\tfrac{1}{2}d_x + 1\tfrac{1}{2}d_{x+1} + \ldots + (w-x+\tfrac{1}{2})d_w)/l_x \qquad (10.13)$$

for $x = 0, 1, \ldots, w$ ($e_w = \frac{1}{2}$). Now if the distribution of deaths in each interval $[x, x+1)$ is uniform, so that, as far as expected values are concerned, the d_x individuals may be regarded as having lived for half a year in the interval, then Y_x, the further length of life lived by an individual of age x, may be treated as a random variable taking values $\frac{1}{2}, 1\frac{1}{2}, \ldots, w-x-\frac{1}{2}$, $w-x+\frac{1}{2}$, with respective probabilities $Q_x, P_x Q_{x+1}, \ldots, P_x P_{x+1} \ldots P_{w-2}Q_{w-1}, P_x P_{x+1} \ldots P_{w-1}$. Hence

$$E_x = E[Y_x]$$

$$= \tfrac{1}{2}Q_x + 1\tfrac{1}{2}P_x Q_{x+1} + \ldots + (w-x+\tfrac{1}{2})P_x P_{x+1} \ldots P_{w-1}.$$

Therefore, since

$$E[d_y | l_x] = l_x P_x P_{x+1} \ldots P_{y-1} Q_y, \quad (y = x, x+1, \ldots, w-1)$$

and

$$E[d_w | l_x] = l_x P_x P_{x+1} \ldots P_{w-1},$$

we have from (10.13) that

$$E[e_x] = \underset{l_x}{E} E[e_x | l_x]$$

$$= \underset{l_x}{E}[\tfrac{1}{2}Q_x + \ldots + (w-x+\tfrac{1}{2})P_x P_{x+1} \ldots P_{w-1}]$$

$$= E_x,$$

and e_x is an unbiased estimate of E_x. Also from (10.13) we see that e_x is a sample mean based on the frequencies d_y for l_x $(= d_x + d_{x+1} + \ldots + d_w)$ individuals. Hence, by the central limit theorem, e_x is asymptotically normal for large l_x and, given an estimate of $V[e_x]$, a large-sample confidence interval for E_x can be calculated in the usual manner (cf. (1.5)).

Chiang [1960a] proves that

$$V[e_x] = E[1/l_x] \sum_{y=x}^{w-1} \{\phi_{xy}[E_{y+1} + \tfrac{1}{2}]^2 P_y Q_y\}, \quad x = 0, 1, \ldots, w-1,$$

where ϕ_{xy} is the probability that an individual of age x survives to age y (i.e. $\phi_{xx} = 1$ and $\phi_{xy} = P_x P_{x+1} \ldots P_{y-1}$ for $y > x$). Hence estimating $E[1/l_x]$ by $1/l_x$, and ϕ_{xy} by l_y/l_x, we see that $V[e_x]$ can be estimated by

$$\hat{V}[e_x] = (1/l_x^2) \sum_{y=x}^{w-1} \{l_y[e_{y+1} + \tfrac{1}{2}]^2 \hat{P}_y \hat{Q}_y\}.$$

However, since e_x is a sample mean, an unbiased estimate of $V[e_x]$ is given by (cf. (1.9))

$$v[e_x] = \frac{(\tfrac{1}{2} - e_x)^2 d_x + (1\tfrac{1}{2} - e_x)^2 d_{x+1} + \ldots + (w - x + \tfrac{1}{2} - e_x)^2 d_w}{l_x(l_x - 1)}$$

$$= \left\{ \sum_{y=x}^{w} [(y + \tfrac{1}{2} - x) - e_x]^2 d_y \right\} \Big/ \{l_x(l_x - 1)\}$$

$$= \left\{ \sum_{y=x}^{w} [(y + \tfrac{1}{2})^2 d_y] - l_x(e_x + x)^2 \right\} \Big/ \{l_x(l_x - 1)\}.$$

If the term $l_x - 1$ in the above expression is replaced by l_x, we find from Chiang [1960b: 225] that $v[e_x] = \hat{V}[e_x]$.

Chiang's formulation of the above problem is slightly more general than the one given here. He considers the more general situation when the sequence $\{l_x\}$ is defined for the arbitrary sequence $x_0, x_1, \ldots, x_w$. For this this case let

l_i = number surviving to age x_i,

d_i = $l_i - l_{i+1}$ (the number dying in the age interval $[x_i, x_{i+1})$),

t_i = $x_{i+1} - x_i$,

P_i = probability that an individual of age x_i survives to age x_{i+1},

ϕ_{ij} = $P_i P_{i+1} \ldots P_{j-1}$ $(j > i; \phi_{ii} = 1)$

and

E_i = expected value of the further expectation of life for an individual of age x_i.

Suppose that the distribution of deaths in each interval is such that, on the average, each of the d_i individuals lives $a_i t_i$ years in the interval $[x_i, x_{i+1})$, and define

$$c_j = (1 - a_{j-1}) t_{j-1} + a_j t_j.$$

Then Chiang shows that

$$e_i = \sum_{j=i}^{w} (x_j - x_i + a_j t_j) d_j / l_i$$

$$= a_i t_i + \left(\sum_{j=i+1}^{w} c_j l_j \right) / l_i$$

is an unbiased estimate of E_j with variance

$$V[e_i] = E[1/l_i] \sum_{j=i}^{w-1} \{\phi_{ij}[E_{j+1} + (1-a_j) t_j]^2 P_j Q_j\}.$$

This variance can be estimated by

$$\hat{V}[e_i] = (1/l_i^2) \sum_{j=i}^{w-1} \{l_j [e_{j+1} + (1-a_j) t_j]^2 \hat{P}_j \hat{Q}_j\}$$

$$= (1/l_i^2)[\sum_{j=i}^{w} (x_j + a_j t_j)^2 d_j - l_i (e_i + x_i)^2],$$

$$(i = 0, 1, \ldots, w-1).$$

10.2.2 Constant probability of survival

When P_x, the probability of surviving from age x to age $x + 1$, is constant $(= P$, say), then using (10.8) we find that (10.9) reduces to

$$\frac{l_0!}{\prod\limits_{x=0}^{w} d_x!} P^{l_1 + l_2 + \ldots + l_w} Q^{l_0 - l_w}.$$

Hence the maximum-likelihood estimate of Q $(= 1 - P)$ is readily shown to be

$$\hat{Q} = (l_0 - l_w)/(l_0 + l_1 + \ldots + l_{w-1})$$

$$= \left(\sum_{x=0}^{w-1} d_x \right) \bigg/ \left(\sum_{x=0}^{w-1} l_x \right),$$

which is the same as q_{pool} of (10.7). Using standard maximum-likelihood theory, we find that the asymptotic variance of $\hat{Q}$ is estimated by (cf. (1.27))

$$\hat{V}[\hat{Q}] = \left(\sum_{x=1}^{w} l_x \right) (l_0 - l_w) \bigg/ \left(\sum_{x=0}^{w-1} l_x \right)^3.$$

Alternatively one can use the average estimate of (10.6), namely

$$q_{ave} = \sum_{x=0}^{w-1} \hat{Q}_x / w.$$

Then, from (10.12) and (1.9), an unbiased estimate of $V[q_{ave}]$ is given by

$$v[q_{ave}] = \frac{1}{w(w-1)} \sum_{x=0}^{w-1} (\hat{Q}_x - q_{ave})^2.$$

REGRESSION MODELS. Since

$$E[d_x] = l_0 P^x Q, \quad (x = 0, 1, \ldots, w-1),$$

we can take logarithms and consider the linear model

$$\log_{10} (d_x) = \log_{10} (l_0 Q) + x \log_{10} P + \epsilon_x. \tag{10.14}$$

Here the ϵ_x are correlated, though the degree of correlation will usually be small. Therefore, assuming the ϵ_x to be independently normally distributed with zero mean and constant variance, least-squares estimates of $\log_{10} P$

and $\log_{10}(l_0 Q)$ can be obtained in the usual manner. If l_0 is unknown, as is sometimes the case, then it may be estimated from this model.

Another regression model which is often used can be obtained from the equation

$$E[l_x] = l_0 P^x$$

i.e., taking logarithms,

$$\log_{10} l_x = \log_{10} l_0 + x \log_{10} P + \epsilon_x.$$

However, the degree of correlation between the l_x variables is much greater than that between the d_x variables, so that, from a least-squares point of view, the model (10.14) is preferred.

10.3   TIME-SPECIFIC DATA: CONSTANT SURVIVAL RATE

10.3.1   Geometric model

Life-table methods for calculating mortality rates (and the complementary survival rates) for time-specific data are discussed in some detail in **10.1.2**(2). But when the mortality rate is constant over all age-classes we can either combine the rates to give q_{ave} or q_{pool}, or else use the following techniques of Chapman and Robson [1960] and Robson and Chapman [1961].

1   Chapman–Robson survival estimate

We shall assume that:

(1) S, the proportion of a given age-group surviving for one year, is the same for each age-group and remains constant from year to year.
(2) N_x, the number of animals of age x in the population at the time of the year when reproduction occurs (called time "zero") is constant from year to year (N_0 = annual number of births).
(3) Sampling from the population is random with respect to age.

Then, from assumption (2), the number of animals N_x growing out of age x will be balanced by the number SN_{x-1} growing into this age-group. Hence

$$N_x = SN_{x-1} = S^x N_0,$$

and since

$$N = \sum_{x=0}^{\infty} N_x = N_0(1 + S + S^2 + \ldots) = N_0/(1-S),$$

we have

$$N_x = (1-S)S^x N, \quad (x = 0, 1, 2, \ldots).$$

If an animal is chosen at random at time zero, then the probability that it is of age x is N_x/N, i.e. x is a random variable with probability function

$$f(x) = (1-S)S^x \quad (x = 0, 1, 2, \ldots). \tag{10.15}$$

In fact, even if the animal is chosen at random at some other time of the year, so that it is found to be of age $x+$, x will still have probability function (10.15), provided that the survival rate is the same for all age-groups throughout the year (though this rate may vary with the time of the year), i.e. provided that N_x/N remains constant. Therefore, given a random sample of size n $(= \sum_x n_x)$ in which n_x are observed to be of age x $(x = 0, 1, 2, \ldots, r)$, then, ignoring the complications of sampling without replacement in each age-group, the likelihood function for the sample is

$$(1-S)^n \, S^{\sum x n_x} \;=\; (1-S)^n S^X,$$

where $X = \sum_{x=0}^{r} x n_x$ is a sufficient statistic for S. Since X is a complete statistic, a uniformly minimum-variance unbiased estimate $\hat{S}$ of S exists, and this estimate is a unique function of X, say $h(X)$. From the identity $E[h(X)] = S$ in S, Chapman and Robson show that

$$\hat{S} \;=\; X/(n+X-1).$$

Although the exact variance of $\hat{S}$ is not expressible in closed form,

$$v[\hat{S}] \;=\; \hat{S}\left(\hat{S} - \frac{X-1}{n+X-2}\right)$$

is the minimum-variance unbiased estimate of this exact variance. But when n is large, $\hat{S}$ is approximately equal to $X/(n+X)$, the maximum-likelihood estimate, which has asymptotic variance $[S(1-S)^2]/n$; hence, for large n,

$$V[\hat{S}] \;\approx\; S(1-S)^2/n. \tag{10.16}$$

Although the equations leading up to (10.15) are deterministic, the above theory can be justified on other grounds using a stochastic model (cf. Chapman and Robson [1960: 357]).

ALTERNATIVE ESTIMATES. Various alternative estimates have been have been used in practice. For example, Jackson's [1939] estimate

$$S' \;=\; (n_1 + n_2 + \ldots + n_r)/(n_0 + n_1 + \ldots + n_{r-1})$$
$$\;=\; (n-n_0)/(n-n_r)$$

is widely used although it is biased, can be greater than unity, and depends on the behaviour of the two extreme age-classes n_0 and n_r in the sample. A more useful modification (Heincke [1913]), which differs only slightly from S' when n_r is small compared with n, is

$$\hat{S}_1 \;=\; (n-n_0)/n.$$

Since n_0 is binomially distributed with parameters n and $(1-S)$, $\hat{S}_1$ is unbiased with variance

$$V[\hat{S}_1] \;=\; S(1-S)/n.$$

We note from (10.16) that $V[\hat{S}] \approx (1-S)V[\hat{S}_1]$, so that $V[\hat{S}]$ will be much smaller than $V[\hat{S}_1]$ when $1-S$ is small.

Chapman and Robson [1960] have considered a more general class of estimates, namely

$$\hat{S}_p \;=\; [(n-n_0-n_1 \ldots -n_{p-1})/n]^{1/p}, \qquad p = 1, 2, \ldots,$$

where, to terms of order $1/n$,

$$E[\hat{S}_p] \;=\; S + \frac{(1-p)(1-S^p)}{np^2 S^{p-1}}$$

and

$$V[\hat{S}_p] \;=\; \frac{(1-S^p)}{np^2 S^{p-2}}.$$

This variance, considered as a function of p, is minimised for the value of p, say $\hat{p}$, satisfying

$$1 - S^p \;=\; \tfrac{1}{2} |\log S| p.$$

If this were to be used as a guide in practice, assuming some prior information on the general magnitude of S, p would be chosen as one of the integers near $\hat{p}$. Since $\hat{p}$ is an increasing function of S, $\hat{S}_1$ is preferred among this class when S is small, while p should be chosen fairly large when S is large. For example, when $S = 0\cdot5$, $\hat{p}$ lies between 2 and 3. Choosing $p = 2$ we note that the bias of $\hat{S}_2$ is approximately $-(1-S^2)/4Sn$, which is negligible for S near 1, but may not be otherwise. Also $V[\hat{S}_2] < V[\hat{S}_1]$ $(= S(1-S)/n)$ when $S > \tfrac{1}{3}$. However it should be stressed that when the geometric probability model (10.15) is valid, $\hat{S}$ is the most efficient estimate.

THRESHOLD AGE. In many populations there is a "threshold" age above which S is independent of age. For example, in fisheries there are many types of fishing gear which are selective against smaller sizes and consequently younger ages. In this case the exploitation rate for the younger fish will be smaller, and this in turn will affect the probability of survival of the younger classes. For this situation it is not unreasonable to assume that there is some age A, such that for all ages $x \geqslant A$, $N_{x+1} = S N_x$ and $N_x = S^{x-A} N_A$. Thus by simply relabelling the ages so that $A = 0$ (i.e. using a "coded" age) the above theory can still be applied. However, there may be no way of determining A in practice, so that the data in the youngest age-class may be suspect. Chapman and Robson [1960] give a useful test of the validity of the geometric probability model which gives particular emphasis to this zero class. They show that for large n (say greater than 100)

$$z \;=\; (\hat{S}_1 - \hat{S}) \Big/ \left\{ \frac{X(X-1)(n-1)}{n(n+X-1)^2(n+X-2)} \right\}^{\frac{1}{2}} \tag{10.17}$$

is asymptotically distributed as the unit normal when (10.15) is valid. Since $\hat{S}$ and $\hat{S}_1$ are both unbiased estimates of S under the above model, any significant difference in these estimates will indicate that, compared with the older age-classes, the $x = 0$ class is not properly represented. If a two-tailed test is required then we can use z^2 as χ_1^2. The authors also give a small-sample test based on the hypergeometric distribution (or its binomial approximation).

Example 10.3   Rock bass: Robson and Chapman [1961]

For a sample of rock bass trap-netted from Cayuga Lake, New York, during a single summer season, the threshold age A was arbitrarily chosen as VI and Table 10.7 was obtained. From this table we have

$$\hat{S} = \frac{X}{n + X - 1} = \frac{196}{243 + 196 - 1} = 0.4475$$

or, expressed as a percentage survival rate, 45 per cent. Also,

$$v[\hat{S}] = \frac{X}{n + X - 1}\left(\frac{X}{n+X-1} - \frac{X-1}{n+X-2}\right)$$

$$= \frac{196}{438}\left(\frac{196}{438} - \frac{195}{437}\right) = 5.66 \times 10^{-4}$$

and hence

$$\sqrt{v[\hat{S}]} = 0.0238.$$

TABLE 10.7

Age frequencies, n_x, from a sample of rock bass: from Robson and Chapman [1961: 182].

Age	Coded age (x)	n_x	$x\,n_x$
VI+	0	118	0
VII+	1	73	73
VIII+	2	36	72
IX+	3	14	42
X+	4	1	4
XI+	5	1	5
Total		$n = 243$	$X = 196$

An approximate 95 per cent confidence interval for S is given by $0.4475 \pm 1.96\,(0.0238)$ or $(0.40, 0.50)$.

Heincke's estimate

$$\hat{S}_1 = \frac{n - n_0}{n} = \frac{243 - 118}{243} = 0.5144$$

is greater than $\hat{S}$, which suggests that there is a deficit in the zero age-group relative to the older age-group. To test whether the difference is significant, we calculate the chi-squared statistic

$$z^2 = (0.4475 - 0.5144)^2 \Bigg/ \left\{\frac{(196)(195)(242)}{243(438^2)(437)}\right\}$$

$$= 9.858,$$

which is highly significant ($P < 0.002$). This leads us to suspect the validity of the geometric model. Unfortunately the chi-squared test cannot tell us which of the three basic assumptions on p. 414 is at fault, as it merely establishes whether or not there is a reasonable agreement between the observed frequency in the age-group 0 and the frequency expected on the basis of the data in the older age-groups. If we suspect that age-group 0 is at fault we can eliminate it and recode the remaining age-groups as in Table 10.8. For this recoded set of data

$$\hat{S} = 71/(125+71-1) = 0 \cdot 3641,$$

$$\hat{S}_1 = (125-13)/125 = 0 \cdot 4160,$$

and

$$z^2 = (0 \cdot 3641 - 0 \cdot 4160)^2 \bigg/ \left\{ \frac{71\,(70)(124)}{125\,(195^2)\,194} \right\} = 4 \cdot 032$$

TABLE 10.8

Recoded age frequencies, n_x, from Table 10.7.

Age	Recoded age (x)	n_x	$x\,n_x$
VII+	0	73	0
VIII+	1	36	36
IX+	2	14	28
X+	3	1	3
XI+	4	1	4
Total		$n = 125$	$X = 71$

which is still significant at the 5 per cent level. Therefore, eliminating age-class VII+ and recoding, we finally get a non-significant test.

For a further application of the above theory see Johnson [1968: male fur seal population].

INSTANTANEOUS MORTALITY RATE. Suppose that mortality is a Poisson process with parameter Z, then using 1 year, say, as the time unit, $S = \exp(-Z)$ or

$$Z = -\log S.$$

A reasonable estimate of Z is $\hat{Z} = -\log \hat{S}$ and, using the delta method, we have asymptotically

$$E[\hat{Z}] = S + \tfrac{1}{2}(1-S)^2/(nS)$$

and

$$V[\hat{Z}] = (1-S)^2/(nS).$$

Chapman and Robson [1960] suggest a modified estimate

$$Z^* = -\log \hat{S} - \frac{(n-1)(n-2)}{n(n+X-1)(X+1)}$$

which has negligible bias.

We note that in fisheries the instantaneous mortality rate Z is generally the instantaneous natural mortality rate plus the instantaneous exploitation rate.

418

2 Older age-groups pooled

In addition to the random sampling errors of the n_x there is usually an error in the process of measuring age. For example, in fishery work with commonly used procedures such as scale or otolith reading, the age determination becomes more time-consuming and more subject to error with increasing age. To alleviate both these problems Robson and Chapman [1961] suggest that exact ageing should be attempted for just the younger age-groups and that the remaining age-groups be combined. Although the pooling of older age-groups represents a loss in information, the time saved in the ageing process can be utilised for larger samples.

Suppose that animals J years old or less are aged exactly, and all animals $J + 1$ years old or more are pooled. Let

$$n_{(J)} = \sum_{x=J+1}^{r} n_x$$

and

$$X = \sum_{x=0}^{J} xn_x + (J+1)n_{(J)} ,$$

then the maximum-likelihood estimate of S is now

$$\hat{S}_{\text{pool}} = \frac{X}{n - n_{(J)} + X}$$

with asymptotic variance

$$V[\hat{S}_{\text{pool}}] = S(1-S)^2/[n(1-S^{J+1})] .$$

In this case no minimum-variance unbiased estimate of S exists. When $J = 0$, the above estimate reduces to simply Heincke's estimate $\hat{S}_1$.

We note that, for large n (cf. (10.16)),

$$\frac{V[\hat{S}]}{V[\hat{S}_{\text{pool}}]} \approx 1 - S^{J+1}$$

is the efficiency of this method relative to ageing all the fish exactly. Thus, for every 100 fish aged exactly, the present pooling method would require

$$n = 100/(1-S^{J+1}) \tag{10.18}$$

fish to obtain the same accuracy of estimation. Robson and Chapman present a table of this relationship (Table 10.9) and conclude that when S is in the neighbourhood of 0·50, little is gained by attempting ageing on more than the first two or three age-groups if the above method of estimation is to be used.

Example 10.4   Using the data of Table 10.7 with $J = 3$ we have Table 10.10. Thus

$$\hat{S}_{\text{pool}} = 193/(243 - 16 + 193) = 0\cdot46 ,$$

and

$$\hat{V}[\hat{S}_{\text{pool}}] = 0\cdot000\ 612$$

which is only slightly larger than the value of 0·000 566 for the case when all the fish are aged exactly.

TABLE 10.9

The sample size n given by equation (10.18) for different values of S and J: from Robson and Chapman [1961: Table 1].

J \ S	0·25	0·50	0·75
0	133	200	400
1	107	133	228
2	101	114	173
3	100	107	146
4	100	103	132
5	100	101	122
...	...	...	...
∞	100	100	100

TABLE 10.10

Age data from a rock bass sample with age IX+ and older pooled: from Robson and Chapman [1961: 184].

Age	Coded age (x)	n_x	$x\,n_x$
VI+	0	118	0
VII+	1	73	73
VIII+	2	36	72
IX+ and older	$\geqslant 3$	16	48
Total		$n = 243$	$X = 193$

3 Age range truncated at both ends

In some experimental circumstances it may be necessary to truncate the age data on the right as well as on the left. For example, in fisheries the sampling gear may be effective only for a limited range of fish size, and therefore age, so that different gear is used for different age ranges. In addition the exploitation rate may vary for different size-classes, so that the probability of survival may vary with age. Such circumstances therefore lead to a partitioning of the age range into segments and a separate analysis for each segment, as follows.

Suppose that the age in a particular segment runs from 0 to K on the *coded* scale and assume that the assumptions on p. 414 are valid for this segment. Then the age distribution for the segment is given by the truncated probability function (cf. (10.15))

$$f(x) = S^x \bigg/ \sum_{j=0}^{K} S^j, \quad x = 0, 1, 2, \dots, K.$$

Chapman and Robson [1960: 361] mention that an unbiased estimate of S does not exist for this model, and the asymptotically unbiased maximum-likelihood estimate, $\hat{S}_{seg}$, say, is the solution of

$$X/n = \sum_{k=0}^{K} kS^k \Big/ \sum_{k=0}^{K} S^k$$
$$= [S/(1-S)] - (K+1)S^{K+1}/(1-S^{K+1}), \qquad (10.19)$$

where $X = \sum_{x=0}^{K} xn_x$ and $n = \sum_{x=0}^{K} n_x$. Methods for solving the above equation, using a table reproduced from Robson and Chapman [1961], are given in **A6**. The asymptotic variance of $\hat{S}_{seg}$ is given by

$$V[\hat{S}_{seg}] = \frac{1}{n}\left[\frac{1}{S(1-S)^2} - \frac{(K+1)^2 S^{K-1}}{(1-S^{K+1})^2}\right]^{-1}$$

which decreases to the variance of the "non-truncated" estimate $S(1-S)^2/n$ as $K \to \infty$. For small K there can be a considerable loss in efficiency through an unnecessary truncation of the data.

Heincke's estimate $\hat{S}_1$ can also be used, though with a different mean and variance; n_0 is now binomial with parameters n and $(1-S)/(1-S^{K-1})$, so that

$$E[\hat{S}_1] = S - S^{K+1}(1-S)/(1-S^{K+1})$$
$$= S - b(\hat{S}_1), \text{ say,}$$

and

$$V[\hat{S}_1] = S(1-S)(1-S^K)/[n(1-S^{K+1})^2].$$

Although $\hat{S}_1$ is now biased the bias will usually be small. Unfortunately a simple test statistic similar to (10.17) is not available, though a comparison of the confidence interval based on $\hat{S}_1 + b(\hat{S}_1)$ with that based on $\hat{S}_{seg}$ will give some idea as to the reliability of the zero class.

Example 10.5 Using the data of Table 10.7 with $K = 3$, we have Table 10.11. Here,

$$X/n = 187/241 = 0\cdot775\,93,$$

and entering Table **A6** with $K = 3$ and using linear interpolation leads to

$$\hat{S}_{seg} = 0\cdot5247,$$

TABLE 10.11

Age data for a segment of the age-range only: from Robson and Chapman [1961: 188].

Age	Coded age (x)	n_x	$x\,n_x$
VI+	0	118	0
VII+	1	73	73
VIII+	2	36	72
IX+	3	14	42
Total		$n = 241$	$X = 187$

with variance estimated by

$$\frac{1}{241}\left[\frac{1}{(0\cdot525)(1-0\cdot525)^2} - \frac{16(0\cdot525)^2}{[1-(0\cdot525)^4]^2}\right]^{-1} = 0\cdot001\ 266.$$

Without truncation on the right, the variance estimate obtained in Example 10·3 is 0·000 566, which is less than half the above estimate, thus indicating a considerable loss in precision through truncation.

An approximate 95 per cent confidence interval for S is given by $0\cdot525 \pm 1\cdot96\ (0\cdot001\ 266)^{\frac{1}{2}}$ or $(0\cdot45, 0\cdot60)$.

Heincke's estimate is

$$\hat{S}_1 = (241-118)/241 = 0\cdot510,$$

and since

$$\hat{S}_1 + \hat{b}(\hat{S}_1) = 0\cdot510 + 0\cdot035$$
$$= 0\cdot545$$

lies within the above confidence interval, there is no reason for suspecting the zero class.

4   Use of age–length data

When n, the sample size, is large, as in commercial fishing, or when age determinations are time-consuming, it may be impractical or inefficient to age the whole sample. If age and length are correlated, a simpler method would be to stratify the sample by length, and then age just a subsample from each length-class (Ketchen [1950]). The sum of the coded ages for the subsample could then be used to provide an estimate X' say of X, the sum for the whole sample. In this case Robson and Chapman [1961] show that an almost unbiased estimate of S is

$$\hat{S}_L = \frac{X'}{n + X' - 1} + \frac{(n-1)\,V[X']}{(n+X'-1)(n+X'-2)(n+X'-3)}$$

with asymptotic variance

$$V[\hat{S}_L] = \frac{S(1-S)^2}{n} + \frac{(1-S)^4}{(n-1)^2}\,V[X']. \qquad (10.20)$$

The estimate X' and its variance can be obtained by one of the following two methods, depending on whether individual lengths in the sample are known or not.

GROUPED DATA. Consider the simpler case where the lengths are not determined accurately but are merely assigned to one of I length-classes. For $i = 1, 2, \ldots, I$ let

L_i = number in the sample belonging to the ith length-class,
l_i = number in the ith class subsampled for age determination,
X_i = (unknown) total coded age for the L_i individuals in the ith class,
$\bar{X}_i = X_i/L_i$,

x_{ij} = jth coded age determination in the subsample from the ith class
$\quad (j = 1, 2, \ldots, l_i)$,

$x_i = \sum_j x_{ij}$ (total coded age for ith subsample),

and

$\quad \bar{x}_i = x_i / l_i$.

Then

$$X = \sum_{i=1}^{I} X_i = \sum_{i=1}^{I} L_i \bar{X}_i, \qquad (10.21)$$

and from Cochran [1963: Chapter 5] we have the stratified-sample estimate of X, namely

$$X'_{\text{strat}} = \sum_{i=1}^{I} L_i \bar{x}_i.$$

The variance of X'_{strat} is estimated by

$$v[X'_{\text{strat}}] = \sum_{i=1}^{I} L_i (L_i - l_i) \, \hat{\sigma}_i^2 / l_i, \qquad (10.22)$$

where $\hat{\sigma}_i^2$ is the usual unbiased estimate of the variance σ_i^2 of the age of individuals in the ith class, namely

$$\hat{\sigma}_i^2 = \sum_{j=1}^{l_i} (x_{ij} - \bar{x}_i)^2 / (l_i - 1).$$

Example 10.6 Lemon soles: Ricker [1958]

To demonstrate the calculations, the first few rows of a table from Ricker [1958] (cf. Table 10.15) are reproduced in Table 10.12. Thus, for

TABLE 10.12

Extract from a table of age-length data given by Ricker [1958: 79].

Length-class (cm)	L_i	l_i	Coded age				
			0	1	2	3	4
27	6	6	5	1	0	0	0
28	9	9	3	4	2	0	0
29	30	10	4	4	1	1	0
30	51	10	1	5	4	0	0

example,

$$L_2 \bar{x}_2 = x_2 = 3(0) + 4(1) + 2(2) = 8$$

and

$$L_3 \bar{x}_3 = 3 x_3 = 3[4(0) + 4(1) + 1(2) + 1(3)] = 27.$$

The contribution of the first two length-classes to $v[X'_{\text{strat}}]$ is zero as $L_i = l_i$ ($i = 1, 2$). The contribution of the third class is $30(30-10)\hat{\sigma}_3^2/10$ or

$60\,\hat{\sigma}_3^2$, where

$$\hat{\sigma}_3^2 = \tfrac{1}{9}\{\Sigma\, x_{3j}^2 - x_3^2/l_3\}$$
$$= \tfrac{1}{9}\{4\,(0) + 4\,(1^2) + 1\,(2^2) + 1\,(3^2) - 81/10\}$$
$$= 0\cdot99.$$

Example 10.7   Halibut: Robson and Chapman [1961]

A sample from a halibut population was stratified by length and sub-sampled for age determinations; the data are set out in Table 10.13. Here,

$$n = 1612, \quad X'_{\text{strat}} = 5660\cdot95$$

TABLE 10.13

Sample from a halibut population stratified by length and subsampled for age determinations: adapted from Robson and Chapman [1961: Table 2].

Length-class (i)	L_i	l_i	x_i	$\hat{\sigma}_i^2$	$L_i\bar{x}_i$	$\hat{\sigma}_i^2 L_i(L_i - l_i)/l_i$
1	25	13	28	3·31	53·84	76·38
2	37	18	42	6·59	86·33	257·38
3	79	40	57	3·79	112·57	291·92
4	146	75	90	3·51	175·20	485·13
5	204	103	191	5·95	378·29	1 190·23
6	259	132	296	7·15	580·78	1 781·70
7	224	103	291	7·89	632·85	2 076·22
8	177	81	389	4·74	850·03	994·35
9	163	74	340	8·76	748·91	1 717·32
10	116	61	363	12·78	690·29	1 336·66
11	87	41	266	16·01	564·43	1 562·73
12	37	20	168	20·46	310·80	643·47
13	26	16	107	21·83	173·87	354·74
14	32	13	123	20·02	302·76	936·32
Total	1.612	790			5 660·95	13 704·55

$$v[X'_{\text{strat}}] = 13\,704\cdot55,$$

$$\hat{S}_L = \frac{5660\cdot95}{1612 + 5660\cdot95 - 1} + \frac{1611\,(13\,704\cdot55)}{7270\cdot67\,(7269\cdot67)\,(7268\cdot67)}$$
$$= 0\cdot7785$$

and

$$\hat{V}[\hat{S}_L] = \frac{(0\cdot7785)\,(1 - 0\cdot7785)^2}{1612} + \frac{(1 - 0\cdot7785)^4\,(13\,704\cdot55)}{1611^2}$$
$$= 0\cdot000\,036.$$

An approximate 95 per cent confidence interval for S is

$$0\cdot779 \pm 1\cdot96\,(0\cdot006) \quad \text{or} \quad (0\cdot767, 0\cdot791).$$

INDIVIDUAL MEASUREMENTS. Suppose now that the lengths of all the members of the sample are carefully measured and let $\bar{Z}_i$ be the average of all the lengths in the ith length-class. Let z_{ij} be the length of the individual with coded age x_{ij}, and let

$$\bar{Z}_i = \sum_{j=1}^{l_i} z_{ij}/l_i.$$

Then if the regression of x_{ij} on z_{ij} is approximately linear within the ith class, a simple regression estimate of X_i is available, namely (Cochran [1963: Chapter 7])

$$\bar{X}_i'' = \bar{x}_i + b_i(\bar{Z}_i - \bar{z}_i),$$

where $b_i = \sum_{j=1}^{l_i} (x_{ij} - \bar{x}_i)(z_{ij} - \bar{z}_i) \Big/ \sum_{j=1}^{l_i} (z_{ij} - \bar{z}_i)^2$

is the usual estimate of slope. Thus X can now be estimated by (cf. (10.21))

$$X'_{\text{reg}} = \sum_{i=1}^{l} L_i \bar{X}_i''$$

and the variance of this estimate is again estimated by (10.22) but with

$$\hat{\sigma}_i^2 = \frac{1}{l_i - 2} \left\{ \sum_j (x_{ij} - \bar{x}_i)^2 - \frac{[\sum_j (x_{ij} - \bar{x}_i)(z_{ij} - \bar{z}_i)]^2}{\sum_j (z_{ij} - \bar{z}_i)^2} \right\}.$$

OPTIMUM ALLOCATION OF EFFORT. Let c_A and c_L be the costs of making an age determination and a length reading, respectively, and suppose that c is the total finance available for the experiment, i.e.

$$c = nc_L + n'c_A,$$

where $n' = \sum_{i=1}^{l} l_i$. If proportional subsampling is used, so that $l_i/L_i = f$, then

$$c = n(c_L + fc_A) \tag{10.23}$$

and one would choose n and f subject to (10.23), which minimise $V[\hat{S}_L]$ of (10.20). To do this, Robson and Chapman put the two components of variance in (10.20) on a per-unit basis, so that

$$V[\hat{S}_L] = \frac{V_1}{n} + \frac{V_2}{n'}(1-f) = \frac{1}{n}\left(V_1 - V_2 + \frac{V_2}{f}\right), \tag{10.24}$$

where $V_1 = S(1-S)^2$
and

$$V_2 = n' \frac{(1-S)^4}{(n-1)^2} \sum_{i=1}^{l} \frac{L_i^2 \sigma_i^2}{l_i},$$

which is approximately constant as $n'/(n-1) \approx l_i/L_i$ and $L_i/(n-1)$ ($\approx L_i/\sum L_i$) is approximately equal to the population proportion in the ith length-class when the sampling is random. Thus, minimising (10.24) subject to (10.23) leads to

$$f_{\text{opt}} = \left(\frac{V_2}{V_1 - V_2} \cdot \frac{c_L}{c_A}\right)^{\frac{1}{2}};$$

if $V_2 \geqslant V_1$, then 100 per cent subsampling is required and $f_{opt} = 1$. Here σ_i^2 (the ith class variance), V_1 and V_2 can be estimated from a pilot sample or from previous experiments, using either of the two methods given above (provided that proportional subsampling is used). For example, using the data of Example 10.7, we have $f \approx \frac{1}{2}$,

$$\hat{V}_1 = (0{\cdot}7785)(1-0{\cdot}7785)^2 = 0{\cdot}038\ 159,$$

$$\hat{V}_2 = \frac{790\,(1-0{\cdot}7785)^4}{1611^2} \sum_i \hat{\sigma}_i^2 \frac{L_i^2}{l_i} = 0{\cdot}019\ 244,$$

and hence

$$f_{opt} = 1{\cdot}0077\,(c_L/c_A)^{\frac{1}{2}}.$$

Thus for $c_A/c_L = 10$, the optimum sampling ratio for future experiments is one age-reading to every three length-readings.

10.3.2 Regression model

If the probability p of catching an individual is the same for each individual irrespective of age, then

$$E[n_x | N_x] = pN_x. \qquad (10.25)$$

Also, if S is now interpreted as the probability of survival, we have that

$$E[N_x] = N_0 S^x. \qquad (10.26)$$

Combining (10.25) and (10.26), and taking logarithms, leads to the linear regression model

$$E[y_x] \approx \log\,(pN_0) + x\log S, \qquad (10.27)$$

where $y_x = \log n_x$ (logarithms to the base 10 can also be used). In fishery research the plot of y_x versus x is called a "catch curve", and for a full discussion on the interpretation of such curves the reader is referred to Ricker [1958: Chapter 2]; further examples of catch curves are given in Kennedy [1954] and Tester [1955].

Assuming the y_x to be independent with constant variance, the usual least-squares estimate of $\log S$ (and hence of S) can be obtained. In the case of fisheries there is some empirical evidence to suggest that for haul data, the log transformation stabilizes the variance (Jones [1956], Winsor and Clark [1940]).

If the n_x are assumed to be Poisson random variables, Chapman and Robson [1960: 366] suggest a modification of (10.27) to allow for bias, namely $y_x = \log n_x - [1/(n_x+1)]$ and recommend that the sample be truncated on the right for values of n_x less than 5. They also show that under certain conditions the regression method can still be used when S and N_0 are *independent* random variables, as follows.

Let $S_1, S_2, \ldots, S_x$ be the x realisations of the random variable S associated with N_x, then

$$E[N_x | N_0, S_1, \ldots, S_x] = N_0\,S_1\,S_2 \ldots S_x. \qquad (10.28)$$

Combining (10.28) and (10.25), and assuming the variances of the n_x, N_0 and S to be small compared with their means, it transpires that (10.27) now becomes

$$E[\log n_x] \approx \log (pE[N_0]) + x \log (E[S]),$$

so that $E[S]$ can be estimated by least squares.

The above regression methods are particularly useful when the age distribution in the population is geometric but either the sampling is not random — when for example there is age segregation — or length data are used to obtain n_x (cf. Example 10.9). However, in any circumstances a graph is always useful as a check on the underlying assumptions; in particular the threshold age A above which S is approximately constant can usually be determined empirically from the graph.

Example 10.8   Herring: Tester [1955]

The age data in Table 10.14 are from Tester [1955: Table 5]. Here the graph of $\log n_x$ against x (Fig. 10.3) indicates that the threshold age is IV, so that S appears to be approximately constant for fish of age IV or older.

TABLE 10.14

Age-composition data for a herring population: from Tester [1955: Table 5].

Age (x)	I	II	III	IV	V	VI	VII	VIII	IX
n_x	34	608	6141	2607	497	91	17	4	1
$\log_{10} n_x$	1·53	2·78	3·79	3·42	2·70	1·96	1·23	0·60	0·00
Coded age				0	1	2	3	4	5

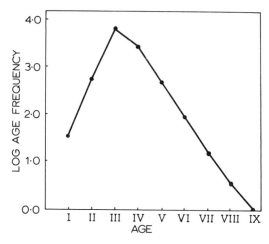

Fig. 10.3   Catch curve for a sample of herring: redrawn from Tester [1955].

TABLE 10.15
Age–length data for a population of lemon soles: from Ricker [1958: 79].

Length-class (i)	L_i	l_i	Age-class IV	V	VI	VII	VIII	IX
1	6	6	5 (5·0)	1 (1·0)	—	—	—	—
2	9	9	3 (3·0)	4 (4·0)	2 (2·0)	—	—	—
3	30	10	4 (12·0)	4 (12·0)	1 (3·0)	1 (3·0)	—	—
4	51	10	1 (5·1)	5 (25·5)	4 (20·4)	—	—	—
5	54	10	—	8 (43·2)	2 (10·8)	—	—	—
6	48	10	1 (4·8)	7 (33·6)	1 (4·8)	1 (4·8)	—	—
7	41	10	1 (4·1)	3 (12·3)	3 (12·3)	2 (8·2)	1 (4·1)	—
8	27	10	—	2 (5·4)	6 (16·2)	1 (2·7)	1 (2·7)	—
9	13	10	—	1 (1·3)	4 (5·2)	3 (3·9)	—	2 (2·6)
10	6	6	—	—	1 (1·0)	3 (3·0)	2 (2·0)	—
11	3	3	—	—	1 (1·0)	1 (1·0)	1 (1·0)	—
12	1	1	—	—	—	—	1 (1·0)	—
Total ($\hat{n}_x$) =			(34·0)	(138·3)	(76·7)	(26·6)	(10·8)	(2·6)
$\log_{10} \hat{n}_x$ =			1·53	2·14	1·88	1·42	1·03	0·41
x =			0	1	2	3	4	5

Example 10.9   Lemon sole: Ricker [1958]

Table 10.15 is taken from Ricker [1958: 79] and the numbers in brackets for the ith length-class, say, are simply the observed age frequencies in the subsample from the ith class multiplied by L_i/l_i. Summing over each length-class leads to the estimates $\hat{n}_x$, and a plot of log $\hat{n}_x$ versus x is given in Fig. 10.4. From this graph we see that the regression method can be used for the four age-classes V–VIII inclusive; the last age-class is not used as $\hat{n}_5 < 5$.

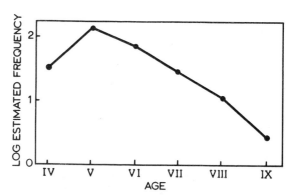

Fig. 10.4   Catch curve for a sample of lemon soles using subsampling for age determinations: data from Ricker [1958].

10.4　TIME-SPECIFIC DATA: AGE-DEPENDENT SURVIVAL

Suppose that S is no longer constant with respect to age, and let S_x be the proportion of the group of age x which survives to age $x + 1$ $(x = 1, 2, \ldots,$ $K - 1)$. Assuming stationarity, N_x will remain constant from year to year, so that

$$N_{x+1} = S_x N_x = S_0 S_1 \ldots S_x N_0 \quad (x = 0, 1, \ldots, K-1)$$

and

$$\sum_{x=0}^{K} N_x = (1 + S_0 + S_0 S_1 + \ldots + S_0 S_1 S_2 \ldots S_{K-1}) N_0 .$$

If K is chosen in advance, then, by considering $N_x \big/ \sum_{x=0}^{K} N_x$, the probability function for x, the age of an individual selected at random (conditional on $0 \leqslant x < K$), is (Chapman and Robson [1960])

$$f(x) = \frac{S_0 S_1 \ldots S_{x-1}}{(1 + S_0 + S_0 S_1 + \ldots + S_0 S_1 S_2 \ldots S_{K-1})} \quad (x = 0, 1, \ldots, K),$$

$$= \theta_x, \text{ say.}$$

Therefore if a random sample is taken and n_x are found to be of age x, then the joint probability function of $n_0, n_1, \ldots, n_K$, given n $\left(= \sum_{x=0}^{K} n_x \right)$, is

$$f(\{n_x\} \mid n) = \frac{n!}{\prod_{x=0}^{K} n_x!} \prod_{x=0}^{K} \theta_x^{n_x}.$$

It is readily shown that the maximum-likelihood estimate of S_x is the moment estimate

$$\hat{S}_x = n_{x+1}/n_x, \quad (x = 0, 1, \ldots, K-1),$$

and using the delta method we find that asymptotically

$$V[\hat{S}_x] = \frac{S_x^2}{n} \left[\frac{1}{\theta_{x+1}} + \frac{1}{\theta_x} \right],$$

$$\text{cov}[\hat{S}_x, \hat{S}_{x+1}] = -S_x S_{x+1}/(n\theta_{x+1}),$$

and

$$\text{cov}[\hat{S}_x, \hat{S}_y] = 0, \quad (y > x+1).$$

Chapman and Robson suggest a slight modification of the maximum-likelihood estimate, namely

$$S_x^* = n_{x+1}/(n_x + 1),$$

which is almost unbiased since

$$E[S_x^*] = S_x[1 - (1 - \theta_x)^n].$$

Thus an almost unbiased estimate of the average survival rate

$$S_{\text{ave}} = (S_0 + S_1 + \ldots + S_{K-1})/K$$

is

$$S_{\text{ave}}^* = (S_0^* + S_1^* + \ldots + S_{K-1}^*)/K$$

with asymptotic variance

$$V[S_{ave}^*] = \frac{1}{K^2}\left\{\sum_{x=0}^{K-1} V[\hat{S}_x] + 2\sum_{x=0}^{K-2} cov[\hat{S}_x, \hat{S}_{x+1}]\right\}.$$

To estimate this variance we replace each S_x by S_x^* and each θ_x by n_x/n. Alternatively we can use the estimate given by (1.10).

One other estimate of interest is

$$\hat{S}_{geo} = (\hat{S}_0 \hat{S}_1 \dots \hat{S}_{K-1})^{1/K} = (n_k/n_0)^{1/K},$$

an estimate of the geometric mean. If Z_x is the annual instantaneous mortality rate, so that $S_x = \exp(-Z_x)$, then $\sum_x Z_x/K$, the average instantaneous mortality rate, can be estimated by $-\log \hat{S}_{geo}$. However, $\hat{S}_{geo}$ depends only on the two extreme age-classes, so that it will be subject to greater fluctuations than S_{ave}^*.

To test the hypothesis $H : S_x = S$ ($x = 0, 1, \dots, K-1$) we can calculate a standard multinomial goodness-of-fit statistic

$$T = \sum_{x=0}^{K} (n_x - n\hat{\theta}_x)^2/(n\hat{\theta}_x)$$

which is asymptotically χ_{K-1}^2 when H is true. Here

$$\hat{\theta}_x = \hat{S}_{seg}^x \Big/ \sum_{x=0}^{K} \hat{S}_{seg}^x$$

$$= \hat{S}_{seg}^x (1 - \hat{S}_{seg})/(1 - \hat{S}_{seg}^{K+1})$$

$$= \frac{1}{K+1}\left[1 - \frac{X(1-\hat{S}_{seg})}{n\hat{S}_{seg}}\right],$$

where $\hat{S}_{seg}$ is given by equation (10.19).

In conclusion we note that estimates of the form n_{x+1}/n_x have sometimes been used in analysing age data from bird-banding experiments. However, as pointed out in **5.4.3** (with $D_x = n_x$), such estimates are only valid when survival is independent of age, i.e. when $S_x = S$.

CHAPTER 11

POPULATIONS STRATIFIED GEOGRAPHICALLY

11.1 CLOSED POPULATION

11.1.1 The Petersen method

Very often a population is stratified geographically, so that the total population may be regarded as consisting of separate strata living in different areas. The experimenter may then be interested not only in the sizes of the separate strata but also in the degree of mixing between strata. One method of estimating the various population parameters is to use a generalisation of the Petersen method in which marked animals are released in each stratum, using a different mark for each stratum. After allowing for the dispersal of the marked, a sample is then taken from each stratum and the recaptures recorded. This model has been investigated by Schaefer [1951: population stratified temporally rather than spatially], Chapman and Junge [1956: moment estimation], Beverton and Holt [1957: deterministic approach with a mark release in just one stratum], Darroch [1961: extensive maximum-likelihood theory], and Overton and Davis [1969: 437–41]. The theory in this chapter is based on Darroch's article.

Let

s = number of strata,
A = total number of marked animals released,
a_i = number of marked released in stratum i,
U = total number in the unmarked population,
$\overset{*}{U}_i$ = number of unmarked in stratum i at the time of the mark release,
U_j = number of unmarked in stratum j at the time of the sampling,
N = $U + A$ (total population size),
n_j = size of sample from stratum j,
m_{ij} = marked members of n_j which were released in stratum i,
u_j = number of unmarked in n_j,
θ_{ij} = probability that a member of a_i moves to stratum j, and
ψ_{ij} = probability that a member of a_i is caught in stratum j.

We shall assume that:

(i) The population is closed, i.e. there is no mortality, so that

$$\sum_j \theta_{ij} = 1 \quad (i = 1, 2, \ldots, s).$$

(ii) All individuals in the jth stratum, whether marked or unmarked, have the same probability p_j of being caught in the sample.

(iii) Marked individuals behave independently of one another in regard to moving between strata and being caught.

(iv) $\psi_{ij} = \theta_{ij} p_j$.

(v) The matrix $\hat{\Theta} = [(\theta_{ij})]$ is non-singular: this implies that the transformation in (iv) subject to (i) is one-to-one, so that maximum-likelihood estimates of the θ_{ij} and p_j can be obtained from estimates of the ψ_{ij}.

MAXIMUM-LIKELIHOOD ESTIMATES. Given the above assumptions, the joint probability function of the random variables $\{m_{ij}\}$, $\{u_j\}$ is the product of

$$f(\{m_{ij}\}|\{a_i\}) = \prod_{i=1}^{s} \left\{ \frac{a_i!}{(a_i - m_{i.})! \prod_{j=1}^{s} m_{ij}!} (1 - \sum_j \theta_{ij} p_j)^{a_i - m_{i.}} \cdot \prod_{j=1}^{s} (\theta_{ij} p_j)^{m_{ij}} \right\}$$

(11.1)

and

$$f(\{u_j\}) = \prod_{i=1}^{s} \left\{ \binom{U_j}{u_j} p_j^{u_j}(1 - p_j)^{U_j - u_j} \right\},$$

(11.2)

where $m_{i.} = \sum_j m_{ij}$. As the $\{m_{ij}\}$ and $\{u_j\}$ are independent, maximum-likelihood estimation for the joint distribution amounts to obtaining maximum-likelihood estimates $\hat{\theta}_{ij}$, $\hat{p}_j$, from (11.1) and then using (11.2) to write down the moment equations $\hat{U}_j \hat{p}_j = u_j$. Therefore, if Θ is non-singular, we find that

$$\hat{\theta}_{ij}\hat{p}_j = \hat{\psi}_{ij} = m_{ij}/a_i, \quad (i, j = 1, 2, \ldots, s),$$

(11.3)

$$\sum_{j=1}^{s} \hat{\theta}_{ij} = 1, \quad (i = 1, 2, \ldots, s),$$

(11.4)

and

$$\hat{U}_j = u_j/\hat{p}_j, \quad (j = 1, 2, \ldots, s).$$

(11.5)

Cross-multiplying in (11.3), these equations can be written in matrix form, namely

$$D_a \hat{\Theta} \hat{D}_p = [(m_{ij})] = M,$$

$$\Theta 1 = 1,$$

and

$$\hat{U} = D_u \hat{p}.$$

(11.6)

Here 1 is a column vector of ones; $U = [(U_j)]$, $a = [(a_i)]$, $\rho = [(p_j^{-1})]$ are column vectors; $D_a = [(\delta_{ij} a_j)]$, $D_u = [(\delta_{ij} u_j)]$ and $\hat{D}_p = [(\delta_{ij} \hat{p}_j)]$ are diagonal matrices. If Θ is non-singular and the sample sizes n_j large, a singular matrix M will be extremely unlikely. Therefore, assuming M to be non-singular,

$$\hat{D}_p^{-1} = M^{-1} D_a \hat{\Theta}$$

and

$$\begin{aligned} \hat{\rho} &= \hat{D}_p^{-1} 1 \\ &= M^{-1} D_a \hat{\Theta} 1 \\ &= M^{-1} D_a 1 \\ &= M^{-1} a. \end{aligned}$$

(11.7)

This means that the $\hat{p}_i$ are given by

$$
\begin{bmatrix} \hat{p}_1^{-1} \\ \hat{p}_2^{-1} \\ \cdots \\ \hat{p}_s^{-1} \end{bmatrix} = \begin{bmatrix} m_{11} & m_{12} & \cdots & m_{1s} \\ m_{21} & m_{22} & \cdots & m_{2s} \\ \cdots & \cdots & \cdots & \cdots \\ m_{s1} & m_{s2} & \cdots & m_{ss} \end{bmatrix}^{-1} \begin{bmatrix} a_1 \\ a_2 \\ \cdots \\ a_s \end{bmatrix}
$$

and the $\hat{\theta}_{ij}$, $\hat{U}_j$ then follow from equations (11.3) and (11.5), namely

$$\hat{\theta}_{ij} = a_i^{-1} m_{ij} \hat{p}_j^{-1}, \quad \hat{U}_j = u_j \hat{p}_j^{-1}$$

or, in matrix notation,

$$\hat{\Theta} = D_a^{-1} M \hat{D}_\rho \tag{11.8}$$

and

$$\hat{U} = D_u \hat{\rho} = D_u M^{-1} a. \tag{11.9}$$

Estimates of U and N are also available; thus

$$\hat{U} = 1'\hat{U} = 1'D_u M^{-1} a = u'M^{-1}a \quad (= u'\hat{\rho}) \tag{11.10}$$

and

$$
\begin{aligned}
\hat{N} &= \hat{U} + A \\
&= u'M^{-1}a + 1'a \\
&= (u'M^{-1} + 1'MM^{-1})a \\
&= n'M^{-1}a,
\end{aligned}
$$

where $n = [(n_j)]$. We note that the estimate $\hat{N}$ was given by Chapman and Junge [1956: their $\hat{N}_3$] and has been applied, for example, to the Alaska fur-seal herd (Kenyon *et al.* [1954]).

VARIANCES AND COVARIANCES. Darroch proves that the above estimates are consistent (for Θ non-singular) and approximately unbiased (for large a_i). Also $\hat{\rho}$ has negligible bias so that its variance–covariance matrix is approximately

$$E[(\hat{\rho} - \rho)(\hat{\rho} - \rho)'] = \Sigma, \quad \text{say.}$$

Using the delta method, Darroch shows that

$$\Sigma \approx D_\rho \Theta^{-1} D_\mu D_a^{-1} \Theta'^{-1} D_\rho$$

and

$$E[(\hat{U} - U)(\hat{U} - U)'] \approx D_U \Theta^{-1} D_\mu D_a^{-1} \Theta'^{-1} D_U + D_U(D_\rho - I),$$

where I is the unit matrix, and the diagonal elements of D_μ are

$$\mu_i = (\sum_j \theta_{ij}/p_j) - 1.$$

Hence

$$
\begin{aligned}
E[(\hat{N} - N)^2] &= E[(\hat{U} - U)^2] \\
&= 1'E[(\hat{U} - U)(\hat{U} - U)']1 \\
&\approx U'\Theta^{-1} D_\mu D_a^{-1} \Theta'^{-1} U + U'(\rho - 1) \\
&= \sum_i \eta_i^2 \mu_i/a_i + \sum_j U_j(p_j^{-1} - 1),
\end{aligned}
$$

where
$$\boldsymbol{\eta}' = [(\eta_i)]' = \mathbf{U}'\boldsymbol{\Theta}^{-1}. \tag{11.11}$$

The vector $\boldsymbol{\eta}$ has a simple interpretation when the movement pattern for marked and unmarked is the same. Assuming that θ_{ij} applies also to the unmarked individuals, we have the approximate deterministic equations

$$U_j \approx \sum_i \overset{*}{U_i}\theta_{ij}. \tag{11.12}$$

Writing these in matrix form and post-multiplying by $\boldsymbol{\Theta}^{-1}$, we have

$$\overset{*}{\mathbf{U}}' \approx \mathbf{U}'\boldsymbol{\Theta}^{-1}$$

which, together with (11.11), implies that $\eta_i \approx \overset{*}{U_i}$, the number of unmarked in stratum i at the time of the mark release.

11.1.2 Validity of assumptions

Although assumption (i) that there is no natural mortality cannot be tested for the above model, the theory is readily generalised in **11.2**, below, to deal with the case $\sum_j \theta_{ij} < 1$. Assumption (ii) of constant p_j will be satisfied if marking and point of release do not affect catchability and if the stratum area is sufficiently small, so that either the sampling effort is uniform over it and every member has an equal chance of capture, or there is uniform mixing of marked and unmarked, regardless of sampling uniformity (cf. **3.1.2**).

In the definition of ψ_{ij} and the derivation of (11.1) it was assumed that the a_i individuals released in the ith stratum move and are caught (or not caught) (1) independently of those released in any other stratum, and (2) independently of each other. (1) is very reasonable, but (2) is less likely to hold true in practice. For example, if the marked individuals are released close together at some random point in the stratum, they may tend to move to the same stratum (e.g. schooling in fish). On the other hand, if they are released over a carefully spaced grid of points so that they start with maximum possible distances between them, they may tend to move in different strata.

As well as "contagious" movement there is the possibility of "contagious" catching. For, example, if sampling in the jth stratum is not uniform but is concentrated in one or more subareas and if, having a common stratum of origin, the marked individuals from the ith stratum are not uniformly distributed in the jth stratum, there will be a certain amount of positive dependence on their being caught or not caught. However, Darroch [1961: 250–2] shows that the estimates are still consistent when there is both contagious movement and catching and that contagious movement has little effect on the variance–covariance matrix of $\hat{\boldsymbol{\rho}}$. Positive dependence in catchability will tend to increase the variances, though this effect will also be small if every effort is made to ensure that both the release and the sampling procedures are as nearly random as possible.

If Θ is singular, the above theory is no longer valid. Instead of working through the parameters $\{\psi_{ij}\}$ we must now work with the $\{\theta_{ij}\}$ and $\{p_j\}$ directly. Unfortunately this direct approach of maximising the likelihood subject to the constraints $\Theta 1 = 1$ does not yield simple solutions of the resulting maximum-likelihood equations, so that a singular Θ should be avoided. In fact, the further Θ is from singularity the more reliable are the estimates. This is reflected in the variance formulae which depend on Θ^{-1}; the elements of Θ^{-1} will be large if the determinant of Θ is near zero. However, when (11.12) is true, $E[(\hat{U}-U)^2]$ will be insensitive to near singularity, for then η_i, although defined in terms of Θ^{-1}, is approximately equal to $\overset{*}{U}_i$ and is therefore virtually independent of Θ^{-1}. Even if (11.12) is not true, the η_i will tend to be positive and therefore not too large, as $\Sigma \eta_i = U$. (This last equation follows from the sequence $\Sigma \eta_i = \eta'1 = \eta'\Theta 1 = U'1 = U$.)

Although it is impossible to predetermine the strata so that Θ is guaranteed to be non-singular, one or two trivial cases may be foreseeable. For example, if it is suspected that $\theta_{hj} = \theta_{ij}$ for all j, then the hth and ith strata can be combined. However, any near singularity in Θ will generally be reflected in the near singularity of M, and M can be used to decide which strata to pool.

Finally it is noted that the estimate $\hat{U}$ does not depend on the assumption $\psi_{ij} = \theta_{ij}p_j$ when ψ_{ij} is the same for both marked and unmarked, i.e. when marked and unmarked have the same behaviour patterns. To see this we consider the moment equations

$$\sum_i \overset{*}{U}_i \psi_{ij} = u_j \quad \text{and} \quad a_i \psi_{ij} = m_{ij},$$

which lead to

$$\sum_i \overset{*}{U}_i m_{ij}/a_i = u_j \tag{11.13}$$

or

$$\overset{*}{U}'D_a^{-1}M = u'. \tag{11.14}$$

Thus a moment estimate of U is

$$\overset{*}{U}'1 = u'M^{-1}D_a 1$$

$$= u'M^{-1}a$$

which, from (11.10), is simply $\hat{U}$ once again. It is also readily shown from Darroch that $E[(\hat{U}-U)^2]$ can be written in terms of the ψ_{ij} and p_j only.

11.1.3   Variable number of strata

In some situations the number of strata may change between the times of the mark release and of sampling. Suppose that the marked individuals are released into s strata, and sampling is carried out in t strata. Then (11.1) and (11.2) are still valid, provided we define $i = 1, 2, \ldots, s$ and $j = 1, 2, \ldots, t$.

When $s < t$, the parameters $\{\theta_{ij}\}$, $\{p_j\}$ are no longer identifiable and the U_j cannot be estimated. But an estimate of U is still available if ψ_{ij} is the same for marked and unmarked. In this case, with $j = 1, 2, \ldots, t$ we have, from (11.13), t equations for the s unknowns $\{\overset{*}{U}_i\}$. We can then replace these t equations by s linear combinations of them, giving, from (11.14),

$$\overset{*}{\mathbf{U}}'\mathbf{D}_a^{-1}\mathbf{MQ} = \mathbf{u}'\mathbf{Q}$$

where $\mathbf{Q}$ is a t-by-s matrix of rank s such that $\mathbf{MQ}$ is non-singular. Thus our estimate of U, namely

$$\begin{aligned}
\overset{*}{\mathbf{U}}'\mathbf{1} &= (\mathbf{Q}'\mathbf{u})'[\mathbf{MQ}]^{-1}\mathbf{D}_a\mathbf{1} \\
&= (\mathbf{Q}'\mathbf{u})'[\mathbf{MQ}]^{-1}\mathbf{a}
\end{aligned}$$

takes the same form as $\hat{U}$ (cf. (11.10)). Its asymptotic variance can be found, using Darroch's technique [1961: 83], by writing $\hat{\mathbf{p}} = [\mathbf{MQ}]^{-1}\mathbf{a}$.

When $s > t$ the maximum-likelihood estimates of the $\{\theta_{ij}\}$, $\{p_j\}$ are not readily found. However, we can once again obtain a moment-type estimate of U using a method similar to one given by Chapman and Junge [1956]. Consider the moment equations

$$U_j p_j = u_j \quad \text{and} \quad m_{ij}/p_j = a_{ij},$$

where a_{ij} is the number from the group of a_i released in the ith stratum which move to stratum j. Then, substituting for p_j in the second equation and summing on j gives us s equations

$$\sum_j m_{ij}U_j/u_j = a_i, \quad (i = 1, 2, \ldots, s),$$

or

$$\mathbf{MD}_u^{-1}\mathbf{U} = \mathbf{a} \tag{11.15}$$

in the t unknowns $\{U_j\}$. Once again when $s = t$, (11.15) leads to $\hat{U}$, while for $s > t$ we look for a t-by-s matrix $\mathbf{R}$ such that $\mathbf{RM}$ is non-singular. Then $\mathbf{U}$ is estimated by $\mathbf{D}_u(\mathbf{RM})^{-1}\mathbf{Ra}$ and U by $\mathbf{u}'(\mathbf{RM})^{-1}\mathbf{Ra}$; expressions for asymptotic variances and covariances can be obtained using the methods of Darroch. In conclusion we mention that more general methods for dealing with the case $s > t$ (which also allow for mortality) are given in **11.2** below.

11.1.4   The Petersen estimate using pooled data

CONSISTENCY. If the experimenter does not know how the population is stratified, or he is unable to use different marks for different strata, he will pool the data and use the Petersen estimate of N, namely

$$\hat{N}_P = An/m,$$

where $u = \sum_{j=1}^{t} u_j$, $m = \sum_{i=1}^{s}\sum_{j=1}^{t} m_{ij}$ and $n = u + m$. In discussing the properties of this estimate it is convenient to consider

$$\hat{U}_P = \hat{N}_P - A = Au/m,$$

the Petersen estimate of the unmarked population size. The first property we consider is that of consistency. Letting $a_i \to \infty$, $U_j \to \infty$ in such a way that a_i/A, U_j/U and A/U are all constant, we find that $\hat{U}_P$ is a consistent estimate of

$$U_P = AE[u]/E[m]$$
$$= AB/C, \quad \text{say},$$
$$= \frac{A \, \Sigma \, U_j p_j}{U \, \Sigma\Sigma \, a_i \theta_{ij} p_j} \cdot U. \tag{11.16}$$

Although the equation $U_P = U$ can be satisfied in an infinite number of ways, only the following four special cases are listed because of their simple physical interpretations:

(1) $p_j = p$ for $j = 1, 2, \ldots, t$.
(2) $a_i/\overset{*}{U}_i = A/U$ $(i = 1, 2, \ldots, s)$, \tag{11.17}

and the movement pattern for marked and unmarked is the same, i.e.

$$U_j = \sum_i \overset{*}{U}_i \theta_{ij}. \tag{11.18}$$

(3) Equation (11.18) holds and $\theta_{ij} = \theta_j$ for all i, j.
(4) $\sum_i a_i \theta_{ij} = k U_j$, $(j = 1, 2, \ldots, t)$.

Condition (1) implies that a constant proportion of each stratum is sampled, while (11.17) implies that a constant proportion of each stratum is marked. (11.17) is the marking analogue of (1) and corresponds to a constant probability of capture over the whole population when the animals are initially caught for marking. If (3) is true, we have a complete mixing of the whole population and, when $s = t$, Θ is singular. Condition (4) is equivalent to saying that the expected number of marked in the jth stratum is proportional to the number of unmarked; when $\sum_j \theta_{ij} = 1$, $k = A/U$. We note that conditions (1)–(3) were first considered by Chapman and Junge [1956].

EFFICIENCY. Using the delta method it can be shown that

$$E[(\hat{U}_P - U_P)^2] \approx (A^2 B^2/C^4) V[m] + (A^2/C^2) V[u].$$

Now from the inequality $(\Sigma \, x_i y_i)^2 \leqslant (\Sigma \, x_i^2)(\Sigma \, y_i^2)$ with $x_i = \sqrt{a_i}$, $y_i = \sqrt{a_i} \sum_j \theta_{ij} p_j$ we find that (cf. (11.1))

$$V[m] = \sum_i V[m_{i.}]$$
$$= \sum_i a_i (\sum_j \theta_{ij} p_j)(1 - \sum_j \theta_{ij} p_j)$$
$$\leqslant (\sum_i \sum_j a_i \theta_{ij} p_j)(1 - \sum_i \sum_j a_i \theta_{ij} p_j/A)$$
$$= A(C/A)(1 - C/A)$$
$$= \text{var}[m], \quad \text{say}.$$

Similarly,
$$V[u] = \sum_j V[u_j]$$
$$= \sum_j U_j p_j(1 - p_j)$$

$$\leqslant \; (\sum_j U_j p_j)(1 - \sum_j U_j p_j/U)$$
$$= \; U(B/U)(1 - B/U)$$
$$= \; \mathrm{var}\,[u], \quad \text{say.}$$

Therefore, if the experimenter uses the Petersen estimate (or its approximately unbiased version $(A+1)u/(m+1)$, cf. (5.21)) and treats u and m as binomial variables, his expression for the mean-square error, namely

$$(A^2 B^2/C^4)\,\mathrm{var}\,[m] + (A^2/C^2)\,\mathrm{var}\,[u]$$

will be conservative in that it will slightly overestimate the true mean-square error $E[(\hat{U}_P - U_P)^2]$.

When $s = t$, and $\hat{U}_P$ and the maximum-likelihood estimate $\hat{U}$ are both consistent, Darroch shows that asymptotically

$$E[(\hat{U}_P - U)^2] \; = \; Y \leqslant Z \; = \; E[(\hat{U} - U)^2], \tag{11.19}$$

with equality when $p_i = p$ and $\eta_i/a_i = U/A$ for $i = 1, 2, \dots, s$ (see (11.11) for definition of η_i). These two conditions for equality are satisfied if, for example, (1) and (2) above are true.

TESTS FOR CONSISTENCY. A goodness-of-fit test for the hypothesis $H_1 : \theta_{ij} = \theta_j$ is equivalent to a test of homogeneity for the rows of the s-by-$(t+1)$ contingency table:

m_{11}	m_{12}	$\dots$	m_{1t}	$a_1 - m_{1.}$	a_1
m_{21}	m_{22}	$\dots$	m_{2t}	$a_2 - m_{2.}$	a_2
$\dots$	$\dots$	$\dots$	$\dots$	$\dots$	$\dots$
m_{s1}	m_{s2}	$\dots$	m_{st}	$a_s - m_{s.}$	a_s
$m_{.1}$	$m_{.2}$	$\dots$	$m_{.t}$	$A - m$	A

A "pooled" version of this test given by both Chapman and Junge, and Darroch, is a test of homogeneity on the columns of the 2-by-s contingency table:

$m_{1.}$	$m_{2.}$	$\dots$	$m_{s.}$	m
$a_1 - m_{1.}$	$a_2 - m_{2.}$	$\dots$	$a_s - m_{s.}$	$A - m$
a_1	a_2	$\dots$	a_s	A

using

$$T \; = \; \frac{\sum\limits_i (m_{i.} - a_i m/A)^2}{a_i \dfrac{m}{A}\left(1 - \dfrac{m}{A}\right)}.$$

Here T is asymptotically distributed as chi-squared with $s - 1$ degrees of freedom when the hypothesis $H_2 : \sum_j \theta_{ij} p_j = d$ is true; in this case $\hat{U}_P$ is consistent when (11.18) is also true.

Two special cases of H_2 are H_1 and $H_3 : p_j = p$ (all j). When T is not significant, one can test H_1 against H_2 using the s-by-t contingency table obtained from the s-by-$(t+1)$ table above by deleting the last column. Unfortunately no straightforward test of H_3 against H_2 is available.

$H_4 : \sum_i a_i \theta_{ij} = k U_j$ can be tested by a test of homogeneity on the columns of the 2-by-t contingency table:

$m_{.1}$	$m_{.2}$	$\cdots$	$m_{.t}$	m
u_1	u_2	$\cdots$	u_t	u
n_1	n_2	$\cdots$	n_t	n

In conclusion we mention briefly another estimate of N due to Schaefer [1951], namely

$$\hat{N}_S = \sum_{i=1}^{s} \sum_{j=1}^{t} \frac{n_j a_i m_{ij}}{m_{.i} m_{.j}}.$$

We find, as first pointed out by Chapman and Junge, that $\hat{N}_S$ is consistent when either of the above conditions (1) or (2) is satisfied. The asymptotic variance of $\hat{N}_S$, which would require considerable computation, does not seem to be given in the literature.

11.2   OPEN POPULATION: NATURAL MORTALITY

CASE I: $s = t$. If natural mortality is taking place between the times of marking and sampling, $\sum_j \theta_{ij}$ ($= \phi_i$ say) will be the probability of survival for members of a_i. In this case (11.1) is still valid but $\{\theta_{ij}\}$, $\{p_j\}$ are non-identifiable, as $\{\lambda \theta_{ij}\}$, $\{\lambda^{-1} p_j\}$ also lead to the same probability function (11.1). To tie down this non-identifiability, Darroch assumes that $\phi_i = \phi$ ($i = 1, 2, \ldots, s$) and defines $\beta_{ij} = \theta_{ij}/\phi$ and $P_j = \phi p_j$. Here β_{ij} is the probability that a member of a_i is in stratum j at the time of sampling, given that it is alive at that time, and P_j is the probability of an animal surviving and, if it is in the jth stratum, of being caught there.

This re-parametrisation leads to the same distribution (11.1) but with θ_{ij} and p_j replaced by β_{ij} and P_j. Therefore, when $\mathbf{B} = [(\beta_{ij})]$ is non-singular, the maximum-likelihood estimates of β_{ij} and P_j will satisfy (cf. (11.3) and (11.4))

$$\hat{\beta}_{ij} \hat{P}_j = m_{ij}/a_i \quad (= \hat{\psi}_{ij})$$

and

$$\sum_j \hat{\beta}_{ij} = 1, \quad (i = 1, 2, \ldots, s).$$

From (11.7) and (11.8) these equations have solutions

$$\hat{\mathbf{p}} = \mathbf{M}^{-1} \mathbf{a}$$

and
$$\hat{\mathbf{B}} = \mathbf{D}_a^{-1}\mathbf{M}\hat{\mathbf{D}}_\rho,$$

where $\hat{\rho}_i = 1/\hat{P}_i$. However, although (11.1) is re-parametrised, (11.2) remains unchanged, and we now have the problem of relating P_j to p_j so that U_j can be estimated. Fortunately we can get round this problem if we proceed by analogy with (11.5) and consider

$$\hat{U}_j \hat{p}_j = u_j$$

or, writing $W_j = U_j/\phi$,

$$\hat{W}_j \hat{P}_j = u_j.$$

This means that we can estimate W_j and $W = \Sigma\, W_j$ using the same methods as those used in **11.1.1**; for example, from (11.9) we have

$$\hat{\mathbf{W}} = \mathbf{D}_u\mathbf{M}^{-1}\mathbf{a}$$

and

$$\hat{W} = \mathbf{u}'\mathbf{M}^{-1}\mathbf{a}.$$

We note that $\hat{\mathbf{W}}$ is the same as $\hat{\mathbf{U}}$, though $\hat{W}$ is now an estimate of W (which is approximately the total unmarked population size at the time of the mark release) rather than of U, the final unmarked population size. This fact could have been deduced directly from (11.14) and the following equations where it was shown that, irrespective of any assumptions about the ψ_{ij}, $\hat{\overset{*}{U}}$ is an estimate of $\mathbf{U}'\mathbf{1}$, the initial size of the unmarked population.

As far as variance estimation is concerned, the only changes in the variance–covariance formulae are the addition of non-estimable second-order correction terms in $E[(\hat{\mathbf{W}}-\mathbf{W})(\hat{\mathbf{W}}-\mathbf{W})']$ and $E[(\hat{W}-W)^2]$ involving $1 - \phi^{-1}$. The appropriate formulae can be obtained from equations (11.20) to (11.23) below by dropping out $\mathbf{X}$, and setting $\bar{\phi} = \phi$ and $\gamma_i = 1$. However, when ϕ is not too small these non-estimable corrections will be negligible.

We note that although the estimates $\hat{\psi}_{ij} (= m_{ij}/a_i)$ are less than unity, in transforming from $\hat{\psi}_{ij}$ to $\hat{\beta}_{ij}$ it is possible that $\hat{P}_j$ may be greater than unity. If this anomaly takes place, Darroch points out that as the transformation is made by imposing the constraints $\underset{j}{\Sigma}\,\beta_{ij} = 1$ it is these constraints which are the root of the trouble. He suggests two diagnoses: (i) the ϕ_i are not equal, or (ii) the ϕ_i are equal but have large sampling errors. Here (i) can be allowed for by reducing the number of sample strata and applying the theory of case II below when $s > t$ (see Example 11.1 below). (ii) means that the $\hat{P}_j$ and $\hat{W}_j$ are virtually useless as estimates of P_j and W_j, and this would be confirmed by their having large variances. But this does not mean that the variance of $\hat{W}$ is also large, as the covariances of $\hat{W}_j$ may be large and negative.

Replacing θ_{ij}, p_j and U_j by β_{ij}, P_j and W_j, we find that the general theory of **11.1.4** still holds, though the Petersen estimate $\hat{U}_P$ is now an estimate of W rather than of U. Also the comments in **11.1.2** still apply here.

CASE II: $s > t$. When $s > t$ an alternative approach to that given in

11.1.2 is available. Allowing the ϕ_i to differ and defining $\bar{\phi} = \Sigma\ \phi_i/s$, we now redefine the parameters $\beta_{ij} = \theta_{ij}/\bar{\phi}$, $P_j = \bar{\phi}\,p_j$, $W_j = U_j./\bar{\phi}$ and introduce $\gamma_i = \phi_i/\bar{\phi}$. This means that we have $st - 1 + t$ independent parameters $\{\beta_{ij}\}$, $\{P_j\}$, the subtraction of 1 being due to the single constraint $\Sigma\Sigma\ \beta_{ij} = s$. However, instead of imposing $s - 1$ constraints $\gamma_1 = \gamma_2 = \dots = \gamma_s$, as in the case $s = t$ above, Darroch suggests imposing just $t - 1$ constraints, thus reducing the number of independent parameters to st, the number of $\{\psi_{ij}\}$. The $t - 1$ constraints will usually take the form $\gamma_i - \gamma_j = 0$ or possibly $\gamma_i + \gamma_j - 2\gamma_k = 0$, and they can be written in the general form

$$\sum_{k=1}^{s} x_{jk}\gamma_k = 0, \quad (j = 1, 2, \dots, t-1).$$

To this set we add $\underset{k}{\Sigma}\ \gamma_k = s$ in the form

$$\sum_{k=1}^{s} x_{tk}\gamma_k = 1, \quad (x_{tk} = 1/s).$$

For a suitable choice of constraints the matrix $\mathbf{X} = [(x_{jk})]$ will be a t-by-s matrix of rank t, and

$$\mathbf{XB1} = \mathbf{X\gamma} = \mathbf{v},$$

where $\mathbf{v}'$ is the row vector $(0, 0, \dots, 0, 1)$. If the s-by-t matrices $\mathbf{B}$ and $\mathbf{M}$ are also of rank t, Darroch shows that the maximum-likelihood estimates are now

$$\hat{\mathbf{\rho}} = [\mathbf{X}\mathbf{D}_a^{-1}\mathbf{M}]_a^{-1}\mathbf{v}$$
$$\hat{\mathbf{B}} = \mathbf{D}_a^{-1}\mathbf{M}\hat{\mathbf{D}}_\rho, \quad \hat{\mathbf{\gamma}} = \hat{\mathbf{B}}\mathbf{1}$$
$$\hat{\mathbf{W}} = \mathbf{D}_u[\mathbf{X}\mathbf{D}_a^{-1}\mathbf{M}]^{-1}\mathbf{v}$$

and

$$\hat{W} = \mathbf{v}'[\mathbf{X}\mathbf{D}_a^{-1}\mathbf{M}]^{-1}\mathbf{v}$$

where

$$\hat{\mathbf{\rho}} = [(\hat{P}_j^{-1})].$$

Also if

$$E[(\hat{\mathbf{\rho}} - \mathbf{\rho})(\hat{\mathbf{\rho}} - \mathbf{\rho})'] = \mathbf{\Sigma},$$

Darroch proves that

$$\mathbf{\Sigma} \approx \mathbf{D}_\rho(\mathbf{XB})^{-1}\mathbf{X}\mathbf{D}_\mu\mathbf{D}_a^{-1}\mathbf{X}'(\mathbf{XB})'^{-1}\mathbf{D}_\rho,$$

$$E[(\hat{\mathbf{W}} - \mathbf{W})(\hat{\mathbf{W}} - \mathbf{W})'] \approx \mathbf{D}_W(\mathbf{XB})^{-1}\mathbf{X}\mathbf{D}_\mu\mathbf{D}_a^{-1}\mathbf{X}'(\mathbf{XB})'^{-1}\mathbf{D}_W + \mathbf{D}_W(\mathbf{D}_\rho - \mathbf{I}\bar{\phi}^{-1}), \quad (11.20)$$

and

$$E[(\hat{W} - W)^2] \approx \sum_i \eta_i^2 \mu_i/a_i + \sum_j W_j(\rho_j - \bar{\phi}^{-1}) \tag{11.21}$$

where $\quad \mu_i = \sum_j \beta_{ij}P_j^{-1} - \gamma_i^2$ $\qquad\qquad\qquad\qquad\qquad\qquad$ (11.22)

and

$$\mathbf{\eta}' = \mathbf{W}(\mathbf{XB})^{-1}\mathbf{X}. \tag{11.23}$$

As far as the Petersen estimate $\hat{U}_P$ is concerned, the general theory of **11.1.4** still applies if we replace θ_{ij}, p_j, U_j and U by β_{ij}, P_j, W_j and W. For example, U_P can now be written in the form

$$U_P = \frac{A\ \Sigma\ W_j P_j}{W\ \Sigma\Sigma\ a_i\beta_{ij}P_j}\ \cdot W.$$

However, Darroch shows that of the four conditions given in **11.1.4** (suitably interpreted in terms of our new parameters) only (3) implies that $U_P = W$. For the other three conditions $U_P \neq W$, though if the ϕ_i are not too different, the difference $U_P - W$ will be small. This means that the goodness-of-fit tests given in **11.1.4** will now test for approximate consistency. Once again, when $\hat{U}_P$ and $\hat{W}$ are both consistent, the Petersen estimate will generally have a smaller asymptotic variance; the asymptotic variances will be equal if $P_j = P$ and $\eta_i/a_i = W/A_i$, where η_i is now defined by (11.23).

Example 11.1   Sockeye salmon: Darroch [1961]

Darroch considers an example given by Schaefer [1951] in which both stratifications are with respect to time instead of place. The population consisted of all adult sockeye salmon passing a certain point of a river during a period of $s = 8$ weeks on their way upstream to their spawning grounds. The fish were sampled and tagged according to the week in which they passed this point. Provided they succeeded in reaching the spawning grounds, most adult salmon died after spawning. In this case, the deaths took place over a period of $t = 9$ weeks and during each of these weeks, a number of dead fish were recorded, presumably very soon after death. The data from the experiment are set out in Table 11.1. Here m_{ij} is the number of fish tagged in the

TABLE 11.1

Schaefer's data on sockeye salmon: m_{ij}, a_i and u_j for $s = 8$, $t = 9$
(from Darroch [1961: Table 1]).

Week of tagging (i)	\multicolumn{9}{c}{Week of recovery (j)}									Total	a_i
	1	2	3	4	5	6	7	8	9		
1	1	–	2	–	–	–	–	–	–	3	15
2	1	3	7	–	–	–	–	–	–	11	59
3	1	11	33	24	5	1	–	1	–	76	410
4	–	5	29	79	52	3	2	7	3	180	695
5	–	–	11	67	77	2	16	7	3	183	773
6	–	–	–	14	25	3	10	6	2	60	335
7	–	–	–	–	–	–	1	5	–	6	59
8	–	–	–	–	–	–	1	–	–	1	5
Total	3	19	82	184	159	9	30	26	8	520	2351
u_j	16	113	718	2664	3317	635	1217	904	368	9952	

ith sampling week and recovered dead on the spawning ground in the jth recovery week; p_j is the probability of being recovered on the spawning ground in the jth recovery week; θ_{ij} is the probability that a fish tagged in the ith week dies on the spawning ground in the jth recovery week, and ϕ_i is the probability that a fish tagged in the ith week dies on the spawning ground during the 9-week period. $(1 - \phi_i)$ represents the probability of dying before reaching the grounds or of surviving until after this period. Evidently a small percentage of salmon do manage to reach the sea alive and return to spawn again (Jones [1959]).

Since the m_{ij} in some of the outer weeks are too small to be used in what is essentially a large-sample theory, Darroch reduces s and t to four by grouping the first three and last three weeks of tagging into single strata and the first three and last four weeks of recovery into single strata. The new values of $\{m_{ij}\}$, $\{a_i\}$ and $\{u_j\}$ are given in Table 11.2, and assuming that $\phi_i = \phi$ we proceed as in case I above.

TABLE 11.2

Schaefer's data on sockeye salmon: reduced to $s = 4$, $t = 4$ (from Darroch [1961: Table 2]).

	m_{i1}	m_{i2}	m_{i3}	m_{i4}	$m_{i.}$	$a_i - m_{i.}$	a_i	$m_{i.}/a_i$
m_{1j}	59	24	5	2	90	394	484	0·186
m_{2j}	34	79	52	15	180	515	695	0·259
m_{3j}	11	67	77	28	183	590	773	0·237
m_{4j}	0	14	25	28	67	332	399	0·168
$m_{.j}$	104	184	159	73				
u_j	847	2664	3317	3124				

As a first step in the calculations, we investigate the consistency of the Petersen estimate using the tests outlined in **11.1.4**. We find that $T = 16\cdot91$, which is significant at $0\cdot1$ per cent level of significance, and the vectors $[(m_{.j})]$ and $[(u_j)]$ are so obviously not proportional that there is no need to apply a chi-squared test. Since H_1 and H_4 are both rejected, it would appear that the Petersen estimate is unsatisfactory.

Evaluating $\hat{\rho} = \mathbf{M}^{-1}\mathbf{a}$, and recalling that $\hat{P}_i = \hat{\rho}_i^{-1}$, we find that

$$\hat{P}_1 = 0\cdot1318, \quad \hat{P}_2 = 1\cdot9461, \quad \hat{P}_3 = 0\cdot1947, \quad \hat{P}_4 = 0\cdot1063$$

and

$$\hat{W} = \mathbf{u}'\hat{\rho} = 54\ 200.$$

The unsatisfactory value of $\hat{P}_2$ may be just a symptom of the general inaccuracy of the estimates, or it may indicate that the model is incorrect in assuming the ϕ_i equal. As Darroch points out, both of these explanations are probably correct, and although nothing can be done about the first we can act on the second. The $m_{i.}/a_i$ indicate where the possible differences in the ϕ_i lie, the middle two being appreciably larger than the outer two (cf. Table 11.2). Darroch therefore suggests estimating subject to just the two constraints: $\phi_1 = \phi_4$, $\phi_2 = \phi_3$. In order to apply the general theory of case II above, we must reduce t from four to three by pooling two of the recovery periods. It is permissible to group the jth and kth periods if (i) $(\beta_{ij}P_j + \beta_{ik}P_k)/(\beta_{ij} + \beta_{ik})$ is independent of i, in particular if (ii) $P_j = P_k$ or (iii) β_{ij}/β_{ik} is

443

independent of i. (ii) cannot be tested, but (iii) can as it implies proportionality of the jth and kth columns of $\mathbf{M}$. In this case, the columns which are nearest to being proportional are the third and fourth, which we now pool. (Although the hypothesis of proportionality is rejected at the 0·1 per cent level, (i) will still be approximately true if P_3 is not too different from P_4.) Proceeding as in case II above, we have

$$\mathbf{X} = \begin{bmatrix} 1 & 0 & 0 & -1 \\ 0 & 1 & -1 & 0 \\ \frac{1}{4} & \frac{1}{4} & \frac{1}{4} & \frac{1}{4} \end{bmatrix}, \qquad \mathbf{M} = \begin{bmatrix} 59 & 24 & 7 \\ 34 & 79 & 67 \\ 11 & 14 & 105 \\ 0 & 14 & 53 \end{bmatrix},$$

$$\mathbf{a}' = (484, 695, 773, 399), \quad \mathbf{u}' = (847, 2664, 6441),$$

and evaluating $\hat{\boldsymbol{\rho}} = [\mathbf{XD}_a^{-1}\mathbf{M}]^{-1}\mathbf{v}$, we obtain

$$\hat{\boldsymbol{\rho}}' = (6·021, 1·607, 6·397) \quad \text{or} \quad \hat{\mathbf{P}}' = (0·1661, 0·6223, 0·1563).$$

Although $\hat{P}_2$ now lies in $[0, 1]$ it is still rather high. The other estimates are

$$\hat{\mathbf{B}} = \mathbf{D}_a^{-1}\mathbf{M}\hat{\mathbf{D}}_\rho = \begin{bmatrix} 0·7339 & 0·0797 & 0·0925 \\ 0·2945 & 0·1827 & 0·6167 \\ 0·0857 & 0·1393 & 0·8689 \\ 0·0000 & 0·0564 & 0·8497 \end{bmatrix}$$

and, summing the rows,

$$\hat{\gamma}_1 = \hat{\gamma}_4 = 0·9061, \quad \hat{\gamma}_2 = \hat{\gamma}_3 = 1·0939.$$

Also,

$$\hat{\mathbf{W}}' = \hat{\boldsymbol{\rho}}'\mathbf{D}_u = (5099, 4282, 41\ 204)$$

and

$$\hat{W} = 50\ 585.$$

The estimated variance–covariance matrix of $\hat{\boldsymbol{\rho}}$ is

$$\hat{\boldsymbol{\Sigma}} = \begin{bmatrix} 9·96 & -14·84 & 6·31 \\ -14·85 & 23·58 & -10·32 \\ 6·31 & -10·32 & 4·78 \end{bmatrix}$$

and we note that the variance of $\hat{\rho}_2$ is large. It is unlikely that there was any catch dependence in this experiment, so that we need not make any mental reservations about $\hat{\boldsymbol{\Sigma}}$ underestimating the true variance–covariance matrix (see **11.1.2**). Finally the variance–covariance matrix of $\hat{W}$ is estimated by

$$\hat{E}[(\hat{\mathbf{W}} - \mathbf{W})(\hat{\mathbf{W}} - \mathbf{W})'] = 10^6 \begin{bmatrix} 7·168 & -33·474 & 34·441 \\ -33·474 & 167·347 & -177·113 \\ 34·441 & -177·113 & 198·694 \end{bmatrix},$$

and hence
$$\hat{E}[(\hat{W}-W)^2] = 20 \cdot 916 \times 10^6.$$

It is interesting to note that the estimates of $V[\hat{W}_2]$ and $V[\hat{W}_3]$ are both much larger than the estimate of $V[\hat{W}]$.

Although the Petersen estimate is invalid, it is of interest to evaluate it and its variance. Using the approximately unbiased version,
$$\hat{U}_P = (A+1)u/(m+1)$$
$$= 2352(9952)/521$$
$$= 44\ 927$$

and
$$\hat{E}[(\hat{U}_P - U_P)^2] = 3 \cdot 181 \times 10^6.$$

The latter is a good deal smaller than $\hat{E}[(\hat{W}-W)^2]$, which might be expected from (11.19) since the $\hat{P}_j$ differ and the $\hat{\eta}_i/a_i$ vary considerably; here

$$\hat{\eta}' = u'[\mathbf{XD}_a^{-1}\mathbf{M}]^{-1}\mathbf{X} = (9896, -20\ 728, 46\ 020, 15\ 397).$$

CHAPTER 12

CONCLUSION

12.1 COMPARISON OF METHODS

12.1.1 General comments

A wide variety of models have been discussed in this book, and the question now arises as to which method should be used in a given situation. Obviously the choice of method will depend very much on the nature of the population, its distribution over the population area, and the method of sampling the population. Where possible, the experiment should be designed so that more than one method of estimation can be used. For example, the closest-distance methods of Chapter 2 can be used along with quadrat sampling; the removal methods of Chapter 7 can sometimes be used in conjunction with the capture—recapture methods of Chapter 4, where animals are "removed" by tagging; the Petersen method can be incorporated in the CIR method of Chapter 9; and if the population is stationary, the time-specific methods of Chapter 10, using an age analysis of the removal, can be used with the CIR method. The following is a selection of articles which obtain estimates by more than one method: Dunnet [1957, rabbits: Hayne's regression method and variations of the Petersen method], Flyger [1959, squirrels: Schnabel method using trapping and sighting records], Wood [1959, foxes: relative density estimates], Huber [1962, rabbits: Petersen and Schnabel methods], Ammann and Ryel [1963, ruffed grouse: relative density estimates], Eberhardt *et al.* [1963, rabbits: Petersen, CIR, frequency of capture, catch-effort, and other miscellaneous methods], Muir [1964, fish: catch-effort, Schumacher—Eschmeyer and age—composition methods], Chapman [1964, fur seals: age—composition, CIR, and tag release methods], Fischler [1965, crabs: catch-effort and single tag release methods], Bergerud and Mercer [1966, ptarmigan: direct count, King, Petersen, relative density and aerial methods], Mosby [1969, squirrels: Petersen, Schnabel, frequency of capture, age—composition methods], and Phillips and Campbell [1970, whelks: Schnabel, Schumacher and Eschmeyer, Jolly—Seber, frequency of capture methods].

In the past, little attention has been devoted to the problem of designing an experiment to yield an estimate with a given minimum accuracy or precision. For example, in many of the early applications of the Petersen method, too few individuals were tagged, so that the number of recaptures was too small and the resulting confidence intervals too wide. However, for many of the models the variance formulae are complicated, so that it is not easy to plan for a given precision. In such cases a small pilot experiment may be appropriate; much more research is needed, however, on this question of experimental design.

Where possible, the robust but less efficient regression estimates should be calculated along with the more efficient maximum-likelihood estimates. The regression method is particularly useful when expected values appear to be correct (as indicated by the appropriate graph(s)), but the variances predicted by the model underlying the maximum-likelihood theory are open to question because of possible departures from the underlying assumptions, e.g. sampling is not strictly random (cf. **4.1.3** (*1*)). However, in all cases, the assumptions underlying a particular model should be studied carefully and, where possible, appropriate tests carried out. If there is likely to be any question about the validity of the underlying assumptions, the sample data should be collected in such a way that empirical variance estimates are available from replicated samples. A comparison of the sampling variance with the estimated theoretical variance predicted by the model will often throw some light on the validity of the model.

When samples are large enough, it is sometimes possible to use inter-penetrating subsamples. Here the total sample is split up randomly into subsamples from which separate estimates of population size, etc. can be calculated and then averaged (e.g. **3.4.3**). But this technique has so far received little attention in the literature.

Davis [1963] has stated that "the failure of wildlife investigators to check population estimates against known number is a deplorable situation". In spite of the fact that some methods have been widely used for estimating population size, these methods are seldom checked against a known population.

12.1.2 Individual methods

1   Absolute density

SAMPLE PLOTS. Total counts over the whole population are rarely possible, and one must usually resort to counting on sample plots. This method, however, has a limited usefulness for animal populations because of (i) the problem of locating and marking out the sample plots, particularly if the terrain is difficult and the animals mobile, and (ii) the problem of finding all the animals on the plot. When the terrain is difficult and the population area large, strip transects will generally be more appropriate than quadrats.

We recall from Chapter 2 that accuracy of a population-density estimate depends very much on the proportion p of the total area sampled, and some care should be given to the design of such experiments. It was shown in **2.1.1**(*1*) that when $p < 0\cdot2$, $C^2 \approx 1/n$, and the accuracy depends only on the total number n of animals seen when less than 20 per cent of the population area is sampled. However, the validity and usefulness of this formula when the population is not randomly distributed needs further investigation; this could be done using simulation or by working with populations of known size. When non-randomly distributed populations are to be compared, the data can sometimes be transformed to stabilise the variance and achieve some measure of normality (cf. Southwood [1966: 8–12]). For example, the logarithmic transformation (cf. Koch [1969]) is often used for contagious populations; this leads to a study of ratios of population densities rather than of differences.

If the population density is known to vary considerably over different but well defined parts of the population area then stratified sampling should be used. As considerable gains in precision can be achieved if the allocation of sampling effort to each stratum or domain is optimal (i.e. the fraction of each domain area sampled is proportional to the square root of the domain density), one should endeavour to obtain rough estimates of domain densities before the experiment is carried out. Even very approximate estimates will lead to more precise estimates than those obtained using proportional allocation (i.e. the same fraction of each domain area sampled).

LINE TRANSECTS. This method is particularly useful when the animals are too mobile to be sampled using sample plots, or when the animals are difficult to locate and must be flushed into the open. But the particular density estimate to be used for a given species depends very much on the nature of the species and its "flushing" behaviour. For this reason the line transect method is still in its infancy as far as animal populations are concerned, and it should be used with caution until we know more about the behaviour of the various estimates described in **2.1.3**. In this respect further research is needed on known populations: the simulated experiments of Robinette *et al.* [1954, 1956] and Gates [1969] are a step in the right direction

It is not easy to make a direct comparison between the line transect method and quadrat sampling as the reliability of the former is still unknown. However, assuming that the model given in **2.1.3**(2) leading to the estimate $\hat{N}_5$ is valid, some general comments can be made. To begin with, it was noted in that section that $C^2 \approx 2/n$, so that approximately twice the number of animals must be seen to give the same coefficient of variation C as the sample plot method. To compare the costs of the two methods, we note from (2.12) that

$$C^2 = V[\hat{N}_5]/N^2$$

$$\approx \frac{1}{NP}\left(1-P+NP\,E\left[\frac{1}{n-2}\right]\right)$$

$$\approx \frac{2-P}{NP},$$

and since $P = 2L/(A\lambda_1)$, where L is the length of line transect, A the population area, and $1/\lambda_1$ the average distance at which animals are seen from the transect, we find that

$$L \approx A\lambda_1/(1+NC^2).$$

Hence the cost of obtaining an estimate with a given C is approximately Lc_L, where c_L is the cost per unit length of transect. On the other hand, for the sample plot method, the proportion p of the population area which must be sampled for a given C is (cf. (2.2)) $1/(1+NC^2)$, and if quadrats of area a are used, each costing c_q, then the total cost is $c_q Ap/a$. Hence the ratio of cost of the line transect method to the sample plot method is

$$c_L \lambda_1/c_q'$$

where $c_q' = c_q/a$, the quadrat cost on a per-unit-area basis. Therefore, as one would expect $c_L < c_q'$ and $\lambda_1 < 1$, the above ratio will generally be less than unity, so that the line transect method will give more value for money, provided the underlying assumptions are satisfied.

CLOSEST-INDIVIDUAL TECHNIQUES. These methods can be used for animals which are relatively immobile and readily seen. In **2.1.4**(2) a promising estimate $\widetilde{D}$ (cf. (2.31)) of the population density D ($= N/A$) was considered which is unbiased, even when the population is not randomly distributed according to a binomial or Poisson law but rather follows a negative-binomial law, as in the case of contagious populations. A comparison of $\widetilde{D}$ with the maximum-likelihood estimate $\hat{D}$ of (2.29) may shed some light on the validity of the Poisson model.

When the main effort involved is in searching for the animals rather than in locating sample points or quadrat boundaries it was shown on p. 44 that the closest-individual technique using $\hat{D}$ has approximately the same efficiency as quadrat sampling. Otherwise, suppose that c_q is the cost of locating and setting out a quadrat of area a and let c_p be the cost of locating a sampling point and measuring the rth closest individual. Then, from (2.30),

$$C[\hat{D}] = 1/\sqrt{(sr-2)}$$

or

$$s = \frac{1}{r}\left(\frac{1}{C^2} + 2\right).$$

Hence, for a given C, the closest-individual method costs

$$\frac{c_p}{r}\left(\frac{1}{C^2} + 2\right),$$

which may be compared with

$$\frac{c_q A}{a(1+NC^2)},$$

the total cost for the sample plot method.

2   Relative density

The direct methods discussed in **2.2.1** for estimating relative density are particularly useful when one is interested either in detecting changes in population density with time or in comparing populations in different areas. However, it was stressed there that if any comparisons are to be made, then the censusing should be carried out under as nearly identical conditions as possible. As changing conditions have different effects on different density indices, one should endeavour to obtain more than one type of index. For example, Ammann and Ryel [1963] used correlation co- efficients to compare a number of indices for ruffed grouse based on mail- carrier counts along their routes, drumming counts, opinion surveys, brood counts, and kill records.

Where possible, replicated samples should be used for determining indices, so that sampling estimates of variance can be calculated (cf. (1.9)). For example, instead of just counting the total number n of animals seen on a roadside count for a road of length l units, one should count the number n_i $(i = 1, 2, \ldots, l)$ for each unit of road-length; although the index $\bar{n} = \Sigma\, n_i/l = n/l$ is still based on the total n, an estimate of $V[\bar{n}]$ is now available, namely $\Sigma\, (n_i - \bar{n})^2/l\,(l-1)$. In this example we note that the general comments and theory of **2.1.2**(1) still apply, provided we think in terms of sample length rather than sample plot. For example, in choosing the road-length unit we want l to be as large as possible, so that by the central limit theorem $\bar{n}$ is approximately normal; however, the shorter the unit the more crucial is the "end effect" where animals may be allocated to the wrong unit.

Catches from trap-lines have been widely used as an index of population density, particularly for small-mammal populations. The basis of the method is that the catch is always proportional to population density, provided the trapping is standardised (Calhoun [1948]). Hansson [1967, 1968], however, has reviewed this method and concludes that the proportion caught may vary considerably, thus suggesting that the trap-line method can be very un- reliable (particularly when there is a wide variation in home range: cf. Stickel [1948]). On the other hand, Southern [1965] points out that the method is simple and very suitable for naturalists who have little time to spare; both he and Ashby [1967: 411 ff.] demonstrate that the method can be a useful one for detecting population changes.

3   Mark—recapture methods: closed population

PETERSEN METHOD. Of all the methods considered in this book the Petersen method appears to be the most useful, provided that the assumptions underlying the method are satisfied and there are sufficient recaptures in the second sample. With regard to the latter proviso it was noted in **3.1.1** that the coefficient of variation C of the Petersen estimate is given by

$$C^2 \approx N/(n_1 n_2) \approx 1/m_2\,, \tag{12.1}$$

so that for $C < 0\cdot25$ we must have at least 16 recaptures. Many of the early applications of the Petersen method fell down on this score, as too few individuals were marked, so that less than 10 were recaptured. By studying Figs. 3.1–3.6 on pages 65–69 we see that for population sizes commonly encountered in unexploited populations a considerable proportion of the population must be marked for a reasonable accuracy. For example, setting $n_1 = n_2 = n$, say, and defining the accuracy A as on p. 64, we obtain the following values of $100n/N$, the percentage of the population that must be marked for a given A:

A	0·5	0·25	0·1	0·5	0·25	0·1	0·5	0·25	0·1
N	50	50	50	100	100	100	1000	1000	1000
$100n/N$	40	54	74	32	43	64	13	20	40

A number of authors have commented on the need to mark at least 50 per cent of the population, e.g. Huber [1962: 185], Strandgaard [1967: 650] and Mosby [1969: 61].

The main assumption underlying the Petersen estimate is that marked and unmarked have the same probability of being caught in the second sample. Unfortunately it is not always easy to detect departures from this assumption, so that even when all precautions are taken and the assumptions appear to be satisfied, the Petersen estimate may be biased. For example, Buck and Thoits [1965] carried out extensive Petersen experiments in large fishponds, and after draining the ponds found that the Petersen estimates were considerably biased, with errors much larger than one would expect from chance. Thus "in spite of normally accepted indications to the contrary", the assumptions underlying the Petersen method were not valid: recruitment and mortality were ruled out, so that the errors were apparently due to a variable catchability amongst the fish. Further examples from Buck and Thoits [1965: Table 1], Cormack [1968: 470–3], Robinette et al. [1954], Dunnet [1957], Huber [1962], Smith [1968], Mosby [1969], etc., in which estimates are compared with known population sizes, indicate that in some cases the Petersen estimate is satisfactory while in others it is most un-reliable. If the Petersen estimate is to be used extensively for a given species then it should be compared with other estimates and, where possible, tested on a known population. One model, given by Eberhardt [1969a], which allows for a variable trap response is described on p. 178.

If the second sample can be taken in stages then the regression method of **3.7** (p. 125) can be used for testing the assumption that marked and un-marked have the same probability of capture.

BINOMIAL MODEL. Various methods for overcoming the problem of variable catchability, such as pre-baiting, using different sampling methods for the two samples, changing trap positions, etc., were discussed in **3.2.2**. However, the most promising approach to the problem is to avoid recapturing altogether and to obtain an estimate of the proportion marked by simply observing the animals or, if toe-clipping is used, by counting their tracks (Marten [1971]); tagged animals can also be detected using remote sensing techniques (Marten [1970b]).

If sight records are used, as in Example 3.9 (p. 110), then the second sample is obtained by sampling *with* replacement and the binomial model

(3.3) is applicable. In this case we have from Bailey [1951]

$$C^2 \approx (N - n_1)/(n_1 n_2), \tag{12.2}$$

and since animals can be seen more than once, n_2, the number of animals sighted in the second sample, can even be larger than N, thereby increasing the precision of the experiment. Therefore, given a rough estimate of N, and given n_1, n_2 can be determined from (12.2) for a given C with no restriction on n_2 other than that of time and cost of sampling. If the experiment is limited by funds then we can determine the optimum allocation of effort as follows.

Let c_m be the cost of marking an animal in the first sample and let c_0 be the cost of observing an animal in the second sample. Then B, the total cost, is given by

$$B = n_1 c_m + n_2 c_0,$$

and minimising C^2 in (12.2), subject to the above constraint, leads to

$$n_1 = N \left[1 - \left(1 - \frac{B}{N c_m} \right)^{\frac{1}{2}} \right]$$

or, when n_1/N is small,

$$n_1 \approx B/(2c_m).$$

In using the sighting method for the second sample, some animals will be more active than others, so that they will be more likely to be seen. However, this will not affect the Petersen estimate or its modified version $\hat{N}_1$, provided the probability of sighting a given animal is independent of its mark status.

If tracking is used to "observe" the animals, then, from Marten [1970b], the Petersen estimate for the unmarked population size is given by

$$\hat{U} = \hat{N} - n_1 = n_1 u_2/m_2 = u_2/\hat{k},$$

where $\hat{k}$ is the number of tracks per marked individual and u_2 is the number of tracks from unmarked individuals. Therefore, since u_2 and $\hat{k}$ are independent, we have, using the delta method,

$$V[\hat{U}] \approx U^2 \left[\frac{V[u_2]}{(E[u_2])^2} + \frac{V[\hat{k}]}{(E[\hat{k}])^2} \right].$$

Following Marten, it is not unreasonable to assume that the number of tracks per individual is Poisson, so that

$$V[\hat{U}] \approx U^2 \left[\frac{1}{E[u_2]} + \frac{1}{E[m_2]} \right]$$

and

$$C^2[\hat{U}] \approx \frac{1}{E[u_2]} + \frac{1}{E[m_2]}.$$

Marten points out that if animals vary in their activity, the above estimate of U will not be affected, provided activity is independent of mark status. Further extensions of the above tracking method are given in Marten [1971].

SCHNABEL METHOD. Sometimes it is not possible to catch enough individuals on the first sampling occasion for a satisfactory application of the Petersen estimate, so that the Schnabel method must be used (e.g. Williams [1965]). However, the Petersen method is to be preferred as it is less affected by the failure of the underlying assumptions. On the other hand, if the assumption of a closed population is open to question, then the Schnabel method is appropriate using the methods of Chapter 5.

In comparing the Petersen and Schnabel estimates from the point of view of efficiency, we note that the respective coefficients of variation C_1 and C_2 are both of the form $C_i = \sqrt{(N/\lambda_i)}$, where $\lambda_1 = n_1 n_2$ and $\lambda_2 = \sum_{i=2}^{s} n_i M_i$ (cf. p. 130 for notation). We shall assume that for each method the probability of capture is the same, and we shall denote these probabilities by P_1 and P_2 respectively. Then, replacing the n_i and M_i by their expected values leads to

$$r^2 = C_1^2/C_2^2$$
$$= \lambda_2/\lambda_1$$
$$\approx NP_2 \sum_{i=2}^{s} N(1-Q_2^{i-1})/N^2 P_1^2$$
$$= (sP_2 - 1 + Q_2^s)/P_1^2,$$

where $Q_2 = 1 - P_2$. One basis for comparison is to assume that both methods have the same expected total sample size, i.e. $E[n_1 + n_2]$ for the Petersen method equals $E[n_1 + n_2 + \ldots + n_s]$ for the Schnabel method; then $2NP_1 = sNP_2$ and

$$r = \sqrt{4(sP_2 - 1 + Q_2^s)/(s^2 P_2^2)}. \qquad (12.3)$$

From Table 12.1 we see that for the values of P_2 and s commonly used in practice, the Schnabel estimate is more efficient than the Petersen estimate, though the difference is not great. For small P_2, r increases with s up to a maximum value and then decreases to zero as $s \to \infty$; the maximum is reached more quickly for larger values of P_2.

TABLE 12.1

Ratio (r) of the efficiencies of the Petersen and Schnabel estimates for different values of s (the number of samples in the Schnabel census) and P_2 (the probability of capture in each sample of the Schnabel census).

P_2		Ratio (r)			
	$s = 3$	4	5	6	12
0·05	1·14	1·20	1·23	1·25	1·25
0·1	1·13	1·18	1·20	1·21	1·16
0·2	1·11	1·14	1·14	1·13	1·01
0·3	1·10	1·11	1·09	1·06	0·89
0·5	1·05	1·03	0·99	0·95	0·75

If one wishes to compare the costs of the two methods for the same efficiency (i.e. $r = 1$), suppose that c_T is the cost of setting up a single trap each day and c_m is the cost of marking an animal. Assuming that $P_i = aT_i$, where T_i is the number of traps used, and using individual marks or tags (i.e. marked individuals are not re-marked), the ratio of the cost of the Petersen method to the Schnabel method is

$$\frac{2T_1 c_T + NP_1 c_m}{sT_2 c_T + N(1-Q_2^{s-1})c_m}.$$

If rough estimates of N and a are available then the above ratio can be calculated for different values of s and T_2; T_1 is calculated from

$$P_1^2 = sP_2 - 1 + Q_2^s.$$

REGRESSION METHODS. The basic equation underlying the regression method of Schumacher and Eschmeyer (**4.1.3**(*1*)) is $E[m_i/n_i | M_i] = M_i/N$, which holds when marked and unmarked have the same probability of capture in the ith sample. However, another useful equation is $E[u_i] = p_i U_i$, where p_i is the probability that an unmarked individual is caught in the ith sample. If p_i is the same for marked individuals, then $1/p_i$ can be estimated by $(M_i + 1)/(m_i + 1)$, and this leads to the regression model

$$E\left[\frac{u_i(M_i+1)}{(m_i+1)} | M_i\right] \approx U_i = N - M_i.$$

Alternatively, if p_i is k times the probability of catching a marked individual, then $1/p_i$ can be replaced by $(M_i + 1)/k(m_i + 1)$ and

$$E\left[\frac{u_i(M_i+1)}{(m_i+1)} | M_i\right] \approx k(N-M_i), \tag{12.4}$$

which is Marten's model (4.18). However, departures from the underlying assumption that marked and unmarked have the same probability of capture take different forms. For example, Tanaka and Kanamori [1967] suggest that, with trap addiction, the traps will first be taken up by marked animals so that p_i will be proportional to some function g of the traps available for unmarked individuals, i.e. if T is the number of traps and $T_i (= T - m_i)$ is the number of traps not occupied by marked when the ith sample is complete, then $p_i = a g(T_i)$. Therefore, since $m_1 = 0$,

$$p_i = p_1 g(T_i)/g(T)$$

and, given an estimate ρ_i of $g(T)/g(T_i)$, we have the regression model

$$E[\rho_i u_i | M_i] = p_1(N-M_i)$$

which is of the same form as (12.4). Tanaka gives an example where, on the basis of empirical information, he chooses $g(T) = T^x$. Leslie and Davis [1939] and Tanaka [1963b], using the analogy of the random movement of molecules, suggest an alternative model

$$T\{\log T_i - \log (T_i - u_i)\} = p_1(N - M_i) + e_i,$$

which is again of the same form as (12.4).

4   Mark–recapture methods: open population

The single tag–release models of Chapter 6 are mainly applicable to commercially exploited populations where numbers are large and tagging experiments costly. However, since the information from a single release is minimal, strong assumptions have to be built into the models in order that the population size may be estimated. For this reason it is preferable to release several small batches of tagged individuals at suitable intervals rather than one large batch at the beginning, and then use the "multi-sample single-recapture" models discussed on pages 212–14. Up till now these models have not been used much, despite their wide applicability and the simplicity of their estimates.

For the small unexploited populations where multiple recaptures are possible, the Schnabel method described in **5.1.1** can be used, provided there is no trap-shyness or trap addiction and any emigration is permanent; when there is trap response the models of **4.1.6**(4) based on the frequency of capture may be useful. The special cases discussed in **5.1.3**, sections 1 and 2, should not be used unless there is definite biological evidence that the conditions for their applicability are satisfied. When the conditions are satisfied the more restrictive models will yield more efficient estimates.

The large-sample variance formulae of Chapter 5 were derived using the delta method. However, the question of how large is a "large" sample still requires further investigation, and several authors are at present working on this problem, using simulation. In the absence of any direct information I suggest that the recommendations given on p. 204 be followed, i.e. m_i and r_i should both be greater than 10.

5   Catch-effort methods

These are widely used for commercial fisheries and are particularly useful when tagging experiments are unsatisfactory. For example, if the population is very big and the fishing grounds extensive, the recaptures in each sample may be too few, even in spite of large tag–releases; also, when recaptures are few, the non-reporting of tags can be a major problem. Volume 155 of *Rapports et Procès-Verbaux des Réunions du Conseil International pour L'Exploration de la Mer* contains reviews of experience with catch-effort methods in marine fisheries: see especially Gulland [1964] and Paloheimo and Dickie [1964].

For small closed populations, the so-called removal method of **7.2** has been used with mixed success. For example, Gentry *et al.* [1968] found that the probability of capture p did not remain constant from sample to sample as required, while Grodzinksi *et al.* [1966] found that the method can be satisfactory for some species, particularly with prebaiting followed by intensive trapping (see also Babinska and Bock [1969]). The main departures from the assumption of constant p are (i) variation in trap efficiency from sample to sample due to environmental changes such as weather, etc., and (ii) variation in the trap response of individuals: both departures are discussed in some detail in **7.2.2**(*1*).

To compare the two-sample removal method with the Petersen method we shall assume that for the Petersen method the probability of capture in each sample is p. Then, from p. 324, the ratio of the coefficient of variation of the removal estimate to that of the Petersen estimate is $\sqrt{(1+q)/p}$. Values of this ratio are:

p	0·1	0·2	0·3	0·4	0·5	0·6	0·9
ratio	4·36	3·00	2·39	2·00	1·73	1·53	1·10

thus showing that the Petersen estimate is much more efficient (though of course more costly).

For the three-sample removal method we find, using (7.23), that the corresponding ratio is

$$\sqrt{\left[\frac{(1-q^3)\,qp^2}{(1-q^3)^2 - 9p^2 q^2}\right]}$$

which takes values:

p	0·1	0·2	0·3	0·4	0·5
ratio	2·14	1·42	1·09	0·88	0·73

Thus, for $p = 0·3$ there is little difference between the two methods as far as precision is concerned. However, for this value of p the removal method would remove about 66 per cent of the population, while the Petersen method would lead to a marking of 30 per cent of the population.

6   *Change in ratio method*

To compare the CIR method with the Petersen method, suppose that n animals are classified according to x- and y-type before and after a removal of R_x x-type animals (see p. 354 for notation; here $n_1 = n_2 = n$, $R_y = 0$ and $R = R_x$). Then, if θ is the ratio of the cost of tagging to the cost of classification, Chapman [1955: $\theta = 1/\lambda$] considers the Petersen estimate that could be made if $2n/\theta$ animals were tagged and a second sample of R animals was inspected for tags. For this situation we find from Chapman that the ratio of the coefficient of variation of the Petersen estimate to that of the CIR method is approximately

$$\sqrt{\left[\left(\frac{1-P_1}{2P_1 - u}\right)\left(\frac{u}{1-u}\right)\cdot\frac{\theta}{2}\right]},$$

where $P_1 = X_1/N_1$ (the initial proportion of x-types) and $u = R/N_1$. The values of θ required to make the above ratio unity are given in Table 12.2 for different values of P_1 and u. Since the comparison of the two methods is made under the conditions most favourable to the CIR method, i.e. the

TABLE 12.2

The value of θ, the relative cost of tagging to cost of classification, required to give the same coefficients of variation for the Petersen and CIR estimates: from Chapman [1955: Table 3].

X_1/N_1				$R/N_1 = u$			
$(= P_1)$	0·25	0·20	0·15	0·10	0·05	0·02	0·01
0·1	—	—	—	—	6·34	19·60	41·80
0·2	—	—	3·54	6·76	16·62	46·55	96·53
0·3	3·00	4·58	7·28	12·86	29·85	81·20	166·90
0·4	5·51	8·00	12·28	21·00	47·49	127·40	260·70
0·5	9·00	12·80	19·27	32·40	71·20	192·08	392·04
0·6	14·25	20·00	29·75	49·50	109·25	289·10	589·05
0·7	23·00	32·00	47·22	78·00	171·00	450·80	917·33
0·8	40·51	56·00	81·16	135·00	294·51	774·20	1574·10
0·9	93·00	128·00	87·00	306·00	332·51	1744·40	3544·20

allocation of sampling effort is almost optimal (page 363), this table suggests that unless tagging is prohibitively costly, the Petersen method is better value for money.

If considerations of cost are not important then $\sqrt{\theta}$ will represent the ratio of the coefficient of variation of the CIR estimate to that of the Petersen estimate. Therefore, since the Petersen estimate itself is not very precise unless a large fraction of the population is tagged, we find that CIR estimates are generally not very accurate unless a large fraction of the population is classified, or there is a big change in the proportion of x-types through the removal R; from p. 370 we recall that $\Delta P = P_1 - P_2$ should be at least greater than 0·10 (or possibly 0·05 if we have full confidence in the underlying assumptions).

7 Age composition methods

Although life tables are widely used, the assumptions behind the construction of such tables are not always appreciated. For example, the time-specific method of **10.1.2**(2) is only valid when the population is stationary, so that any such analysis should be accompanied by biological and statistical evidence for stationarity.

Age-specific methods have also been applied to tag returns from dead animals. For example, if D_j is the number of birds tagged at birth that die in their jth year of life, then the D_j series is often treated as a d_x series. But we saw on p. 252 that this method is only valid when the tag recovery rate is constant from year to year and there are no tagged birds alive at the end of the experiment. It was also shown there that another series D'_j, based on the percentage tag returns, can be used when the same conditions are satisfied. Occasionally the D'_j are treated as an l_x series, so that D'_{j+1}/D'_j is an estimate of the survival rate (cf. **10.1.1**). However, it was shown at the end of Chapter 5 that this method is only valid when the survival rate is constant from year to year.

In conclusion I would stress that more care be exercised in the construction and interpretation of life tables and that the maximum-likelihood methods of **5.4** and Chapter 10 should be more widely used.

12.2 OTHER TOPICS

When I first planned this book I had hoped to cover a wider range of topics. However, the more I searched the literature the more I realised that I had to restrict myself to just the problem of estimating animal abundance, without considering the nature of the factors underlying population change. This means that a wide range of topics relating to population dynamics have not been covered. For example, there have been many applications of the theory of stochastic processes to population systems (cf. Bartlett [1960], Keyfitz [1968], Pielou [1969]), while in the study of fishery dynamics numerous deterministic models have been developed (cf. Beverton and Holt [1957], Ricker [1958], Cushing [1968]); in recent years simulation (e.g. Eleveback and Varma [1965], King and Paulik [1967], Walters [1969]) and systems analysis (e.g. Watt [1966], Dale [1970]) have provided useful tools for population analysis. Also, such topics as density-dependence (Andrewartha [1961], Varley and Gradwell [1970]), population cycles (Lack [1954]), the natural regulation of population numbers (Lack [1954]), association between species (Greig-Smith [1964], Southwood [1966], Pielou [1969]), and estimating numbers of species (Williams [1964], Holgate [1969], Pielou [1969]), etc., have generated much research. However, I hope that this book will encourage statisticians to become increasingly interested in ecology, with its broad spectrum of challenging research problems.

APPENDIX

A1 Shortest 95% confidence interval for N/λ based on the Poisson distribution

		Entering variable m_2 (or m)			
m_2	Lower limit	Upper limit	m_2	Lower limit	Upper limit
0	0·088 5				
1	0·072 0	19·489	26	0·024 78	0·056 3
2	0·076 7	2·821	27	0·024 08	0·053 9
3	0·073 6	1·230	28	0·023 42	0·051 6
4	0·069 0	0·738	29	0·022 79	0·049 5
5	0·064 4	0·513	30	0·022 21	0·047 5
6	0·060 0	0·388	31	0·021 65	0·045 7
7	0·056 1	0·309	32	0·021 12	0·044 0
8	0·052 6	0·256	33	0·020 61	0·042 5
9	0·049 5	0·217	34	0·020 14	0·041 0
10	0·046 8	0·188	35	0·019 68	0·039 6
11	0·044 3	0·165	36	0·019 25	0·038 4
12	0·042 0	0·147	37	0·018 83	0·037 2
13	0·040 0	0·133	38	0·018 43	0·036 0
14	0·038 2	0·121	39	0·018 05	0·035 0
15	0·036 5	0·111	40	0·017 69	0·033 96
16	0·035 0	0·1020	41	0·017 33	0·033 00
17	0·033 62	0·0945	42	0·017 00	0·032 10
18	0·032 33	0·0880	43	0·016 68	0·031 24
19	0·031 14	0·0823	44	0·016 36	0·030 43
20	0·030 04	0·0773	45	0·016 06	0·029 66
21	0·029 01	0·0729	46	0·015 78	0·028 92
22	0·028 06	0·0689	47	0·015 50	0·028 22
23	0·027 16	0·0653	48	0·015 23	0·027 55
24	0·026 32	0·0620	49	0·014 98	0·026 91
25	0·025 52	0·0591	50	0·014 75	0·026 25

(Reproduced from Chapman [1948].)

Applications of the above table are given on p. 63 and pp. 139–40.

A2 Tag recoveries needed for prescribed probabilities of detecting incomplete tag-reporting with various levels of catch inspection

The parameters are defined in **3.2.4.**

$a = 0.10$

p	$1-\beta$	p_0									
		.05	.10	.15	.20	.25	.30	.40	.50	.70	.90
.25	.50	6	4	4	3	3	4	4	5	9	28
	.80	24	14	11	10	10	9	10	11	17	51
	.90	39	23	18	15	14	14	14	15	23	66
	.95	54	31	24	20	19	18	18	19	28	80
	.99	88	50	37	32	29	27	26	28	40	109
.50	.50	39	23	18	15	14	14	14	15	23	66
	.80	136	76	56	47	42	39	38	39	55	147
	.90	210	116	85	71	63	58	55	56	76	201
	.95	284	156	114	94	83	76	71	73	97	252
	.99	453	246	178	146	128	117	108	109	143	365
.60	.50	84	47	36	30	27	26	25	27	38	106
	.80	276	151	110	91	80	74	69	71	95	247
	.90	421	229	166	136	120	110	101	102	134	344
	.95	564	305	221	180	158	144	132	133	172	436
	.99	888	478	344	280	243	221	201	200	256	639
.70	.50	197	109	80	66	59	55	52	53	72	191
	.80	615	332	240	196	171	156	143	143	186	469
	.90	925	498	359	291	253	230	209	208	266	662
	.95	1230	660	474	384	333	302	273	271	342	847
	.99	1917	1026	734	593	513	464	416	410	513	1256
.75	.50	322	176	128	106	93	86	80	81	108	278
	.80	981	527	380	308	268	244	221	220	280	696
	.90	1468	787	564	456	395	358	323	319	402	990
	.95	1943	1040	744	601	519	470	422	415	519	1272
	.99	3016	1609	1148	925	797	720	643	630	781	1895
.80	.50	568	307	222	182	159	145	133	134	173	439
	.80	1688	904	647	523	453	410	369	364	457	1121
	.90	2512	1341	958	772	667	602	539	529	658	1603
	.95	3314	1767	1260	1015	875	789	704	689	853	2065
	.99	5120	2725	1939	1558	1340	1207	1073	1046	1285	3093
.85	.50	1130	607	436	354	307	279	252	250	317	787
	.80	3290	1754	1251	1007	868	783	699	684	847	2052
	.90	4866	2590	1844	1482	1275	1148	1021	997	1225	2950
	.95	6401	3403	2420	1943	1670	1502	1333	1298	1590	3814
	.99	9845	5227	3712	2975	2554	2294	2031	1972	2402	5733
.90	.50	2832	1512	1079	869	750	677	605	593	736	1789
	.80	8074	4289	3048	2444	2100	1887	1672	1626	1985	4748
	.90	11881	6305	4475	3585	3076	2761	2442	2369	2880	6859
	.95	15579	8261	5859	4691	4022	3608	3186	3086	3743	8894
	.99	23864	12643	8959	7166	6138	5501	4850	4691	5669	13423
.95	.50	12548	6658	4724	3785	3246	2914	2576	2499	3036	7227
	.80	35080	18573	13151	10513	8999	8060	7097	6855	8262	19515
	.90	51367	27181	19237	15369	13149	11770	10352	9989	12013	28320
	.95	67153	35522	25132	20072	17167	15362	13503	13020	15641	36822
	.99	102459	54175	38312	30586	26148	23389	20540	19791	23732	55778

Table A.2 (*continued*)

$$a = 0.05$$

ρ	1−β	\(p_0\) .05	.10	.15	.20	.25	.30	.40	.50	.70	.90
.25	.50	9	6	5	5	5	6	7	8	14	46
	.80	30	19	15	13	13	13	14	15	25	75
	.90	47	28	22	19	18	18	18	20	32	93
	.95	63	37	29	25	23	22	23	25	38	109
	.99	101	57	44	37	34	33	32	35	51	143
.50	.50	63	37	29	25	23	22	23	25	38	109
	.80	180	100	75	63	57	53	51	54	76	207
	.90	264	146	108	90	80	75	71	74	102	271
	.95	346	191	140	116	103	95	90	92	125	331
	.99	530	289	211	174	153	141	131	133	177	458
.60	.50	137	77	58	49	45	42	41	44	63	174
	.80	368	202	148	123	109	101	95	97	132	346
	.90	533	291	212	175	154	142	131	134	178	460
	.95	693	376	273	224	197	180	166	168	221	566
	.99	1048	566	409	334	291	266	243	244	315	794
.70	.50	324	178	131	109	97	90	85	87	119	315
	.80	827	448	325	266	233	213	196	197	257	653
	.90	1183	638	461	375	327	298	272	272	350	879
	.95	1524	820	590	479	417	379	344	343	437	1091
	.99	2281	1223	877	710	615	558	503	498	628	1549
.75	.50	530	289	211	174	153	141	131	133	177	458
	.80	1325	714	515	419	364	332	302	301	386	968
	.90	1883	1011	726	589	511	464	420	416	528	1310
	.95	2417	1296	929	751	651	590	532	525	662	1631
	.99	3600	1924	1375	1110	959	867	777	764	953	2328
.80	.50	935	506	366	299	261	239	219	220	285	722
	.80	2289	1227	880	712	617	560	505	499	630	1554
	.90	3234	1730	1237	999	864	781	701	690	863	2114
	.95	4137	2209	1578	1272	1098	992	888	872	1084	2641
	.99	6132	3268	2329	1874	1614	1455	1297	1269	1567	3788
.85	.50	1862	1000	718	582	506	459	415	412	523	1297
	.80	4473	2388	1705	1374	1185	1070	957	939	1166	2836
	.90	6287	3350	2387	1920	1654	1491	1329	1299	1604	3877
	.95	8016	4267	3037	2441	2100	1891	1682	1641	2018	4859
	.99	11827	6286	4468	3585	3080	2769	2456	2390	2922	7000
.90	.50	4666	2490	1777	1432	1235	1115	997	977	1213	2947
	.80	11012	5854	4162	3340	2871	2582	2291	2230	2730	6544
	.90	15398	8176	5807	4655	3997	3590	3179	3088	3764	8989
	.95	19572	10386	7371	5905	5066	4548	4022	3902	4743	11299
	.99	28752	15243	10808	8651	7415	6649	5870	5684	6886	16345
.95	.50	20671	10967	7782	6234	5348	4799	4243	4116	5000	11905
	.80	47983	25412	18000	14393	12324	11041	9727	9402	11345	26830
	.90	66770	35343	25021	19997	17114	15324	13487	13022	15682	27011
	.95	84619	44776	31689	25317	21660	19389	17053	16456	19791	46652
	.99	123797	65478	46319	36989	31632	28303	24871	23978	28788	67742

Table A.2 (concluded)

$$a = 0.01$$

ρ	$1-\beta$	p_0 .05	.10	.15	.20	.25	.30	.40	.50	.70	.90
.25	.50	17	12	10	10	10	11	13	16	28	92
	.80	45	28	23	21	20	20	22	25	42	131
	.90	65	39	31	28	27	26	28	32	51	155
	.95	84	50	39	35	33	32	33	37	59	175
	.99	126	73	57	49	46	44	45	49	75	218
.50	.50	126	73	57	49	46	44	45	49	75	218
	.80	278	157	118	100	90	85	83	88	127	350
	.90	382	213	158	133	119	112	108	113	159	432
	.95	479	266	197	164	147	137	130	136	188	506
	.99	692	381	280	232	205	190	179	184	250	661
.60	.50	274	154	116	98	89	84	82	87	125	347
	.80	578	319	235	195	174	162	153	158	217	579
	.90	781	428	314	259	230	212	199	204	276	723
	.95	971	531	387	319	281	259	242	246	329	855
	.99	1385	753	546	448	393	360	332	336	441	1131
.70	.50	648	357	262	218	193	179	169	174	238	629
	.80	1312	713	518	425	373	342	316	320	422	1083
	.90	1751	948	686	561	490	449	412	414	539	1368
	.95	2162	1167	843	687	599	547	500	500	645	1629
	.99	3048	1640	1180	959	834	758	688	685	874	2181
.75	.50	1060	578	422	347	305	281	261	266	353	914
	.80	2110	1140	823	671	586	535	489	490	633	1597
	.90	2802	1509	1087	883	769	700	636	634	811	2029
	.95	3446	1852	1331	1080	938	852	772	767	975	2424
	.99	4835	2591	1857	1503	1302	1180	1063	1050	1323	3261
.80	.50	1869	1011	732	597	522	477	437	439	570	1444
	.80	3661	1966	1413	1145	994	903	817	811	1029	2555
	.90	4834	2591	1857	1502	1301	1179	1063	1050	1323	3261
	.95	5927	3172	2270	1834	1586	1435	1290	1271	1593	3909
	.99	8775	4419	3156	2544	2196	1984	1775	1743	2169	5282
.85	.50	3724	2000	1437	1165	1011	918	830	824	1045	2593
	.80	7184	3840	2744	2214	1913	1729	1550	1524	1902	4647
	.90	9441	5038	3595	2896	2498	2255	2015	1976	2452	5957
	.95	11537	6150	4383	3528	3040	2741	2445	2393	2958	7161
	.99	16035	8534	6074	4882	4201	3782	3364	3283	4036	9719
.90	.50	9332	4980	3554	2863	2470	2229	1993	1954	2424	5894
	.80	17752	9444	6720	5398	4643	4178	3714	3621	4445	10689
	.90	23221	12342	8773	7040	6049	5438	4824	4695	5742	13759
	.95	28291	15027	10674	8560	7351	6604	5851	5687	6938	16586
	.99	39150	20774	14743	11812	10134	9096	8044	7804	9488	22601
.95	.50	41348	21937	15566	12469	10696	9600	8487	8232	10002	23813
	.80	77628	41128	29142	23312	19969	17897	15779	15263	18444	43682
	.90	101102	53539	37919	30318	25958	23254	20483	19795	23878	56448
	.95	122823	65021	46037	36797	31496	28206	24830	23981	28893	68223
	.99	169262	89565	63386	50642	43326	38784	34111	32916	39588	93317

(Reproduced from Paulik [1961].)

A3 Solution of $\exp(-x) + ax = 1$

Note: For $a < 0.05$ take $x = 1/a$.

a	x	a	x	a	x	a	x	a	x
0·050	20·0000000	0·100	9·9995458	0·150	6·6581095	0·200	4·9651142	0·250	3·9206904
·051	19·6078431	·101	9·9004937	·151	6·6136295	·201	4·9395120	·251	3·9037180
·052	19·2307691	·102	9·8033798	·152	6·5697225	·202	4·9141487	·252	3·8868671
·053	18·8679244	·103	9·7081477	·153	6·5263771	·203	4·8890207	·253	3·8701363
·054	18·5185184	·104	9·6147429	·154	6·4835823	·204	4·8641246	·254	3·8535241
0·055	18·1818180	0·105	9·5231129	0·155	6·4413272	0·205	4·8394568	0·255	3·8370292
·056	17·8571425	·106	9·4332073	·156	6·3996012	·206	4·8150140	·256	3·8206502
·057	17·5438592	·107	9·3449775	·157	6·3583940	·207	4·7907929	·257	3·8043857
·058	17·2413787	·108	9·2583768	·158	6·3176956	·208	4·7667902	·258	3·7882344
·059	16·9491518	·109	9·1733599	·159	6·2774962	·209	4·7430026	·259	3·7721949
0·060	16·6666657	0·110	9·0898836	0·160	6·2377864	0·210	4·7194272	0·260	3·7562660
·061	16·3934414	·111	9·0079060	·161	6·1985568	·211	4·6960608	·261	3·7404463
·062	16·1290307	·112	8·9273666	·162	6·1597983	·212	4·6729003	·262	3·7247347
·063	15·8730138	·113	8·8482865	·163	6·1215021	·213	4·6499430	·263	3·7091299
·064	15·6249974	·114	8·7705681	·164	6·0836596	·214	4·6271857	·264	3·6936307
0·065	15·3846122	0·115	8·6941952	0·165	6·0462625	0·215	4·6046258	0·265	3·6782358
·066	15·1515112	·116	8·6191326	·166	6·0093024	·216	4·5822605	·266	3·6629441
·067	14·9253682	·117	8·5453466	·167	5·9727714	·217	4·5600869	·267	3·6477545
·068	14·7058763	·118	8·4728044	·168	5·9366617	·218	4·5381024	·268	3·6326663
·069	14·4927463	·119	8·4014745	·169	5·9009657	·219	4·5163044	·269	3·6176767
0·070	14·2857054	0·120	8·3313262	0·170	5·8656758	0·220	4·4946903	0·270	3·6027863
·071	14·0844963	·121	8·2623301	·171	5·8307850	·221	4·4732575	·271	3·5879935
·072	13·8888760	·122	8·1944575	·172	5·7962859	·222	4·4520036	·272	3·5732972
·073	13·6986147	·123	8·1276809	·173	5·7621718	·223	4·4309262	·273	3·5586962
·074	13·5134952	·124	8·0619734	·174	5·7284359	·224	4·4100228	·274	3·5441897
0·075	13·3333117	0·125	7·9973091	0·175	5·6950714	0·225	4·3892910	0·275	3·5297765
·076	13·1578693	·126	7·9336629	·176	5·6620721	·226	4·3687286	·276	3·5154556
·077	12·9869832	·127	7·8710106	·177	5·6294315	·227	4·3483333	·277	3·5012261
·078	12·8204781	·128	7·8093287	·178	5·5971436	·228	4·3281028	·278	3·4870869
·079	12·6581876	·129	7·7485942	·179	5·5652022	·229	4·3080351	·279	3·4730370
0·080	12·4999534	0·130	7·6887851	0·180	5·5336015	0·230	4·2881278	0·280	3·4590756
·081	12·3456253	·131	7·6298800	·181	5·5023357	·231	4·2683789	·281	3·4452017
·082	12·1950603	·132	7·5718581	·182	5·4713992	·232	4·2487864	·282	3·4314143
·083	12·0481222	·133	7·5146991	·183	5·4407865	·233	4·2293482	·283	3·4177125
·084	11·9046814	·134	7·4583837	·184	5·4104922	·234	4·2100622	·284	3·4040955
0·085	11·7646144	0·135	7·4028927	0·185	5·3805110	0·235	4·1909266	0·285	3·3905624
·086	11·6278033	·136	7·3482078	·186	5·3508378	·236	4·1719393	·286	3·3771122
·087	11·4941358	·137	7·2943110	·187	5·3214675	·237	4·1530985	·287	3·3637441
·088	11·3635044	·138	7·2411851	·188	5·2923952	·238	4·1344024	·288	3·3504572
·089	11·2358068	·139	7·1888131	·189	5·2636161	·239	4·1158490	·289	3·3372508
0·090	11·1109450	0·140	7·1371786	0·190	5·2351255	0·240	4·0974366	0·290	3·3241240
·091	10·9888254	·141	7·0862658	·191	5·2069187	·241	4·0791634	·291	3·3110759
·092	10·8693583	·142	7·0360593	·192	5·1789912	·242	4·0610277	·292	3·2981058
·093	10·7524581	·143	6·9865438	·193	5·1513386	·243	4·0430277	·293	3·2852129
·094	10·6380427	·144	6·9377049	·194	5·1239566	·244	4·0251617	·294	3·2723964
0·095	10·5260334	0·145	6·8895283	0·195	5·0968408	0·245	4·0074282	0·295	3·2596555
·096	10·4163548	·146	6·8420002	·196	5·0699872	·246	3·9898254	·296	3·2469895
·097	10·3089347	·147	6·7951072	·197	5·0433917	·247	3·9723518	·297	3·2343976
·098	10·2037037	·148	6·7488362	·198	5·0170503	·248	3·9550058	·298	3·2218790
·099	10·1005954	·149	6·7031744	·199	4·9909591	·249	3·9377858	·299	3·2094331
0·100	9·9995458	0·150	6·6581095	0·200	4·9651142	0·250	3·9206904	0·300	3·1970591

Table A.3 (*continued*)

a	x	a	x	a	x	a	x	a	x
0·300	3·1970591	0·350	2·6566127	0·400	2·2316119	0·450	1·8847348	0·500	1·5936243
·301	3·1847564	·351	2·6471411	·401	2·2240016	·451	1·8784240	·501	1·5882633
·302	3·1725241	·352	2·6377143	·402	2·2164218	·452	1·8721351	·502	1·5829188
·303	3·1603617	·353	2·6283321	·403	2·2088724	·453	1·8658679	·503	1·5775904
·304	3·1482684	·354	2·6189939	·404	2·2013532	·454	1·8596224	·504	1·5722783
0·305	3·1362436	0·355	2·6096996	0·405	2·1938638	0·455	1·8533984	0·505	1·5669823
·306	3·1242866	·356	2·6004487	·406	2·1864042	·456	1·8471958	·506	1·5617022
·307	3·1123968	·357	2·5912409	·407	2·1789740	·457	1·8410145	·507	1·5564381
·308	3·1005735	·358	2·5820758	·408	2·1715732	·458	1·8348542	·508	1·5511898
·309	3·0888160	·359	2·5729530	·409	2·1642015	·459	1·8287149	·509	1·5459572
0·310	3·0771238	0·360	2·5638723	0·410	2·1568586	0·460	1·8225965	0·510	1·5407403
·311	3·0654961	·361	2·5548333	·411	2·1495444	·461	1·8164989	·511	1·5355389
·312	3·0539325	·362	2·5458356	·412	2·1422588	·462	1·8104218	·512	1·5303531
·313	3·0424323	·363	2·5368790	·413	2·1350014	·463	1·8043652	·513	1·5251826
·314	3·0309949	·364	2·5279631	·414	2·1277721	·464	1·7983290	·514	1·5200275
0·315	3·0196198	0·365	2·5190875	0·415	2·1205707	0·465	1·7923131	0·515	1·5148876
·316	3·0083062	·366	2·5102520	·416	2·1133971	·466	1·7863172	·516	1·5097628
·317	2·9970537	·367	2·5014563	·417	2·1062510	·467	1·7803413	·517	1·5046532
·318	2·9858617	·368	2·4927000	·418	2·0991322	·468	1·7743854	·518	1·4995585
·319	2·9747297	·369	2·4839828	·419	2·0920406	·469	1·7684491	·519	1·4944788
0·320	2·9636570	0·370	2·4753044	0·420	2·0849759	0·470	1·7625325	0·520	1·4894139
·321	2·9526432	·371	2·4666645	·421	2·0779381	·471	1·7566355	·521	1·4843637
·322	2·9416876	·372	2·4580629	·422	2·0709268	·472	1·7507578	·522	1·4793282
·323	2·9307898	·373	2·4494991	·423	2·0639420	·473	1·7448995	·523	1·4743074
·324	2·9199493	·374	2·4409730	·424	2·0569834	·474	1·7390603	·524	1·4693011
0·325	2·9091655	0·375	2·4324843	0·425	2·0500510	0·475	1·7332402	0·525	1·4643092
·326	2·8984379	·376	2·4240326	·426	2·0431444	·476	1·7274391	·526	1·4593317
·327	2·8877660	·377	2·4156176	·427	2·0362635	·477	1·7216568	·527	1·4543685
·328	2·8771494	·378	2·4072391	·428	2·0294083	·478	1·7158932	·528	1·4494195
·329	2·8665874	·379	2·3988969	·429	2·0225784	·479	1·7101483	·529	1·4444847
0·330	2·8560797	0·380	2·3905906	0·430	2·0157738	0·480	1·7044219	0·530	1·4395640
·331	2·8456257	·381	2·3823199	·431	2·0089942	·481	1·6987139	·531	1·4346573
·332	2·8352249	·382	2·3740846	·432	2·0022395	·482	1·6930242	·532	1·4297645
·333	2·8248770	·383	2·3658845	·433	1·9955095	·483	1·6873527	·533	1·4248856
·334	2·8145814	·384	2·3577192	·434	1·9888042	·484	1·6816993	·534	1·4200205
0·335	2·8043377	0·385	2·3495885	0·435	1·9821232	0·485	1·6760639	0·535	1·4151691
·336	2·7941455	·386	2·3414922	·436	1·9754665	·486	1·6704464	·536	1·4103313
·337	2·7840042	·387	2·3334300	·437	1·9688339	·487	1·6648467	·537	1·4055072
·338	2·7739134	·388	2·3254016	·438	1·9622253	·488	1·6592647	·538	1·4006965
·339	2·7638728	·389	2·3174068	·439	1·9556405	·489	1·6537002	·539	1·3958994
0·340	2·7538818	0·390	2·3094453	0·440	1·9490792	0·490	1·6481533	0·540	1·3911155
·341	2·7439401	·391	2·3015170	·441	1·9425415	·491	1·6426237	·541	1·3863450
·342	2·7340472·	·392	2·2936214	·442	1·9360271	·492	1·6371114	·542	1·3815878
·343	2·7242027	·393	2·2857585	·443	1·9295360	·493	1·6316164	·543	1·3768437
·344	2·7144062	·394	2·2779280	·444	1·9230678	·494	1·6261384	·544	1·3721128
0·345	2·7046573	0·395	2·2701297	0·445	1·9166226	0·495	1·6206774	0·545	1·3673948
·346	2·6949555	·396	2·2623632	·446	1·9102001	·496	1·6152333	·546	1·3626899
·347	2·6853006	·397	2·2546284	·447	1·9038002	·497	1·6098061	·547	1·3579979
·348	2·6756921	·398	2·2469251	·448	1·8974228	·498	1·6043955	·548	1·3533187
·349	2·6661296	·399	2·2392530	·449	1·8910677	·499	1·5990016	·549	1·3486523
0·350	2·6566127	0·400	2·2316119	0·450	1·8847348	0·500	1·5936243	0·550	1·3439987

Table A.3 (*continued*)

a	x	a	x	a	x	a	x	a	x
0·550	1·3439987	0·600	1·1262612	0·650	0·9336939	0·700	0·7614337	0·750	0·6058600
·551	1·3393577	·601	1·1221820	·651	·9300635	·701	·7581691	·751	·6028986
·552	1·3347293	·602	1·1181127	·652	·9264412	·702	·7549111	·752	·5999427
·553	1·3301135	·603	1·1140533	·653	·9228268	·703	·7516598	·753	·5969924
·554	1·3255101	·604	1·1100038	·654	·9192204	·704	·7484150	·754	·5940475
0·555	1·3209191	0·605	1·1059641	0·655	0·9156220	0·705	0·7451767	0·755	0·5911081
·556	1·3163405	·606	1·1019342	·656	·9120315	·706	·7419450	·756	·5881742
·557	1·3117742	·607	1·0979141	·657	·9084489	·707	·7387198	·757	·5852457
·558	1·3072201	·608	1·0939037	·658	·9048742	·708	·7355010	·758	·5823227
·559	1·3026782	·609	1·0899029	·659	·9013073	·709	·7322887	·759	·5794050
0·560	1·2981485	0·610	1·0859117	0·660	0·8977481	0·710	0·7290829	0·760	0·5764927
·561	1·2936307	·611	1·0819301	·661	·8941968	·711	·7258834	·761	·5735858
·562	1·2891250	·612	1·0779581	·662	·8906531	·712	·7226903	·762	·5706842
·563	1·2846312	·613	1·0739955	·663	·8871172	·713	·7195036	·763	·5677880
·564	1·2801493	·614	1·0700424	·664	·8835889	·714	·7163232	·764	·5648970
0·565	1·2756792	0·615	1·0660987	0·665	0·8800683	0·715	0·7131491	0·765	0·5620114
·566	1·2712209	·616	1·0621643	·666	·8765553	·716	·7099813	·766	·5591310
·567	1·2667743	·617	1·0582393	·667	·8730498	·717	·7068198	·767	·5562558
·568	1·2623394	·618	1·0543236	·668	·8695519	·718	·7036645	·768	·5533859
·569	1·2579161	·619	1·0504171	·669	·8660616	·719	·7005154	·769	·5505213
0·570	1·2535043	0·620	1·0465198	0·670	0·8625787	0·720	0·6973725	0·770	0·5476618
·571	1·2491040	·621	1·0426317	·671	·8591032	·721	·6942358	·771	·5448075
·572	1·2447151	·622	1·0387527	·672	·8556352	·722	·6911053	·772	·5419584
·573	1·2403377	·623	1·0348828	·673	·8521746	·723	·6879808	·773	·5391144
·574	1·2359715	·624	1·0310219	·674	·8487214	·724	·6848625	·774	·5362756
0·575	1·2316167	0·625	1·0271701	0·675	0·8452755	0·725	0·6817503	0·775	0·5334419
·576	1·2272730	·626	1·0233272	·676	·8418370	·726	·6786441	·776	·5306132
·577	1·2229405	·627	1·0194933	·677	·8384057	·727	·6755440	·777	·5277897
·578	1·2186192	·628	1·0156683	·678	·8349816	·728	·6724499	·778	·5249712
·579	1·2143089	·629	1·0118521	·679	·8315648	·729	·6693618	·779	·5221578
0·580	1·2100097	0·630	1·0080447	0·680	0·8281552	0·730	0·6662796	0·780	0·5193494
·581	1·2057214	·631	1·0042461	·681	·8247528	·731	·6632035	·781	·5165460
·582	1·2014440	·632	1·0004563	·682	·8213575	·732	·6601332	·782	·5137476
·583	1·1971774	·633	0·9966752	·683	·8179694	·733	·6570689	·783	·5109542
·584	1·1929217	·634	0·9929027	·684	·8145883	·734	·6540104	·784	·5081658
0·585	1·1886768	0·635	0·9891389	0·685	0·8112143	0·735	0·6509579	0·785	0·5053823
·586	1·1844426	·636	·9853837	·686	·8078473	·736	·6479112	·786	·5026037
·587	1·1802190	·637	·9816371	·687	·8044874	·737	·6448703	·787	·4998301
·588	1·1760060	·638	·9778989	·688	·8011344	·738	·6418352	·788	·4970613
·589	1·1718036	·639	·9741693	·689	·7977884	·739	·6388059	·789	·4942975
0·590	1·1676118	0·640	0·9704481	0·690	0·7944493	0·740	0·6357824	0·790	0·4915365
·591	1·1634304	·641	·9667354	·691	·7911171	·741	·6327646	·791	·4887843
·592	1·1592594	·642	·9630310	·692	·7877918	·742	·6297526	·792	·4860350
·593	1·1550988	·643	·9593350	·693	·7844733	·743	·6267462	·793	·4832905
·594	1·1509485	·644	·9556473	·694	·7811617	·744	·6237456	·794	·4805508
0·595	1·1468085	0·645	0·9519679	0·695	0·7778568	0·745	0·6207506	0·795	0·4778159
·596	1·1426788	·646	·9482967	·696	·7745587	·746	·6177612	·796	·4750858
·597	1·1385592	·647	·9446338	·697	·7712674	·747	·6147775	·797	·4723604
·598	1·1344498	·648	·9409790	·698	·7679828	·748	·6117994	·798	·4696398
·599	1·1303505	·649	·9373324	·699	·7647049	·749	·6088269	·799	·4669239
0·600	1·1262612	0·650	·9336939	0·700	0·7614337	0·750	0·6058600	0·800	0·4642129

a	x	a	x	a	x	a	x
0·800	0·4642128	0·850	0·3343447	0·900	0·2145557	0·950	0·1034788
·801	·4615063	·851	·3318552	·901	·2122529	·951	·1013381
·802	·4588045	·852	·3293697	·902	·2099535	·952	·0992005
·803	·4561073	·853	·3268862	·903	·2076576	·953	·0970658
·804	·4534148	·854	·3244107	·904	·2053652	·954	·0949342
0·805	0·4507270	0·855	0·3219371	0·905	0·2030762	0·955	0·0928056
·806	·4480438	·856	·3194675	·906	·2007906	·956	·0906799
·807	·4453652	·857	·3170019	·907	·1985084	·957	·0885573
·808	·4426911	·858	·3145402	·908	·1962296	·958	·0864376
·809	·4400217	·859	·3120824	·909	·1939542	·959	·0843209
0·810	0·4373568	0·860	0·3096286	0·910	0·1916824	0·960	0·0822071
·811	·4346965	·861	·3071786	·911	·1894137	·961	·0800963
·812	·4320407	·862	·3047325	·912	·1871484	·962	·0779885
·813	·4293894	·863	·3022903	·913	·1848866	·963	·0758836
·814	·4267426	·864	·2998520	·914	·1826281	·964	·0737816
0·815	0·4241004	0·865	0·2974176	0·915	0·1803729	0·965	0·0716825
·816	·4214626	·866	·2949869	·916	·1781211	·966	·0695864
·817	·4188292	·867	·2925601	·917	·1758725	·967	·0674932
·818	·4162003	·868	·2901372	·918	·1736274	·968	·0654028
·819	·4135759	·869	·2877180	·919	·1713855	·969	·0633154
0·820	0·4109558	0·870	0·2853027	0·920	0·1691469	0·970	0·0612308
·821	·4083402	·871	·2828911	·921	·1669116	·971	·0591492
·822	·4057290	·872	·2804833	·922	·1646795	·972	·0570704
·823	·4031222	·873	·2780793	·923	·1624508	·973	·0549944
·824	·4005197	·874	·2756790	·924	·1602253	·974	·0529213
0·825	0·3979215	0·875	0·2732825	0·925	0·1580031	0·975	0·0508511
·826	·3953278	·876	·2708897	·926	·1557841	·976	·0487837
·827	·3927383	·877	·2685007	·927	·1535683	·977	·0467191
·828	·3901532	·878	·2661153	·928	·1513558	·978	·0446574
·829	·3875723	·879	·2637337	·929	·1491465	·979	·0425985
0·830	0·3849957	0·880	0·2613557	0·930	0·1469404	0·980	0·0405424
·831	·3824235	·881	·2589815	·931	·1447374	·981	·0384891
·832	·3798554	·882	·2566109	·932	·1425377	·982	·0364386
·833	·3772916	·883	·2542439	·933	·1403412	·983	·0343909
·834	·3747321	·884	·2518806	·934	·1381478	·984	·0323460
0·835	0·3721768	0·885	0·2495210	0·935	0·1359576	0·985	0·0303038
·836	·3696257	·886	·2471649	·936	·1337706	·986	·0282644
·837	·3670787	·887	·2448125	·937	·1315867	·987	·0262278
·838	·3645360	·888	·2424637	·938	·1294060	·988	·0241939
·839	·3619974	·889	·2401185	·939	·1272283	·989	·0221628
0·840	0·3594630	0·890	0·2377769	0·940	0·1250538	0·990	0·0201345
·841	·3569327	·891	·2354389	·941	·1228825	·991	·0181088
·842	·3544066	·892	·2331044	·942	·1207142	·992	·0160859
·843	·3518846	·893	·2307735	·943	·1185490	·993	·0140657
·844	·3493667	·894	·2284461	·944	·1163869	·994	·0120482
0·845	0·3468528	0·895	0·2261222	0·945	0·1142279	0·995	0·0100335
·846	·3443431	·896	·2238019	·946	·1120720	·996	·0080214
·847	·3418374	·897	·2214851	·947	·1099191	·997	·0060120
·848	·3393358	·898	·2191718	·948	·1077693	·998	·0040053
·849	·3368383	·899	·2168620	·949	·1056226	·999	·0020013
0·850	0·3343447	0·900	0·2145557	0·950	0·1034788	1·000	0·0000000

A4 Table for finding the maximum-likelihood estimate of P, the parameter of a truncated geometric distribution

In the following table the function

$$f(Q) = \frac{sQ^{s+1} - (s+1)Q^s + 1}{Q^{s+1} - Q^s - Q + 1}$$

is evaluated for different values of $P\ (= 1 - Q)$ and s. Using linear interpolation, this table can be used for solving the equation $f(Q) = \bar{x}$ as described in **4.1.6** (4).

P

s	0·001	0·10	0·20	0·30	0·40	0·50	0·60	0·70	0·80	0·90	0·999
2	1·500	1·474	1·444	1·412	1·375	1·333	1·286	1·231	1·167	1·091	1·001
3	1·999	1·930	1·852	1·767	1·673	1·571	1·462	1·345	1·226	1·108	1·001
4	2·499	2·369	2·225	2·069	1·904	1·733	1·562	1·396	1·244	1·111	1·001
5	2·998	2·790	2·563	2·323	2·078	1·839	1·615	1·416	1·248	1·111	1·001
6	3·497	3·195	2·868	2·533	2·206	1·905	1·642	1·424	1·250	1·111	1·001
7	3·996	3·582	3·142	2·705	2·298	1·945	1·655	1·427	1·250	1·111	1·001
8	4·495	3·953	3·387	2·844	2·363	1·969	1·661	1·428	1·250	1·111	1·001
9	4·993	4·308	3·605	2·955	2·408	1·982	1·664	1·428	1·250	1·111	1·001
10	5·492	4·647	3·797	3·043	2·439	1·990	1·666	1·429	1·250	1·111	1·001
11	5·990	4·969	3·966	3·111	2·460	1·995	1·666	1·429	1·250	1·111	1·001
12	6·488	5·277	4·115	3·165	2·474	1·997	1·666	1·429	1·250	1·111	1·001
13	6·986	5·569	4·244	3·206	2·483	1·998	1·667	1·429	1·250	1·111	1·001
14	7·484	5·847	4·356	3·238	2·489	1·999	1·667	1·429	1·250	1·111	1·001
15	7·981	6·111	4·453	3·262	2·493	2·000	1·667	1·429	1·250	1·111	1·001
16	8·479	6·360	4·537	3·280	2·495	2·000	1·667	1·429	1·250	1·111	1·001
17	8·976	6·597	4·608	3·294	2·497	2·000	1·667	1·429	1·250	1·111	1·001
18	9·473	6·821	4·670	3·304	2·498	2·000	1·667	1·429	1·250	1·111	1·001
19	9·970	7·033	4·722	3·312	2·499	2·000	1·667	1·429	1·250	1·111	1·001
20	10·467	7·232	4·767	3·317	2·499	2·000	1·667	1·429	1·250	1·111	1·001
21	10·963	7·429	4·805	3·322	2·500	2·000	1·667	1·429	1·250	1·111	1·001
22	11·460	7·597	4·836	3·325	2·500	2·000	1·667	1·429	1·250	1·111	1·001
23	11·956	7·763	4·863	3·327	2·500	2·000	1·667	1·429	1·250	1·111	1·001
24	12·452	7·920	4·886	3·329	2·500	2·000	1·667	1·429	1·250	1·111	1·001
25	12·948	8·066	4·905	3·330	2·500	2·000	1·667	1·429	1·250	1·111	1·001
26	13·444	8·204	4·921	3·331	2·500	2·000	1·667	1·429	1·250	1·111	1·001

(Reproduced from Thomasson and Kapadia [1968].)

A5 Tabulation of

$$f(x) = \frac{1}{x} - \frac{1}{\exp(x) - 1}$$

Using linear interpolation, the following table can be used for solving the equation $f(x) = a$ (e.g. see equation (6.5)).

x	·0	·1	·2	·3	·4	·5	·6	·7	·8	·9
0		·4916	·4832	·4750	·4668	·4584	·4504	·4422	·4340	·4260
1	·4180	·4102	·4024	·3946	·3870	·3794	·3720	·3648	·3576	·3504
2	·3434	·3366	·3300	·3234	·3168	·3106	·3044	·2984	·2924	·2866
3	·2810	·2754	·2700	·2648	·2596	·2546	·2496	·2450	·2402	·2358
4	·2314	·2270	·2228	·2188	·2148	·2110	·2072	·2036	·2000	·1966
5	·1932	·1900	·1868	·1836	·1806	·1778	·1748	·1720	·1694	·1668
6	·1642	·1616	·1592	·1568	·1546	·1524	·1502	·1480	·1460	·1440
7	·1420	·1400	·1382	·1364	·1346	·1328	·1310	·1294	·1278	·1262
8	·1246	·1232	·1216	·1202	·1188	·1174	·1160	·1148	·1134	·1122

(Reproduced from Deemer and Votaw [1955].)

A6 Tabulation of

$$A_K(S) = \sum_{k=0}^{K} kS^k \Bigg/ \sum_{k=0}^{K} S^k$$

s	K = 2	3	4	5	6	7	8
.01	.0101	.0110	.0110	.0110	.0110	.0110	.0110
.02	.0204	.0204	.0204	.0204	.0204	.0204	.0204
.03	.0308	.0309	.0309	.0309	.0309	.0309	.0309
.04	.0415	.0416	.0416	.0416	.0416	.0416	.0416
.05	.0523	.0526	.0527	.0527	.0527	.0527	.0527
.06	.0632	.0638	.0638	.0638	.0638	.0638	.0638
.07	.0742	.0742	.0752	.0752	.0752	.0752	.0752
.08	.0854	.0854	.0869	.0869	.0869	.0869	.0869
.09	.0967	.0986	.0989	.0989	.0989	.0989	.0989
.10	.1081	.1107	.1111	.1111	.1111	.1111	.1111
.11	.1196	.1230	.1235	.1236	.1236	.1236	.1236
.12	.1312	.1355	.1363	.1364	.1364	.1364	.1364
.13	.1428	.1483	.1493	.1494	.1495	.1495	.1495
.14	.1545	.1612	.1625	.1627	.1628	.1628	.1628
.15	.1663	.1744	.1761	.1764	.1765	.1765	.1765
.16	.1781	.1879	.1900	.1904	.1905	.1905	.1905
.17	.1900	.2015	.2041	.2047	.2047	.2047	.2047
.18	.2019	.2153	.2186	.2193	.2195	.2195	.2195
.19	.2138	.2294	.2333	.2343	.2345	.2346	.2346
.20	.2258	.2436	.2484	.2496	.2499	.2499	.2499
.21	.2378	.2580	.2638	.2653	.2657	.2658	.2658
.22	.2498	.2727	.2795	.2814	.2818	.2819	.2819
.23	.2618	.2875	.2955	.2978	.2985	.2986	.2987
.24	.2737	.3025	.3118	.3147	.3155	.3158	.3158
.25	.2857	.3177	.3285	.3319	.3329	.3332	.3333
.26	.2977	.3330	.3454	.3495	.3508	.3512	.3513
.27	.3096	.3485	.3627	.3675	.3691	.3696	.3698
.28	.3216	.3641	.3803	.3860	.3879	.3886	.3888
.29	.3335	.3800	.3982	.4049	.4072	.4080	.4083
.30	.3453	.3959	.4164	.4242	.4270	.4281	.4284
.31	.3572	.4120	.4349	.4439	.4474	.4486	.4491
.32	.3690	.4282	.4538	.4642	.4682	.4698	.4704
.33	.3807	.4445	.4729	.4848	.4895	.4914	.4921
.34	.3924	.4610	.4923	.5058	.5114	.5137	.5145
.35	.4041	.4775	.5121	.5274	.5340	.5367	.5377
.36	.4157	.4942	.5321	.5494	.5570	.5603	.5616
.37	.4272	.5109	.5524	.5718	.5806	.5845	.5861
.38	.4387	.5277	.5730	.5948	.6049	.6094	.6114
.39	.4502	.5446	.5938	.6181	.6297	.6350	.6374
.40	.4615	.5616	.6149	.6420	.6552	.6614	.6643
.41	.4728	.5786	.6363	.6663	.6813	.6885	.6920
.42	.4841	.5957	.6579	.6910	.7080	.7164	.7205
.43	.4953	.6128	.6798	.7162	.7353	.7450	.7499
.44	.5064	.6299	.7019	.7418	.7633	.7745	.7801
.45	.5174	.6471	.7242	.7679	.7919	.8047	.8114
.46	.5284	.6644	.7467	.7945	.8212	.8358	.8435
.47	.5392	.6816	.7694	.8214	.8512	.8677	.8767
.48	.5500	.6988	.7923	.8488	.8817	.9005	.9109

470

.49	.5608	.7161	.8154	.8766	.9130	.9341	.9461
.50	.5714	.7333	.8387	.9048	.9449	.9686	.9824
.51	.5820	.7506	.8621	.9333	.9774	1.0041	1.0198
.52	.5925	.7678	.8857	.9623	1.0106	1.0403	1.0583
.53	.6029	.7850	.9094	.9917	1.0445	1.0775	1.0978
.54	.6132	.8022	.9333	1.0214	1.0789	1.1157	1.1387
.55	.6235	.8193	.9573	1.0514	1.1140	1.1547	1.1806
.56	.6336	.8365	.9813	1.0818	1.1497	1.1946	1.2237
.57	.6437	.8535	1.0055	1.1125	1.1860	1.2354	1.2681
.58	.6537	.8705	1.0297	1.1435	1.2229	1.2772	1.3136
.59	.6636	.8875	1.0540	1.1748	1.2604	1.3198	1.3604
.60	.6735	.9044	1.0784	1.2064	1.2984	1.3634	1.4084
.61	.6832	.9213	1.1028	1.2382	1.3370	1.4077	1.4576
.62	.6929	.9380	1.1273	1.2702	1.3761	1.4530	1.5081
.63	.7025	.9548	1.1518	1.3025	1.4157	1.4991	1.5598
.64	.7119	.9714	1.1763	1.3350	1.4557	1.5461	1.6127
.65	.7214	.9880	1.2009	1.3677	1.4963	1.5938	1.6668
.66	.7307	1.0044	1.2254	1.4006	1.5373	1.6423	1.7220
.67	.7399	1.0208	1.2499	1.4336	1.5787	1.6917	1.7786
.68	.7491	1.0371	1.2743	1.4667	1.6205	1.7417	1.8362
.69	.7581	1.0534	1.2988	1.5000	1.6626	1.7925	1.8950
.70	.7671	1.0695	1.3232	1.5333	1.7051	1.8439	1.9549
.71	.7760	1.0855	1.3476	1.5667	1.7479	1.8960	2.0158
.72	.7848	1.1014	1.3719	1.6002	1.7910	1.9487	2.0778
.73	.7936	1.1173	1.3961	1.6338	1.8343	2.0019	2.1407
.74	.8022	1.1330	1.4202	1.6773	1.8779	2.0557	2.2046
.75	.8108	1.1486	1.4443	1.7009	1.9217	2.1100	2.2694
.76	.8193	1.1641	1.4683	1.7345	1.9656	2.1647	2.3350
.77	.8277	1.1795	1.4921	1.7680	2.0097	2.2199	2.4014
.78	.8360	1.1947	1.5159	1.8015	2.0539	2.2754	2.4686
.79	.8443	1.2099	1.5395	1.8350	2.0981	2.3312	2.5364
.80	.8525	1.2249	1.5631	1.8683	2.1424	2.3872	2.6048
.81	.8605	1.2399	1.5865	1.9016	2.1868	2.4436	2.6737
.82	.8686	1.2547	1.6097	1.9348	2.2311	2.5001	2.7432
.83	.8765	1.2694	1.6328	1.9678	2.2754	2.5567	2.8131
.84	.8843	1.2839	1.6558	2.0007	2.3197	2.6135	2.8833
.85	.8921	1.2984	1.6786	2.0335	2.3638	2.6702	2.9538
.86	.8998	1.3127	1.7013	2.0662	2.4079	2.7270	3.0245
.87	.9075	1.3269	1.7238	2.0986	2.4518	2.7838	3.0953
.88	.9150	1.3409	1.7461	2.1309	2.4955	2.8405	3.1662
.89	.9225	1.3549	1.7683	2.1629	2.5391	2.8970	3.2372
.90	.9299	1.3687	1.7903	2.1948	2.5824	2.9534	3.3080
.91	.9372	1.3824	1.8121	2.2264	2.6255	3.0096	3.3788
.92	.9445	1.3960	1.8337	2.2578	2.6684	3.0655	3.4493
.93	.9517	1.4094	1.8552	2.2890	2.7110	3.1212	3.5196
.94	.9588	1.4227	1.8765	2.3200	2.7533	3.1765	3.5896
.95	.9658	1.4359	1.8975	2.3506	2.7953	3.2315	3.6593
.96	.9728	1.4490	1.9184	2.3811	2.8369	3.2861	3.7285
.97	.9797	1.4619	1.9391	2.4112	2.8783	3.3402	3.7972
.98	.9865	1.4747	1.9596	2.4411	2.9192	3.3940	3.8654
.99	.9933	1.4874	1.9799	2.4707	2.9598	3.4472	3.9332
1.00	1.0000	1.5000	2.0000	2.5000	3.0000	3.5000	4.0000

The equation

$$A_K(S) = a \qquad (1)$$

can be solved by using linear interpolation in the above table. However, when $K = 1$, $S = a/(1-a)$, and when $K = 2$:

$$S = \frac{-(1-a) + \sqrt{1 + 6a - 3a^2}}{2(2-a)}.$$

If greater accuracy is needed, or for $K > 8$, (1) must be solved iteratively. In this case $A_K(S)$ can be more conveniently written in the form

$$A_K(S) = \frac{S}{1-S} - \frac{(K+1)S^{K+1}}{1-S^{K+1}},$$

and (1) solved by repeated substitution in the right-hand side of

$$S = a(1-S) + \frac{(K+1)S^{K+1}(1-S)}{1-S^{K+1}}.$$

A7 Uniqueness of a certain polynomial root

If r and n_i $(i = 1, 2, \ldots, s)$ are positive integers such that r is greater than each n_i and $r < \sum_{i=1}^{s} n_i$, then the $(s-1)$th degree polynomial

$$N^{s-1}(N-r) = \prod_{i=1}^{s} (N-n_i)$$

has a unique root greater than r.

Proof:

Let

$$\phi(N) = \frac{N}{N-r} \prod_{i=1}^{s} \left(1 - \frac{n_i}{N}\right),$$

then we wish to prove that $\phi(N) = 1$ has a unique root greater than r. Now $\phi(N)$ is continuous for $N > r$, $\phi(r+0) = \infty$ and

$$\phi(N) = 1 - (\Sigma n_i - r)/N + O(1/N^2)$$
$$< 1$$

for large N. Hence $\phi(N) = 1$ has at least one root greater than r. If we can now show that $\phi'(N) = 0$ for only *one* value of N greater than r then this root is unique.

By taking logarithms we see that

$$\phi'(N) = \frac{\phi(N)}{N}\left[\sum_{i=1}^{s}\frac{n_i}{N-n_i} - \frac{r}{N-r}\right]$$
$$= \phi(N)\,\theta(N), \text{ say,}$$

and we wish to prove that $\theta(N) = 0$ has a unique root greater than r. To do this we use a similar argument to that given above in considering $\phi(N) = 1$.

Thus $\theta(N)$ is continuous for $N > r$, $\theta(r+0) = -\infty$,

$$\theta(N) = \frac{1}{N}\left[\frac{(\Sigma\, n_i - r)}{N} + O\left(\frac{1}{N^2}\right)\right] = 0+$$

for large N and $\theta(N) = 0$ has at least one root N_0, say, greater than r. Now

$$\theta'(N_0) = \frac{r}{(N_0 - r)^2} - \sum\frac{n_i}{(N_0 - n_i)^2}$$

$$= \sum\frac{n_i}{(N_0 - n_i)(N_0 - r)} - \sum\frac{n_i}{(N_0 - n_i)^2}$$

$$> 0,$$

since $r > n_i$ and

$$\sum\frac{n_i}{N_0 - n_i} = \frac{r}{N_0 - r}.$$

Therefore $\theta'(N)$ is positive at every root of $\theta(N) = 0$ greater than r, and this can only be true if there is just one root. This in turn implies that $\phi(N) = 1$ has a unique root greater than r.

(The above proof was motivated by some unpublished work of Professor J.N. Darroch.)

REFERENCES

Adams, L. (1951). Confidence limits for the Petersen or Lincoln index used in animal population studies. *J. Wildl. Manag.* 15, 13–19. [63–4]

Adams, L. (1959). An analysis of a population of snowshoe hares in northwestern Montana. *Ecol. Monogr.* 29, 141–70. [54]

Adams, L. and Davis, S.D. (1967). The internal anatomy of home range. *J. Mammal.* 48, 529–36. [20, 94]

Allee, W.C., Emerson, A.E., Park, O., Park, T. and Schmidt, K.P. (1949). *Principles of Animal Ecology.* W.B. Saunders Co.: Philadelphia. [393]

Allen, D.L. (1942). A pheasant inventory method based upon kill records and sex ratios. *Trans. Nth Amer. Wildl. Conf.* 7, 329–33. [353]

Allen, K.R. (1966). Some methods for estimating exploited populations. *J. Fish. Res. Bd. Canada* 23, 1553–74. [336–40]

Allen, K.R. (1968). Simplification of a method of computing recruitment rates. *J. Fish. Res. Bd. Canada* 25, 2701–2. [336–40]

Amman, G.D. and Baldwin, P.H. (1960). A comparison of methods for censusing woodpeckers in spruce-fir forests of Colorado. *Ecology* 41, 699–706. [21, 35]

Ammann, G.A. and Ryel, L.A. (1963). Extensive methods of inventorying ruffed grouse in Michigan. *J. Wildl. Manag.* 27, 617–33. [20, 446, 450]

Andersen, J. (1962). Roe-deer census and population analysis by means of modified marking and release technique. *In* E.D. Le Cren and M.W. Holdgate (Editors), *The Exploitation of Natural Animal Populations,* 72–82. Blackwell: Oxford. [110]

Anderson, D.R. and Pospahala, R.S. (1970). Correction of bias in belt transect studies of immotile objects. *J. Wildl. Manag.* 34, 141–6. [28]

Andrewartha, H.G. (1961). *Introduction to the Study of Animal Populations.* Methuen: London. [25, 458]

Andrzejewski, R. (1963). Processes of incoming, settlement and disappearance of individuals and variations in the numbers of small rodents. *Acta theriologica* 7, 169–213. [19]

Andrzejewski, R. (1967). Estimation of the abundance of small rodent populations for the use of biological productivity investigations. *In* K. Petrusewicz (ed.), *Secondary Productivity of Terrestrial Ecosystems,* vol. 1, 275–81. Institute of Ecology, Polish Academy of Sciences. [19]

Andrzejewski, R., Bujalska, G., Ryszkowski, L. and Ustyniuk, J. (1966). On a relation between the number of traps in a point of catch and trappability of small rodents. *Acta theriologica* 11, 343–9. [83]

Andrzejewski, R. and Jezierski, W. (1966). Studies on the European hare. XI: Estimation of population density and attempt to plan the yearly take of hares. *Acta theriologica* 11, 433–48. [327]

Andrzejewski, R. and Wierzbowska, T. (1961). An attempt at assessing the duration of residence of small rodents in a defined forest area and the rate of interchange between individuals. *Acta theriologica* 5, 153–72. [181]

Ashby, K.R. (1967). Studies in the ecology of field mice and voles (*Apodemus sylvaticus, Clethrionomys glareolus and Microtus agrestis*) in Houghall Wood, Durham. *J. Zool., Lond.* 152, 389–513. [450]

Ashford, J.R., Read, K.L.Q. and Vickers, G.G. (1970). A system of stochastic models applicable to studies of animal population dynamics. *J. Animal Ecol.* 39, 29–50. [399]

REFERENCES

Atwood, E.L. (1956). Validity of mail survey data on bagged waterfowl. *J. Wildl. Manag.* **20**, 1—16.   [360]

Atwood, E.L. and Geis, A.D. (1960). Problems associated with practices that increase the reported recoveries of waterfowl bands. *J. Wildl. Manag.* **24**, 272—9.   [97]

Ayre, G.L. (1962). Problems in using the Lincoln Index for estimating the size of ant colonies (*Hymenoptera: Formicidae*). *J.N.Y. Ent. Soc.* **70**, 159—66.   [72]

Babińska, J. and Bock, E. (1969). The effect of pre-baiting on captures of rodents. *Acta theriologica* **14**, 267—70.

Backiel, T. (1964). Tag detachment in *Vimba vimba* L. (in Polish). *Roczniki Nauk Rolniczych* **84—B**, 241—53.   [281]

Bailey, G.N.A. (1968). Trap-shyness in a woodland population of Bank voles (*Clethrionomys glareolus*). *J. Zool., Lond.* **156**, 517—21.   [20, 84]

Bailey, J.A. (1969). Trap responses of wild cottontails. *J. Wildl. Manag.* **33**, 48—58.   [84]

Bailey, N.T.J. (1951). On estimating the size of mobile populations from capture—recapture data. *Biometrika* **38**, 293—306.   [61, 118, 220, 452]

Bailey, N.T.J. (1952). Improvements in the interpretation of recapture data. *J. Animal Ecol.* **21**, 120—7.   [61, 220]

Balph, D.F. (1968). Behavioural responses of unconfined Uinta ground squirrels to trapping. *J. Wildl. Manag.* **32**, 778—94.   [84]

Banks, C.J. and Brown, E.S. (1962). A comparison of methods of estimating population density of adult sunn pest, *Eurygaster integriceps* Put. (Hemiptera, Scutelleridae) in wheat fields. *Ent. Exp. and Appl.* **5**, 255—60.   [83, 114—15]

Baranov, T.I. (1918). On the question of the biological basis of fisheries. *Rep. Div. Fish Management and Scientific Study of the Fishing Industry* **1**, 81—128.   [329]

Barbehenn, K.R. (1958). Spatial and population relationships between *Microtus* and *Blarina*. *Ecology* **39**, 293—304.   [327]

Barkalow, F.S. Jr., Hamilton, R.B. and Soots, R.F. Jr. (1970). The vital statistics of an unexploited gray squirrel population. *J. Wildl. Manag.* **34**, 489—500.   [383]

Barnett, V.D. (1966). Evaluation of the maximum-likelihood estimator when the likelihood equation has multiple roots. *Biometrika* **53**, 151—65.   [17]

Bartlett, M.S. (1960). *Stochastic Population Models in Ecology and Epidemiology.* Methuen's Monographs on Applied Probability and Statistics: London.   [458]

Barton, D.E., David, F.N. and Merrington, M. (1960). Tables for the solution of the exponential equation, $\exp(-a) + ka = 1$. *Biometrika* **47**, 439—45.   [464—7]

Batcheler, C.L. and Bell, D.J. (1970). Experiments in estimating density from joint point- and nearest-neighbour distance samples. *Proc. N.Z. Ecol. Soc.* **17**, 111—17.   [51]

Beaver, R.A. (1966). The development and expression of population tables for the bark beetle *Scolytus scolytus* (F.). *J. Animal Ecol.* **35**, 27—41.   [398]

Becker, G.F. and Van Orstrand, C.E. (1924). *Smithsonian Mathematical Tables: Hyperbolic Functions.* Smithsonian Institute: Washington.   [136]

Bellrose, F.C. (1955). A comparison of recoveries from reward and standard bands. *J. Wildl. Manag.* **19**, 71—5.   [97, 250]

Bellrose, F.C. and Chase, E.B. (1950). Population losses in the mallard, black duck, and blue-winged teal. *Illinois Nat. Hist. Survey Biol. Notes* **22**, 1—27.   [253]

Bergerud, A.T. and Mercer, W.E. (1966). Census of the willow ptarmigan in Newfoundland. *J. Wildl. Manag.* **30**, 101—13.   [446]

Berryman, A.A. (1968). Development of sampling techniques and life tables for the fir engraver *Scolytus ventralis*. *Canadian Entomologist* **100**, 1138—47.   [398]

Beverton, R.J.H. (1954). *Notes on the use of theoretical models in the study of exploited fish populations.* Misc. Contr. No. 2, U.S. Fishery Lab., Beaufort, N.C., 1—159.   [276]

475

REFERENCES

Beverton, R.J.H. and Holt, S.J. (1957). *On the Dynamics of Exploited Fish Popu-Lations*. London: Her Majesty's Stationery Office. [2, 94, 281, 306, 329, 431, 458]

Birch, M.W. (1964). A new proof of the Pearson—Fisher theorem. *Ann. Math. Statist.* **35**, 817—24. [15]

Blackith, R.E. (1958). Nearest-neighbour distance measurements for the estimation of animal populations. *Ecology* **39**, 147—50. [41, 48, 50]

Blackith, R.E., Siddorn, J.W., Waloff, N. and Emden, H.F. Van (1963). Mound nests of the yellow ant, *Lasius flavus* L., on waterlogged pasture in Devonshire. *Ent. Mon. Mag.* **99**, 48—9. [41]

Bliss, C.I. and Fisher, R.A. (1953). Fitting the negative binomial distribution to biological data; note on the efficient fitting of the negative binomial. *Biometrics* **9**, 176—96; 197—9. [175]

Boguslavsky, G.W. (1955). A mathematical model for conditioning. *Psychometrika* **20**, 125—38. [190]

Boguslavsky, G.W. (1956). Statistical estimation of the size of a small population. *Science* **124**, 317—18. [191—2]

Bohrnstedt, G.W. and Goldberger, A.S. (1969). On the exact covariance of products of random variables. *J. Amer. Statist. Assoc.* **64**, 1439—42. [9]

Bole, B.P. (1939). The quadrat method of studying small mammal populations. *Cleveland Mus. Nat. Hist. Sci. Publ.* No. 5. [21—2]

Bouck, G.R. and Ball, R.C. (1966). Influence of capture methods on blook characteristics and mortality in the rainbow trout (*Salmo gairdneri*). *Trans. Amer. Fish. Soc.* **95**, 170—6. [83]

Boyd, H. (1956). Statistics of the British population of the pink-footed goose. *J. Animal Ecol.* **25**, 253—73. [244—5]

Braaten, D.O. (1969). Robustness of the De Lury population estimator. *J. Fish. Res. Bd. Canada* **26**, 339—55. [304]

Brander, R.B. and Cochran, W.W. (1969). Radio location telemetry. *In* R.H. Giles, Jr. (ed.), *Wildlife Management Techniques*, 3rd edn, 95—103. The Wildlife Society: Washington. [94]

Brass, W. (1958). Simplified methods of fitting the truncated negative binomial distribution. *Biometrika* **45**, 59—68. [175]

Breakey, D.R. (1963). The breeding season and age structure of feral mouse populations near San Francisco Bay, California. *J. Mammal.* **44**, 153—68. [403]

Brock, V.E. (1954). A preliminary report on a method of estimating reef fish populations. *J. Wildl. Manag.* **18**, 297—308. [28]

Brooks, G.R. Jr. (1967). Population ecology of the ground skink, *Lygosoma laterale* (Say). *Ecol. Monogr.* **37**, 71—87. [400]

Brown, L.E. (1954). Small mammal populations at Silwood Park Field Centre, Berkshire, England. *J. Mammal.* **35**, 161—76. [84]

Brunk, H.D. (1965). *An Introduction to Mathematical Statistics*, 2nd edn. Blaisdell: Waltham, Massachusetts. [186]

Buchalczyk, T. and Pucek, Z. (1968). Estimation of the numbers of *Microtus oeconomus* using the Standard Minimum method. *Acta theriologica* **14**, 461—82. [327]

Buck, D.H. and Thoits, C.F. 3rd (1965). An evaluation of Petersen estimation procedures employing seines in 1-acre ponds. *J. Wildl. Manag.* **29**, 598—621. [19, 86, 451]

Buckland, W.R. (1964). *Statistical Assessment of the Life Characteristic: a Bibliographic Guide*. Griffin's Statistical Monographs and Courses No. 13. Griffin: London. [287]

REFERENCES

Buckner, C.H. (1957). Population studies on small mammals of southeastern Manitoba. *J. Mammal.* **38**, 87—97. [83]

Burkitt, J.P. (1926). A study of the robin by means of marked birds. *Brit. Birds* **20**, 91—101. [397]

Calhoun, J.B. (1948). Announcement of program. North American census of small mammals. Release No. 1 (Rodent Ecology Project, Johns Hopkins University), 8 pp. [450]

Calhoun, J.B. and Casby, J.U. (1958). Calculation of home range and density of small mammals. U.S. Dept. of Health, Education, and Welfare, *Public Health Monogr.* **55**, 1—24. [171]

Carney, S.M. and Petrides, G.A. (1957). Analysis of variation among participants in pheasant cock-crowing censuses. *J. Wildl. Manag.* **21**, 392—7. [54]

Carothers, A.D. (1971). An examination and extension of Leslie's test of equal catchability. *Biometrics* **27**, 615—30. [161, 162, 228]

Caughley, G. (1966). Mortality patterns in mammals. *Ecology* **47**, 906—18. [393, 400—7]

Caughley, G. (1970). Eruption of ungulate populations, with emphasis on the Himalayan thar in New Zealand. *Ecology* **51**, 53—72. [407]

Chakravarti, I.M. and Rao, C.R. (1959). Tables for some small sample tests of significance for Poisson distributions and 2×3 contingency tables. *Sankhyā* **21**, 315—26. [170]

Chapman, D.G. (1948). A mathematical study of confidence limits of salmon populations calculated from sample tag ratios. *Internat. Pac. Salmon Fisheries Comm. Bull.* **2**, 69—85. [63—4, 460]

Chapman, D.G. (1951). Some properties of the hypergeometric distribution with applications to zoological censuses. *Univ. Calif. Public. Stat.* **1**, 131—60. [60, 100, 122, 125]

Chapman, D.G. (1952). Inverse, multiple and sequential sample censuses. *Biometrics* **8**, 286—306. [71, 119—21, 131—3, 138—9, 157—8, 187]

Chapman, D.G. (1954). The estimation of biological populations. *Ann. Math. Statist.* **25**, 1—15. [189, 238, 298, 309, 353, 355—6]

Chapman, D.G. (1955). Population estimation based on change of composition caused by selective removal. *Biometrika* **42**, 279—90. [353, 358, 362—3, 372—3, 377—9, 456—7]

Chapman, D.G. (1956). Estimating the parameters of a truncated Gamma distribution. *Ann. Math. Statist.* **27**, 498—506. [277]

Chapman, D.G. (1961). Statistical problems in the dynamics of exploited fish populations. *Proc. 4th Berkeley Symp. 1960*, vol. 4, 153—68. [272—3, 304, 328—31, 344—5]

Chapman, D.G. (1964). A critical study of Pribilof fur seal population estimates. *Fishery Bulletin* **63**, 657—69. [446]

Chapman, D.G. (1965). The estimation of mortality and recruitment from a single-tagging experiment. *Biometrics* **21**, 529—42. [266—71, 347—52]

Chapman, G.G. and Johnson, A.M. (1968). Estimation of fur seal pup populations by randomised sampling. *Trans. Amer. Fish. Soc.* **97**, 264—70. [117]

Chapman, D.G. and Junge, C.O. (1956). The estimation of the size of a stratified animal population. *Ann. Math. Statist.* **27**, 375—89. [431—8]

Chapman, D.G. and Murphy, G.I. (1965). Estimates of mortality and population from survey—removal records. *Biometrics* **21**, 921—35. [353, 356, 376—7, 384—92]

Chapman, D.G. and Overton, W.S. (1966). Estimating and testing differences between population levels by the Schnabel estimation method. *J. Wildl. Manag.* **30**, 173—80. [122—5, 140]

477

REFERENCES

Chapman, D.G. and Robson, D.S. (1960). The analysis of a catch curve. *Biometrics* **16**, 354—68. [171, 414—30]

Chesness, R.A. and Nelson, W.M. (1964). Illegal kill of hen pheasants in Minnesota. *J. Wildl. Manag.* **28**, 249—53. [361]

Chew, R.M. and Butterworth, B.B. (1964). Ecology of rodents in Indian Cove (Mojave Desert), Joshua Tree National Monument, California. *J. Mammal.* **45**, 203—25. [314]

Chiang, C.L. (1960a). A stochastic study of the life table and its applications. I: Probability distributions of the biometric functions. *Biometrics* **16**, 618—35. [408—12]

Chiang, C.L. (1960b). A stochastic study of the life table and its applications. II: Sample variance of the observed expectation of life and other biometric functions. *Human Biology* **32**, 221—38. [408—12]

Chitty, D. and Kempson, D.A. (1949). Prebaiting small mammals and a new design of live-trap. *Ecology* **30**, 536—42. [84]

Chitty, D. and Shorten, M. (1946). Techniques for the study of the Norway rat (*Rattus norvegicus*). *J. Mammal.* **27**, 63—78. [82—3]

Chung, J.H. and De Lury, D.B. (1950). *Confidence Limits for the Hypergeometric Distribution*, University of Toronto Press. [62]

Clancy, D.W. (1963). The effect of tagging with Petersen disc tags on the swimming ability of fingerling Steelhead Trout (*Salmo gairdneri*). *J. Fish. Res. Bd. Canada* **20**, 969—81. [83]

Clapham A.R. (1932). The form of the observational unit in quantitative ecology. *J. Ecology* **20**, 192—7. [21]

Clark, P.J. and Evans, F.C. (1954). Distance to nearest neighbor as a measure of spatial relationships in populations. *Ecology* **35**, 445—53. [43, 45, 49, 50]

Clark, P.J. and Evans, F.C. (1955). Some aspects of spatial pattern in biological populations. *Science* **121**, 397—8. [43]

Cochran, W.G. (1954). Some methods for strengthening the common chi-squared tests. *Biometrics* **10**, 417—51. [163]

Cochran, W.G. (1963). *Sampling Techniques*, 2nd edn. John Wiley and Sons: New York. [5, 20, 23, 27, 55, 63, 64, 112—15, 134, 425, 426]

Cohen, A., Peters, H.S. and Foote, L.E. (1960). Calling behaviour of mourning doves in two midwest life zones. *J. Wildl. Manag.* **24**, 203—12. [54]

Comfort, A. (1957). Survival curves of mammals in captivity. *Proc. Zool. Soc. London* **128**, 349—64. [396, 400, 409]

Comrie, L.J. (1959). *Chambers' Six-figure Mathematical Tables*, vol. 2 — *Natural Values*. Chambers Ltd.: London. [136]

Corbet, P.S. (1952). An adult population study of *Pyrrhosoma Nymphula* (Sulzer): (Odonata: Coenagrionidae). *J. Animal Ecol.* **21**, 206—22. [82]

Cormack, R.M. (1964). Estimates of survival from the sighting of marked animals. *Biometrika* **51**, 429—38. [214—17]

Cormack, R.M. (1966). A test for equal catchability. *Biometrics* **22**, 330—42. [88—93]

Cormack, R.M. (1968). The statistics of capture—recapture methods. *Oceanogr. Mar. Bio. Ann. Rev.* **6**, 455—506. [2, 187, 204, 208, 451]

Cormack, R.M. (1972). The logic of capture—recapture estimates. *Biometrics* **28**, (to appear). [205, 223]

Cottam, C. (1956). Uses of marking animals in ecological studies: marking birds for scientific purposes. *Ecology* **37**, 674—81. [93]

Cottam, G. and Curtis, J.T. (1949). A method for making rapid surveys of woodlands by means of pairs of randomly selected trees. *Ecology* **30**, 101—4. [50]

Cottam, G. and Curtis, J.T. (1955). Correction for various exclusion angles in the random pairs methods. *Ecology* **36**, 767. [50]

REFERENCES

Cottam, G. and Curtis, J.T. (1956). The use of distance measures in phytosocio-logical sampling. *Ecology* 37, 451—60. [44, 50]

Cottam, G., Curtis, J.T. and Hale, B.W. (1953). Some sampling characteristics of a population of randomly dispersed individuals. *Ecology* 34, 741—57. [44]

Coulson, J.C. (1960). A study of the mortality of the starling based on ringing re-coveries. *J. Animal Ecol.* 29, 251—71. [245, 250]

Cox, D.R. and Lewis, P.A.W. (1966). *The Statistical Analysis of Series of Events.* Methuen: London. [46]

Craig, C.C. (1953a). On a method of estimating biological populations in the field. *Biometrika* 40, 216—18. [47—9]

Craig, C.C. (1953b). Use of marked specimens in estimating populations. *Biometrika* 40, 170—6. [136—7, 170]

Crow, E.L. (1956). Confidence limits for a proportion. *Biometrika* 43, 423—35. [123]

Crow, E.L. and Gardner, R.S. (1959). Confidence intervals for the expectation of a Poisson variable. *Biometrika* 46, 441—53. [63, 123]

Crowcroft, P. and Jeffers, J.N.R. (1961). Variability in the behaviour of wild house mice (*Mus musculus L.*) towards live-traps. *Proc. Zool. Soc. London* 137, 573—82. [84, 162, 165]

Cucin, D. and Regier, H.A. (1966). Dynamics and exploitation of lake whitefish in Southern Georgian Bay. *J. Fish. Res. Bd. Canada* 23, 221—74. [283]

Cushing, D.H. (1964). The counting of fish with an echo-sounder. *Rapp. P.-v. Réun. Cons. perm. int. Explor. Mer.* 155, 190—5. [19]

Cushing, D.H. (1968). *Fisheries Biology: a Study in Population Dynamics.* University of Wisconsin Press. [19, 458]

Czen Pin (1962). O minimaksowym estymatorze liczności populacji. *Zastosow. Mat.* 6, 137—48. [121]

Dacey, M.F. (1963). Order neighbor statistics for a class of random patterns in multidimensional space. *Ann. Assoc. Amer. Geographers,* 53, 505—15. [42]

Dacey, M.F. (1964a). Two-dimensional random point pattern: a review and an inter-pretation. *Papers, The Regional Science Association,* 13, 41—55. [24, 42]

Dacey, M.F. (1964b). Modified Poisson probability law for point pattern more regular than random. *Ann. Assoc. Amer. Geographers,* 54, 559—65. [41]

Dacey, M.F. (1965). Order distance in an inhomogeneous random point pattern. *The Canadian Geographer* 9, 144—53. [41]

Dacey, M.F. (1966). A compound probability law for a pattern more dispersed than random and with areal inhomogeneity. *Economic Geography* 42, 172—9. [41]

Dale, M.B. (1970). Systems analysis and ecology. *Ecology* 51, 2—16. [458]

Darling, D.A. and Robbins, H. (1967). Finding the size of a finite population. *Ann. Math. Statist.* 38, 1392—8. [191—3]

Darroch, J.N. (1958). The multiple-recapture census. I: Estimation of a closed population. *Biometrika* 45, 343—59. [131—4, 136, 139, 154, 164, 189—91, 234]

Darroch, J.N. (1959). The multiple-recapture census. II: Estimation when there is immigration or death. *Biometrika* 46, 336—51. [15, 218—19]

Darroch, J.N. (1961). The two-sample capture—recapture census when tagging and sampling are stratified. *Biometrika* 48, 241—60. [431—45]

Dasmann, R.F. (1952). Methods for estimating deer populations from kill data. *Calif. Fish and Game* 38(2), 225—33. [353]

David, F.N. and Johnson, N.L. (1952). The truncated Poisson. *Biometrics* 8, 275—85. [170]

Davis, D.E. (1957). Observations on the abundance of Korean mice. *J. Mammal.* 38, 374—7. [315]

Davis, D.E. (1963). Estimating the numbers of game populations. *In* H.S. Mosby (ed.), *Wildlife Investigational Techniques*, 2nd edn, 89—118. The Wildlife Society: Washington.   [19, 53—5, 57, 146, 376, 447]

Davis, D.E., Christian, J.J. and Bronson, F. (1964). Effect of exploitation on birth, mortality and movement rates in a woodchuck population. *J. Wildl. Manag.* **28**, 1—9.   [233]

Davis, W.S. (1964). Graphic representation of confidence intervals for Petersen population estimates. *Trans. Amer. Fish. Soc.* **93**, 227—32.   [64]

Debauche, H.R. (1962). The structural analysis of animal communities of the soil. *In* P.W. Murphy (ed.), *Progress in Soil Zoology*, 10—25. Butterworths: London.   [25]

Deemer, W.L. and Votaw, D.V. Jr. (1955). Estimation of parameters of truncated or censored exponential distributions. *Ann. Math. Statist.* **26**, 498—504.   [273, 345, 469]

Deevey, E.S. Jr. (1947). Life tables for natural populations of animals. *Quart. Rev. Biol.* **22**, 283—314.   [393, 395]

Delong, K.T. (1966). Population ecology of feral house mice: interference by *Microtus*. *Ecology* **47**, 481—4.   [160, 205]

De Lury, D.B. (1947). On the estimation of biological populations. *Biometrics* **3**, 145—67.   [298—300, 303]

De Lury, D.B. (1951). On the planning of experiments for the estimation of fish populations. *J. Fish. Res. Bd. Canada* **8**, 281—307.   [298, 305, 307]

De Lury, D.B. (1954). On the assumptions underlying estimates of mobile populations. *In* O. Kempthorne (ed.), *Statistics and Mathematics in Biology*. Iowa State College Press: Ames, 287—93.   [2]

De Lury, D.B. (1958). The estimation of population size by marking and recapture procedure. *J. Fish. Res. Bd. Canada* **15**, 19—25.   [126, 142]

Dice, L.R. (1938). Some census methods for mammals. *J. Wildl. Manag.* **2**, 119—30.   [19, 51]

Dice, L.R. (1952). *Natural Communities*. University of Michigan Press: Ann Arbor.   [22, 52—6]

Dickie, L.M. (1955). Fluctuations in abundance of the giant scallop, *Placopecten magellanicus* (Gmelin), in the Digby area of the Bay of Fundy. *J. Fish. Res. Bd. Canada* **12**, 797—857.   [307—8]

Dobson, R.M. (1962). Marking techniques and their application to the study of small terrestrial animals. *In* P.W. Murphy (ed.), *Progress in Soil Zoology*, 228—39. Butterworths: London.   [93—4]

Dobzhansky, T., Cooper, D.M., Phaff, H.J., Knapp, E.P. and Carson, H.L. (1956). Studies on the ecology of Drosophila in the Yosemite region of California. IV: Differential attraction of species of Drosophila to different species of yeast. *Ecology*, **37**, 544—50.   [82]

Dole, J.W. (1965). Summer movements of adult leopard frogs, *Rana pipiens* (Schreber), in Northern Michigan. *Ecology* **46**, 236—55.   [20]

Dorney, R.S. (1958). Ruffed grouse roosts as a spring-census technique. *J. Wildl. Manag.* **22**, 97—9.   [54]

Dorney, R.S., Thompson, D.R., Hale, J.B. and Wendt, R.F. (1958). An evaluation of ruffed grouse drumming counts. *J. Wildl. Manag.* **22**, 35—40.   [54]

Draper, N.R. and Smith, H. (1966). *Applied Regression Analysis*. Wiley: New York.   [10, 11, 268, 305]

Dunnet, G.M. (1957). A test of the recapture method of estimating the number of rabbits, *Oryctolagus cuniculus* (L). *C.S.I.R.O. Wildl. Res.* **2**, 90—100.   [84, 446, 451]

REFERENCES

Dunnet, G.M. (1963). A population study of the quokka, *Setonix brachyurus* Quoy and Gaimard (Marsupialia). *C.S.I.R.O. Wildl. Res.* **8**, 78—117. [71, 160]

Dyer, M.I. (1967). Photo-electric cell technique for analyzing radar film. *J. Wildl. Manag.* **31**, 484—91. [19]

Eberhardt, L.L. (1967). Some developments in 'distance sampling'. *Biometrics* **23**, 207—16. [41—9]

Eberhardt, L.L. (1968). A preliminary appraisal of line transects. *J. Wildl. Manag.* **32**, 82—8. [29, 34—5]

Eberhardt, L.L. (1969a). Population estimates from recapture frequencies. *J. Wildl. Manag.* **33**, 28—39. [62, 88, 171, 174, 178, 451]

Eberhardt, L.L. (1969b). Population analysis. *In* R.H. Giles, Jr. (ed.), *Wildlife Management Techniques*, 3rd edn. 457—95. The Wildlife Society: Washington. [393]

Eberhardt, L.L., Peterle, T.J. and Schofield, R. (1963). Problems in a rabbit population study. *Wildl. Monogr.* **10**, 1—51. [84, 170, 174, 299, 383, 446]

Edwards, W.R. and Eberhardt, L.L. (1967). Estimating cottontail abundance from live-trapping data. *J. Wildl. Manag.* **31**, 87—96. [84, 165, 171—2]

Einarsen, A.S. (1945). Quadrat inventory of pheasant trends in Oregon. *J. Wildl. Manag.* **9**, 121—31. [21]

Eleveback, L. and Varma, A. (1965). A selected bibliography on simulation and empirical sampling. *N.Y. Statistician* 16: No. VI. [458]

Elling, C.H. and Macy, P.T. (1955). Pink salmon tagging experiments in Icy Strait and Upper Chatham Strait, 1950. *U.S. Dept. Int. Fish. and Wildlife Ser. Fish. Bull.* **100**, 331—71. [290, 294]

Ellis, J.A., Westemeier, R.L., Thomas, K.P. and Norton, H.W. (1969). Spatial relationships among quail coveys. *J. Wildl. Manag.* **33**, 249—54. [41]

Emlen, J.T. Jr., Hine, R.L., Fuller, W.A. and Alfonso, P. (1957). Dropping boards for population studies of small mammals. *J. Wildl. Manag.* **21**, 300—14. [54]

Epstein, B. (1954). Truncated life tests in the exponential case. *Ann. Math. Statist.* **25**, 555—64. [287]

Epstein, B. (1960a). Tests for the validity of the assumption that the underlying distribution of life is exponential: Part I. *Technometrics* **2**, 83—101. [46, 287]

Epstein, B. (1960b). Tests for the validity of the assumption that the underlying distribution of life is exponential: Part II. *Technometrics* **2**, 167—83. [46, 237]

Epstein, B. (1960c). Statistical life test acceptance procedures. *Technometrics* **2**, 435—46. [287]

Epstein, B. (1960d). Estimation from life test data. *Technometrics* **2**, 447—54. [287]

Epstein, B. and Sobel, M. (1953). Life testing. *J. Amer. Statist. Assoc.* **48**, 486—502. [287]

Epstein, B. and Sobel, M. (1954). Some theorems relevant to life testing from an exponential distribution. *Ann. Math. Statist.* **25**, 373—81. [287]

Evans, C.D.,. Troyer, W.A. and Lensink, C.J. (1966). Aerial census of moose by quadrat sampling units. *J. Wildl. Manag.* **30**, 767—76. [28]

Evans, F.C. (1951). Notes on a population of the striped ground squirrel (*Citellus tridecemlineatus*) in an abandoned field in southeastern Michigan. *J. Mammal.* **32**, 437—49. [84]

Farner, D.S. (1945). Age groups and longevity in the American robin. *Wilson Bull.* **57**, 56—74. [247]

Farner, D.S. (1949). Age groups and longevity in the American robin: comments, further discussion and certain revisions. *Wilson Bull.* **61**, 68—81. [247]

Farner, D.S. (1955). Birdbanding in the study of population dynamics. *In* A. Wolfson (ed.), *Recent Studies in Avian Biology*, 397—449. University of Illinois Press, Urbana. [247, 251, 397]

Feller, W. (1957). *An Introduction to Probability Theory and its Applications*, vol. 1, 2nd edn. Wiley: New York. [3, 15, 260]

Fieller, E.C. (1940). The biological standardisation of insulin. *J. Roy. Statist. Soc. Suppl.* **7**, 1—65. [11]

Fischler, K.J. (1965). The use of catch—effort, catch—sampling, and tagging data to estimate a population of blue crabs. *Trans. Amer. Fish. Soc.* **94**, 287—310. [261—3, 300—2, 335—6, 446]

Fisher, H.I., Hiatt, R.W. and Bergeson, W. (1947). The validity of the roadside census as applied to pheasants. *J. Wildl. Manag.* **11**, 205—31. [53]

Fisher, R.A. (1954). *Statistical Methods for Research Workers*, 12th edn. Oliver and Boyd: Edinburgh. [55]

Flyger, V.F. (1959). A comparison of methods for estimating squirrel populations. *J. Wildl. Manag.* **23**, 220—3. [84, 172]

Flyger, V.F. (1960). Movements and home range of the gray squirrel *Sciurus Carolinensis*, in two Maryland woodlots. *Ecology* **41**, 365—9. [85, 446]

Foote, L.E., Peters, H.S. and Finkner, A.L. (1958). Design tests for mourning dove call-count sampling in seven southeastern states. *J. Wildl. Manag.* **22**, 402—8. [28, 53, 54]

Fordham, R.A. (1970). Mortality and population change of dominican gulls in Wellington, New Zealand; with a statistical appendix by R.M. Cormack. *J. Animal Ecol.* **39**, 13—27. [252]

Gangwere, S.K., Chavin, W. and Evans, F.C. (1964). Methods of marking insects, with special reference to Orthoptera (Sens. Lat.). *Ann. Entomol. Soc. America* **57**, 662—9. [93]

Gates, C.E. (1969). Simulation study of estimators for the line transect sampling method. *Biometrics* **25**, 317—28. [29, 32, 36—8, 448]

Gates, C.E., Marshall, W.H. and Olson, D.P. (1968). Line transect method of estimating grouse population densities. *Biometrics* **24**, 135—45. [29—32, 40]

Gates, C.E. and Smith, W.B. (1972). Estimation of density of mourning doves from aural information. *Biometrics* **28** (June). [53—4]

Gates, J.M. (1966). Crowing counts as indices to cock pheasant populations in Wisconsin. *J. Wildl. Manag.* **30**, 735—44. [54]

Geis, A.D. (1955). Trap response of the cottontail rabbit and its effect on censusing. *J. Wildl. Manag.* **19**, 466—72. [84, 162, 165, 167—8]

Geis, A.D. and Atwood, E.L. (1961). Proportion of recovered waterfowl bands reported. *J. Wildl. Manag.* **25**, 154—9. [250]

Geis, A.D. and Taber, R.D. (1963). Measuring hunting and other mortality. *In* H.S. Mosby (ed.), *Wildlife Investigational Techniques*, 2nd edn, 284—98. The Wildlife Society: Washington. [361, 383]

Gentry, J.B. and Odum, E.P. (1957). The effect of weather on the winter activity of old-field rodents. *J. Mammal.* **38**, 72—7. [318]

Gentry, J.B., Golley, F.B. and Smith, M.H. (1968). An evaluation of the proposed International Biological Program census method for estimating small mammal populations. *Acta theriologica* **13**, 313—27. [455]

Gerrard, P.J. and Chiang, H.C. (1970). Density estimation of corn rootworm egg populations based on frequency of occurrence. *Ecology* **51**, 237—45. [55—6]

Getz, L.L. (1961). Responses of small mammals to live-traps and weather conditions. *Amer. Midl. Nat.* **66**, 160—70. [84]

Gilbert, P.F. and Grieb, J.R. (1957). Comparison of air and ground deer counts in Colorado. *J. Wildl. Manag.* **21**, 33—7. [21]

Giles, R.H. Jr. (1963). Instrumentation in wildlife investigations. *In* H.S. Mosby (ed.), *Wildlife Investigational Techniques*, 2nd edn, 1—21. The Wildlife Society: Washington. [20, 94]

Giles, R.H. Jr. (ed.) (1969). *Wildlife Management Techniques*, 3rd edn. The Wildlife Society: Washington. [2]

Golley, F.B., Gentry, J.B., Caldwell, L.D. and Davenport, L.B. Jr. (1965). Number and variety of small mammals on the A.E.C. Savannah River Plant. *J. Mammal.* **46**, 1–18. [327]

Goodman, L.A. (1953). Sequential sampling tagging for population size problems. *Ann. Math. Statist.* **24**, 56–69. [120, 188–9]

Goodman, L.A. (1960). On the exact variance of products. *J. Amer. Statist. Assoc.* **55**, 708–13. [9]

Goodman, L.A. (1969). The analysis of population growth when the birth and death rates depend on several factors. *Biometrics* **25**, 659–81. [393]

Green, R.G. and Evans, C.A. (1940). Studies on a population cycle of snowshoe hares on the Lake Alexander area. I: Gross annual census, 1932–1939. *J. Wildl. Manag.* **4**, 220–38. [106]

Greig-Smith, P. (1964). *Quantitative Plant Ecology*, 2nd edn. Butterworths: London. [22–5, 41, 51, 56, 458]

Grodzinski, W., Pucek, Z. and Ryszkowski, L. (1966). Estimation of rodent numbers by means of prebaiting and intensive removal. *Acta theriologica* **11**, 297–314. [315, 455]

Gulland, J.A. (1955a). Estimation of growth and mortality in commercial fish populations. *U.K. Ministry Agric. and Fish, Fish. Invest.*, Ser. 2, **18**, 1–46. [306]

Gulland, J.A. (1955b). On the estimation of population parameters from marked members. *Biometrika* **42**, 269–70. [272]

Gulland, J.A. (1963). On the analysis of double-tagging experiments. *In* North Atlantic Fish Marking Symposium, *I.C.N.A.F. Special Publication No. 4*, 228–9. [94, 281]

Gulland, J.A. (1964). Catch per unit effort as a measure of abundance. *Rapp. P.-v. Réun. Cons. perm. int. Explor. Mer* **155**, 8–14. [455]

Gullion, G.W. (1966). The use of drumming behaviour in ruffed grouse population studies. *J. Wildl. Manag.* **30**, 717–29. [54]

Gunderson, D.R. (1968). Foodplain use related to stream morphology and fish populations. *J. Wildl. Manag.* **32**, 507–14. [82]

Guthrie, D.R., Osborne, J.C. and Mosby, H.S. (1967). Physiological changes associated with shock in confined gray squirrels. *J. Wildl. Manag.* **31**, 102–8. [83]

Hald, A.H. (1952). *Statistical Tables and Formulas*. Wiley: New York. [263]

Haldane, J.B.S. (1945). On a method of estimating frequencies. *Biometrika* **33**, 222–5. [121]

Haldane, J.B.S. (1953). Some animal life tables. *J. Inst. Actuaries* **79**, 83–9. [247]

Haldane, J.B.S. (1955). The calculation of mortality rates from ringing data. *Proc. XIth Int. Orn. Congr. Basel*, 454–8. [246, 248–9]

Hancock, D.A. (1963). Marking experiments with the commercial whelk (*Buccinum undatum*). *In* North Atlantic Fish Marking Symposium, *I.C.N.A.F.*, Special Publication No. 4, 167–87. [82]

Hanson, W.R. (1963). Calculation of productivity, survival and abundance of selected vertebrates from sex and age ratios. *Wildl. Monogr.* **9**, 1–60. [353, 375, 381]

Hansson, L. (1967). Index line catches as a basis of population studies on small mammals. *Oikos* **18**, 261–76. [450]

Hansson, L. (1968). Population densities of small mammals in open field habitats in South Sweden in 1964–1967. *Oikos* **19**, 53–60. [450]

Hansson, L. (1969). Home range, population structure and density estimates at removal catches with edge effect. *Acta theriologica* **14**, 153–60. [52]

Harke, D.T. and Stickley, A.R. Jr. (1968). Sensitive resettable odometer aids roadside census of red-winged blackbirds. *J. Wildl. Manag.* **32**, 635–6. [54]

REFERENCES

Hartley, H.O. (1958). Maximum likelihood estimation from incomplete data. *Biometrics* **14**, 174—94. [169, 175]

Hartley, H.O., Homeyer, P.G. and Kozicky, E.L. (1955). The use of log transformations in analysing fall roadside pheasant counts. *J. Wildl. Manag.* **19**, 495—6. [53]

Harvard Computation Laboratory (1955). *Tables of the Cumulative Binomial Probability Distribution.* Harvard University Press. [64, 122]

Harvey, J.M. and Barbour, R.W. (1965). Home range of *Microtus Ochrogaster* as determined by a modified minimum area method. *J. Mammal.* **46**, 398—402. [20]

Havey, K.A. (1960). Recovery, growth and movement of hatchery-reared Lake Atlantic salmon at Long Pond, Maine. *Trans. Amer. Fish. Soc.* **89**, 212—17. [304]

Hayne, D.W. (1949a). An examination of the strip census method for estimating animal populations. *J. Wildl. Manag.* **13**, 145—57. [36, 38]

Hayne, D.W. (1949b). Two methods for estimating populations from trapping records. *J. Mammal.* **30**, 399—411. [142, 146—50, 325, 327]

Healy, M.J.R. (1962). Some basic statistical techniques in soil zoology. *In* P.W. Murphy (ed.), *Progress in Soil Zoology*, 3—9. Butterworths: London. [20]

Heezen, K.L. and Tester, J.R. (1967). Evaluation of radio-tracking by triangulation with special reference to deer movements. *J. Wildl. Manag.* **31**, 124—41. 94]

Heincke, F. (1913). Investigations on the plaice. General report. 1. The plaice fishery and protective measures. Preliminary brief summary of the most important points of the report. *Cons. int. Explor. Mer, Rapp. et P.-v.* **16**, 1—67. [415]

Henderson, H.F., Hasler, A.D. and Chipman, G.G. (1966). An ultrasonic transmitter for use in studies of movements of fishes. *Trans. Amer. Fish. Soc.* **95**, 350—6. [94]

Henny, C.J. (1967). Estimating band-reporting rates from banding and crippling loss data. *J. Wildl. Manag.* **31**, 533—8. [250]

Hessler, E., Tester, J.R., Siniff, D.B. and Nelson, M.M. (1970). A biotelemetry study of survival of pen-reared pheasants released in selected habitats. *J. Wildl. Manag.* **34**, 267—74. [94]

Hewitt, O.H. (1967). A road-count index to breeding populations of red-winged blackbirds. *J. Wildl. Manag.* **31**, 39—47. [53]

Hickey, F. (1960). Death and reproductive rate of sheep in relation to flock culling and selection. *New Zeal. J. Agric. Res.* **3**, 332—44. [401]

Hickey, J.J. (1952). Survival studies of banded birds. *U.S. Fish and Wildl. Serv., Spec. Sci. Rept.: Wildl.* No. 15, 1—177. [245, 247]

Hjort, J.G. and Ottestad, P. (1933). The optimum catch. *Hvalradets Skrifter, Oslo* **7**, 92—127. [296]

Holgate, P. (1964a). The efficiency of nearest neighbour estimators. *Biometrics* **20**, 647—9. [42—3]

Holgate, P. (1964b). A modified geometric distribution arising in trapping studies. *Acta theriologica* **9**, 353—6. [180—2]

Holgate, P. (1965). Tests of randomness based on distance methods. *Biometrika* **52**, 345—53. [41]

Holgate, P. (1966). Contributions to the mathematics of animal trapping. *Biometrics* **22**, 925—36. [182—4]

Holgate, P. (1969). Species frequency distributions. *Biometrika* **56**, 651—60. [458]

Hopkins, B. (1954). A new method for determining the type of distribution of plant individuals. *Annals of Botany* **18**, 213—27. [43]

Houser, A. and Dunn, J.E. (1967). Estimating the size of the threadfin shad population in Bull Shoals Reservoir from midwater trawl catches. *Trans. Amer. Fish. Soc.* **96**, 176—84. [25]

Howell, J.C. (1951). The roadside census as a method of measuring bird populations. *Auk* **68**, 334—57. [53]

REFERENCES

Huber, J.J. (1962). Trap response of confined cottontail populations. *J. Wildl. Manag.* **26**, 177–85. [84, 162, 446, 451]

Hughes, R.D. (1965). On the composition of a small sample of individuals from a population of the banded hare wallaby, *Lagostrophus fasciatus* (Peron and Lesneur). *Austral. J. Zool.* **13**, 75–95. [403]

Hughes, R.D. and Gilbert, N. (1968). A model of an aphid population — a general statement. *J. Animal Ecol.* **37**, 553–63. [399]

Hunter, G.N. and Yeager, L.E. (1949). Big game management in Colorado. *J. Wildl. Manag.* **13**, 392–411. [21]

Hunter, W.R. and Grant, D.C. (1966). Estimates of population density and dispersal in the Natacid Gastropod, *Polinices duplicatus*, with a discussion of computational methods. *Biolog. Bull.* **131**, 292–307. [152]

Imber, M.J. and Williams, G.R. (1968). Mortality rates of a Canada population in New Zealand. *J. Wildl. Manag.* **32**, 256–67. [253]

International Commission for the Northwest Atlantic Fisheries (1963). The selectivity of fishing gear. *I.C.N.A.F., Special Publication No. 5*, 1–225. [82]

International Pacific Halibut Commission (1960). *Utilisation of Pacific Halibut Stocks; Yield per Recruitment*. Seattle University. [351]

Jackson, C.H.N. (1933). On the true density of tsetse flies. *J. Animal Ecol.* **2**, 204–9. [1]

Jackson, C.H.N. (1937). Some new methods in the study of *Glossina morsitans*. *Proc. Zool. Soc. London 1936*, 811–96. [1]

Jackson, C.H.N. (1939). The analysis of an animal population. *J. Animal Ecol.* **8**, 238–46. [1, 415]

Jackson, C.H.N. (1940). The analysis of a tsetse-fly population: I. *Ann. Eugen. London* **10**, 332–69. [1]

Jackson, C.H.N. (1944). The analysis of a tsetse-fly population: II. *Ann. Eugen. London* **12**, 176–205. [1]

Jackson, C.H.N. (1948). The analysis of a tsetse-fly population: III. *Ann. Eugen. London* **14**, 91–108. [1]

Jenkins, D., Watson, A. and Miller, G.R. (1967). Population fluctuations in the red grouse (*Lagopus lagopus scoticus*). *J. Animal Ecol.* **36**, 97–122. [255]

Jennrich, R.I. and Turner, F.B. (1969). Measurement of non-circular home range. *J. Theoret. Biol.* **22**, 227–37. [20]

Johnson, A.M. (1968). Annual mortality of territorial male fur seals and its management significance. *J. Wildl. Manag.* **32**, 94–9. [418]

Johnson, M.G. (1965). Estimates of fish populations in warmwater streams by the removal method. *Trans. Amer. Fish. Soc.* **94**, 350–7. [315]

Jolly, G.M. (1965). Explicit estimates from capture–recapture data with both death and immigration — stochastic model. *Biometrika* **52**, 225–47. [196–205, 225]

Jones, J.W. (1959). *The Salmon*. Collins: London. [443]

Jones, R. (1956). The analysis of trawl haul statistics with particular reference to the estimation of survival rates. *Cons. Int. Explor. Mer, Rapp. et P.-v.* **140** (*1*), 30–9. [426]

Junge, C.O. (1963). A quantitative evaluation of the bias in population estimates based on selective samples. *In* North Atlantic Fish Marking Symposium, *I.C.N.A.F., Special Publication No. 4*, 26–8. [86]

Junge, C.O. and Libosvárský, J. (1965). Effects of size selectivity on population estimates based on successive removals with electrical fishing gear. *Zool. Listy* **14**, 171–8. [315]

Justice, K.E. (1961). A new method for measuring home ranges of small mammals. *J. Mammal.* **42**, 462–70. [20, 54]

Kale, B.K. (1961). On the solution of the likelihood equation by iteration processes. *Biometrika* **48**, 452—6. [17]

Kale, B.K. (1962). On the solution of the likelihood equations by iteration processes: the multiparametric case. *Biometrika* **49**, 479—86. [17]

Kathirgamatamby, N. (1953). Note on the Poisson index of dispersion. *Biometrika* **40**, 225—8. [14, 45]

Kaye, S.V. (1961). Movements of harvest mice tagged with gold-198. *J. Mammal.* **42**, 323—37. [20]

Keeping, E.S. (1962). *Introduction to Statistical Inference*. D. Van Nostrand: Princeton, New Jersey. [38, 82, 87, 98, 268]

Keith, L.B. and Meslow, E.C. (1968). Trap response by snowshoe hares. *J. Wildl. Manag.* **32**, 795—801. [165, 170]

Keith, L.B., Meslow, E.C. and Rongstad, O.J. (1968). Techniques for snowshoe hare population studies. *J. Wildl. Manag.* **32**, 801—12. [83]

Kelker, G.H. (1940). Estimating deer.populations by a differential hunting loss in the sexes. *Proc. Utah Acad. Sci., Arts and Letters*, **17**, 6—69. [353]

Kelker, G.H. (1944). Sex-ratio equations and formulas for determining wildlife populations. *Proc. Utah Acad. Sci., Arts and Letters*, **19—20**, 189—98. [353]

Kelker, G.H. and Hanson, W.R. (1964). Simplifying the calculation of differential survival of age classes. *J. Wildl. Manag.* **28**, 411. [353]

Kelley, T.L. (1948). *The Kelley Statistical Tables*. (Revised 1948). Harvard Univ. Press: Cambridge, Massachusetts. [365]

Kelly, G.F. and Barker, A.M. (1963). Estimation of population size and mortality rates from tagged redfish, *Sebastes marinus* L., at Eastport, Maine. *In* North Atlantic Fish Marking Symposium, *I.C.N.A.F., Special Publication No. 4*, 204—9. [172]

Kendall, M.G. and Moran, P.A.P. (1963). *Geometrical Probability*. Griffin's Statistical Monographs and Courses No. 10. Griffin: London. [41, 44, 50]

Kendall, M.G. and Stuart, A. (1969). *The Advanced Theory of Statistics*, vol. 1, 3rd edn. Griffin: London. [32, 36, 161, 174]

Kendall, M.G. and Stuart, A. (1973). *The Advanced Theory of Statistics*, vol. 2, 3rd edn. Griffin: London. [14, 38, 46, 286]

Kendall, M.G. and Stuart, A. (1968). *The Advanced Theory of Statistics*, vol. 3, 2nd edn. Griffin: London. [9]

Kendeigh, S.C. (1944). Measurement of bird populations. *Ecol. Monogr.* **14**, No. 1, 67—106. [21]

Kennedy, W.A. (1954). Tagging returns, age studies and fluctuations in abundance of Lake Winnipeg whitefish, 1931—1951. *J. Fish. Res. Bd. Canada* **11**, 284—309. [426]

Kenyon, K.W., Scheffer, V.B. and Chapman, D.G. (1954). A population study of the Alaska fur seal herd. *U.S. Fish and Wildl. Serv., Spec. Sci. Report: Wildl. No. 12*, 1—77. [433]

Kershaw, K.A. (1964). *Quantitative and Dynamic Ecology*. Edward Arnold: London. [25]

Ketchen, K.S. (1950). Stratified subsampling for determining age distributions. *Trans. Amer. Fish. Soc.* **79**, 205—12. [422]

Ketchen, K.S. (1953). The use of catch-effort and tagging data in estimating a flatfish population. *J. Fish. Res. Bd. Canada* **10**, 459—83. [298, 308]

Keuls, M., Over, H.J. and De Wit, C.T. (1963). The distance method for estimating densities. *Statistica Neerlandica* **17**, 71—91. [41]

Keyfitz, N. (1968). *Introduction to the Mathematics of Population*. University of Chicago. [393, 458]

REFERENCES

Kikkawa, J. (1964). Movement, activity and distribution of the small rodents *Clethrionomys glareolus and Apodemus sylvaticus* in woodland. *J. Animal Ecol.* **33**, 259—99. [20, 81, 205, 315]

Kimball, A. (1960). Estimation of mortality intensities in animal experiments. *Biometrics* **16**, 505—21. [410—11]

Kimball, J.W. (1948). Pheasant population characteristics and trends in the Dakotas. *Trans. Nth Amer. Wildl. Conf.* **13**, 291—311. [375]

King, C.E. and Paulik, G.J. (1967). Dynamic models and the simulation of ecological systems. *J. Theoret. Biol.* **16**, 251—67. [458]

King, L.J. (1969). *Statistical Analysis in Geography*. Prentice-Hall: Englewood Cliffs, New Jersey. [25]

Kiritani, K. and Nakasuji, F. (1967). Estimation of the stage-specific survival rate in the insect population with overlapping stages. *Res. Popul. Ecol.* **9**, 143—52. [399]

Kiritani, K., Hokyo, N., Sasaba, T. and Nakasuji, F. (1970). Studies on population dynamics of the green rice leafhopper. *Res. Popul. Ecol.* **12**, 137—53. [399]

Kline, P.D. (1965). Factors influencing roadside counts of cottontails. *J. Wildl. Manag.* **29**, 665—71. [53]

Knowlton, F.F., Martin, P.E. and Haug, J.C. (1968). A telemetric monitor for determining animal activity. *J. Wildl. Manag.* **32**, 943—8. [94]

Koch, A.L. (1969). The logarithm in biology. II: Distributions simulating the lognormal. *J. Theoret. Biol.* **23**, 251—68. [448]

Kozicky, E.L. (1952). Variations in two spring indices of male ring-necked pheasant populations. *J. Wildl. Manag.* **16**, 429—37. [54]

Kozicky, E.L., Bancroft, T.A. and Homeyer, P.G. (1954). An analysis of woodcock singing ground counts, 1948—1952. *J. Wildl. Manag.* **18**, 259—66. [53—4]

Kozicky, E.L., Hendrickson, G.O., Homeyer, P.G. and Speaker, E.B. (1952). The adequacy of the fall roadside pheasant census in Iowa. *Trans. Nth Amer. Wildl. Conf.* **17**, 293—305. [53]

Krebs, C.J. (1966). Demographic changes in fluctuating populations of *Microtus californicus*. *Ecol. Monogr.* **36**, 239—73. [160]

Kurtén, B. (1953). On the variation and population dynamics of fossil and recent mammal populations. *Acta Zool. Fennica* **76**, 1—122. [403]

Lack, D. (1943). The age of the blackbird. *Brit. Birds* **36**, 166—75. [247]

Lack, D. (1951). Population ecology in birds. *Proc. Xth Int. Orn. Congr. Uppsala*, 409—48. [247]

Lack, D. (1954). *The Natural Regulation of Animal Numbers*. Clarendon Press: Oxford. [458]

Lagler, K.F. (1968). Capture, sampling and examination of fishes. *In* W.E. Ricker (ed.), *Methods for Assessment of Fish Production in Fresh Waters*, IBP Handbook No. 3, 7—45. Blackwell Scientific Publications. [82]

Lander, R.H. (1962). A method of estimating mortality rates from change in composition. *J. Fish. Res. Bd. Canada* **19**, 159—68. [353, 387]

Laplace, P.S. (1786). Sur les naissances, les mariages et les morts. *Histoire de l'Académie Royale des Sciences, Année 1783*, Paris, p. 693. [104]

Lauckhart, J.B. (1950). Determining the big-game population from the kill. *Trans. Nth Amer. Wildl. Conf.* **15**, 644—9. [353]

Laughlin, R. (1965). Capacity for increase: a useful population statistic. *J. Animal Ecol.* **34**, 77—91. [396]

Le Cren, E.D. (1965). A note on the history of mark—recapture population estimates. *J. Animal Ecol.* **34**, 453—4. [59]

REFERENCES

Leedy, D.L. (1949). Ohio pheasant nesting surveys based on farmer interviews. *J. Wildl. Manag.* **13**, 274—86. [54]

Leopold, A. (1933). *Game Management.* Charles Scribner's Sons: New York. [36]

Leslie, P.H. (1952). The estimation of population parameters from data obtained by means of the capture—recapture method. II: the estimation of total numbers. *Biometrika* **39**, 363—88. [160]

Leslie, P.H. (1958). Statistical appendix. *J. Animal Ecol.* **27**, 84—6. [161]

Leslie, P.H. and Chitty, D. (1951). The estimation of population parameters from data obtained by means of the capture—recapture method. I: The maximum likelihood equations for estimating the death rate. *Biometrika* **38**, 269—92. [201, 205]

Leslie, P.H., Chitty, D. and Chitty, H. (1953). The estimation of population parameters from data obtained by means of the capture—recapture method. III: An example of the practical applications of the method. *Biometrika* **40**, 137—69. [81, 160, 201, 224—5]

Leslie, P.H. and Davis, D.H.S. (1939). An attempt to determine the absolute number of rats on a given area. *J. Animal Ecol.* **8**, 94—113. [298, 454]

Leslie, P.H. and Ranson, R.M. (1940). The mortality, fertility and rate of natural increase of the vole (*Microtus agrestis*) as observed in the laboratory. *J. Animal Ecol.* **9**, 27—52. [400]

Leslie, P.H., Tener, T.S., Vizoso, M. and Chitty, H. (1955). The longevity and fertility of the Orkney vole, *Microtus orcadensis*, as observed in the laboratory. *Proc. Zool. Soc. Lond.* **125**, 115—25. [400]

Lewis, J.C. and Farrar, J.W. (1968). An attempt to use the Leslie census method on deer. *J. Wildl. Manag.* **32**, 760—4. [299]

Libosvársky, J. (1962). Application of De Lury method in estimating the weight of fish stock in small streams. *Int. Revue ges. Hydrobiol.* **47**, 515—27. [303, 314—15]

Lidicker, W.Z. Jr. (1966). Ecological observations on a feral house mouse population declining to extinction. *Ecol. Monogr.* **36**, 27—50. [205]

Lieberman, G.J. and Owen, D.B. (1961). *Tables of the Hypergeometric Distribution.* Stanford University Press. [63, 66]

Lincoln, F.C. (1930). Calculating waterfowl abundance on the basis of banding returns. *U.S. Dept. Agric. Circ. No. 118*, 1—4. [1, 104]

Lloyd, M. (1967). Mean crowding. *J. Animal Ecol.* **36**, 1—29. [21]

Lowe, J.I. (1956). Breeding density and productivity of mourning doves on a county-wide basis in Georgia. *J. Wildl. Manag.* **20**, 428—33. [54]

Ludwig, J.P. (1967). Band loss — its effect on banding data and apparent survivorship in the ringed-billed gull populations of the Great Lakes. *Bird-Banding* **38**, 309—23. [251]

McClure, H.E. (1945). Comparison of census methods for pheasants in Nebraska. *J. Wildl. Manag.* **9**, 38—45. [54]

MacDonald, D. and Dillman, E.G. (1968). Techniques for estimating non-statistical bias in big game harvest surveys. *J. Wildl. Manag.* **32**, 119—29. [360]

McLaren, A.D. (1967). Appendix: statistical analysis of the spacing of the settled aphids over the leaf and over the glass. *J. Animal Ecol.* **36**, 163—70. [41]

McLaren, G.F. and Pottinger, R.P. (1969). A technique for studying the population dynamics of the cabbage aphid *Brevicoryne brassicae* (L.). *N.Z.J. agric. Res.* **12**, 757—70. [399]

McLeese, D.W. and Wilder, D.G. (1958). The activity and catchability of the lobster (*Homarus americanus*) in relation to temperature. *J. Fish. Res. Bd. Canada* **15**, 1345—54. [305]

REFERENCES

MacLulich, D.A. (1951). A new technique of animal census, with examples. *J. Mammal.* **32**, 318–28. [20, 51]

Manly, B.F.J. (1969). Some properties of a method of estimating the size of mobile animal populations. *Biometrika* **56**, 407–10. [233]

Manly, B.F.J. (1970). A simulation study of animal population estimation using the capture–recapture method. *J. appl. Ecol.* **7**, 13–39. [237]

Manly, B.F.J. (1971a). A simulation study of Jolly's method for analysing capture–recapture data. *Biometrics* **27**, 415–24. [204]

Manly, B.F.J. (1971b). Estimates of a marking effect with capture–recapture sampling. *J. appl. Ecol.* **8**, 181–9. [231]

Manly, B.F.J. and Parr, M.J. (1968). A new method of estimating population size, survivorship, and birth rate from capture–recapture data. *Trans. Soc. Brit. Ent.* **18**, 81–9. [226, 233–5]

Margetts, A.R. (1963). Measurement of the efficiency of recovery and reporting of tags from recaptured fish. *In* North Atlantic Fish Marking Symposium, *I.C.N.A.F., Special Publication No. 4*, 255–7. [100]

Marsden, H.M. and Baskett, T.S. (1958). Annual mortality in a banded bobwhite population. *J. Wildl. Manag.* **22**, 414–19. [400]

Marten, G.G. (1970a). A regression method for mark–recapture estimates with unequal catchability. *Ecology* **51**, 291–5. [128, 150–2]

Marten, G.G. (1970b). The remote-sensing approach to censusing deer mice and monitoring their activity. (In preparation.) [451–2]

Marten, G.G. (1971). Censusing mouse populations by means of smoked-paper tracking. (In preparation.) [54, 451–2]

Martinson, R.K. (1966). Proportion of recovered duck bands that are reported. *J. Wildl. Manag.* **30**, 264–8. [250]

Martinson, R.K. and Grondahl, C.R. (1966). Weather and pheasant populations in Southwestern Dakota. *J. Wildl. Manag.* **30**, 74–81. [54]

Martinson, R.K. and McCann, J.A. (1966). Proportion of recovered goose and brant brands that are reported. *J. Wildl. Manag.* **30**, 856–8. [250]

Mech, L.D. (1967). Telemetry as a technique in the study of predation. *J. Wildl. Manag.* **31**, 492–6. [94]

Menhinick, E.F. (1963). Estimation of insect population density in herbaceous vegetation, with emphasis on removal sweeping. *Ecology* **44**, 617–21. [315, 327]

Meslow, E.C. and Keith, L.B. (1968). Demographic parameters of a snowshoe hare population. *J. Wildl. Manag.* **32**, 812–34. [233, 383]

Michener, C.D., Cross, E.A., Daly, H.V., Rettermeyer, C.W. and Wille, A. (1955). Additional techniques for studying the behaviour of wild bees. *Insectes Sociaux* **2**, 237–46. [96]

Miller, L.S. (1957). Tracing vole movements by radioactive excretory products. *Ecology* **38**, 132–6. [20]

Miller, R.G. Jr. (1966). *Simultaneous Statistical Inference.* McGraw-Hill: New York. [11]

Mitra, S.K. (1958). On the limiting power function of the frequency chi-squared test. *Ann. Math. Statist.* **29**, 1221–33. [195, 243]

Mohr, C.O. and Stumpf, W.A. (1966). Comparison of methods for calculating areas of animal activity. *J. Wildl. Manag.* **30**, 293–304. [20]

Mood, A.M. (1940). The distribution theory of runs. *Ann. Math. Statist.* **11**, 367–92. [185]

Moore, G. and Wallis, W.A. (1943). Time series significance tests based on signs of differences. *J. Amer. Statist. Assoc.* **38**, 153–64. [158]

REFERENCES

Moore, P.G. (1954). Spacing in plant populations. *Ecology* **35**, 222–7. [42–3]

Moore, W.H. (1954). A new type of electrical fish-catcher. *J. Animal Ecol.* **23**, 373–5. [315]

Moran, P.A.P. (1951). A mathematical theory of animal trapping. *Biometrika* **38**, 307–11. [312]

Morisita, M. (1954). Estimation of population density by spacing method. *Mem. Fac. Sci. Kyushu Univ.* E1, 187–97. [42, 44]

Morisita, M. (1957). A new method for the estimation of density by the spacing method applicable to non-randomly distributed populations. *Physiology and Ecology* **7**, 134–44 (Japanese): *U.S.D.A.* Forest Service translation no. 11116. [42]

Morris, R.F. (1955). Population studies on some small forest mammals in Eastern Canada. *J. Mammal.* **36**, 21–35. [84]

Morris, R.F. and Miller, C.A. (1954). The development of life tables for the spruce budworm. *Canadian J. Zool.* **32**, 283–301. [399]

Mosby, H.S. (ed.), (1963). *Wildlife Investigational Techniques*, 2nd edn. The Wildlife Society: Washington. [2]

Mosby, H.S. (1967). Population dynamics. *In* O.H. Hewitt (ed.), *The Wild Turkey and its Management*, 113–36. The Wildlife Society: Washington. [398]

Mosby, H.S. (1969). The influence of hunting on the population dynamics of a wood-lot gray squirrel population. *J. Wildl. Manag.* **33**, 59–73. [446, 451]

Moyle, J.B. and Lound, R. (1960). Confidence limits associated with means and medians of series of net catches. *Trans. Amer. Fish. Soc.* **89**, 53–8. [25]

Muir, B.S. (1963). Vital statistics of *Esox masquinongy* in Nogies Creek, Ontario. I: Tag loss, mortality due to tagging, and the estimate of exploitation. *J. Fish. Res. Bd. Canada* **20**, 1213–30. [333]

Muir, B.S. (1964). Vital statistics of *Esox masquinongy* in Nogies Creek, Ontario. II: Population size, natural mortality and effect of fishing. *J. Fish. Res. Bd. Canada* **21**, 727–46. [333–4, 446]

Muir, B.S. and White, H. (1963). Application of the Paloheimo linear equation for estimating mortalities to a seasonal fishery. *J. Fish. Res. Bd. Canada* **20**, 839–40. [331–2]

Murphy, G.I. (1952). An analysis of silver salmon counts at Benkow Dam, South Fork of Eel River, California. *Calif. Fish and Game* **38**, 105–12. [358]

Murphy, G.I. (1960). Estimating abundance from longline catches. *J. Fish. Res. Bd. Canada* **17**, 33–40. [306]

Murphy, G.I. (1965). A solution of the catch equation. *J. Fish. Res. Bd. Canada* **22** 191–202. [333]

Murphy, P.W. (ed.), (1962). *Progress in Soil Zoology.* Butterworths: London. [21]

Murton, R.K. (1966). A statistical evaluation of the effect of wood-pigeon shooting as evidenced by the recoveries of ringed birds. *The Statistician* **16**, 183–202. [247, 249]

Myres, M.T. (1969). Uses of radar in wildlife management. *In* R.H. Giles, Jr. (ed.), *Wildlife Management Techniques*, 3rd edn, 105–8. The Wildlife Society: Washington. [19]

Neff, D.J. (1968). The pellet-group count technique for big game trend, census, and distribution: a review. *J. Wildl. Manag.* **32**, 597–614. [54]

Nelson, R.D., Buss, I.O. and Baines, G.A. (1962). Daily and seasonal crowing frequency of ring-necked pheasants. *J. Wildl. Manag.* **26**, 269–72. [54]

New, J.G. (1958). Dyes for studying the movements of small mammals. *J. Mammal.* **39**, 416–29. [20, 93]

New, J.G. (1959). Additional uses of dyes for studying the movements of small mammals. *J. Wildl. Manag.* **23**, 348–51. [93]

REFERENCES

Newman, D.E. (1959). Factors influencing the winter roadside count of cottontails. *J. Wildl. Manag.* **23**, 290–4. [53]

Nixon, C.M., Edwards, W.R. and Eberhardt, L.L. (1967). Estimating squirrel abundance from live-trapping data. *J. Wildl. Manag.* **31**, 96–101. [84, 172–3]

North Atlantic Fish Marking Symposium (1963). *International Commission for Northwest Atlantic Fisheries, Special Publication No. 4*, 1–370. [83, 85]

Northcote, T.G. and Wilkie, D.W. (1963). Underwater census of stream fish populations. *Trans. Amer. Fish. Soc.* **92**, 146–51. [58]

Nunneley, S.A. (1964). Analysis of banding records of local populations of Blue Jays and Redpolls at Granby, Mass. *Bird-Banding* **35**, 8–22. [109]

Odum, E.P. and Pontin, A.J. (1961). Population density of the underground ant *Lasius flavus*, as determined by tagging with P^{32}. *Ecology* **42**, 186–8. [105]

Omand, D.N. (1951). A study of populations of fish based on catch-effort statistics. *J. Wildl. Manag.* **15**, 88–98. [299, 304]

Orians, G.H. (1958). A capture–recapture analysis of a shearwater population. *J. Animal Ecol.* **27**, 71–84. [205, 228]

Overton, W.S. (1965). A modification of the Schnabel estimator to account for removal of animals from the population. *J. Wildl. Manag.* **29**, 392–5. [154–6]

Overton, W.S. and Davis, D.E. (1969). Estimating the numbers of animals in wildlife populations. *In* R.H. Giles, Jr. (ed.), *Wildlife Management Techniques*, 3rd edn, 403–55. The Wildlife Society: Washington. [36, 53–4, 58, 118, 170, 431]

Owen, D.B. (1962). *Handbook of Statistical Tables*. Addison-Wesley: Reading, Massachusetts. [123]

Pahl, P.J. (1969). On testing for goodness-of-fit of the negative binomial distribution when expectations are small. *Biometrics* **25**, 143–51. [25]

Paloheimo, J.E. (1961). Studies on estimation of mortalities.1: Comparison of a method described by Beverton and Holt and a new linear formula. *J. Fish. Res. Bd. Canada* **18**, 645–62. [304, 330]

Paloheimo, J.E. (1963). Estimation of catchabilities and population sizes of lobsters. *J. Fish. Res. Bd. Canada* **20**, 59–88. [125–8, 305]

Paloheimo, J.E. and Dickie, L.M. (1964). Abundance and fishing success. *Rapp. P.-v. Réun. Cons. perm. int. Extor. Mer* **155**, 152–63. [455]

Paloheimo, J.E. and Kohler, A.C. (1968). Analysis of the Southern Gulf of St Lawrence cod population. *J. Fish. Res. Bd. Canada* **25**, 555–78. [334]

Paludan, K. (1951). Contributions to the breeding biology of *Larus argentatus* and *Larus fuscus*. *Dansk naturh. Foren. Vidensk. Medd.* **114**, 1–128. [253]

Parker, R.A. (1955). A method for removing the effect of recruitment on Petersen-type population estimates. *J. Fish. Res. Bd. Canada* **12**, 447–50. [263–6]

Parker, R.A. (1963). On the estimation of population size, mortality and recruitment. *Biometrics* **19**, 318–23. [256–60]

Parker, R.R. (1965). Estimation of sea mortality rates for the 1961 brood-year pink salmon of the Bella Coola Area, British Columbia. *J. Fish. Res. Bd. Canada* **22**, 1523–54. [383]

Parker, R.R. (1968). Marine mortality schedules of pink salmon of the Bella Coola River, Central British Columbia. *J. Fish. Res. Bd. Canada* **25**, 757–94. [205]

Parr, M.J. (1965). A population study of a colony of imaginal *Ischnura elegaris* (van der Linden) (Odonata: Coenagriidae) at Dale, Pembrokeshire. *Field Studies* **2**, 237–82. [204–5, 208]

Parr, M.J., Gaskell, T.J. and George, B.J. (1968). Capture–recapture methods of estimating animal numbers. *J. Biol. Educ.* **2**, 95–117. [205, 221, 229]

491

REFERENCES

Parrish, B.B. (1962). Problems concerning the population dynamics of the Atlantic herring (*Clupea Harengus L.*) with special reference to the North Sea. *In* E.D. Le Cren and M.W. Holdgate (Editors), *The Exploitation of Natural Animal Populations*, 3–28. Blackwell: Oxford. [306]

Paulik, G.J. (1961). Detection of incomplete reporting of tags. *J. Fish. Res. Bd. Canada* 18, 817–29. [97–100, 346–7, 461–3]

Paulik, G.J. (1962). Use of the Chapman–Robson survival estimate for single- and multi-release tagging experiments. *Trans. Amer. Fish. Soc.* 91, 95–8. [275, 290–2]

Paulik, G.J. (1963a). Estimates of mortality rates from tag recoveries. *Biometrics* 19, 28–57. [83, 238, 272–80, 286, 288, 293–5, 386]

Paulik, G.J. (1963b). Exponential rates of decline and type (1) losses for populations of tagged pink salmon. *In* North Atlantic Fish Marking Symposium, *I.C.N.A.F., Special Publication No. 4*, 230–7. [279, 294–5]

Paulik, G.J. and Robson, D.S. (1969). Statistical calculations for change-in-ratio estimators of population parameters. *J. Wildl. Manag.* 33, 1–27. [103, 353–84]

Paynter, R.A. (1966). A new attempt to construct life tables for Kent Island Herring Gulls. *Bull. Mus. of Comp. Zool., Harvard Univ.* 133, (11), 489–528. [251]

Pearson, P.G. (1955). Population ecology of the spade foot toad, *Scaphiopus h. holbrooki* (Harlan). *Ecol. Monogr.* 25, 233–67. [150]

Pearson, E.S. and Hartley, H.O. (1966). *Biometrika Tables for Statisticians*, vol. 1 (3rd edn). Cambridge University Press. [63–4, 123]

Pelikán, J., Zejda, J. and Holišová, V. (1964). On the question of investigating small mammal populations by the quadrate method. *Acta theriologica* 9, 1–24. [21]

Peterle, T.J. (1969). Radio isotopes and their use in wildlife research. *In* R.H. Giles, Jr. (ed.), *Wildlife Management Techniques*, 3rd edn, 109–18. The Wildlife Society: Washington. [94]

Petraborg, W.H., Wellein, E.G. and Gunvalson, V.E. (1953). Roadside drumming counts: a spring census method for ruffed grouse. *J. Wildl. Manag.* 17, 292–5. [54]

Petrides, G.A. (1949). View points on the analysis of open season sex and age ratios. *Trans. Nth Amer. Wildl. Conf.* 14, 391–410. [353, 376]

Petrides, G.A. (1954). Estimating the percentage kill in ringnecked pheasants and other game species. *J. Wildl. Manag.* 18, 294–7. [380]

Petrusewicz, K. and Andrzejewski, R. (1962). Natural history of a free-living population of house mice (*Mus musculus* Linnaeus) with particular reference to groupings within the population. *Ekol. Pol. (A)* 10, 1–122. [19]

Phillips, B.F. and Campbell, N.A. (1970). Comparison of methods of estimating population size using data on the whelk. *J. Animal Ecol.* 39, 753–9. [211, 446]

Pielou, E.C. (1969). *An Introduction to Mathematical Ecology*, Wiley-Interscience: New York. [25, 41, 50, 56, 458]

Pielowski, Z. (1969). Belt assessment as a reliable method of determining the number of hares. *Acta theriologica* 14, 133–40. [28]

Pottinger, R.P. and Le Roux, E.J. (1971). The biology and dynamics of *Lithocolletis blancardella* on apple in Quebec. *Mem. Ent. Soc. Can.* 77, 1–437. [399]

Progulske, D.R. and Duerre, D.C. (1964). Factors influencing spotlighting counts of deer. *J. Wildl. Manag.* 28, 27–34. [54]

Pucek, Z. (1969). Trap response and estimation of numbers of shrews in removal catches. *Acta theriologica* 14, 403–26. [81, 83–4, 327]

Pyburn, W.F. (1958). Size and movements of a local population of cricket frogs (*Acris crepitans*). *Tex. J. Sci.* 10, 325–42. [144]

Quick, H.F. (1963). Animal population analysis. *In* H.S. Mosby (ed.), *Wildlife Investigational Techniques*, 2nd edn, 190–228. The Wildlife Society: Washington. [393, 397–9, 402–3]

REFERENCES

Raff, M.S. (1956). On approximating the point binomial. *J. Amer. Statist. Assoc.* **51**, 293–303. [98, 122, 140]

Rao, C.R. and Chakravarti, I.M. (1956). Some sample tests of significance for a Poisson distribution. *Biometrics* **12**, 264–82. [14, 170]

Rasmussen, D.I. and Doman, E.R. (1943). Census methods and their applications in the management of mule deer. *Trans. Nth Amer. Wildl. Conf.* **8**, 369–79. [353–4]

Read, K.L.Q. and Ashford, J.R. (1968). A system of models for the life cycle of a biological organism. *Biometrika* **55**, 211–21. [399]

Regier, H.A. and Robson, D.S. (1967). Estimating population number and mortality rates. *In* S.D. Gerking (ed.), *The Biological Basis of Freshwater Fish Production*, 31–66. Blackwell Scientific Publications: Oxford. [58]

Reid, V.H., Hansen, R.M. and Ward, A.L. (1966). Counting mounds and earth plugs to census mountain pocket gophers. *J. Wildl. Manag.* **30**, 327–34. [54]

Ricker, W.E. (1940). Relation of catch per unit effort to abundance and rate of exploitation. *J. Fish. Res. Bd. Canada* **5**, 43–70. [329]

Ricker, W.E. (1944). Further notes on fishing mortality and effort. *Copeia* **1**, 23–44. [329]

Ricker, W.E. (1956). Uses of marking animals in ecological studies: the marking of fish. *Ecology* **37**, 665–70. [93]

Ricker, W.E. (1958). Handbook of computations for biological statistics of fish populations. *Bull. Fish. Res. Bd. Canada* **119**, 1–300. [2, 62, 83–4, 86, 127, 142–3, 222–3, 298, 302–6, 325–6, 350, 380, 383, 423, 428, 458]

Riordan, L.E. (1948). The sexing of deer and elk by airplane in Colorado. *Trans. Nth Amer. Wildl. Conf.* **13**, 409–28. [21, 353, 359–60]

Robinette, W.L. (1949). Winter mortality among mule deer in the Fishlake National Forest, Utah. *U.S. Fish and Wildl. Serv., Spec. Sci. Report No.65*, 1–15. [380]

Robinette, W.L., Jones, D.A., Gashwiler, J.S. and Aldous, C.M. (1954). Methods of censusing winter-lost deer. *Trans. Nth Amer. Wildl. Conf.* **19**, 511–25. [361, 448, 451]

Robinette, W.L., Jones, D.A., Gashwiler, J.S. and Aldous, C.M. (1956). Further analysis of methods of censusing winter-lost deer. *J. Wildl. Manag.* **20**, 75–8. [40, 361, 448]

Robson, D.S. (1960). An unbiased sampling and estimation procedure for creel censuses of fishermen. *Biometrics* **16**, 261–77. [304]

Robson, D.S. (1961). On the statistical theory of a roving creel census of fishermen. *Biometrics* **17**, 415–37. [304]

Robson, D.S. (1963). Maximum likelihood estimation of a sequence of annual survival rates from a capture–recapture series. *In* North Atlantic Fish Marking Symposium, *I.C.N.A.F., Special Publication No. 4*, 330–5. [214]

Robson, D.S. (1969). Mark–recapture methods of population estimation. *In* N.L. Johnson and H. Smith Jr. (Editors), *New Developments in Survey Sampling*, 120–40. Wiley–Interscience, Wiley and Sons: New York. [71, 82, 87, 205, 231]

Robson, D.S. and Chapman, D.G. (1961). Catch curves and mortality rates. *Trans. Amer. Fish. Soc.* **90**, 181–9. [414–30, 470–2]

Robson, D.S. and Flick, W.A. (1965). A non-parametric statistical method for culling recruits from a mark–recapture experiment. *Biometrics* **21**, 936–47. [74–81, 127]

Robson, D.S. and Regier, H.A. (1964). Sample size in Petersen mark–recapture experiments. *Trans. Amer. Fish. Soc.* **93**, 215–26. [60, 63, 64–70, 119]

Robson, D.S. and Regier, H.A. (1966). Estimates of tag loss from recoveries of fish tagged and permanently marked. *Trans. Amer. Fish. Soc.* **95**, 56–9. [282–6]

Robson, D.S. and Regier, H.A. (1968). Estimation of population number and mortality rates. *In* W.E. Ricker (ed.), *Methods for Assessment of Fish Production in Fresh Waters*, IBP Handbook No. 3, 124—58. Blackwell Scientific Publications: Oxford. [16, 71—3, 153—5, 160, 319]

Robson, D.S. and Whitlock, J.H. (1964). Estimation of a truncation point. *Biometrika* **51**, 33—9. [58]

Roessler, M. (1965). An analysis of the variability of fish populations taken by otter trawl in Biscayne Bay, Florida. *Trans. Amer. Fish. Soc.* **94**, 311—18. [25]

Rogers, G., Julander, P. and Robinette, W.L. (1958). Pellet-group counts for deer census and range-use index. *J. Wildl. Manag.* **22**, 193—9. [54]

Rose, C.D. and Hassler, W.W. (1969). Application of survey techniques to the dolphin, *Coryphaena hippurus*, fishery of North Carolina. *Trans. Amer. Fish. Soc.* **98**, 94—103. [304]

Rupp, R.S. (1966). Generalised equation for the ratio method of estimating population abundance. *J. Wildl. Manag.* **30**, 523—6. [354—5, 372]

Ryszkowski, L., Andrzejewski, R. and Petrusewicz, K. (1966). Comparison of estimates of numbers obtained by the methods of release of marked individuals and complete removal of rodents. *Acta theriologica* **11**, 329—41. [19]

Sadleir, R.M.F.S. (1965). The relationship between agonistic behaviour and population changes in the deermouse, *Peromyscus maniculatus* (Wagner). *J. Animal Ecol.* **34**, 331—53. [205]

Sampford, M.R. (1955). The truncated negative binomial distribution. *Biometrika* **42**, 58—69. [174]

Sampford, M.R. (1962). *An Introduction to Sampling Theory: with Applications to Agriculture*. Oliver and Boyd: London. [22, 43, 174]

Samuel, E. (1968). Sequential maximum likelihood estimation of the size of a population. *Ann. Math. Statist.* **39**, 1057—68. [189, 191, 193]

Samuel, E. (1969). Comparison of sequential rules for estimation of the size of a population. *Biometrics* **25**, 517—35. [136, 193]

Sanderson, G.C. (1950). Small-mammal population of a prairie grove. *J. Mammal.* **31**, 17—25. [51]

Sanderson, G.C. (1966). The study of mammal movements — a review. *J. Wildl. Manag.* **30**, 215—35. [20, 94]

Sands, W.A. (1965). Termite distribution in man-modified habitats in West Africa, with special reference to species segregation in the genus *Trinervitermes* (Isoptera, Termitidae, Nasutitermitinae). *J. Animal Ecol.* **34**, 557—71. [21, 41]

Scattergood, L.W. (1954). Estimating fish and wildlife populations: a survey of methods. *In* O. Kempthorne (ed), *Statistics and Mathematics in Biology*. Iowa State College Press: Ames, 273—85. [2, 19, 54]

Schaefer, M.B. (1951). Estimation of the size of animal populations by marking experiments. *U.S. Fish and Wildlife Service Fisheries Bulletin* **69**, 191—203. [64, 431, 439, 442]

Schnabel, Z.E. (1938). The estimation of the total fish population of a lake. *Amer. Math. Mon.* **45**, 348—52. [130, 139]

Schnell, J.H. (1968). The limiting effects of natural predation on experimental cotton rat populations. *J. Wildl. Manag.* **32**, 698—711. [400]

Schultz, V. (1959). A contribution toward a bibliography on game kill and creel census procedures. *Misc. Publ. No. 359*, Md. Agric. Exp. Sta. [304]

Schultz, V. (1961). *An Annotated Bibliography of the Uses of Statistics in Ecology — a Search of 31 Periodicals*. Off. Tech. Ser., Dept. Comm., TID — 3908, 1—315. [1]

Schultz, V. and Brooks, S.H (1958). Some statistical aspects of the relationship of quail density to farm composition. *J. Wildl. Manag.* **22**, 283—91. [53]

REFERENCES

Schultz, V. and Byrd, M.A. (1957). An analysis of covariance of cottontail rabbit population data. *J. Wildl. Manag.* **21**, 315—19. [53]

Schultz, V. and Muncy, R.J. (1957). An analysis of variance applicable to transect population data. *J. Wildl. Manag.* **21**, 274—8. [53]

Schumacher, F.X. and Eschmeyer, R.W. (1943). The estimation of fish populations in lakes and ponds. *J. Tennessee Acad. Sci.* **18**, 228—49. [141—2]

Seal, H.L. (1954). The estimation of mortality and other decremental probabilities. *Skandinavisk Aktuarietidskrift* **37**, 137—62. [411]

Sealander, J.A., Griffin, D.N., De Costa, J.J. and Jester, D.B. (1958). A technique for studying behavioural responses of small mammals to traps. *Ecology* **39**, 541—2. [84]

Seber, G.A.F. (1962). The multi-sample single recapture census. *Biometrika* **49**, 339—49. [194—5, 212, 273]

Seber, G.A.F. (1965). A note on the multiple—recapture census. *Biometrika* **52**, 249—259. [196, 199]

Seber, G.A.F. (1967). Asymptotic variances in multinomial distributions. *New Zealand Statistician* **2(1)**, 23—4. [9]

Seber, G.A.F. (1970a). The effects of trap resonse on tag—recapture estimates. *Biometrics* **26**, 13—22. [60, 85, 151]

Seber, G.A.F. (1970b). Estimating time-specific survival and reporting rates for adult birds from band returns. *Biometrika* **57**, 313—18. [213, 241—5]

Seber, G.A.F. (1971). Estimating age-specific survival rates from bird-band returns when the reporting rate is constant. *Biometrika* **58**, 491—7. [252]

Seber, G.A.F. (1972). Estimating survival rates from bird-band returns. *J. Wildl. Manag.* **36**, 405—13. [245, 255]

Seber, G.A.F. and Le Cren, E.D. (1967). Estimating population parameters from catches large relative to the population. *J. Animal Ecol.* **36**, 631—43.

Seber, G.A.F. and Whale, J.F. (1970). The removal method for two and three samples. *Biometrics* **26**, 393—400. [312, 315]

Selleck, D.M. and Hart, C.M. (1957). Calculating the percentage of kill from sex and age ratios. *Calif. Fish and Game* **43(4)**, 309—16.

Sen, A.R. (1970). Relative efficiency of sampling systems in the Canadian waterfowl harvest survey, 1967—68. *Biometrics* **26**, 315—26. [28, 53]

Sen, A.R. (1971a). Some recent developments in waterfowl sample survey techniques. *Applied Statistics* **20**, 139—47. [360]

Sen, A.R. (1971b). Increased precision in Canadian waterfowl harvest survey through successive sampling. *J. Wildl. Manag.* **35**, 664—8. [360]

Seshadri, V., Csorgo, M. and Stevens, M.A. (1969). Tests for the exponential distribution using Kolmogorov-type statistics. *J. Roy. Statist. Soc.*, B, **31**, 499—509. [46]

Severinghaus, C.W. and Maguire, H.F. (1955). Use of age composition data for determining sex ratios among adult deer. *N.Y. Fish and Game J.* **2 (2)**, 242—6. [375]

Sheppe, W. (1965). Characteristics and uses of *Peromyscus* tracking data. *Ecology* **46**, 630—4. [20, 54]

Shetter, D.S. (1967). Effect of jaw tags and fin excision upon the growth, survival and exploitation of hatchery rainbow trout fingerlings in Machigan. *Trans. Amer. Fish. Soc.* **96**, 394—9. [83]

Shiraishi, Y. and Furuta, Y. (1963). Estimation of the distribution and the stock number in the Lake Ashinoko, Kanagawa Prefecture, from the records of the fish-finder. *Bull. Fresh. Fish. Res. Lab.* **13**, 57—75. [19]

Sindermann, C.J. (1961). Parasite tags for marine fish. *J. Wildl. Manag.* **25**, 41—7. [94]

Siniff, D.B. and Skoog, R.O. (1964). Aerial censusing of caribou using stratified random sampling. *J. Wildl. Manag.* **28**, 391—401. [21, 28]

Skellam, J.G. (1948). A probability distribution derived from the binomial distribution by regarding the probability of success as variable between the sets of trials. *J. Roy. Statist. Soc., B*, **10**, 257—61. [62, 88, 177]

Skellam, J.G. (1952). Studies in statistical ecology. I: Spatial pattern. *Biometrika* **39**, 346—62. [43]

Skellam, J.G. (1958). The mathematical foundations underlying the use of line transects in animal ecology. *Biometrics* **14**, 385—400. [40]

Slater, L.E. (ed.), (1963). *Bio-telemetry: the Use of Telemetry in Animal Behaviour and Physiology in Relation to Ecological Problems*. Macmillan: New York. [94]

Slobodkin, L.B. (1962). *Growth and Regulation of Animal Populations*. Holt, Rinehart and Winston: New York. [395, 403]

Smirnov, V.S. (1967). The estimation of animal numbers based on the analysis of population structure. *In* K. Petrusewicz (ed.), *Secondary Productivity of Terrestrial Ecosystems*, vol. 1, 199—223. Institute of Ecology, Polish Academy of Sciences: Warsaw. [353]

Smith, A.D. (1964). Defecation rates of mule deer. *J. Wildl. Manag.* **28**, 435—44. [54]

Smith, M.H. (1968). A comparison of different methods of capturing and estimating numbers of mice. *J. Mammal.* **49**, 455—62. [451]

Snedecor, G.W. (1946). *Statistical Methods*, 4th edn. Iowa State Univ. Press.: Ames (Iowa). [263]

Sonleitner, F.J. and Bateman, M.A. (1963). Mark—recapture analysis of a population of Queensland fruit-fly, *Dacus tryoni* (Frogg.), in an orchard. *J. Animal Ecol.* **32**, 259—69. [205]

South, A. (1965). Biology and ecology of *Agriolamax reticulatus* (Müll.) and other slugs: spatial distribution. *J. Animal Ecol.* **34**, 403—17. [25]

Southern, H.N. (1965). The trap-line index to small mammal populations. *J. Zool., Lond.* **147**, 217—21. [450]

Southwick, C.H. and Siddiqi, M.R. (1968). Population trends of Rhesus monkeys in villages and towns of Northern India, 1959—65. *J. Animal Ecol.* **37**, 199—204. [53]

Southwood, T.R.E. (1966). *Ecological Methods*. Methuen: London. [20, 21, 25—6, 52, 84—5, 93—4, 96, 327, 393, 397, 399, 448, 458]

Spenceley, G.W., Spenceley, R.M. and Epperson, E.R. (1952). *Smithsonian Logarithmic Tables to Base e and Base 10*. Smithsonian Institute: Washington. [136]

Spitz, F. (1963). Les techniques d'échantillonage utilisées dans l'étude des populations de petits mammifères. *La Terre et la Vie* **2**, 203—37. [20]

Stair, J. (1957). Dove investigations. *Arizona Game and Fish Dept. Job Completion Rept., P.-R. Proj. W—53—R—7, WP 3, J—9*, 1—32. [251]

Stickel, L.F. (1948). The trap line as a measure of small mammal populations. *J. Wildl. Manag.* **12**, 153—61. [450]

Stoddart, L.C. (1970). A telemetric method for detecting jackrabbit mortality. *J. Wildl. Manag.* **34**, 501—7. [94]

Stott, B. (1968). Marking and tagging. *In* W.E. Ricker (ed.), *Methods for Assessment of Fish Production in Fresh Waters*, IBP Handbook No. 3, 78—92. Blackwell Scientific Publications: Oxford. [93]

Stradling, D.J. (1970). The estimation of worker ant populations by the mark—release—recapture method: an improved marking technique. *J. Animal Ecol.* **39**, 575—91. [106]

Strandgaard, H. (1967). Reliability of the Petersen method tested on a roe-deer population. *J. Wildl. Manag.* **31**, 643—51. [82, 84, 111, 451]

Swed, F.S. and Eisenhart, C. (1943). Tables for testing randomness of grouping in a sequence of alternatives. *Ann. Math. Statist.* 14, 66—87. [186]

Swinebroad, J. (1964). Net-shyness and wood thrush populations. *Bird-Banding* 35, 196—202. [180]

Taber, R.D. (1956). Uses of marking animals in ecological studies: marking of mammals; standard methods and new developments. *Ecology* 37, 681—5. [93]

Taber, R.D. (1969). Criteria of sex and age. *In* R.H. Giles, Jr. (ed.), *Wildlife Management Techniques*, 3rd edn, 325—402. The Wildlife Society: Washington. [393]

Taber, R.D. and Cowan, I.M. (1969). Capturing and marking wild animals. *In* R.H. Giles, Jr. (ed.), *Wildlife Management Techniques*, 3rd edn, 277—317. The Wildlife Society: Washington. [85, 93]

Takahasi, K. (1961). Model for the estimation of the size of a population by using the capture—recapture method. *Ann. Inst. Statist. Math.* 12, 237—48. [160]

Tanaka, R. (1951). Estimation of vole and mouse populations on Mount Ishizuchi and on the uplands of Southern Shikoku. *J. Mammal.* 32, 450—8. [84, 145—8]

Tanaka, R. (1952). Theoretical justification of the mark-and-release index for small mammals. *Bull. Kochi Women's College* 1, 38—47. [145]

Tanaka, R. (1956). On differential response to live traps of marked and unmarked small mammals. *Annot. Zool. Jap.* 29, 44—51. [84]

Tanaka, R. (1963a). On the problem of trap-response types of small mammal populations. *Res. Popul. Ecol.* 5, 139—46. [84]

Tanaka, R. (1963b). Examination of the routine census equation by considering multiple collisions with a single-catch trap in small mammals. *Jap. J. Ecol.* 13, 16—21. [454]

Tanaka, R. (1970). A field study on the effect of prebaiting on censusing by the capture—recapture method in a vole population. *Res. Popul. Ecol.* 12, 111—25. [84]

Tanaka, R. and Kanamori, M. (1967). New regression formula to estimate the shole population for recapture-addicted small mammals. *Res. Popul. Ecol.* 9, 83—94. [454]

Tanaka, R. and Teramura, S. (1953). A population of the Japanese field vole infested with Tsutsugamuchi disease. *J. Mammal.* 34, 345—52. [146]

Tanaka, S. (1953). Precision of age-composition of fish estimated by double sampling method using the length for stratification. *Bull. Jap. Soc. Sci. Fish.* 19(5), 657—70. [33]

Tanton, M.T. (1965). Problems of live-trapping and population estimation for the wood mouse, *Apodemus sylvaticus* (L.). *J. Animal Ecol.* 34, 1—22. [174, 176—7]

Tanton, M.T. (1969). The estimation and biology of populations of the bank vole and wood mouse. *J. Animal Ecol.* 38, 511—29. [174, 184]

Taylor, R.H. and Williams, R.M. (1956). The use of pellet counts for estimating the density of populations of the wild rabbit, *Oryctolagus cuniculus* (L.). *N.Z. J. Sci. and Technol.* 38B, 236—56. [54]

Taylor, S.M. (1966). Recent quantitative work on British bird populations: a review. *The Statistician* 16, 119—70. [84, 174]

Tesch, F.W. (1968). Age and growth. *In* W.E. Ricker (ed.), *Methods for Assessment of Fish Production in Fresh Waters*, IBP Handbook No. 3, 93—123. Blackwell Scientific Publications: Oxford. [393]

Tester, A.L. (1955). Estimation of recruitment and natural mortality rate from age-composition and catch data in British Columbia herring populations. *J. Fish. Res. Bd. Canada* 12, 649—81. [426—7]

Tester, J.R. and Siniff, D.B. (1965). Aspects of animal movement and home range data obtained by telemetry. *Trans. Nth Amer. Wildl. Conf.* 30, 379—92. [174]

Thomasson, R.L. and Kapadia, C.H. (1968). On estimating the parameter of a trun- cated geometric distribution. *Ann. Inst. Statist. Math.* **20**, 519—23.   [171, 468]

Thompson, H.R. (1956). Distribution of distance to *n*th neighbour in a population of randomly distributed individuals. *Ecology* **37**, 391—4.   [42, 44—5]

Tomlinson, R.E. (1968). Reward banding to determine reporting rate of recovered mourning dove bands. *J. Wildl. Manag.* **32**, 6—11.   [250, 360]

Trippensee, R.E. (1948). *Wildlife Management, Upland Game and General Principles.* McGraw-Hill: New York.   [21]

Turček, F.J. (1958). Zonologische Arbeitsmethoden für Wirbeltiere. *In* J. Balogh (ed.), *Lebensgemeinschaften der Landtiere*, 415—50.   [21]

Turner, F.B. (1960a). Size and dispersion of a Louisiana population of the cricket frog, *Acris gryllus. Ecology* **41**, 258—68.   [41, 134, 144, 160, 205]

Turner, F.B. (1960b). Tests of randomness in recaptures of *Rana pipretiosa. Ecology* **41**, 237—9.   [228]

Turner, F.B., Hoddenbach, G.A., Medica, P.A. and Lannom, J.R. (1970). The demo- graphy of the lazard, *Uta stansburiana* Baird and Girard, in Southern Nevada. *J. Animal Ecol.* **39**, 505—20.   [397]

Van Etten, R.C., Switzenberg, D.F. and Eberhardt, L. (1965). Controlled deer hunting in a square-mile enclosure. *J. Wildl. Manag.* **29**, 59—73.   [299]

Varley, G.C. and Gradwell, G.R. (1970). Recent advances in insect population dyna- mics. *Ann. Rev. Ent.* **15**, 1—24.   [393, 458]

Waloff, N. and Blackith, R.E. (1962). The growth and distribution of the mounds of *Lasius flavus* (Fabricius)(Hym.: Formicidae) in Silwood Park, Berkshire. *J. Animal Ecol.* **31**, 421—37.   [21, 41]

Walters, C.J. (1969). A generalised computer simulation model for fish population studies. *Trans. Amer. Fish. Soc.* **98**, 505—12.   [458]

Waters, T.F. (1960). The development of population estimate procedures in small trout lakes. *Trans. Amer. Fish. Soc.* **89**, 287—94.   [86]

Watt, K. (ed.), (1966). *Systems Analysis in Ecology.* Academic Press: New York.   [458]

Webb, W.L. (1942). Notes on a method of censusing snowshoe hare populations. *J. Wildl. Manag.* **6**, 67—9.   [36]

Webb, W.L. (1965). Small mammal populations on islands. *Ecology* **46**, 479—88.   [327]

Welch, H.E. (1960). Two applications of a method of determining the error of popu- lation estimates of mosquito larvae by the mark and recapture technique. *Ecology* **41**, 228—9.   [112]

Westerskov, K. (1963). Superior survival of black-necked over ring-necked pheasants in New Zealand. *J. Wildl. Manag.* **27**, 239—45.   [253]

Wetherill, G.B. (1967). *Elementary Statistical Methods.* Methuen: London.   [308]

White, E.G. (1970). A self-checking coding technique for mark—recapture studies. *Bull. Ent. Res.*, **60**, 303—7.   [93, 96]

White, E.G. (1971). A versatile Fortran computer program for the capture—recapture stochastic model of G.M. Jolly. *J. Fish. Res. Bd. Canada* **28**, 443—5.   [205]

Whitlock, S.C. and Eberhardt, L.L. (1956). Large-scale dead deer surveys: methods, results and management implications. *Trans. Nth Amer. Wildl. Conf.* **21**, 555—66.   [361]

Widrig, T.M. (1954). Method of estimating fish populations with application to Pacific sardine. *U.S. Fish and Wildl. Serv., Fish. Bull.* **56**, 141—66.   [329]

Wiegert, R.G. (1964). Population energetics of meadow spittlebugs (*Philaenus spumarius* L.) as affected by migration and habitat. *Ecol. Monogr.* **34**, 217—41.   [28]

REFERENCES

Wierzbowska, T. and Petrusewicz, K. (1963). Residency and rate of disappearance of two free-living populations of the house mouse (*Mus musculus* L.). *Ekol. Pol.* (*A*), **11**, 557—74. [181]

Wight, H. (1959). Eleven years of rabbit-population data in Missouri. *J. Wildl. Manag.* **23**, 34—9. [53]

Williams, C.B. (1964). *Patterns in the Balance of Nature*. Academic Press: London. [458]

Williams, W.P. (1965). The population density of four species of freshwater fish, roach (*Rutilus rutilus* (L.)), bleak (*Alburnus alburnus* (L.)), dace (*Leuciscus leuciscus* (L.)), and perch (*Perca fluviatilis* (L.)) in the River Thames at Reading. *J. Animal Ecol.* **34**, 173—85. [453]

Winsor, C.P. and Clark, G.L. (1940). A statistical study of variation in the catch of plankton nets. *J. Mar. Res.* **3**, 1—34. [276, 426]

Wood, G.W. (1963). The capture—recapture technique as a means of estimating populations of climbing cutworms. *Canadian J. Zool.* **41**, 47—50. [108]

Wood, J.E. (1959). Relative estimates of fox population levels. *J. Wildl. Manag.* **23**, 53—63. [56, 446]

Woodbury, A.M. (1956). Uses of marking animals in ecological studies: marking amphibians and reptiles. *Ecology* **37**, 670—4. [93]

Yapp, W.B. (1956). The theory of line transects. *Bird Study* **3**, 93—104. [40]

Young, H. (1958). Some repeat data on the Cardinal. *Bird-Banding* **29**, 219—23. [180]

Young, H. (1961). A test for randomness in trapping. *Bird-Banding* **32**, 160—2. [184—6]

Young, H., Nees, J. and Emlen, J.T. Jr. (1952). Heterogeneity of trap response in a population of house mice. *J. Wildl. Manag.* **16**, 169—80. [84, 160, 165—6, 179]

Young, H., Strecker, R.L. and Emlen, J.T. Jr. (1950). Localisation of activity in two indoor populations of house mice, *Mus musculus*. *J. Mammal.* **31**, 403—10. [166]

Zippin, C. (1956). An evaluation of the removal method of estimating animal populations. *Biometrics* **12**, 163—9. [312—15]

Zippin, C. (1958). The removal method of population estimation. *J. Wildl. Manag.* **22**, 82—90. [312—14]

Zubrzycki, S. (1963). O Minimaksowym szacowaniu licznosci populacji. *Zastosow. Mat.* **7**, 183—94. [121]

Zubrzycki, S. (1966). Explicit formulas for minimax admissible estimators in some cases of restrictions imposed on the parameter. *Zastosow. Mat.* **9**, 31—52. [121]

INDEX

Absolute density, *see* Density
Accidental deaths, 70, 83, 152, 196, 342
Accuracy of estimation, 64, 367
Aerial census, 21
   compared to ground count, 21, 359
Age-class, 251, 332, 338
Age-composition, 332, 457
Age-dependent survival, *see* Survival
Age—length data, 333, 422
Age-specific
   data, 399
   fertility, 396
   life table, *see* Life tables
Ageing animals
   error in, 402
   method of, 393
   subsampling from length-classes, 333, 422
Analysis of variance
   applications of, 53, 54, 268, 270
Angle of sighting, 28
Animal signs, 54
   signs per animal, 55, 376
Ant mounds, *see* Insects
Antelopes, *see* Mammals
Ants, *see* Insects
Anurans
   cricket frogs, 41, 134, 144
   toads, 150
Arcsin transformation, 263
Association between species, *see* Species
Asymptotic variance, *see* Maximum-likelihood estimate
Asymptotic variance—covariance matrix, *see* Maximum-likelihood estimate

Bag checks, 360, 402
Bailey's binomial model for the Petersen method, 61, 111, 451
Band
   loss, 251
   recovery rate, 240, 245
   reporting rate, 97, 250
   wear, 251
Banding nestlings, 247, 251
Bass, *see* Fish
Beta distribution
   applications of, 62, 177, 182, 316
Bias, definition, 5
   proportional, 5

Big game, *see* Mammals
Binomial coefficient of catchability, *see* Catchability
Binomial distribution
   applications of, 22, 46, 61, 85, 97—8, 164, 355, 451
   approximation to the hypergeometric distribution, *see* Hypergeometric distribution
   confidence interval for proportion, 123
   goodness-of-fit tests, 12
   normal approximation, 123
   Poisson approximation, 123, 139—40
   test for proportion, 122—3
   zero-truncated, 169
Binomial Index of Dispersion, 13
Birds
   blackbirds, 53
   bluetits, 84
   fulmars, 214
   geese, 243
   juncos, 185
   lapwings, 248
   mourning doves, 53, 54, 250
   pheasants, 21, 53, 54, 356, 358, 362, 371, 376, 383
   quail, 41, 381
   redpolls, 109
   ruffed grouse, 20, 32, 54
   starlings, 250
   waterfowl, 1, 250
   woodcocks, 53, 54
   woodpeckers, 21, 35
   wood-pigeons, 247
Blackbirds, *see* Birds
Black-kneed capsids, *see* Insects
Bluegills, *see* Fish
Bluetits, *see* Birds
Bounded counts method, 58
Budget, *see* Life equation
Butterflies, *see* Insects

Calendar of captures method, 19
Call counts, 54
Capture history, 130, 197
Catch curve, 426
Catch-effort methods
   closed population, 296
   open population, 328
   general comments on, 455
Catch equations, 328

500